Contents

Part 2.2: Human factors – advanced

Appendices

Foreword

My predecessor, Sir Bill Callaghan, provided the foreword to the two earlier editions of this book. I am delighted to welcome this latest (third) edition of *Health and safety: risk management* on behalf of the Health and Safety Executive.

Despite the passage of more than 30 years since the introduction of the Health and Safety at Work etc Act, the Act and the principles enshrined therein are as relevant today as they were back in the 1970s. While the world of work has changed in many ways, it is still absolutely the case that:

- the responsibility for managing risk rests with those who create the risk
- good health and safety performance goes hand in hand with good business and organisational performance
- building a health and safety culture requires strong leadership
- workforce engagement and involvement is essential to truly embedding health and safety in the fabric of an organisation.

In short, delivering strong health and safety performance in any organisation requires the commitment of hearts and minds, not a 'tick the box' compliance culture.

Health and safety professionals have a pivotal role to play in helping organisations to achieve this. Knowing what is required by the law is essential but basic stuff – it is just as important to understand:

- how to apply that knowledge in a proportionate and common-sense way and share good practice
- how people think and behave and the role this plays in causing accidents and ill health
- how hazards and risks change in the changing world of work and the broader societal change that is going on around us. Smaller businesses bring different risks but not no risk.

If you are reading this foreword it is probable that you are studying to become a health and safety professional or maybe you have already become one. Congratulations! You have chosen an exciting and dynamic profession which really makes a difference to people's lives. Our mission is not to eliminate risk or wrap people in cotton wool while at work – but to enable them all to enjoy safe and healthy working lives.

Judith Hackitt CBE
Chair
Health and Safety Executive

Introduction

The process of managing risks to health and safety is no different from that of any other management activity – it is just that bit more difficult because of its complexity. Three groups of people are typically involved:

- Directors and senior managers must decide what is to be achieved and set in train the necessary organisation to achieve it. They must then monitor progress and make changes when the desired objectives are not being achieved. Where necessary, they must investigate when things go wrong.
- Managers are responsible for day-to-day management of health and safety risks and they must implement, maintain and monitor the necessary risk control measures.
- Health and safety practitioners advise all levels of management on how to fulfil their respective roles and, in particular, they provide advice on the hazards in the workplace, the risks they represent and the risk control measures which are required.

Three factors have had an influence on the contents and layout of this book:

- the requirements of the occupational standards in health and safety practice
- the knowledge requirements for Chartered membership of the Institution of Occupational Safety and Health (IOSH)
- the IOSH accreditation requirements for degrees in occupational safety and health.

The book deals with two main topics: risk management and human factors. Each of these topics is divided into two parts, so the overall structure of the book is as follows:

Part 1.1 – Introduction to risk management
Part 1.2 – Introduction to human factors
Part 2.1 – Advanced risk management
Part 2.2 – Advanced human factors.

The rationale for this division is that Parts 1.1 and 1.2 are appropriate for managers and practitioners who require basic guidance on the subjects under discussion. These two parts provide simple solutions to the sometimes complex problems associated with risk management and human factors and will guide people in the practical tasks required to reduce risk in organisations.

However, many aspects of risk management and human factors are the subject of continuing debate and people managing complex risks, or wishing to study the subjects in more detail, should be aware of these debates. Parts 2.1 and 2.2 of the book are, therefore, devoted to descriptions and discussions of these more advanced topics. It is hoped that those who have understood and applied the techniques described in Parts 1.1 and 1.2 will, in the course of time, feel able to move on to the issues dealt with in Parts 2.1 and 2.2.

In the risk management part of the book, the topics covered include identifying hazards, assessing risk, selecting appropriate risk control measures, accident investigation techniques, and performance measurement. Review and audit are dealt with in a separate chapter, as are emergency planning and systems management as it applies to health and safety.

In the human factors part of the book, the topics covered include the sensory and perceptual processes of the individual, perception and decision making, human error and improving human reliability.

Appendix 1 covers the content of certain parts of the NEBOSH National Diploma syllabus. Appendix 2 covers the knowledge requirements identified in the relevant elements of the standards for health, safety and environmental protection practice. Appendix 3 contains the guidance given by IOSH to universities to help them structure academic courses in occupational health and safety. If they follow this guidance then their successful students will have satisfied the knowledge requirements for Chartered membership of IOSH.

Dr Peter Waterhouse
Editor

1: Preliminaries

Introduction

This book is intended to serve the needs of a number of audiences, in particular:

- those studying for university qualifications in health and safety or the National Examination Board in Occupational Safety and Health (NEBOSH) Diploma in Occupational Safety and Health
- those working towards a National or Scottish Vocational Qualification (N/SVQ) in Occupational Health and Safety Practice
- managers and others who have health and safety as part of their responsibilities
- health and safety professionals.

The book is in four main parts, preceded by the present chapter, which describes the material covered in the book and the audiences for which it is most relevant. The four main parts are followed by three appendices.

Part 1.1: Risk management – introduction

Part 1.1 covers the material required for the basic risk management topics in the NEBOSH Diploma syllabus. However, we have presented the material in the best order for learning, rather than the order in which it is presented in the syllabus. Appendix 1 contains a reproduction of the NEBOSH National Diploma in Occupational Safety and Health syllabus,[12] cross-referenced to the chapters in which the relevant material appears in this book.

Part 1.1 also covers certain parts of the knowledge requirements for those studying for the N/SVQ. These are listed in Appendix 2, with references to the chapters in which the relevant material appears in this book.

However, the material in Part 1.1 is primarily concerned with the fundamentals of risk management and, as such, it is relevant for all students of health and safety. Managers and others whose responsibility includes managing health and safety will also find this part valuable.

The final audience for the material in Part 1.1 is health and safety professionals. For most people in this category, the material will already be familiar but they may wish to use parts of it to develop training courses for managers in their organisation. To help with this, there is a chapter on training and communication techniques at the end of Part 1.1. By using the Part 1.1 material in this way, health and safety professionals can encourage the idea that managers are responsible for managing health and safety while health and safety professionals are responsible for providing advice and guidance.

Thus, there are the following main audiences for Part 1.1 of this book:

- those studying for university qualifications or the NEBOSH Diploma
- those trying to obtain the NVQ/SVQ
- managers with health and safety responsibilities
- health and safety professionals.

These are also the audiences for the other parts of the book, but the extent to which they may wish to use the material differs from part to part as described below.

Part 1.2: Human factors – introduction

Part 1.2 covers the basic human factors material dealt with in the NEBOSH Diploma syllabus and certain aspects of the vocational standards (see Appendices 1 and 2 for more details).

As far as managers are concerned, most of the material may be of interest since it deals with how humans function. In practical terms, only those managers who will get involved in detailed accident investigations, or some of the more advanced risk control measures described in Part 2.1, need to be familiar with the contents of Part 1.2.

For health and safety professionals, the material will again be familiar, but they may wish to use it for training purposes.

Part 2.1: Risk management – advanced

Part 2.1 deals with the more advanced aspects of risk management contained in the NEBOSH Diploma syllabus and further aspects of the knowledge requirements for the vocational standards. Details are given in Appendices 1 and 2.

As far as managers are concerned, they should use only those parts of this section which are relevant to their requirements. For example, managers involved in designing complex processes may well wish to make use of the chapter on detailed risk rating techniques.

For health and safety professionals, all of Part 2.1 is relevant, although particular individuals may already have knowledge and skills in some of the areas covered. Many aspects of the more advanced risk management techniques are still actively being researched and developed, and the discussions of these topics, and the recommendations for further reading, will help health and safety professionals in their continuing professional development.

Part 2.2: Human factors – advanced

This part covers the more advanced human factors material contained in the NEBOSH Diploma syllabus and further aspects of the knowledge requirements for the vocational standards (see Appendices 1 and 2 for details).

Much of this material will be beyond most managers' requirements, although, where risks are high, or risk management is already at a high level of expertise, they may wish to study the chapters dealing with human error and human reliability.

For health and safety professionals, all of Part 2.2 is relevant and the suggestions for further reading will be valuable in their continuing professional development.

Appendices

The book ends with three appendices:
- the NEBOSH Diploma syllabus, cross-referenced to the contents of this book
- vocational standards knowledge requirements, cross-referenced to the contents of this book
- IOSH learning objectives, also cross-referenced to the contents of this book.

Conventions

The English language has many words with multiple meanings, and many objects and concepts have numerous different words to describe them. In the hope of reducing ambiguity and repetition, the following conventions have been used throughout the book.

Occupational health and safety

The term occupational health and safety (OH&S) is used in a number of published management systems. The term is variously defined but the definition in BS OHSAS 18001[3] is:

> conditions and factors that affect, or could affect, the health and safety of employees or other workers (including temporary workers and contractor personnel), visitors, or any other person in the workplace.

This definition is rather restricted and the Scope section of BS OHSAS 18001 restricts it further:

> This OHSAS Standard is intended to address occupational health and safety, and it is not intended to address other health and safety areas such as employee wellbeing/wellness programmes, product safety, property damage or environmental impacts.

This book deals with a broader range of risks than OH&S risks and for this reason we have used health and safety, rather than OH&S, and health and safety management system (HSMS), rather than OH&S management system (OH&SMS). However, OH&S and OH&SMS are used when discussing documents, such as BS OHSAS 18001, which use this terminology.

Categories of people

Dozens of words are used to describe the various categories of people that are relevant to risk management. However, certain important categories for the purposes of this book are defined as follows.

Operatives

These are people who have no management responsibilities and whose primary health and safety responsibility is to make sure that they carry out their activities without risks to themselves or others. Everyone has this responsibility, but the other categories of people listed below have additional health and safety responsibilities.

First-line managers

These are people who have operatives reporting to them. They are responsible for making sure that the operatives carry out their activities without risks to themselves or others.

Middle managers

These are people who have first-line managers, or other middle managers, reporting to them. Their main health and safety responsibility is to make sure that those who report to them effectively meet their health and safety management responsibilities.

Senior managers

These people are the 'directing mind' of their part of the organisation and have overall responsibility for health and safety in that part of the organisation. They are responsible for setting policy on health and safety matters and typically control the organisation's resources.

Note that there will be overlaps between these categories. For example, in small and medium-sized enterprises (SMEs), the senior managers may also be first-line managers, since there are no middle managers. In larger organisations, senior managers and middle managers may also have one or more operatives reporting to them directly (for example, secretaries and personal assistants), so that they are also first-line managers as far as these people are concerned. Where we use the term 'managers' without further qualification, it should be taken to include all of the categories listed above.

Self-employed

These people have all the health and safety responsibilities of the other groups although, obviously, they apply over a much narrower range.

Quantification

At various points, expressions such as 'some companies', 'many organisations' or 'most people' have been used. Unless otherwise stated, these should all be understood to mean 'in the author's experience some companies...', 'in the author's experience many organisations...', and so on.

Legislation

The main concern throughout this book has been to describe best practice in risk management techniques. For some aspects of risk management, the use of this best practice is required by United Kingdom (UK) legislation and, where this is the case, details of the relevant legislation are given in clearly identified 'UK legislation' boxes. However, the majority of the topics covered in this book focus on good practice and they are often not referred to in UK legislation.

The majority of health and safety legislation in the UK is divided into legislation applying to England, Wales and Scotland (collectively referred to as Great Britain) and legislation applying only to Northern Ireland. Great Britain's legislation is primarily in the form of Statutory Instruments (SIs) and Northern Ireland's legislation primarily in the form of Statutory Rules (SRs).

Typically, there will be a two pieces of legislation covering each health and safety topic, for example:

- The Management of Health and Safety at Work Regulations 1999 (SI 1999/3242)
- The Management of Health and Safety at Work (Northern Ireland) Regulations 2000 (SR 2000/388)

However, so far as duties imposed by the legislation are concerned, the content of the two pieces of legislation is usually the same. For this reason, only the Great Britain title is given in the text with, in the References section, the SI and SR number for the relevant legislation for Great Britain and Northern Ireland respectively. The exception to this is legislation on fire safety, which is not health and safety legislation, and where there is separate legislation for England and Wales (SI), Scotland (Scottish SI (SSI)) and Northern Ireland (SR).

HSE

In line with UK legislation, the functions of the Health and Safety Executive (HSE) are carried out by separate organisations in Great Britain (the HSE) and Northern Ireland (the HSENI). In this book, the UK HSE is used to describe these two organisations.

More detailed information on these organisations is available at www.hse.gov.uk and www.hseni.gov.uk.

References

At the end of the book there are two lists of documents:

1. **References.** These are provided so that readers can check the accuracy of attributed quotations, summaries of referenced work, and so on. They are identified in the text with a superscript number. **Essential reading.** These texts form part of one or more courses of study and students should be as familiar with these texts as they are with the contents of this book. Items in the list of references that are printed in bold are essential reading.
2. **Further reading.** These texts are intended to provide a guide for those who wish to learn more about a specific topic. Note that these texts are provided for information only; their content is not examinable.

Part 1.1:
Risk management – introduction

2: Part 1.1 – overview

Introduction

Part 1.1 contains a number of chapters which, taken together, form an introduction to all the basic concepts of risk management. The present chapter is intended to provide a route map through this risk management material by describing briefly what is contained in each of the chapters in this part.

Risk management, like most specialist subjects, has its own vocabulary and basic principles. While it would have been possible to begin the description of risk management with a detailed exposition of these, we considered that this would put an unnecessary intellectual burden on the reader. Instead, the next chapter (**Chapter 3: Risk management – setting the scene**) is devoted to an informal introduction to risk management with the minimum of specialist vocabulary. Chapter 3 also deals with the various reasons why risk management is important, including ethical and financial reasons.

Following this informal introduction, **Chapter 4: Key elements of risk management** sets out the various activities which have to take place if risk management is to be effective. This is done in two stages:

- a **risk management model** is described that illustrates the core elements of risk management and how they are linked with each other. This model includes such things as accident investigation and risk assessment
- two commonly used **occupational health and safety management systems** (OH&SMSs) are described, which illustrate the additional activities needed to keep the risk management model operating effectively. These systems include additional elements such as policy, planning and audit.

In the course of Chapter 4, all the aspects of risk management to be dealt with in Chapters 5 to 11 are described in outline, and information is given on which of these chapters contains more detailed information on each aspect. Readers who are not familiar with basic risk management terminology should read Chapters 3 and 4, and then use Chapter 4 as their route map through Chapters 5 to 11. For those readers with a working knowledge of risk management terminology, the main contents of Chapters 5 to 11 are as follows:

- **Chapter 5: Risk assessment.** Risk assessment – that is identifying the nature and magnitude of risks – is a key part of risk management. Chapter 5 describes the basic risk assessment techniques.
- **Chapter 6: Risk control.** There is a variety of ways in which risks can be controlled, some of them more effective than others. This chapter describes the main types of risk control and how their relative effectiveness can be estimated.
- **Chapter 7: Safe systems of work.** Making sure that people work safely is an important aspect of risk management. In Chapter 7 three aspects of getting people to carry out tasks safely are described. These are safe work procedures, permit to work procedures and safety rules.
- **Chapter 8: Monitoring and measuring losses.** This chapter begins with a general description of the sorts of losses people, organisations and the environment can sustain, and identifies which of these losses will be dealt with in the remainder of the book – for example, injury, ill health and asset damage. There is then a discussion of how data on these losses can be collected effectively and analysed.
- **Chapter 9: Identifying causes and patterns.** This chapter deals with two main topics: accident investigation and a specialised form of data analysis used to identify patterns in the occurrence of accidents.
- **Chapter 10: Monitoring and measuring conformity.** Monitoring is checking that, for example, risk control measures are in place and are working as intended. Chapter 10 deals with the purposes and scope of monitoring, the techniques that can be used, and how monitoring conformity can be extended to measuring conformity.

- **Chapter 11: Other elements of occupational health and safety management systems.** This chapter describes those elements of OH&SMSs that have not already been dealt with in previous chapters – for example, review and internal audit.

Taken together, these seven chapters deal with the majority of topics required for effective risk management. However, two other topics are also required – communication and training – and these are dealt with in **Chapter 12: Communication and training**, which is the last chapter in Part 1.1.

3: Risk management – setting the scene

Introduction

This chapter deals with risk management in a fairly informal manner as an introduction to the subject. It is covered more formally in the next chapter. This two-stage approach is desirable because risk management is a potentially complex subject and it is preferable to go for gradual immersion, rather than diving in at the deep end.

Since getting to grips with risk management is not a trivial task, it is worthwhile considering whether it is worth doing. For this reason, the chapter also contains a discussion of the reasons why people would want to manage risk.

Basic terminology

There are four basic terms used in risk management: 'hazard', 'hazardous event', 'risk' and 'loss'. However, the term 'harm' is often used instead of loss, and harm and loss will be used interchangeably in this book.* This part of the chapter deals with how these words are used in the context of risk management, which can be succinctly defined as the management practices put in place to reduce losses.

Hazard, hazardous event, risk and loss

Imagine that you are the owner of a house and that, for some reason, the roof has been temporarily removed. If this were the case, you would naturally be concerned about rain, since this could damage the decoration in your house and its contents.

In these circumstances, rain clouds would be a hazard and a period of rainfall would be a hazardous event.† As illustrated below, you would want to know various things about this hazardous event.

How likely is rain?
Clouds can pass over and rain on someone else!

How much rain will there be, and how fast?

A few seconds of drizzle may not be a problem – a prolonged downpour would be.

---

* There are various definitions of these terms, and we will consider them in Chapter 5, which deals with risk assessment in more detail.

† Hazardous events are sometimes referred to as 'incidents'. The critical feature of both hazardous events and incidents is that they could result in loss, but the loss is not inevitable – the rain could be of such short duration and so light that you suffer no loss.

What the rain clouds constitute is a threat to your assets – that is, your possessions are 'at risk' of water damage and you may sustain a loss. The extent of your loss depends on the value of your assets and the form they are in. For example, if your main assets are ancient Greek statues, a few hours of rain will do little damage. If you are a collector of first edition books, the same amount of rain will create a substantial amount of damage. The important point to note is that the extent of the loss is independent of the hazardous event, since it depends on the nature of the assets. This is summarised in the diagram below.

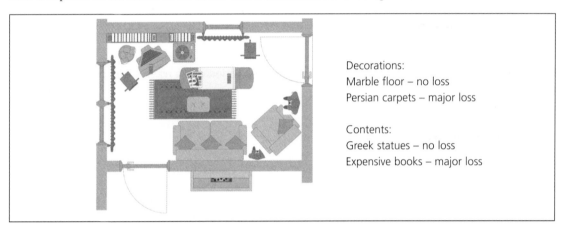

Decorations:
Marble floor – no loss
Persian carpets – major loss

Contents:
Greek statues – no loss
Expensive books – major loss

In risk management, the term 'risk' is used to describe a combination of the hazardous event and the loss, and, in any given set of circumstances, risk takes into account the relevant aspects of both. That is, risk is a combination of the likelihood and nature of the hazardous event and the extent of the loss. In risk management terms, it would be said that:

$$\text{rain} \times \text{statues} = \text{low risk}$$
$$\text{and}$$
$$\text{rain} \times \text{books} = \text{high risk}$$

In everyday English, these two aspects of risk are encapsulated in two distinct grammatical forms:
1. When the reference is to the likelihood of a hazardous event occurring, it is said that 'there is a risk of'. Thus, we could say that 'there is a 10 per cent risk of rain'.
2. When the reference is to the sorts of loss arising as a result of a hazardous event, it is said that 'there is a risk from'. Thus, we could say that 'the risk from rain is property damage or flooding'.

Unfortunately, there is no grammatical construction in English which encapsulates the meaning of risk as used in risk management. In risk management, 'risk' is an abstract concept which takes into account both the hazardous event (risk of) and the loss (risk from). In this book, risk is defined as:

the likelihood of the hazardous event × the severity of the most likely loss in the circumstances

which will be abbreviated to 'likelihood × severity'.

There are many other definitions of risk, and these are described in Chapter 20, which also includes the reasons for using the definition above rather than one of the others.

Using this definition has a major advantage in that it allows a calculation of risk: it is possible to multiply likelihood (in the form of a probability) by severity (in the form of an amount of money) to give a numerical measure of risk. This can then be used in a variety of ways, including to compare one risk with another. However, a full discussion of this aspect of risk will be left until the chapter on risk assessment (Chapter 5).

Another word which is used in discussing risk is 'danger'. However, there is no generally agreed definition of danger in this context and it is not clear how danger and risk differ. The *Oxford English Dictionary* (*OED*) gives eight meanings for danger and identifies the current 'main sense' as:

> liability or exposure to harm or injury; the condition of being exposed to the chance of evil; risk, peril.

So even the *OED* sees risk and danger as synonymous! Nevertheless, danger is a useful general word for informal discussion and will be used as such throughout the book. However, danger will not be used in any technical sense.

The example just used is rather artificial, since it would not be normal to have a house without a roof. However, we will continue with it because it enables us to consider two more important concepts in risk management: risk control measures and residual risk.

Risk control measures

Risk control measures* are, rather obviously, what is done to reduce risk. Having a roof on a house is one of the risk control measures used to deal with the risk from rain. There are three basic categories of risk control measures:

- **Measures which are designed to reduce the likelihood aspect of risk.** The roof on a house does not fall into this category since the presence or absence of a roof does not affect the likelihood of rain. However, many risk control measures are designed to affect likelihood – for example, a manhole cover reduces the likelihood of there being an open manhole down which people can fall.
- **Measures which are designed to reduce the severity aspect of risk.** The roof on a house does fall into this category, and an effective roof reduces the severity to near zero. Many other risk control measures, including safety glasses, car seat belts and hard hats, also fall into this category.
- **Measures which are designed to reduce both aspects of risk.** These are usually technical measures; a fairly common one is the use of low voltage, battery-driven, portable electrical drills instead of 250 V drills with cables to the mains electricity supply. This measure not only reduces the likelihood of electric shock (there are no cables or sockets required where the drill is in use, although they will be required at the point where the drill is recharged) but also reduces the severity of any shock received (a low voltage instead of 250 V).

Whichever category the risk control measures fall into, they have two features in common.

First, they vary in effectiveness. This is fairly obvious in the case of roofs, with some being better at keeping out the rain than others. However, for other sorts of risk control measure, it is not always obvious how their effectiveness can be judged. One of the many skills required in risk management is the ability to choose the most effective risk control measures.

Second, most risk control measures require some degree of maintenance. Again, this is fairly obvious for roofs. All roofs deteriorate with time and, if they are not repaired as necessary, they will eventually let in the rain. It is usual to make a distinction between two types of maintenance:

1. **Reactive maintenance** – waiting until the roof leaks and then having it fixed.

---

* Risk control measures are also known as 'controls', 'risk controls' and 'workplace precautions'. These terms will be described in Chapter 6, which deals with risk control.

2. **Preventive maintenance** – checking the roof regularly and replacing anything which looks likely to cause the roof to leak before the next check is due.

In general, preventive maintenance is preferable, but it is not always possible.

Residual risk

No matter how good a roof may be, or how well it is maintained, there is always the possibility that it will fail and the contents of the house will be damaged by rain. This risk, which will be very small if there is a good, well-maintained roof, is referred to as the residual risk.

There are various ways of dealing with residual risk but the most frequently used methods are to:

- **accept the risk** – that is, do nothing about it and hope it will not materialise
- **use secondary risk control measures** – for example, keep very expensive items in a waterproof cabinet in the house
- **take out insurance so that the financial damage sustained is borne by someone else** – this does nothing to reduce the likelihood or severity of the loss, but it transfers the financial consequences from the owner of the assets to the insurance company.

Note that the level of risk without risk control measures is referred to as 'absolute risk', so:

$$\text{absolute risk} - \text{risk reduction from risk control measures} = \text{residual risk}$$

By now, you may have come to the conclusion that risk management is more trouble than it is worth, despite the fact that this is only the scene-setting stage. To counter any such impression, the reasons for managing risk will be considered before we look more formally at the elements of risk management in the next chapter. However, one common source of loss is accidents and it is necessary to consider what is meant by an accident before moving on.

Accidents and incidents

There are various definitions of 'accident' and two examples are:

> ... any undesired circumstances which give rise to ill health or injury; damage to property, plant, products or the environment; production losses or increased liabilities.[1]

> ... an incident which has given rise to injury, ill health or fatality.[2,3]

These two definitions imply that an accident is something which has given rise to some form of adverse outcome or loss; this is probably what most people would think of as an accident. However, there can still be 'undesired circumstances', 'undesired events' or 'incidents' which do not result in a loss, and in the three sources just quoted these are defined as:

> incidents ... all undesired circumstances and 'near misses' which could cause accidents.[1]

> work-related event(s) in which injury or ill health (regardless of severity) or fatality occurred, or could have occurred.[2,3]

It can be seen from these definitions that there are two distinct uses of the term 'incident':
1. Incidents are those undesired circumstances or hazardous events where no harm occurs[1] and accidents are a separate category.
2. Incidents are events where harm did occur, or could have occurred, and accidents are a subset of incidents.[2,3]

Since it is useful to differentiate between events where harm or loss occurred and events where there was no harm or loss, the following definitions are used for the remainder of this book:

Accident: a hazardous event which resulted in harm
Near miss: a hazardous event which did not result in harm
Incidents: accidents and near misses

The reasons for managing risk

There are various reasons for managing risk, but they can be summarised in four main groups:
• ethical and moral considerations
• legal requirements
• financial matters
• more general business considerations.

This introductory section looks briefly at all of these.

Ethical and moral reasons
Most people would agree that, whatever risks they choose to take themselves, it is unacceptable to put other people at risk, particularly when this is done without their knowledge or consent. Many of the decisions taken by managers and health and safety professionals influence the risks to which other people are exposed and they have, therefore, an ethical and moral duty to make sure that these risks are not unacceptable. In addition, there is an expectation on the part of society in general that organisations will take reasonable care to ensure that the people and activities they manage do not harm other people or their property.

This expectation has changed over the years with general shifts in the attitude of society to health and safety. What was acceptable 20 years ago in many aspects of life is no longer acceptable today. This is perhaps most noticeable in relation to environmental issues, which were not even generally discussed 20 years ago. However, people in general are now less tolerant of a lack of health and safety, and ethically and morally unjustifiable practices such as the payment of 'danger money' and 'dirt money' are now rare. It seems likely that as people's expectations of life in general increase, their expectations of a healthy and safe life also increase.

Legal requirements
Health and safety legislation in most countries places a number of duties on managers* and failure to carry out these duties can result in fines and, in extreme cases, imprisonment.

Action can also be taken even before an injury has been sustained. In the UK, for example, the Health and Safety Executive (HSE) has the power to stop organisations carrying out practices that it considers too risky, and this is done by issuing a Prohibition Notice. The HSE can also require organisations to improve health

---

* More accurately, UK law places duties on employers, and these duties are discharged by managers. However, referring to managers suits the purposes of this book, and will be continued throughout.

and safety practices by issuing an Improvement Notice, which gives an organisation a period of time to make improvements in the relevant areas.

The law also provides mechanisms by which people who have been injured or who have suffered a loss as a result of the unsafe practices of an organisation can claim compensation. Paying this compensation, or having to insure against such payments, can be a major source of financial loss to an organisation – financial losses are dealt with next.

Financial matters

In 2011, the HSE published figures showing that the total cost associated with workplace injuries and ill health (excluding occupational cancers) in Great Britain was £14 billion in 2009 to 2010.[4] Of the total cost, workplace illness cost society an estimated £8.5 billion, and workplace injury (including fatalities) an estimated £5.4 billion. £6.3 billion of the total represents financial costs; the remaining £7.6 billion represents the monetary value given to individuals' 'pain, grief and suffering'.

The financial costs are made up of direct (usually insurable) costs and indirect (usually uninsurable) costs. Examples of direct costs include those associated with injuries, ill health and damage, and examples of indirect costs include time spent on investigations, legal actions and production delays. The HSE has calculated that for every £1 of insured costs, there are between £8 and £36 of uninsured costs, depending on the type of organisation.

For most organisations, accident costs result in a continuing, but often unquantified, drain on their resources. There have been cases where accidents have resulted in companies closing down, either directly because key personnel are killed or injured or key assets are damaged, or indirectly because the costs of recovery from the accident put too great a strain on the organisation's cash flow.

When the true costs of accidents are taken into account, it makes good business sense to minimise risk. Later in the book, the financial aspects of risk management are discussed in more detail (see Chapter 24).

When an organisation is convicted of a breach of health and safety legislation, it can be fined. This, and claims for compensation, are additional costs of failing to manage risk.[4]

Other business reasons

Theoretically, the other business reasons could be considered as costs of accidents but, in practice, they are dealt with separately because they are more difficult to quantify. Essentially, they involve the organisation's image and the effect a poor health and safety record can have on a range of people and organisations, including:

- **employees** – a poor health and safety record can increase costs due to low morale and increased staff turnover
- **customers** – a poor health and safety record can result in customers using alternative products or services
- **shareholders, banks and other sources of finance** – a poor health and safety record can discourage investment in and lending to the organisation
- **insurance companies** – a poor health and safety record can result in increased insurance premiums and, in extreme cases, make insurance impossible to obtain.

In the UK, it is recognised that maintaining a good corporate image is important for many organisations. The following are two examples of how an organisation's image is used as a motivator:

- The HSE devotes a section of its website to listing organisations which have been prosecuted and giving details of the prosecutions.
- Where there has been a fatality at work involving a gross breach of a relevant duty of care, and the way in which the organisation's activities are managed or organised by its senior managers is a substantial element in the breach, the court can issue a publicity order. These publicity orders require the organisation to publicise, in a specified way, a range of information, including the fact that it has been convicted of the offence and the amount of any fine imposed.

Summary

This chapter looked, in an informal way, at what is meant by risk management and how the terms 'hazard', 'hazardous event' 'risk' and 'loss' are used in the context of risk management. We also considered briefly the reasons why managing risk is important. The next chapter outlines a more formal approach to risk management techniques.

4: Key elements of risk management

Introduction

This chapter deals more formally with the key elements of risk management and, in order to provide a structure for this discussion, it will be based on the Plan, Do, Check, Act (PDCA) sequence. This sequence forms the basis of many of the published management systems, including those for quality (ISO 9001[5]) and environment (ISO 14001[6]). These management systems are described in Chapter 18 but, for now, the following summary of the PDCA sequence is all that is required.

- **Plan** – establish the objectives and processes necessary to deliver results in accordance with the organisation's health and safety policy
- **Do** – implement the processes
- **Check** – monitor and measure the processes against the health and safety policy, objectives, and legal and other requirements, and report the results
- **Act** – take actions to continually improve the processes.

This PDCA sequence can be applied at various levels in an organisation. This chapter deals with its application at two levels:
- **Managing specific risks.** The PDCA sequence would be appropriate for managing specific risks, such as those arising from vehicles or electricity. This application of PDCA is described later in this chapter in the form of the risk management model.
- **Managing all risks in an organisation.** This application of PDCA is described in the form of an occupational health and safety management system (OH&SMS), and various such systems are described later in this chapter.

The risk management model

Figure 4.1 shows the elements of the risk management model (RMM) and each of these elements is explained briefly in the notes below. The subsequent chapters in Part 1.1 deal with each of the elements in more detail.

Plan
The purpose of planning is to decide on appropriate risk control measures. This will require a risk assessment, which is the process used to identify the nature and extent of the risks and the possible options for control measures. Incident investigation results should also be taken into account, since these are a record of what has gone wrong in the past and, by implication, what should be prevented in the future. Legal requirements and any other requirements imposed by, for example, customers should also be taken into account, since risk control measures are unlikely to be appropriate if they do not satisfy all relevant legal and other requirements. Risk assessment is dealt with in Chapter 5 and risk control in Chapter 6. The nature of standards and how they are set is examined in Chapter 10.

Do
This part of the sequence consists of implementing and maintaining the risk control measures decided on during planning; there is no separate chapter on this element.

Figure 4.1
The elements in the risk management model

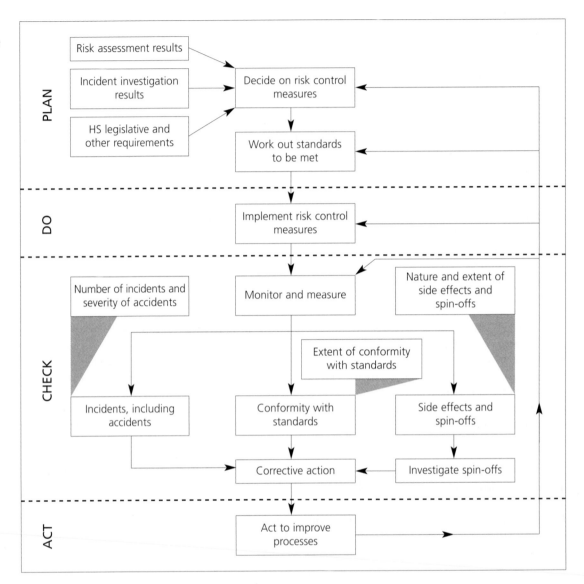

Check

The minimum requirement for checking is monitoring that the risk control measures decided on during planning continue to be implemented and maintained in the ways intended – that is, monitoring conformity with the standards to be met. However, two other types of monitoring are desirable:

- **Monitoring of incidents, including accidents.** This enables an assessment of whether the risk control measures are effective: that is, are they providing the risk reduction they were intended to provide?
- **Monitoring of side effects and spin-offs.** Side effects are detrimental effects of risk control measures, such as reduced productivity or discomfort caused by wearing goggles. Note that incidents are one type of side effect, but because of their use in estimating the effectiveness of the risk control measures, they are treated as a separate category. Spin-offs are beneficial effects of the risk control measures other than their effect in reducing risk. Examples of spin-offs include increased productivity and more accurate working because wearing goggles enables closer inspection.

Monitoring does not usually involve measuring, but it may do in special cases – this can be illustrated by examples.

- The majority of monitoring of car tyre pressure involves visual inspection, without measuring. However, when visual inspection suggests that a tyre is deflating, then the pressure is measured to check this.
- There are various gases, such as carbon monoxide and carbon dioxide, which humans cannot detect. The only way of monitoring the presence of these gases is, therefore, to have instruments which measure their concentration in the atmosphere.

This use of measuring as part of monitoring must not be confused with measuring as a separate process. Most of the things which can be monitored can also be measured and examples include the following:

- **Risk control measures.** If the risk control measure is people wearing hard hats on a building site, monitoring involves checking that people are wearing them. Measuring involves counting the number of people who are wearing their hard hats and the number who are not.
- **Incidents.** Measuring incidents involves, for example, determining how many there are and, in the case of accidents, how severe they are in terms of days lost or extent of injury.
- **Side effects.** If the side effect is reduced productivity, measuring involves determining the extent of the reduced productivity. If the side effect is discomfort cause by wearing goggles, this may be more difficult to measure. However, if people take more breaks because of the discomfort, the effect can be measured by the reduced productivity. In more extreme cases, the effect may be measurable as days away from work.
- **Spin-offs.** As with side effects, measurement involves assessing the extent of the benefit arising from the spin-off rather than just noting that there is a benefit.

When an incident occurs, or a nonconformity,* side effect or spin-off is identified, the actions required are as follows:

- **Nonconformities.** Corrective action should be taken. In the PDCA sequence, 'corrective action' means action taken to correct the causes of the nonconformity, so it usually requires an investigation to identify these causes. The term 'correction' is used in the PDCA sequence when a nonconformity is dealt with but the cause is not. For example, if a fire extinguisher is found off its bracket and being used to prop open a door, correction is to put the fire extinguisher back on its bracket. Corrective action is finding out why the fire extinguisher is being used in this way and either providing a suitable alternative method of propping the door open, or removing the need to prop it open.
- **Incidents.** Where appropriate, corrective action should be taken so that the incident will not happen again.
- **Side effects and spin-offs.** Where necessary, there should be corrective action for side effects, and the causes of spin-offs should be investigated to find out whether the beneficial effect can be achieved elsewhere.

Monitoring and measuring are dealt with in more detail in Chapters 8 and 10 and corrective action is dealt with in Chapter 9.

Act

The Act element of the PDCA sequence involves taking some action to improve the processes used to control or manage risk. The Act element ensures the continual improvement required by modern management systems and the nature of this continual improvement is dealt with in the next section.

---

* In BS OHSAS 18001 (and in ISO 9000 and ISO 14001), nonconformity is defined as 'non-fulfilment of a requirement'. This can be any requirement, including a requirement to have risk control measures.

Occupational health and safety management systems – introduction

Effective management of health and safety risks across an organisation requires the implementation and maintenance of an HSMS. As we mentioned earlier, management systems are also based on the PDCA sequence but they have many more elements than the RMM. There are many published OH&SMSs and a number of these are described later in the chapter. However, BS OHSAS 18001 is used first as an illustration.

BS OHSAS 18001[3]

The diagram used for this OH&SMS is given in Figure 4.2, which is followed by notes on the main elements. Where relevant, the notes identify the chapter in which a more detailed treatment of the element can be found.

Figure 4.2
BS OHSAS 18001: 2007[3] occupational health and safety management system

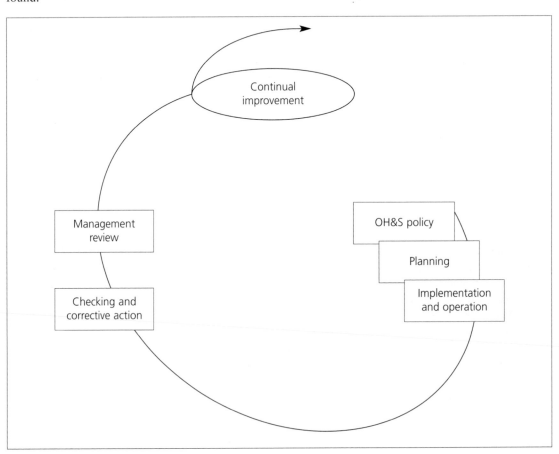

Continual improvement

Continual improvement is defined in BS OHSAS 18001 as the 'recurring process of enhancing the OH&S management system in order to achieve improvements in overall OH&S performance consistent with the organization's policy'.

Note that the focus is on enhancing the management system in order to achieve improvements in occupational health and safety performance, which is defined as 'measurable results of an organization's management of its OH&S risks'. This is consistent with the continual improvement of the processes used to control risk as described in the RMM. In both cases, the assumption is that improvements to the processes or systems will result in improvements in what these processes and systems are intended to achieve.

OH&S policy

BS OHSAS 18001 requires that top management 'shall define and authorize the organization's OH&S policy' and specifies various requirements for this policy. These are discussed in Chapter 11.

Planning

The subsections making up the 'Planning' section are:

- **Hazard identification, risk assessment and determining controls.** This subsection sets out detailed requirements for hazard identification and risk assessment procedures and methodologies, and specifies a hierarchy of controls to be considered. These topics are dealt with in detail in Chapters 5 and 6.
- **Legal and other requirements.** Organisations are required to identify and access the applicable legal and other requirements, for example those imposed by customers or trade associations. They also have to keep this information up to date and communicate pertinent information to relevant interested parties.
- **Objectives and programmes.** This subsection requires organisations to have documented occupational health and safety objectives at relevant functions and levels, and these objectives must be measurable, where practicable. There must be programmes for achieving these objectives and the programmes must include designation of responsibility and authority, means and time frames. Programmes must be reviewed at planned intervals and adjusted as necessary to make sure the objectives are achieved.

Occupational health and safety planning at the organisational level is dealt with in Chapter 11.

Implementation and operation

The subsections making up the 'Implementation and operation' section are as follows:

- **Resources, roles, responsibility, accountability and authority.** This subsection begins by requiring top management to take ultimate responsibility for occupational health and safety and the OH&SMS, and goes on to set requirements for resource allocation and the allocation of occupational health and safety roles, responsibility, accountability and authority. A particular requirement is the appointment of a member of the top management as the 'management appointee' who has specific responsibility for the OH&SMS. All those with management responsibility have to 'demonstrate their commitment to the continual improvement of OH&S performance' and the organisation must ensure that 'persons in the workplace take responsibility for aspects of OH&S over which they have control'.
- **Competence, training and awareness.** The overall requirement in this subsection is to ensure that those performing tasks that can affect occupational health and safety are competent 'on the basis of appropriate education, training or experience'. There are more detailed requirements for training needs analysis and for awareness provision and the delivery of training. These topics are dealt with in more detail in Chapter 12.
- **Communication, participation and consultation.** The communication requirements deal with both internal and external communications. The participation requirements apply only to 'workers' and they include involvement in various activities, consultation on changes, and representation. The consultation requirements involve consultation with contractors and 'relevant external interested parties'.* These topics are dealt with in more detail in Chapter 12.

---

* The inclusion of consultation of workers in 'participation' can lead to confusion in the UK, where such consultation is the subject of specific legislation – see Chapter 12.

- **Documentation.** This subsection lists what should be included in the OH&SMS documentation – for example, the occupational health and safety policy and objectives and a description of the main elements of the OH&SMS and their interaction, and reference to related documents. There is a note to this subsection:

  > It is important that documentation is proportional to the level of complexity, hazard and risks concerned and is kept to the minimum required for effectiveness and efficiency.

  These topics are dealt with in more detail in Chapter 11.
- **Control of documents.** Document control includes approval, review, and ensuring that documents remain legible and readily identifiable. The documents covered include both internally generated documents and documents of external origin. Documentation and document control are dealt with in more detail in Chapter 11.
- **Operational control.** Where controls are needed to manage occupational health and safety risks, there are requirements for operational controls and documented procedures where necessary. The requirements extend to occupational health and safety risks arising from change, contractors and other visitors to the workplace, and purchased goods, equipment and services. These topics are dealt with in more detail in Chapter 11.
- **Emergency preparedness and response.** This subsection requires organisations to have procedures for identifying the potential for emergencies and for responding to them. The emergency response procedures must be tested periodically. Emergency preparedness is dealt with in Chapter 22.

Checking*
The subsections making up the 'Checking' section are as follows.
- **Performance measurement and monitoring.** Occupational health and safety performance has to be monitored and measured regularly and particular items have to be monitored, including the extent to which objectives have been met and the effectiveness of controls. These topics are dealt with in more detail in Chapters 8 and 10.
- **Evaluation of compliance.** Organisations have periodically to evaluate compliance with applicable legal requirements and with the other requirements to which they subscribe. Records of these evaluations have to be kept. Evaluation of compliance is dealt with in more detail in Chapter 10.
- **Incident investigation.** Incidents have to be recorded, investigated and analysed. The purposes of this work include identifying the need for corrective action, identifying opportunities for preventive action,[†] and identifying opportunities for continual improvement. Incident investigation is dealt with in Chapter 9.
- **Nonconformity, corrective action and preventive action.** Nonconformities must be identified and corrected and there must be action to mitigate their consequences for occupational health and safety. Corrective and preventive actions must be subject to risk assessment when they concern new hazards or new controls, and the effectiveness of corrective and preventive actions must be reviewed. These topics are dealt with in more detail in Chapter 10.
- **Control of records.** The purpose of keeping records is to demonstrate conformity to the requirements of BS OHSAS 18001. There are requirements for the identification, storage, protection, retrieval, retention

---

* In the diagram used in BS OHSAS 18001 and reproduced here as Figure 4.2, this section is called 'Checking and corrective action'. However, in the text of the Standard, the heading is just 'Checking'.

† These are defined in BS OHSAS 18001 as follows. Corrective action is 'action taken to eliminate the cause of a detected nonconformity or other undesirable situation' and preventive action is 'action taken to eliminate the cause of a potential nonconformity or other undesirable potential situation'. However, in other OH&SMSs, 'preventive action' is used to mean prevention of recurrence of accidents and incidents.

and disposal of records, and records must remain legible, identifiable and traceable. Control of records is dealt with in more detail in Chapter 11.

- **Internal audit.** The requirement is for audits of the management system and there must be a programme of these audits. The objectivity and impartiality of the audit process must be ensured. Internal audit is dealt with in more detail in Chapter 11.

Management review

The requirement in this element is that the top management of the organisation review various aspects of the organisation's occupational health and safety performance and the organisation's OH&SMS. This type of review is dealt with in Chapter 11.

Other HSMSs

There are four other published HSMSs which we need to consider in this part. Other systems are dealt with in Part 2.1. Two of the systems to be dealt with here have been produced by the HSE, one by the British Standards Institution (BSI) and one by the International Labour Office. The four HSMSs will be referred to as follows:

- HS(G)65.[8] This HSMS was described in a booklet produced by the HSE, *Successful health and safety management*. However, since this title is rather unwieldy, the booklet is usually referred to by its HSE code number, HS(G)65. This booklet was produced in 1991 and it has had a major influence on health and safety management in the UK.
- HSG65.[1] This is a revised version of *Successful health and safety management*, published in 1997. Fortunately, the HSE dropped the brackets around the 'G' in this new version, so there is a convenient way of differentiating the two versions.
- OSH 2001.[9] These *Guidelines on occupational safety and health management systems* are produced by the International Labour Office (ILO) and they cover two main areas. First, they provide 'a national occupational safety and health framework' which is intended to provide guidance for governments wanting to set up a national occupational health and safety standard. This aspect of the guidelines is not within the scope of the present text and is not dealt with further. The second main area is the 'occupational safety and health management system in the organization' and it is this part of the guidelines which is dealt with later in this chapter.
- BS 18004. This HSMS is described in *Guide to achieving effective occupational health and safety performance*. The HSMS described is a slightly modified version of BS OHSAS 18001 and the text of BS 18004 describing this HSMS is taken from BS OHSAS 18001 and OHSAS 18002,[86] which is the guidance on BS OHSAS 18001. The only substantial new material in BS 18004 is in the annexes and this material will be dealt with at appropriate points in later chapters.

Each of the four HSMSs is summarised in its respective source document in the form of a diagram, and these diagrams are reproduced here as Figures 4.3, 4.4, and 4.5, followed by explanatory notes.

The key points to note about these three diagrams is that they all contain roughly the same elements, but in different orders. BS OHSAS 18001 has been used to illustrate these elements but any of the other HSMSs could have been used.

In the UK, HSG65 is widely used and, for this reason, the remainder of this chapter consists of a description of the main elements of this system and a table comparing BS OHSAS 18001 and HSG65. However, there are four introductory points to make:

- Although many organisations claim to have adopted HSG65, few have moved from the HS(G)65 terminology. For example, HSG65 introduced the terms 'workplace precautions' and 'risk control systems' (RCSs), but these terms have not been widely adopted.
- HS(G)65 and HSG65 do not use the term 'occupational health and safety'; they use 'health and safety'.

Figure 4.3
HS(G)65 and HSG65
health and safety
management
systems

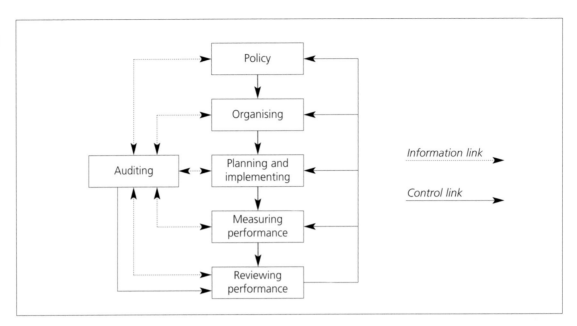

Figure 4.4
BS 18004
occupational health
and safety
management system

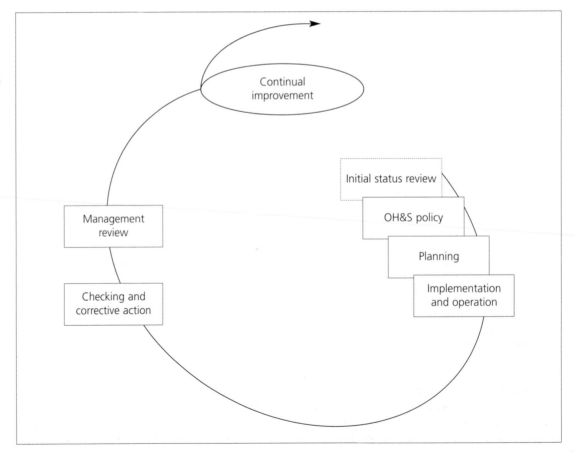

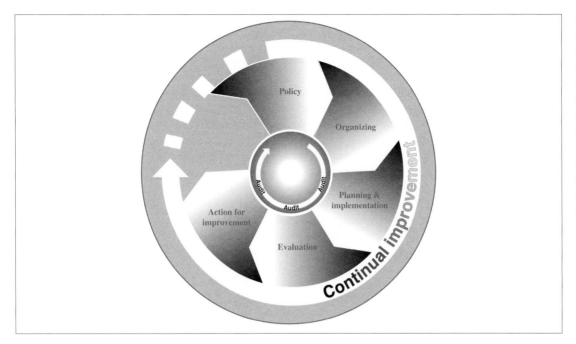

Figure 4.5
OSH 2001 occupational health and safety management system (Acknowledgment: International Labour Organization, *Guidelines on occupational safety and health management systems*, ILO-OSH 2001, figure 2, page 5. Geneva: ILO, 2001. © International Labour Organization, 2001)[9]

- BS 18004 is intended as additional guidance for organisations using BS OHSAS 18001, which can be inferred from the similarity of their respective diagrams.
- ILO 2001 uses much of the HS(G)65 terminology.

HSG65
The notes which follow on the main elements of HSG65 summarise the key points included in each element.

Policy
In the context of health and safety, a policy statement should be a succinct record of what an organisation intends to achieve. The written policy can take many forms, ranging from a commitment to comply with relevant legislation to a commitment to being the safest organisation in the relevant industry sector. HSG65 contains a list of typical statements of 'safety philosophy' and a list of the issues which could form the basis of a policy statement. Further information on the HSG65 view of policy is given in Chapter 11.

In the UK, there is a legal requirement for some organisations to have a written health and safety policy. The UK requirements are given in the *UK legislation – health and safety policy* box overleaf.

Organising
In HSG65, the HSE identifies four important organisational issues, the four 'Cs' – Control, Co-operation, Communication and Competence – and brief notes on each of these are given below. All of these issues will be dealt with in more detail in later chapters.

Control
Control of safety in an organisation revolves around establishing clear responsibilities and making sure that they are met. Since this will entail the assessment of each individual's performance, it is important that any necessary standards are also clearly defined. Responsibilities and standards should be written down and, as appropriate, agreed with the people who will accept the responsibilities or have to meet the standards.

UK legislation – health and safety policy

It is a legal requirement to have a health and safety policy. The relevant legislation states:

> Except in such cases as may be prescribed, it shall be the duty of every employer to prepare, and as often as may be appropriate, revise a written statement of his general policy with respect to the health and safety at work of his employees and the organisation and arrangements for the time being in force for carrying out that policy, and to bring the statement and any revision of it to the notice of all of his employees. (Health and Safety at Work etc Act 1974, section 2(3))[10]

There is a general 'prescribed case' concerning the number of employees:

> Any employer who carries on an undertaking in which for the time being he employs less [sic] than five employees is hereby excepted as respects that undertaking from the provisions of subsection (3) of section 2 of the Health and Safety at Work etc Act 1974. (The Employers' Health and Safety Policy Statements (Exception) Regulations 1975)[11]

The legislation says 'policy, organisation and arrangements', but the NEBOSH syllabus[12] had the following:

> Description of the general components of a health and safety policy document:
> - statement of intent – overview, safety goals and objectives
> - organisation – duties, responsibilities and organisational structure in relation to health and safety
> - arrangements – systems, procedures, standards, cross-reference to key documents.

This teaching, in practice, led to health and safety professionals in the UK referring to the NEBOSH 'policy document' as the policy and attempting to bring this, usually substantial, document to the notice of all employees. This 'policy document' and 'policy' terminology is, however, included in the Control of Major Accident Hazards (COMAH) Regulations.[85]

NEBOSH now has 'legal requirements relating to health and safety policies and arrangements', so things should improve, at least outside the COMAH community.

Although it is not required by legislation, in large organisations it is usually necessary to have more than one health and safety policy. Typically, there is a policy written and revised by the board which covers the whole organisation. Each major part of the organisation then has its own policy. These policies should be consistent with the board policy but be tailored to the part of the organisation concerned. In particular, they should deal in detail with the risks in that part of the organisation. There will also be a need for more focused policies at lower levels, each consistent with those above, but dealing with a defined part of the organisation. The number of levels required in this cascade of policies will depend on the size of the organisation and the nature of the risks to be managed.

The requirements for the arrangements section of the health and safety policy have been extended by the Management of Health and Safety at Work Regulations 1999,[13] regulation 5:

> (1) Every employer shall make and give effect to such arrangements as are appropriate, having regard to the nature of his activities and the size of his undertaking, for the effective planning, organisation, control, monitoring and review of the preventive and protective measures.
> (2) Where the employer employs five or more employees, he shall record the arrangements referred to in paragraph (1).

An important issue in control is rewards and sanctions, in particular the sanctions to be used for non-compliance with safety standards. Ideally, everyone should be motivated to comply voluntarily with safety standards, but there must be effective sanctions as a backup.

Co-operation
Successful organisations are ones in which safety is 'everybody's business'. This means that there must be effective methods for ensuring participation and involvement at all levels. This should include involvement in activities such as setting performance standards, devising and recording systems of work, and dealing with safety problems as and when they arise. These activities should be carried out by groups set up to achieve a particular objective. However, superimposed on these groups there should be a formal structure of health and safety representatives and health and safety committees.

Communication
When considering communications, there are three main flows of information which are relevant. These are shown in Table 4.1.

Information flows		
Into organisation	**Within organisation**	**Out of organisation**
Legal developments Technological developments affecting risk control Developments in safety management practice	Safety policy Plans and standards Ideas for improvement Reports on performance	Accident and ill health data to enforcing authorities and HSE Data to planning authorities and emergency services Emergency communications

Table 4.1
Information flows into, within and out of an organisation[1]

Each of the information flows is maintained using three main techniques and these are summarised in Table 4.2.

Maintaining information flows		
Visible behaviour	**Written communications**	**Face-to-face discussions**
Active monitoring Chairing meetings Involvement in accident and incident investigations	Policy statements Responsibility statements Performance standards Safe systems of work Notices and posters Newsletters	Active monitoring Consultative meetings Safety on agenda of all management meetings Team briefings using the cascade principle 'Tool box talks', particularly for upward communication

Table 4.2
Techniques for maintaining information flows[1]

Competence
All employees, including managers, should be competent to meet their defined responsibilities. Training has a key role in promoting competence but other factors have to be taken into account.

These include having systems in place for competence needs analysis (or training needs analysis), and recruitment and placement procedures which ensure the selection of employees with the necessary mental and physical abilities for the job in question.

UK legislation – health and safety assistance

Regulation 7 of the Management of Health and Safety at Work Regulations 1999[13] requires that:

> (1) every employer shall ... appoint one or more competent persons to assist him in undertaking the measures he needs to take to comply with the requirements and prohibitions imposed on him by or under the relevant statutory provisions;
> (2) a person shall be regarded as competent for the purposes of paragraph (1) above where he has sufficient training and experience or knowledge and other qualities to enable him properly to assist in undertaking the measures referred to in that paragraph.

The Approved Code of Practice published in conjunction with the Regulations[14] provides information on what is meant by competence:

> Competence in the sense it is used in these Regulations does not necessarily depend on the possession of particular skills or qualifications. Simple situations may require only the following:
> (a) an understanding of current best practice;
> (b) an awareness of the limitations of one's own experience and knowledge; and
> (c) the willingness and ability to supplement existing experience and knowledge, when necessary by obtaining external help and advice.

> More complicated situations will require the competent assistant to have a higher level of knowledge and experience. More complex or highly technical situations will call for specific applied knowledge and skills which can be offered by appropriately qualified specialists. Employers are advised to check the appropriate health and safety qualifications (some of which may be competence-based and/or industry-specific), or membership of a professional body or similar organisation (at an appropriate level and in an appropriate part of health and safety) to satisfy themselves that the assistant they appoint has a sufficiently high level of competence. Competence-based qualifications on the Qualifications and Credit Framework and the Scottish Qualifications Framework may also provide a guide.

With these systems in place, training can be provided as appropriate to meet the three main types of need:
- **organisational needs,** for example safety policy and organisational structure
- **job needs,** for example management skills in leadership, managing safety and understanding risk, and knowledge of planning, measuring, reviewing and auditing systems
- **individual needs,** for example induction, maintaining performance and job changes.

In the UK, there is a special requirement for competence and this is dealt with in the *UK legislation – health and safety assistance* box above.

Planning and implementing

In HSG65, the HSE recognises the importance of planning but gives limited guidance on planning techniques. However, in BS 18004 one of the annexes deals in detail with planning and planning techniques, and these will be described in Chapter 22, where they will be dealt with in the context of emergency planning. An important adjunct to the planning process is the identification of the resources required to implement and maintain the plan. Once the resource requirements have been established, appropriate action will be needed to ensure that adequate resources can be made available. If they cannot be made available, the plan should be revised in the light of available resources.

Measuring performance
There are two parts to measuring performance: reactive monitoring, which deals mainly with accidents, and active monitoring, which involves, for example, health and safety inspections. This element of HSG65 also deals with accident investigation.

Audit
Audits should be carried out by people who are independent of the line management being audited and who have special competences in audit techniques and in the subject being audited. Audit should also cover all aspects of the HSMS.

Review
The primary purpose of performance review is to enable organisations to learn by experience and use the lessons learned to improve their health and safety performance. The reviews should be conducted regularly by managers at all levels in the organisation and cover all aspects of health and safety performance.

Comparison of BS OHSAS 18001 and HSG65
The tables below provide a mapping of HSG65 to BS OHSAS 18001 (Table 4.3) and a mapping of BS OHSAS 18001 to HSG65 (Table 4.4). Bold type is used in these tables to identify the main elements in each HSMS, with normal type used to identify sub-elements.

HSG65	BS OHSAS 18001
Policy	Occupational health and safety policy
Organising – control	Roles, responsibility, accountability and authority
Organising – co-operation	Participation and consultation
Organising – communication	Communication
Organising – competence	Competence, training and awareness
Planning and implementation – planning	Planning
Planning and implementation – implementation	Implementation and operation
Measuring performance – active monitoring	Performance measurement and monitoring
Measuring performance – reactive monitoring	Performance measurement and monitoring
Measuring performance – investigation	Incident investigation
Reviewing performance	Management review
Audit	Internal audit

Table 4.3
Comparison of HSG65 and BS OHSAS 18001

BS OHSAS 18001	HSG65
Occupational health and safety policy	Policy
Planning – hazard identification, risk assessment and determining controls	Planning
Planning – legal and other requirements	Planning
Planning – objectives and programme(s)	Planning
Implementation and operation – resources, roles, responsibility, accountability and authority	**Organising** – control
Implementation and operation – competence, training and awareness	**Organising** – competence
Implementation and operation – communication	**Organising** – communication
Implementation and operation – participation and consultation	**Organising** – consultation
Implementation and operation – documentation	
Implementation and operation – control of documents	
Implementation and operation – operational control	
Implementation and operation – emergency preparedness and response	
Checking – performance measurement and monitoring	Measuring performance
Checking – evaluation of compliance	
Checking – incident investigation	Measuring performance
Checking – nonconformity, corrective action and preventive action	Measuring performance
Checking and corrective action – control of records	
Checking and corrective action – internal audit	Audit
Management review	Reviewing performance

Summary

This chapter has dealt with the application of the PDCA sequence for two types of management in an organisation:

- the management of specific risks using the RMM
- the management of all aspects of health and safety using the BS OHSAS 18001 or HSG65 HSMSs.

The chapter also provided diagrams showing the main elements in other published HSMSs and a comparison of BS OHSAS 18001 and HSG65. The following chapters in Part 1.1 deal with these elements in more detail, starting with risk assessment.

5: Risk assessment

Introduction

This chapter will consider the techniques required for basic risk assessment. The more advanced risk assessment techniques will be dealt with in Part 2.1. Although there is much talk about risk assessment, there is no general agreement on terminology. For example, some writers include risk control as part of the risk assessment process, whereas others deal with it as a separate topic. The latter convention will be used in this book and risk control is dealt with separately in the next chapter.

However, even when risk control has been separated out in this way, there are still terminological difficulties with what constitutes risk assessment. To get round this problem with terminology, we will deal with the risk assessment process in separate stages as follows:

1. **Make an inventory of sources of hazards.** If there is no record of exactly what is being managed, there will not be a record of the sources of hazards. This record will be referred to as an inventory and there will also have to be adequate arrangements for keeping this inventory up to date.
2. **Identify hazards and predict hazardous events.** For each item in the inventory, it is necessary to identify the hazards associated with it and predict the hazardous events that may occur.
3. **Rate risks.** Some hazards are more serious than others; this is determined by rating the risk associated with each hazard.
4. **Decide whether the risk is acceptable.** Once the risks have been rated, it is possible to decide whether they are acceptable. If they are not, move to the next stage and select appropriate risk control measures.
5. **Select appropriate risk control measures.** Risk control measures are, rather obviously, the measures put in place to reduce the risk associated with a hazard. However, as was mentioned above, risk control will be dealt with as a separate topic.
6. **Decide on monitoring requirements.** While some risk control measures – for example, fixed barriers – will require very little monitoring, others – such as wearing gloves or hearing protection – may need very frequent monitoring.
7. **Decide on effectiveness criteria.** There should be some method of assessing whether the risk control measure is having the desired effect, and it is useful to identify the criteria that will be used to judge the risk control measure's effectiveness.
8. **Record risk assessments.** It is good practice to record the results of risk assessments, but it is also a legal requirement in some cases.
9. **Review risk assessments.** Things change and these changes can affect the appropriateness of the risk assessment. It is therefore good practice to review risk assessments from time to time and, in some circumstances in the UK, it is also a legal requirement.

This chapter will deal with all the aspects of risk assessment listed above, with the exception of the selection of appropriate risk control measures and deciding on monitoring requirements, which are covered in Chapter 6. It should be remembered, however, that risk assessment is a continuous process – this is shown in Figure 5.1 (overleaf), which is a flow chart summary of the steps listed above.

Figure 5.1

Summary of the risk assessment process

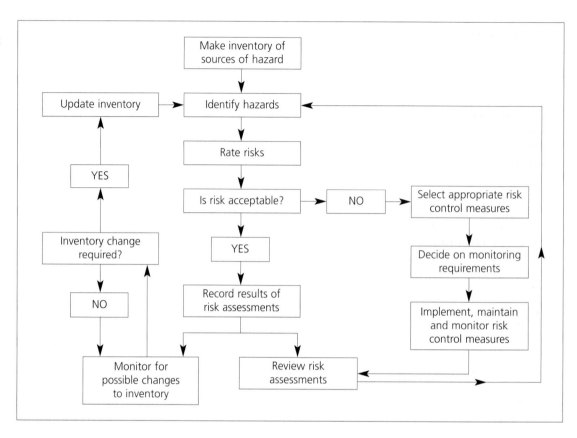

Inventory preparation

All of the people, things, activities and places which have to be managed can, in some circumstances, be hazards. It follows, therefore, that unless you know exactly what is being managed, it will not be possible to identify all the sources of hazards. We refer to the process of identifying what is being managed as inventory preparation. Typical categories which might form the basis for inventories include:

- a description of the locations for which each manager is responsible, including, if appropriate, a sketch map
- a list of the personnel regularly present in the locations
- a list of people who may enter hazardous areas within the locations from time to time, such as visitors and contractors
- a list of the permanent or semi-permanent items of machinery, plant and equipment present in the locations
- a list of the mobile machinery, plant and equipment which enter the locations
- a list of power tools and other tools used at the locations
- a list of the hazardous chemicals and substances stored, transported or used at the locations
- a list of the main energy sources, such as electricity, LPG and radiation, used at the locations
- a list of the tasks carried out at the locations.

In some organisations, particularly those managing very high risks, inventory preparation is a formal, detailed procedure with everything recorded. While this level of detail is not always required, it is always essential to make sure that there is clarity about what each person is managing and that everyone, everything and everywhere is being managed by someone.

For the majority of circumstances, this allocation of responsibility is straightforward, but common sources of ambiguity include:

- managing staff when they are working off site or in other departments
- managing common areas, such as car parks, corridors and stairs
- managing contractors and other people working on site who are carrying out, for example, cleaning and repairs
- managing shared items of equipment, such as photocopiers and lifts.

Inventories also have to be kept up to date – this means that they have to be reviewed periodically to check that they are still valid. A wide variety of ways can be used for establishing inventories and keeping them up to date, including IT-based methods. Whichever method is used, the critical test of its effectiveness is whether it is possible to identify easily all possible sources of hazards.

Identifying hazards and predicting hazardous events

The simplest approach to hazard identification and predicting hazardous events is to visit workplaces and look for unsafe conditions such as trip hazards, blocked fire exits and damaged electrical equipment. In more effective hazard identification procedures, this is supplemented by attempts to observe how people are working, with a view to identifying unsafe acts. However, the direct observation of unsafe acts is notoriously difficult for two reasons.

1. **The acts are usually of short duration and, unless the risk assessor happens to be there at the time they occur, they are not available for observation.** Contrast this with unsafe conditions which remain in place until someone puts them right. One way of getting round this problem is to allocate periods of time to work observation and make detailed records of any unsafe acts which occur during these periods of observation. If the periods for observation are carefully chosen, it may then be possible to draw conclusions about the number and nature of unsafe acts occurring overall. This general technique of observing samples of work activities in order to identify unsafe acts is known by a variety of names including 'safety sampling' and 'job safety sampling'.
2. **The fact that someone is observing usually changes the behaviour of the people being observed.** When it is known that someone is 'inspecting safety', procedures are carried out as they should be until the people carrying out the inspection stop observing. This problem can be overcome by using covert observation but this is usually of dubious ethical standing. A more acceptable alternative is to analyse the task itself, since this can be done either as a 'desk exercise' or under controlled conditions. This technique is considered in detail in Chapter 7.

There are two other problems with straightforward observation as a method of hazard identification.

1. **Not all hazards are perceptible by the unaided human senses.** For example, there are many toxic gases which are invisible and odourless; the only radiations that humans can detect are those in the spectrum of visible light; and humans are unable to detect the presence of viruses, bacteria and other micro-organisms.
2. **Even when the hazards are perceptible, a certain amount of knowledge is required to identify them as hazards.** Most people are familiar with most of the hazards associated with such things as electricity, fire and working at height, but specialised knowledge is required if people are to identify the hazards associated with particular chemicals, radiations, or working in confined spaces.

Thus, although direct observation is the single most important method of hazard identification, it must be carried out by people with the relevant knowledge and it must be a creative process. Passive observation is inadequate and it must be supplemented with questions about the work and the working environment. Two critical types of question are 'How?' and 'What if?', and some examples of such questions are given below.

How...
   do they clean those windows?
   do they get that heavy machine casing off for maintenance?
   do they get fresh supplies of chemicals into that hopper?
   do they isolate that electrical equipment?
   do they ensure that the fire exit does not get blocked?

What if...
   the container leaks?
   the machine stops (or starts) without warning?
   there is a fire?
   the mechanical handling equipment is out of use?
   the lights go out?

Asking the relevant questions about particular work activities and workplaces is a skill which has to be learned, but there are various sources which can be used to identify the sorts of hazards which might be present in any particular circumstances. Two of these sources – interviews and documentation – are described next.

Interviews

Interviews with, for example, the people who do the work, their managers and supervisors, and employee and health and safety representatives,* can produce valuable information. It is often possible to use the critical incident technique (recounting of 'horror stories') in this context. However, much valuable information can be obtained simply by asking people what they do under normal circumstances, and what they do when things go wrong. This basic and fairly obvious technique is too rarely used, but it can be a very effective method for identifying unsafe acts. Good incident investigation interviewing techniques (see Chapter 9) should also be applied in interviews used to identify hazards.

Documentation

There are numerous documentary sources which should be consulted in the course of hazard identification. These include the following:
* **Accident, ill health and other incident records from within the organisation.** These will provide information on what has gone wrong in the past; unfortunately, it is often the case that little has been done to stop it happening again.
* **Legislation, Approved Codes of Practice (ACoPs) and guidance.** Much legislation, and the ACoPs and guidance which accompany it, has been developed in response to known hazards, and these documents provide invaluable information on hazards at work and how they can be identified. Examples of hazards dealt with in UK legislation include chemicals and substances, work equipment, electricity and manual handling.

---

* In the UK, there is separate legislation dealing with safety representatives, who are nominated by a trade union, and representatives of employee safety, who are elected by their work colleagues (see Chapter 12). 'Health and safety representatives' is used in this book to include both categories.

- **Health and Safety Commission (HSC)* and HSE publications.** The HSC and HSE produce a range of publications, many of which concern specific hazards. Examples include guidance on identifying the hazards associated with construction or transport activities.
- **Guidance produced in house.** Many organisations produce guidance for internal use. This guidance may be prepared by in-house experts or specially commissioned from outside consultants, but, whatever its source, it can deal with hazards specific to the organisation or with particular procedures for hazard identification to be used within the organisation concerned.
- **British and international standards.** These standards deal with a wide variety of hazards, including those associated with the materials used in making soft furnishings (including the soft furnishings used in offices, hotels, hospitals and homes for the elderly) and with the use of machinery at work.
- **Industry and trade association guidance.** Many industries produce their own guidance material on hazards relevant to them. Examples include the printing and construction industries. In addition, specific trades such as welding produce guidance on the hazards associated with their work activities.
- **Information from manufacturers or suppliers.** The manufacturers or suppliers of, for example, chemicals, substances, tools and equipment may provide information which can be used to identify any hazards associated with the products they manufacture or supply.

Referring to an appropriate range of the documents listed above should provide the necessary knowledge to identify hazards in an effective manner, although you will still need to learn how to put this knowledge to use. This learning should preferably involve practice in hazard identification under the supervision of a competent person.

Risk rating

The usual method of risk rating is to devise one scale for the likelihood of the hazardous event and another for the severity of the most likely outcome. Individual hazardous events are then rated on each scale and the two ratings are combined to give a measure of risk.

If this methodology is to be effective, however, we have to be clear about what we mean by likelihood and severity, and these terms are discussed next.

Likelihood
There are different views on what constitutes likelihood in the context of risk rating. However, for reasons which will be discussed in detail in Part 2.1, the preferred measure of likelihood is the likelihood of a hazardous event. This is probably best illustrated by example:
- If the hazard is a trailing cable, one hazardous event is someone tripping.
- If the hazard is electricity, one hazardous event is the live conductor and a person coming into contact with each other.
- If the hazard is a fuel, one hazardous event is this fuel coming into contact simultaneously with an ignition source and oxygen.

The likelihood of the hazardous event is defined independently of the possible outcomes. For example, the likelihood of tripping is rated, irrespective of whether the person who trips will be hurt.

Suggested scales for likelihood rating are given in Table 5.1 which includes numerical values and possible qualitative equivalents. Examples of how these scales can be used include:

---

* The HSC and HSE have now merged as the HSE, but there are still many publications with the HSC imprint.

Table 5.1

Scale for rating the likelihood of a hazardous event

Rating	Probability	Qualitative description
1	<0.19	Very unlikely
2	0.2–0.39	Unlikely
3	0.4–0.59	Likely
4	0.6–0.79	Very likely
5	>0.79	Almost certain

- Rating 1 ($p < 0.19$). The likelihood that a person will fall down a manhole with a properly fitted, intact cover. The likelihood that a person will push too hard when opening a conventional room door.
- Rating 3 ($p = 0.4$–0.59). The likelihood that a person will fall down an open manhole protected only by a flimsy barrier. The likelihood that someone will push too hard when pushing a broken-down car with help from three other people.
- Rating 5 ($p > 0.79$). The likelihood that a person will fall down an open manhole which has no barrier and no warning signs. The likelihood that someone will push too hard when pushing a broken-down car uphill unassisted.

A number of factors will have to be taken into account when rating the likelihood of a hazardous event. For example, if the hazardous event was falling down an open manhole, relevant factors would be the size of the manhole, the number of people walking past it and how long the manhole will remain open. More generally, likelihood can be estimated on the basis of duration of exposure where hazards are continuous, or frequency of exposure where hazards are intermittent. However, it is usually the case that likelihood increases with time and it is, therefore, important in all cases to take account of how long the hazard will be in existence.

The examples given illustrate the extremes and a mid-point of the scale. All the points on the scale would, however, be used as appropriate in 'real life' likelihood ratings.

It is important to note that these ratings have to be subjective. They depend on risk assessors' judgment given their knowledge at the time.

Severity
A number of different types of severity could be used in severity ratings and some examples are given below:
- The severity of the most serious harm which could occur. For example, if a person trips, the most severe harm could be a fatality.
- The severity of the most likely harm in the circumstances. For example, if a person trips in an office, the most likely harm is a bruise.
- The combined severity of all the harms which could occur, taking into account their different probabilities.

For reasons which will be discussed in detail in Part 2.1, the second of these is to be preferred in most circumstances and scales for carrying out this type of rating are given in Table 5.2.

The cost scale shown in Table 5.2 would be suitable for the majority of risk ratings, but where more serious accidents are possible, either of the following can be done:
- increase the values in the 1–5 cost scale
- extend the 1–5 cost scale to, say, a 1–9 scale. Where this is done, the likelihood scale can also be changed to a nine-point scale.

Rating	Cost scale	Qualitative description	Time away from work
1	< £100	No injury or trivial injury	< 1 hr
2	£100–£999	Minor injury	1 hr < 1 day
3	£1,000–£4,999	Serious injury	1–5 days
4	£5,000–£24,999	Major injury	6–15 days
5	> £24,999	Disabling injury or fatality	> 15 days

Table 5.2
Scale for rating severity of most likely harm

Remember that the likelihood scale is the likelihood of a hazardous event, irrespective of the harm, and the severity scale is the severity of the most likely harm in the circumstances. Note that there are different factors which influence likelihood and severity.

Factors affecting the likelihood that a person will fall down a manhole include:
- the size of the opening
- the lighting
- whether there is a barrier
- whether there are warning signs.

Factors affecting the severity of harm if a person falls down a manhole include:
- the depth of the manhole
- whether there are opportunities to arrest the fall
- what is at the bottom of the shaft.

Risk

In order to determine the risk associated with a hazard, multiply its likelihood rating by its severity rating. The higher the score, the higher the risk and the higher the hazard's priority for risk control measures. Typical examples are shown in Table 5.3.

Likelihood		Severity		Risk
1	×	1	=	1
3	×	1	=	3
5	×	1	=	5
2	×	4	=	8
5	×	5	=	25

Table 5.3
Sample risk calculation

Where there are existing risk control measures for the hazardous event, it is preferable to rate two risks:
- the risk without the existing risk control measures, that is the absolute risk
- the risk with the existing risk control measures, that is the residual risk.

This provides a check that the existing risk control measures do have an effect.

Risk can also be calculated by multiplying the probability of the hazardous event and the cost of the most likely outcome. How this is done is described in Chapter 6.

The two elements of risk – likelihood and severity – are normally independent of each other. Thus, an open manhole cover would have one severity rating (say 4) but the likelihood rating could range from, say 1 to 4, depending on how many people had to walk around the manhole (that is, were exposed to the hazard).

However, there are circumstances in which likelihood and severity are not independent. For example, when an increasing number of people makes an accident more likely, and when the accident would involve all of the people, then likelihood and severity are not independent. Collapses of seating gantries at music festivals are an example of this. The more people there are on the gantry, the more likely it is to collapse, and the more people there are to be hurt.

However, these sorts of circumstance are fairly unusual and the method described above is suitable for rating 'day-to-day' hazards, remembering that a hazard with a high risk, especially if it has a potential for fatalities or serious injuries, may require a more detailed risk assessment by a specialist.

When risk assessors rate the risk arising from a hazard, they should first do it as things are now, then think what might change and whether this would affect the likelihood rating or severity rating of the hazard.

If foreseeable changes in circumstances – for example different personnel doing a task or different materials being used in a process – could result in an increase in the likelihood or the severity, then they should also record details of this separate rating, including a description of the changes they have foreseen.

Qualitative risk rating

It is possible to apply the principles just described without using numbers and there are health and safety professionals who prefer this approach. The qualitative version of risk rating based on Tables 5.1 and 5.2 is shown in Table 5.4. Using Table 5.4, the qualitative descriptions of likelihood and severity are combined to give a qualitative description of risk from 'very low' to 'very high'.

Table 5.4
Qualitative risk rating

Severity	Likelihood				
	Very unlikely	Unlikely	Likely	Very likely	Almost certain
No injury or trivial injury	Very low	Low	Low	Medium	Medium
Minor injury	Low	Low	Medium	Medium	Medium
Serious injury	Low	Medium	Medium	Medium	High
Major injury	Medium	Medium	Medium	High	High
Disabling injury or fatality	Medium	Medium	High	High	Very high

Deciding whether a risk is acceptable

It will always be debatable whether or not a particular risk is acceptable. However, using a risk rating method of the type just described at least provides the opportunity for a rational approach to such decision making, including the allocation of priorities for implementing additional risk control measures. For example, it enables a table of the type shown in Table 5.5 to be drawn up for managers, or people who have to carry out risk assessments on their behalf. Note that the qualitative risk levels used in Table 5.4 could be used instead of the numbers in Table 5.5.

Risk rating	Action required
1	No action
2–6	Record hazard; record risk ratings; recommend upgrades to risk control measures if the risk rating is not as low as reasonably practicable
8–16	Record hazard; record risk ratings; record the recommended risk control measure needed to reduce the risk rating to below 7 or as low as reasonably practicable
20	Seek further advice
25	Stop the activity immediately and seek further advice

Table 5.5
Sample guidance on action required by different risk levels

Alternatively, the use of risk rating enables attention to be focused on the highest risks, irrespective of their absolute value. This approach is of particular value in organisations where risk levels vary widely from one part of the organisation to another – for example, where there is relatively low-risk office accommodation in association with relatively high-risk production facilities. In these cases, all managers can be instructed to implement a programme designed to reduce their highest risks.

Recording risk assessments

It is good practice to record the results of risk assessments, although detailed records may not be necessary when all the risks identified are trivial, or when the risk assessment can be repeated very easily. In the UK, this good practice is supported by legislation and details of the UK legislative requirements are given later in this chapter.

It is preferable to use a form for recording risk assessments, since this serves to remind you of what has to be recorded, as well as providing a convenient way of structuring the information you gather. There is no general agreement on how a risk assessment form should be laid out or what it should contain. However, the following pieces of information should be recorded as a minimum:

- details of the risk assessor(s) – name(s) and job title(s)
- the date and time of the assessment
- details of what was assessed – for example the location, task, plant or equipment
- any hazards and hazardous events identified, along with, for each:
  - its risk rating without risk control measures in place
  - its risk rating with existing risk control measures in place
- whether any additional risk control measures are required and, if they are:
  - what they should be and how they should be monitored
  - the risk rating if the recommended risk control measures were implemented
- a recommended date for reviewing the risk assessment.

Risk assessment recording systems can be paper-based, but there are now several computer programs which are intended to simplify the recording and analysis of risk assessment data. Magazines directed at risk managers and health and safety professionals often contain advertisements for and evaluations of the latest examples.

Reviewing risk assessments

Risk assessments, like most other things in life, can go out of date. In the case of risk assessments there are two main reasons for this:

1. **There are changes within the organisation which mean that the risk assessment results are no longer valid.** For example, work procedures may change, different components may come into use, or there may be a change in personnel carrying out a particular task.

2. **There are changes outside the organisation which mean that the risk assessment results are no longer valid.** For example, a new hazard associated with a particular chemical or method of work may have been identified, or more effective or cheaper methods of risk control may have been developed, making it possible to implement risk control measures which were not previously cost-effective.

For these reasons, it is necessary to review risk assessments when relevant changes are identified. However, it is also good practice to review risk assessments regularly, for example annually, in case relevant changes within the organisation have not already been taken into account.

UK legislation – legal requirements for risk assessment

The following Regulations contain requirements for risk assessment:
- the Management of Health and Safety at Work Regulations 1999[13]
- the Manual Handling Operations Regulations 1992[15]
- the Personal Protective Equipment at Work Regulations 1992[16]
- the Health and Safety (Display Screen Equipment) Regulations 1992[17]
- the Control of Noise at Work Regulations 2005[18]
- the Control of Substances Hazardous to Health Regulations 2002 (as amended)[19,20]
- the Control of Asbestos Regulations 2012[21]
- the Control of Lead at Work Regulations 2002.[22]

These are the risk assessments that will be required in most organisations, but note that some specialised risks – such as major hazards, genetic manipulation and ionising radiation – have their own legal requirements for risk assessment.

Other uses for risk assessment techniques

The basic techniques of hazard identification, risk rating and so on can also be used in contexts other than formal risk assessment. Indeed, there is a strong argument that managers and health and safety professionals should be employing these techniques continuously as they go about their other activities. Only if managers integrate risk assessment with their other activities will they be able to maintain the constant vigilance required for effective health and safety management.

In the past, a variety of terms has been used to describe management activities specifically focused on hazard identification, including safety inspections, workplace inspections, safety self-audits, housekeeping checks, safety sampling, safety tours and safety surveys. Different organisations define these various activities in different ways but, in general, they can be divided into the following two categories:

1. **Inspections, surveys or tours conducted by personnel with competences in hazard identification, risk rating, and so on.** The primary purposes of this category are to identify circumstances in which required

control measures are not being implemented properly, to identify risks which require additional control measures, and to reach a decision about which types of control measure would be most appropriate. This category is described in more detail in Chapter 10.

2. **Inspections, surveys or tours carried out by personnel with no particular competences in hazard identification, risk rating and so on.** These activities are usually undertaken by middle and senior managers and the primary purpose is to demonstrate these managers' commitment to safety.

The second category is often of limited value since, typically, the senior managers fail to identify hazards that are well known to the workforce. This leads to the accusation that senior managers may say that they are committed to health and safety but they are not sufficiently committed to learn anything about it! When inspections, surveys and tours are to be used in this way, their purpose should be made clear – for example, the senior managers should explain that they have no particular health and safety expertise but they want to demonstrate their commitment by taking time to find out about the health and safety concerns of the workforce.

Since there is no general agreement about what is meant by a safety inspection, safety tour or safety survey, the only sure way of knowing what is meant by such a term in any given organisation is to obtain a detailed description of what is done, by whom, and at what level of competence.

Summary

In this chapter we noted that there is no general agreement about what is meant by risk assessment, although a number of separate aspects of risk assessment can be identified. All of these aspects were mentioned at the beginning of the chapter and the majority were then described in detail. A major item not discussed was risk control, which is the subject of the next chapter.

6: Risk control

Introduction

This chapter deals with the measures which can be taken to control risks, and describes various classifications of risk control measures. There is also a discussion of the relative effectiveness of different types of control measure, and an introduction to the cost implications of using different types of risk control. Note that one important category of risk control measure is safe systems of work. However, this category is so widely used, and so important, that Chapter 7 is devoted solely to this topic.

Risk control is the process of designing, implementing and maintaining measures which will reduce a particular risk. However, because there are many different types of risk, there is a need for a large number of different types of risk control measure. Each type has its own design, implementation and maintenance requirements.

This chapter considers the general principles associated with risk control measures by examining risk control measures from a number of different points of view:
1. effects on likelihood and severity
2. methods of risk reduction
3. hierarchies of risk control measures
4. risk rating and risk control measures.

Effects on likelihood and severity

Remember that risk is defined as:

likelihood of the hazardous event × severity of the most likely outcome

It follows from this definition that there are only three ways a risk control measure can reduce risk. It can:
1. reduce the likelihood of the hazardous event
2. reduce the severity of the most likely outcome
3. reduce both likelihood and severity.

This statement is so obvious that it is often overlooked but it has fundamental implications for determining the effectiveness of a risk control measure.

After you have carried out a risk rating exercise and have decided that some additional or modified risk control measure is required, you must then choose which risk control measure to implement. There are several factors to be taken into account in making this decision, but an important one will be the extent to which the chosen risk control measure will reduce the risk. Unless the effect a particular risk control measure will have on the likelihood and severity of the risk can be established, it is not possible to make an informed decision on which of a choice of risk control measures should be implemented.

We will consider later in this chapter how to measure the effects of risk control measures on the level of risk.

Methods of risk reduction

Although there are large numbers of different types of risk control measure, many of them use similar methods to produce their risk reducing effect. The following basic methods are dealt with in this section:

1. hazard reduction methods
2. separation methods
3. physical barrier methods
4. dose limitation methods.

There is another major category, safe systems of work, that involves carrying out work activities in ways which minimise risk, but this category will be dealt with in Chapter 7.

Hazard reduction methods

These methods include all those which attempt to do something about the hazard itself, rather than its possible effects. The most effective of these methods is to remove the hazard completely – that is, reduce it so far that it is eliminated. Less effective methods reduce the potency of the hazard to a greater or lesser extent.

Table 6.1 gives examples of these hazard elimination and hazard reduction methods. From a risk control point of view, it is preferable to eliminate or reduce hazards, but where this is not possible, you can adopt one or more of the categories described next.

Table 6.1

Hazard elimination and reduction methods

Hazard	Elimination	Reduction
Noisy machine	Replace with a quiet machine	Make the existing machine less noisy by, for example, running it at a slower speed, better maintenance or design modifications
Hazardous chemical	Use a non-hazardous chemical instead	Use a less hazardous chemical instead
Injury from manual handling	Remove the need for manual handling by redesigning the process or using mechanical handling	Use smaller packages, provide mechanical aids to manual handling

Separation methods

These methods all rely for their effectiveness on maintaining a suitable distance between the hazard and the people who could be harmed by it. Typical examples include siting electrical power supply lines overhead and storing flammable materials at remote locations. All of the separation methods suffer from the same weakness: there is nothing to stop people approaching the hazard, either deliberately or inadvertently, and there is usually no backup protection.

Physical barrier methods

There is a wide range of physical barrier methods, but they can be subdivided into three broad categories:
1. **Barriers which are close to the hazard and prevent people coming into contact with it.** This category includes machinery guarding, covers on pits and electrical insulation.
2. **Barriers which are close to the person and prevent the hazard damaging people if it comes into contact with them.** This category includes a whole range of personal protective equipment (PPE) such as gloves, goggles and hard hats.
3. **Barriers which fall into neither of the first two categories.** These include specialised barriers, such as the flow of air used in ventilation systems and sound-absorbing partitions, as well as general fencing and barriers to keep people away from areas containing hazards.

The major problem with barrier methods is that, in the vast majority of cases, the barrier must be removable for a range of legitimate reasons. For example, machine guards have to be removed for maintenance and resetting; PPE, rather obviously, has to be removable; and ventilation systems have to be capable of being switched off for cleaning and maintenance.

Because these barriers have to be designed and made in ways which make them capable of being removed for legitimate reasons, it is also possible for them to be removed for reasons which are not legitimate. For example, machine guards may be removed so that work can be done more quickly; PPE may not be worn because it is uncomfortable; and ventilation systems may be switched off because the noise they make irritates the people working near to them.

Much time and ingenuity is expended on development work intended to produce barrier methods which can only be removed for legitimate reasons with, for example, machinery guards interlocked to the machine's power supply so that removing the guard stops the machine working. Unfortunately, just as much time is spent by operators in finding ever more ingenious ways of defeating these interlocks.

Dose limitation methods

These methods can be used where there is a relationship between the extent to which a person is exposed to a hazard (the dose) and the amount of harm caused to the person. Examples of such hazards include:

- **Noise.** The extent of harm done to people's hearing by noise depends on the 'amount' of noise to which they are exposed, and this 'amount' depends on two factors: how loud the noise is and how long the people are exposed to the noise.* For example, for a given noise level, eight hours' exposure per day for five years could cause irreversible hearing damage, while an exposure of two hours per day to the same noise level would produce little or no long-term damage. It follows from this that jobs which have to be done at this noise level are better done by four people, each working for two hours and suffering little or no long-term hearing damage, than by one person working for eight hours and suffering irreversible hearing damage.

- **Chemicals.** Many chemicals are harmless or even beneficial in small doses, but toxic or harmful in other ways in larger doses. In some circumstances, the length of time spent in an environment is linked to the amount of a chemical which gets into the body, for example when the chemical is airborne and can be breathed in. In such circumstances, having a number of people working in the environment for short periods will result in a low and harmless dose, while having one person work in the same environment for a long period will result in a high and harmful dose.

- **Work-related upper limb disorders (WRULDs).** This is a category of conditions (including tennis elbow and carpal tunnel syndrome) affecting the hands and arms. One cause of such conditions is repetitive movement carried out for too long a period each day. Reallocation of the tasks requiring potentially harmful repetitive movements among a number of people will ensure that no one person has to carry out the task for lengths of time which would result in long-term damage.

The main weakness with all the dose limitation methods is that they rely on people keeping to the job rotation procedures which have been set up. It has been known for groups of operators to decide among themselves that not rotating jobs is more efficient (which it may very well be), forgetting that the job rotation is for their protection.

It can be seen from this description of methods of risk reduction that almost all of the methods have inherent weaknesses. However, some methods are less prone to weakness than others, and this has led to the development of hierarchies of risk control measures, in which types measure are ranked in a preferred order of use. The next section contains a description of a selection of these hierarchies.

---

* See Chapter 14 for further information on the effects of noise on hearing.

Hierarchies of risk control measures

In any given set of circumstances, some risk control measures will be 'better' than others and it is obviously preferable to use the 'best' option. However, there are various different criteria which can be used to define 'best', including:

- **The number of people protected by the risk control measure.** In general, it is better to use a risk control measure which will protect everyone who could be exposed to the hazard, rather than relying on individuals to provide their own protection. For example, it is better to put a soundproof enclosure around a noisy machine than to expect everyone who might be exposed to the machine's noise to wear hearing protection.
- **The extent to which the continuing effectiveness of the risk control measure relies on human behaviour.** In general, it is preferable to have risk control measures which, apart from any necessary maintenance, operate without human intervention. When a risk control measure relies on the actions of people, it is inevitable that on some occasions it will not be used, either deliberately or inadvertently.
- **The extent to which the risk control measure requires testing, maintenance, cleaning, replacement and so on.** All of these required activities rely on human intervention and can, therefore, fail. This reduces the likelihood that the risk control measure will continue to be effective.
- **The cost of the risk control measure.** Ideally, the cost should be calculated over the whole of the time for which risk control is required, since some risk control measures have a low installation cost but are expensive to maintain, while others have higher installation costs but are cheaper to maintain. This aspect of risk control measures is dealt with in more detail later in this chapter.
- **And, last but not least, the extent to which the risk control measure reduces the risk.** Ideally a risk control measure, or combination of measures, will reduce the risk to near zero, but this may not be achievable in practice.

When deciding on the 'best' risk control measures, we need to arrive at a compromise between all the demands listed above, since they are often in competition. However, a critical issue for a measure's long-term effectiveness is the extent to which it relies on human beings continuing to carry out particular activities. For this reason, a number of risk control hierarchies have been developed which are based on this criterion. Six of these hierarchies are described next.

Hierarchy 1: Technical, procedural and behavioural

This hierarchy has three categories.

1. **Technical** risk control measures, including machinery guarding, various forms of fencing, and different types of ventilation. In general, these types of measure can be designed and installed so that they have the minimum reliance on people doing what they are supposed to do.
2. **Procedural** risk control measures, including systems of work and maintaining plant, equipment and so on in a safe condition. Where the procedures are well designed and their implementation is effectively monitored, these measures can be successful, but they do rely on, for example, people keeping to work procedures and remembering to carry out the necessary maintenance work.
3. **Behavioural** risk control measures, including information and training. These measures rely for their effectiveness on a number of things, including the appropriateness of the information or training, the extent to which it is understood, and the extent to which this understanding is put into practice. As a category, therefore, it is the one which is most prone to failure as a result of people not doing what they are supposed to do.

Hierarchy 2: Elimination to PPE

This hierarchy has six categories.

1. **Eliminate hazard at source** – for example:
   - use a non-hazardous substance instead of a hazardous one
   - stop using a noisy machine.
2. **Reduce hazard at source** – for example:
   - use a substance less hazardous than the one used at present
   - replace a noisy machine with a quieter one.
3. **Remove person from hazard** – for example:
   - use unattended robots for paint spraying
   - do not allow people to work near noisy machines.
4. **Contain hazard by enclosure** – for example:
   - do all painting in a proper, enclosed painting bay
   - put soundproofing round a noisy machine.
5. **Reduce employee exposure** – for example:
   - four people exposed for two hours each, not one person for eight hours – applies to exposure to substances or noise.
6. **PPE** – for example:
   - gloves or goggles for substances and hearing protection for noise.

Hierarchy 3: HSG65[1]

The following extract is from HSG65's advice on risk control hierarchies:

> The following is a summary of the preferred hierarchy of risk control principles:
> - Eliminate risks by substituting the dangerous by the inherently less dangerous, eg:
>   - use less hazardous substances;
>   - substitute a type of machine which is better guarded to make the same product;
>   - avoid the use of certain processes, eg by buying from subcontractors.
> - Combat risks at source by engineering controls and giving collective protective measures priority, eg:
>   - separate the operator from the risk of exposure to a known hazardous substance by enclosing the process;
>   - protect the dangerous parts of a machine by guarding;
>   - design process machinery and work activities to minimise the release of, or to suppress or contain, airborne hazards;
>   - design machinery which is remotely operated and to which materials are fed automatically, thus separating the operator from danger areas.
> - Minimise risk by:
>   - designing suitable systems of work;
>   - using personal protective clothing and equipment – this should be only be used as a last resort.

> The hierarchy reflects the fact that eliminating and controlling risk by using physical engineering controls and safeguards is more reliable than relying solely on people.

Hierarchy 4: Airborne hazardous substances

There are various hierarchies which have been devised specifically for controlling risks from airborne hazardous substances. They can be summarised as follows:

1. eliminate the need for the hazardous substances by, for example, redesigning the process
2. substitute a less hazardous substance for the substance currently being used
3. isolate or enclose the process in which the hazardous substance is used
4. local exhaust ventilation
5. general or dilution ventilation
6. PPE
7. reduce exposure time
8. personal hygiene (for example a ban on eating and drinking in contaminated areas, providing washing facilities, and disposing of contaminated PPE).

Hierarchy 5: Management Regulations

The hierarchy given below is quoted from the UK's Management of Health and Safety at Work Regulations 1999.[13] Where an employer implements any preventive and protective measures, he or she must do so on the basis of the principles set out in Schedule 1 of the Regulations, which are:

a) avoiding risks;
b) evaluating the risks which cannot be avoided;
c) combating risks at source;
d) adapting the work to the individual, especially as regards the design of workplaces, the choice of work equipment and production methods, with a view, in particular, to alleviating monotonous work and work at a predetermined work rate and to reducing their effect on health;
e) adapting to technical progress;
f) replacing the dangerous by non-dangerous or less dangerous;
g) developing a coherent overall prevention policy which covers technology, organisation of work, working conditions, social relationships and the influence of factors relating to the working environment;
h) giving collective protection measures priority over individual protective measures; and
i) giving appropriate instructions to employees.

Hierarchy 6: BS OHSAS 18001[3]

BS OHSAS 18001 requires that:

...consideration shall be given to reducing the risks according to the following hierarchy:
a) elimination;
b) substitution;
c) engineering controls;
d) signage/warnings and/or administrative controls;
e) personal protective equipment.

In practice, it is not really important which hierarchy is used in selecting a risk control measure or combination of measures. What is important is that you recognise that some types of risk control measure are more effective in the long term than others, and that you take this into account when deciding which measures to recommend.

Effects of risk control measures

The effects of risk control measures can be expressed in two ways:
- the reduction in risk that they will produce
- the financial savings they will produce because they reduce the number of incidents.

The first term is needed where targets are set for risk reduction, and the second where business cases are used for making decisions on risk reduction. The two methods are the subject of the next two sections.

Risk rating and risk control measures

As we saw at the beginning of this chapter under the heading 'Effects on likelihood and severity', there are three possible ways a risk control measure can reduce risk: it can reduce the likelihood, it can reduce the severity, or it can reduce both.

It is possible to calculate the effect of a risk control measure simply by estimating the effect it will have on likelihood and severity and, since this can be done as a 'desk exercise', the relative merits of different risk control measures can be compared before putting them in place. Some examples of such calculations are given below, using the 1–5 scales for likelihood and severity described in Chapter 5, but the principles underlying these calculations will apply whichever scales are used for rating likelihood and severity.

Putting a guard on a machine
Suppose that the likelihood of someone trapping a hand in an unguarded machine is 4 and that the severity, were this to happen, would also be 4 so that the risk rating is 16. If a guard is fitted to this machine, the likelihood that someone will trap his or her hand should be reduced to 1, so that the risk is reduced to 4.

However, this will be the case only if the guard is used all the time and the guard itself is not a hazard. It might, therefore, be necessary to make adjustments to the calculation of the effectiveness of the guard based on the following assumptions:
1. The guard will not be used for a proportion of the time the machine is in operation so that the likelihood is 2, not 1. Since the severity remains at 4, the risk on this assumption is 8.
2. The guard will be used all of the time but there will be relatively frequent contact between the guard and the operator's knuckles! The new risk is, therefore, likelihood 4 × severity 1 = 4 for operator's knuckles plus the risk of trapping a hand (4) giving a total of 8.
3. The guard will not be used for a proportion of the time (risk 8), and there will be frequent knuckle damage. In these circumstances, the two risk ratings just calculated can be added so that the total is 4 + 8 = 12.*

Getting people to wear hard hats
Suppose that the likelihood of someone being hit by a falling object is 2. If people wear hard hats, it has no influence on the likelihood of them being hit so that likelihood remains at 2, but the severity is reduced. Since the severity will depend on the nature of the object and its velocity, and it can reasonably be expected that these will vary widely, it is only possible to estimate an average severity which might, for example, be 4 on the 1 to 5 scale. Wearing a hard hat might reduce this severity to 1 so that the risk goes from 2 × 4 = 8, to 2 × 1 = 2.

---

* There are alternative methods of calculating risks in these circumstances; these are described in Chapter 24.

If there is reason to believe that the hard hats will not be worn all of the time, then this 6-point reduction is an overestimate and should be adjusted to reflect the extent to which it is thought that hard hats will not be worn.

Not using a toxic chemical

Suppose that the likelihood of significant exposure to a toxic chemical is 2 and that the severity of this significant exposure is 4, giving a risk rating of 8. If it is possible to substitute a non-toxic chemical for the toxic one, then there is no hazard and, therefore, no risk from this hazard. When it is only possible to substitute a less toxic chemical, the likelihood would remain at 2 to but the severity would be reduced to, for example 1, giving a risk rating of 2.

Choosing risk control measures

There is usually more than one way of controlling a risk and the sorts of calculations illustrated above can be used to compare the relative effectiveness in terms of risk reduction of the various options. However, the following will have to be taken into account when choosing between risk control measures:

- **The required level of residual risk.** There may be a requirement that all risks are reduced to a level that is acceptable to the organisation, or the organisation may set risk reduction targets.
- **The cost of the risk control measures.** Where more than one risk control measure will produce the required level of risk reduction, it will be necessary to make a selection. This can be done on the basis of the cheapest option producing the required risk reduction, but is can also be done on the basis of the best value option.

An example of determining a best value option is given below based on three options for controlling the risk associated with the use of a toxic chemical. The three options are:

1. use a non-toxic alternative
2. install suitable ventilation
3. get operators to wear appropriate PPE.

If it is assumed that the likelihood of exposure without any of these risk control measures is 2 and the severity is 4 so that the risk rating is 8, then the calculations would look something like this:

1. Using a non-toxic alternative would eliminate the risk giving an 8-point risk reduction.
2. Installing suitable ventilation would, in theory, reduce the likelihood of exposure to 0 but would have no effect on the severity. In practice, few ventilation systems are 100 per cent effective, so the real risk reduction might be from $2 \times 4 = 8$ to $1 \times 4 = 4$, giving a 4-point risk reduction.
3. Getting people to wear appropriate PPE does not reduce the likelihood of exposure but reduces the severity to 0 if the PPE is 100 per cent effective. However, it is known that PPE is rarely 100 per cent effective because of, for example, leakage round the mask and poor maintenance of filtration systems. For this reason, a more realistic estimate of the risk reduction might be from $2 \times 4 = 8$ to $2 \times 1 = 2$. In addition, it is known that it is unlikely that the PPE will be worn by everyone at all relevant times so that it may be necessary to estimate the risk at 2×3 to take this into account, giving a 2-point risk reduction.

Once the best estimate of risk reduction has been made for each option, it is then appropriate to compare the relative costs of the options being considered, since it is now possible to calculate a cost per unit risk reduction. This can be quite complex in practice, but the principles can be illustrated by using simple cost figures for the three options for risk control just described. The following assumptions will be made:

- The calculations will be based on implementation costs and running costs for a three-year period.
- The non-toxic alternative chemical will cost an additional £8,000 over the three-year period because it is more expensive and less efficient.

- An adequate ventilation system will cost £3,000 to install and £1,000 per year to run and maintain.
- It will cost £1,000 per year to buy, clean, test and maintain suitable PPE, and train people in its use.

Based on these assumptions and the risk reductions calculated earlier, the costs per unit risk reduction are as follows:

Using a non-toxic chemical	£8,000 ÷ 8 = £1,000 per unit reduction.*
Installing ventilation	£5,000 ÷ 4 = £1,250 per unit reduction.
Issuing PPE	£3,000 ÷ 2 = £1,500 per unit reduction.

Obviously, these figures have been chosen to illustrate a point, but they are not necessarily unrealistic. There will be occasions when PPE is the most cost-effective solution – for example, when tasks are carried out infrequently or the necessary ventilation system is extremely expensive. What this method provides, however, is a rational way of deciding between risk control options. This basic technique can be elaborated in a number of ways, and these are dealt with in Chapter 24.

Financial savings

The effects of risk control measures can be expressed in terms of the financial savings they will produce because they reduce the number and/or severity of incidents. This is done by using probability and cost rather than the 1–5 scales we used earlier. For example:
- instead of rating a likelihood as 2, we estimate probability, for example 0.25
- instead of rating cost as 3, we estimate an actual cost, for example £2,000
- to get the saving we multiply the probability and cost, that is £500.

What we are doing with calculations of financial savings is to put a business case for health and safety expenditure, so that by spending money, money is saved. Preventive maintenance can be used as an analogy, and two examples are:
- The present situation is that there are five machine breakdowns a year, each costing the organisation £1,000 – a total of £5,000. If the organisation spends £3,000 per year on preventive maintenance, this will reduce the number of breakdowns to one per year. Thus spending £3,000 per year results in a gain to the organisation of £1,000 per year.
- Relevant personnel in the organisation can agree an estimate of the breakdowns for the coming year and the mean cost per breakdown. They then establish the cost of preventive maintenance and its likely effectiveness. With these data, they can then decide whether it is cost-effective to pay for preventive maintenance.

However, there are practical problems with applying this technique, and these are discussed in Chapter 24.

Other factors

We have discussed the theoretical effect a risk control measure will have on the level of risk, the likely effectiveness in practice of the risk control measure, and its cost. It is likely that we will need to take other

---

* When making these calculations, we may have to take account of the fact that the 1–25 risk rating scale does not have 25 units. For example, there is no 7, 11 or 13 in the scale. This is dealt with in more detail in the discussion of costing in Chapter 24.

factors into account when choosing risk control measures, including the following:

- **The practicability of the risk control measure.** Sometimes risk control measures which do not rely on people doing the right thing, or not doing the wrong thing, are impracticable. Certain high-risk tasks, such as working with live electrical equipment or constructing scaffolding, have to be carried out and in these cases the majority of the risk control measures rely on safe systems of work. These systems are dealt with in Chapter 7.
- **The need for interim control measures.** When risks are high and it will take some time to install an effective long-term risk control measure, then it may be necessary to use interim measures. For example, replacing a noisy machine or installing a ventilation system could take several months and in the interim it will be necessary to use, for example, dose limitation methods and/or PPE.

Whatever factors have to be taken into account in any particular decision about risk control measures, the criteria which have to be satisfied are:

- the risk control measure should reduce the risk to an acceptable level
- any mechanical features of the risk control measure should be reliable
- the risk control measure should place the minimum requirements on people doing the right things and not doing the wrong things.

UK risk terms

Various terms are used to describe risk in the UK health and safety literature, and this can cause confusion. The contents of this box describe the main sources of confusion and how they can be resolved.

Figure 6.1 is used to illustrate the points made. The vertical axis represents the level of risk, which could be presumed to start at 0 and increase to infinity, or some arbitrary maximum figure, such as 100.

The four boxes in the figure represent the risks arising from four different hazards – for example, an unguarded machine, objects falling from a scaffold, or a trailing cable in an office. The layout of these representations is:

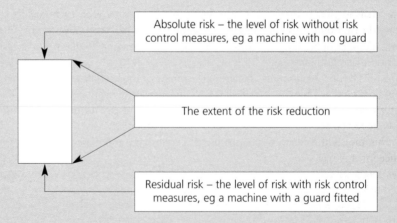

Absolute risk – the level of risk without risk control measures, eg a machine with no guard

The extent of the risk reduction

Residual risk – the level of risk with risk control measures, eg a machine with a guard fitted

Figure 6.1 also shows two levels of risk – 'trivial' and 'tolerable'.

Continued...

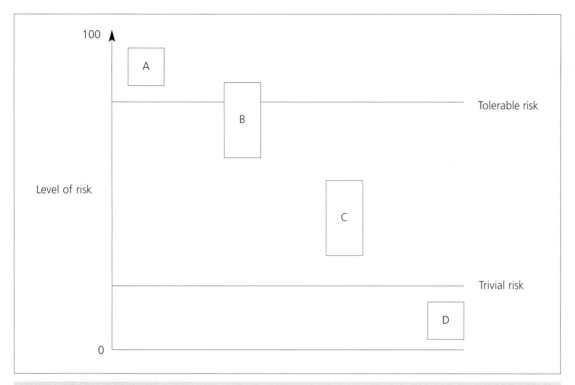

Figure 6.1
Illustration of
various terms used
in discussing risk

... continued

BS OHSAS 18001[3] defines acceptable risk as 'risk that has been reduced to a level that can be tolerated by the organization having regard to its legal obligations and its own OH&S policy', but the term 'acceptable risk' does not appear in the Standard. This would appear to cover all the risk levels below the tolerable risk level in Figure 6.1.

The Management Regulations[13] refer to 'significant' and 'insignificant' risk and state that insignificant risks 'can usually be ignored'. Insignificant risks would appear to be those below the trivial risk level in Figure 6.1.

The idea of 'reasonably practicable' is enshrined in UK legislation and it has been defined in several ways. Essentially, a risk need not be reduced if the extent of the risk reduction that would be obtained is small compared to the resources (in time, money and effort) needed to achieve the reduction.

The clear implication is that a risk should be reduced where it is reasonably practicable to do so. There is no stipulation as to the absolute level of risk which should be subject to the requirement of reasonable practicability. This means that even a risk with a low absolute level (such as example D in Figure 6.1) should be reduced if it is reasonably practicable to do so.

There is the potential for confusion because of the use of the term 'insignificant' risk. There is no logical reason why an insignificant risk cannot be reduced and, therefore, no logical reason why it cannot be subject to the test of reasonable practicability. In the UK, an authoritative statement on this issue is needed.

A more rational approach would be for organisations to decide on the level of risk they deem tolerable and then deal with risk reduction as follows:

Continued...

... continued

- Where it is not reasonably practicable to reduce the risk below the tolerable level (example A in Figure 6.1), the organisation stops the activity. This would not be required under UK law. So long as the risk has been reduced to the lowest level that is reasonably practicable, the activity can continue, irrespective of the level of residual risk.
- Where it is reasonably practicable to reduce the risk below the tolerable level (example B in Figure 8.1), the organisation implements the risk control measures and continues with the activity. This process would also satisfy UK law.

The problem arises with examples C and D in Figure 6.1. If the organisation is operating in the UK, it will have to reduce these risks because it is reasonably practicable to do so, even though the risks are tolerable to the organisation. However, where the requirement for reasonable practicability does not apply, the organisation would do nothing to reduce the risk further – unless, of course, it had adopted this standard for all its operations.

All these confusions arise because, in the UK, the absolute level of risk, the residual risk and reasonably practicable risk reduction are inextricably mixed in the minds of the authorities. It seems unlikely that the situation will improve unless these concepts are disentangled.

A much more straightforward (and logically defensible) approach would be to require two commitments from organisations:

- **To adopt an appropriate tolerable (or acceptable) level of risk or, where organisations have a wide range of activities, a different level for each activity.** The latter would be needed, for example, in the emergency services, where much higher risks are often tolerable during an emergency response than at other times. Guidance could be provided on what is appropriate and the appropriateness of an organisation's choice could be tested in law.
- **To ensure that all their risks are at or below this tolerable (or acceptable) level.** This would not stop organisations adopting other criteria for risk reduction. For example, they could still apply the 'as low as reasonably practicable' principle, or make a business case for reducing risks where they were already below the tolerable level.

Workplace precautions, RCSs and management arrangements

The term 'risk control measure' has been used so far to describe actions taken to reduce risk. However, HSG65 uses alternative terminology:

Workplace precautions
These are not defined in HSG65 but they 'can include machine guards, local exhaust ventilation, safety instructions and systems of work'. Effectively, they are what we have been calling risk control measures.

Risk control systems (RCSs)
These are 'the basis for ensuring that adequate workplace precautions are provided and maintained'. Examples of RCSs include the systems for ensuring that PPE continues to be provided, and the maintenance and repair of local exhaust ventilation systems.

Management arrangements

These are the 'management processes ... necessary to organise, plan, control and monitor the design and implementation of the RCSs'.

In the context of risk management, these distinctions are useful, since they allow for separate discussion of the risk control measures themselves (workplace precautions), the requirements for maintaining the effectiveness of risk control measures (that is RCSs), and more general risk management requirements (management arrangements). However, this terminology has not been widely adopted and other terms will be used where it is necessary to make these distinctions.

Summary

This chapter dealt with the various broad categories of risk control measure and how they can be organised into hierarchies according to their effectiveness. We have also considered how this effectiveness can be measured in terms of adjustments to risk ratings. The chapter ended with a note on the calculation of the cost-effectiveness of risk control measures.

The consideration of risk control measures will now continue with a chapter dealing with a particular type of risk control – safe systems of work.

7: Safe systems of work

Introduction

Safe systems of work are an important aspect of risk control but, as with many other aspects of health and safety management, there is no general agreement on what is meant by the term.* As will be seen in Chapter 21, 'system' can have a very specific meaning, and there is a whole body of research work based on 'system theory'. Unfortunately, safe systems of work, in systems theory terms, are not necessarily systems at all.

The term 'safe system of work' will, therefore, be used only as general description of three different categories of risk control, which are based on attempts to get people to carry out work activities in a particular way:

1. **Safe work procedures.** These are methods of carrying out tasks in ways which minimise the risks associated with the tasks' hazards. They take the form of written descriptions of how work should be carried out.

2. **Permit to work procedures.** These are a way of making tasks less of a risk by specifying procedures to be followed with respect to one or more of the hazards associated with the task. Permit to work procedures do not usually give instructions on how the task itself is to be carried out; rather, they give instructions on precautions to be taken. For example, permits to work are often used for maintenance work on electrical equipment as a way of ensuring that the equipment is switched off and isolated before the work begins and remains isolated for the duration of the work. The permits to work do not usually, however, contain any information on how the actual maintenance work should be carried out.

3. **Safety rules.** These are lists of dos and don'ts and are mainly used for highly variable activities, such as using ladders or driving fork-lift trucks, where it is impossible to specify, in a safe work procedure, all the possible types of task which may have to be carried out.

The rest of this chapter is a more detailed description of these three categories of safe systems of work.

Safe work procedures

It is a fact of life that most tasks can be carried out in a number of different ways and that some ways will be 'better' than others. However, the definition of 'better' depends on the current concerns of the person doing the defining. Some examples are given below.

- For the accountant, 'better' will mean cheaper or at least more cost-effective.
- For the quality manager, 'better' will mean more consistent.
- For the production manager who has to meet deadlines, 'better' will mean faster.
- For the training manager, 'better' will mean simpler.
- For the health and safety professional, 'better' will mean safer.

When a work procedure is being specified, the result will usually be a compromise which meets, in part, all of the relevant demands. This fact should be recognised when developing safe work procedures, since there is little point in designing procedures that reduce risk by creating ways of working that make the procedure unacceptable to managers or to the people who have to do the work.

---

* In the UK, the phrase 'systems of work that are safe' is used in legislation, for example the Health and Safety at Work etc Act 1974. However, there is no legal definition of what constitutes a safe system of work.

The rest of this section deals with the techniques of devising safe work procedures. It is assumed throughout that managers and the workforce are involved in developing the procedures they will have to implement, and that this involvement is extended to other interested parties, including, for example, quality managers and trainers.

Devising a safe working procedure involves the following main stages:

1. specifying the objectives of the work procedure
2. describing the actions necessary to reach the specified objectives
3. identifying the hazards, if any, associated with each action
4. specifying, for each hazard identified, an appropriate risk control measure or measures.

The first two stages are not specific to health and safety, since they are required when developing, for example, quality procedures. The methods for carrying out these stages are, therefore, the general task analysis methodologies such as hierarchical task analysis (HTA). However, the last two stages are specific to health and safety and require specialised techniques such as job safety analysis. General task analysis is, therefore, dealt with first before moving on to the specific health and safety methodologies.

General task analysis

More than 60 task analysis techniques have been developed for a wide range of different purposes. However, the general principles will be illustrated by using HTA. We will use it as an example because it has wide general usefulness for the following reasons:

- It takes into account both mental skills (such as perception and decision making) and motor skills, and the errors associated with each. Some techniques emphasise, or are restricted to, one or other of these aspects.
- It takes into account the required timings of actions – for example, the fact that some actions have to be carried out before others. Some techniques concentrate on what should be done, without adequate facilities for taking into account when they should be done.

However, it should be remembered that for certain task analyses, HTA may not be the most appropriate technique and that other techniques may have to be substituted for, or used as a supplement to, HTA.

The best way of explaining HTA is to work through a simple example – we will use an HTA for taking a bath.

The first stage is, as described above, to specify the objectives. This may appear obvious – getting clean – but taking a bath can have other purposes, including relaxation and an opportunity to read in peace! These different objectives will influence the next stage, deciding on the actions necessary to achieve the objective(s). We will assume for the purposes of this illustration that the objective is simply to get clean.

HTA is so called because it involves breaking the task down into successively more detailed levels in the form of a hierarchy. For the first level of detail, therefore, the units might be as follows:

1	fill bath with water at the required temperature
2	get undressed and get into bath
3	wash and get out of bath
4	dry yourself and get dressed
5	empty and clean bath.

For the second level of detail, each of these main units is further subdivided as necessary. For example, the first unit might be subdivided as follows:

1.1   put plug in
1.2   turn hot tap on
1.3   turn cold tap on
1.2.1   monitor water temperature and adjust flow from hot and cold taps as appropriate
1.2.1.1   monitor water level and turn off both taps when the required level is reached.

For a task as well known as taking a bath, this may be the last level of subdivision required, but for more complex or less familiar tasks, as many levels of detail as are necessary can be added.

Once all of the necessary activities have been specified at the required level of detail, they can be presented in diagrammatic form as shown in Figure 7.1.

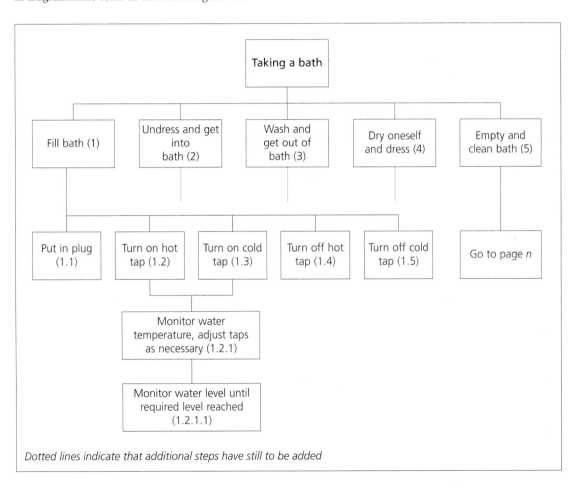

Dotted lines indicate that additional steps have still to be added

Figure 7.1
Partial HTA diagram for taking a bath

Where complex tasks are being analysed and recorded, it is usual to record separate subtasks on different pages, in which case the relevant branch of the hierarchy terminates with a 'Go to page *n*' instruction as indicated in Figure 7.1. It is also desirable, particularly for complex HTA diagrams, to number the various boxes for ease of reference. For example, one numbering system which can be used is in the form '*x.y.z*...', where *x* identifies the position (reading from left to right) of boxes at the first level of detail, *y* identifies second level positions, and so on. We have used this convention in Figure 7.1, with the box numbers given in brackets.

For normal HTA, working through the required steps in detail and recording them in the form of a hierarchy completes the process. However, the hierarchy may not be the best format for instructions on how to carry out the task. For example, it is usually preferable to carry out other tasks while the bath is filling since only periodic monitoring is required. Where this is the case, separate written work instructions should be prepared for the supplementary tasks.*

Task analysis for health and safety

All task analysis for health and safety reasons should begin with a detailed general task analysis using HTA or some other appropriate format. Each stage of the work should then be analysed and any hazards associated with a stage should be recorded and their risks assessed.

Where there are significant risks, consideration should be given to amending the task to remove or reduce these risks. For example, in Figure 7.1 the sequence for turning on the taps is 'hot' then 'cold'. Analysis of this sequence in safety terms would be likely to result in this order being reversed with, perhaps, an additional step in the HTA in the form 'Allow cold tap to run until bath is part full'. This would reduce the risk of scalding and might also reduce the risks associated with high levels of water vapour building up in the bathroom. But where the risk from hot water is particularly high, as may be the case with elderly people or the mentally ill, the risk assessment may recommend thermostatically controlled bath filling.

Only when alteration of the task is not a practical option should appropriate risk control measures be devised and recorded.

The written job instructions for carrying out the tasks, variously referred to as job safety instructions or safe operating procedures, can then be prepared, to take into account the required risk control measures.

Permits to work

A permit to work (also referred to as a work permit) is a document used as part of a permit to work procedure. Permits to work take many forms, but no permit to work is of much value unless it is part of an effective permit to work procedure or, as it is referred to by some authors, a permit to work system. For this reason, we will describe all the elements in a permit to work procedure before returning to the nature of the permit to work itself.

The key elements in a permit to work procedure are as follows:
- a clear statement of the aims of the procedure
- the steps to be taken before a permit to work is issued and the arrangements for issuing it
- the steps to be followed while the permit to work is in force
- the steps to be taken when the permit to work is being withdrawn, either temporarily or permanently.

We will look at each of these elements in detail later, but first it is necessary to consider typical uses for permit to work procedures.

Uses of permit to work procedures

Using permit to work procedures results from the recognition that a high-risk task has to be carried out. Typical high-risk tasks for which permit to work procedures are used include:
- **Work with electrical equipment, particularly high-voltage electrical equipment, or equipment powered by electricity.** The main purposes of the permit to work procedure in these circumstances are to ensure that

---

* Work instructions are one of the types of documentation used in quality management. This documentation is described in Chapter 11.

the equipment is isolated, and if necessary earthed, before the task begins, and that the equipment is re-energised only when it is safe to do so.

- **Work with flammable liquids or gases.** In these cases, the permit to work procedure is used to ensure that there are no leaks during tasks such as work on pipes and valves. In essence, the procedures are similar to those involving electricity, but the flow in these cases is of a gas or liquid. In addition, permit to work procedures for flammable gases and liquids will also deal with the need to ensure the absence of ignition sources.
- **Work in confined spaces.** The main purposes of this type of permit to work procedure are to ensure that the quality of the atmosphere in the confined space is checked before entry and, as necessary, while anyone is in the confined space. The permit to work may also be used to specify any necessary risk control measures and any emergency response arrangements which should be in force for the duration of the work.
- **Hot work, such as welding or burning, which introduces an actual or potential source of ignition.** The first purpose of this type of permit to work procedure is to ensure that the environment in which the work is being done is free from, for example, fire and explosion risks that would be increased by hot work. Where risks are identified, the hot work permit may be used to specify risk control measures to be used during, and on completion of, the hot work.
- **Work at height, for example roof work, particularly where the roof is made from fragile material or has no adequate edge protection.** In these circumstances, the permit to work procedure is used to ensure that relevant risk control measures are in place for the duration of the work.

However, the principles underlying a permit to work procedure can be used for a wide variety of tasks, including work with hazardous chemicals, radiation sources and specific types of machinery such as cranes. They have also been used to control access to hazardous areas and for particular categories of work such as excavation and moving pieces of equipment.

The practical problem with all permit to work procedures is, however, making sure that anyone who has to carry out high-risk tasks knows that they need a permit. It is essential to provide information on permit to work requirements and to monitor their implementation rigorously if the procedure is to be effective.

Aims of permit to work procedures

It is possible to identify a number of aims which could be satisfied by a permit to work procedure:
- granting written permission to carry out a particular task
- providing a procedure for carrying out a risk assessment before undertaking a high-risk task
- providing a procedure to keep essential risk controls in place for the duration of a high-risk task
- providing a procedure for returning the work site to a safe condition when a task is complete
- providing a means of communication between people involved in managing high-risk tasks and those carrying them out.

Although a particular permit to work procedure can be used to satisfy one or more of the aims listed above, the procedure is unlikely to be satisfactory unless the actual aims are clearly identified and written down as part of the procedure's documentation.

Issuing permits to work

There should be a written procedure describing the steps to be taken before a permit to work is issued and the arrangements for the issue of permits. The details required in this procedure will vary depending on the type of activity in question, but you may need to consider the following:
- **The work to be done.** The permit must record accurately what has to be done but need not include details of how it is to be done. Where more information on the method is necessary, it can be provided as a safe working procedure, sometimes referred to as a method statement.

- **Where the work is to be done.** This is particularly important in circumstances where there are several pieces of equipment or plant of the same or similar type. Where items of plant are not numbered, permits should include diagrams or plans to help to ensure that the work is carried out at the correct location and on the correct item of plant or equipment.
- **When the work has to be done.** Some tasks have to be carried out within a specified period of time and, if not completed within this time, they have to be suspended. A common example of this is railway maintenance work, where a 'possession' is granted only for a specified period and any work not completed during that period can only be continued when arrangements have been made for a second possession.
- **The hazards associated with the work and the risks that may be encountered.** In theory, the issue of each permit to work should be preceded by a sufficiently detailed risk assessment – this will be essential for 'one-off' jobs. However, in practice, 'generic' risk assessments are often used, tailored as appropriate to the specific requirements of the task to be carried out.
- **The risk control measures which have to be implemented and maintained.** A wide variety of risk control measures can be specified, ranging from the competence of the personnel who are to carry out the work to using appropriate PPE.

Permits to work should be filled in by a competent person, in consultation with the people who will carry out the work and, if necessary, the person in charge of the location where the work will take place. Effective consultation helps to minimise misunderstandings during the use of the permit.

When all the details of the permit to work have been agreed, it is usual for the people issuing and accepting it to sign and date it and, if appropriate, record the time of issue.

Procedures while the permit to work is in force

The permit to work, or associated documentation, should specify the procedure to be followed while the permit is in force, but it may not be practical to cover all eventualities. Usually, there is some degree of reliance on the competence of the people carrying out the work to deal with unforeseen risks. This is why it is important, at the pre-issue stage, to consider what competences are required.

The permit to work procedure should also specify what should be done if the job has to be abandoned in the case of an emergency or suspended for any reason (for example, a lack of spare parts).

Where work is expected to take place over an extended period, the procedures should specify arrangements for such things as shift changes and overnight suspension of the work.

Procedure for withdrawing the permit

There should be a formal procedure for withdrawing permits to work, also referred to as cancellation or handback. This procedure should include a check that the work has been completed to specification and that normal work can be resumed without exposing anyone to risks arising from the work done under the permit. Usually, the person who issued the permit to work, or someone acting in the same capacity, will make the necessary checks and sign the permit as a confirmation that normal work can be resumed.

General comments on permit to work procedures

Like any procedure, permit to work procedures are only effective if they:
- have clearly specified aims
- are well designed
- are operated by people with the relevant competences
- are enforced rigorously.

Too often, organisations pay lip service to permit to work procedures so that they become a 'paper chase' rather than an effective risk control procedure. Failure to audit the permit to work procedure adequately can result in poor practices continuing for long periods of time; for example, see the report on the Piper Alpha disaster.[23]

These weaknesses in permit to work procedures, together with organisations' failures to deal properly with the work done by contractors, cast serious doubt on the effectiveness of such procedures as a risk control measure. In too many organisations, it is assumed that, because the forms are available and filled in, the procedures are effective. However, as Piper Alpha showed, this is not necessarily the case and frequent checks on the permit to work procedures, with corrective action when necessary, are essential if these procedures are to achieve their desired aims.

Safety rules

Safety rules take a wide variety of forms, but the following general categories can be identified:
1. **Single rules for specific circumstances** cover the sorts of safety rules typically displayed as signs or notices:
   - 'hearing protection must be worn in this area'
   - 'no smoking'
   - 'access only to authorised personnel'
   - 'fire door – keep shut'.
2. **Small sets of rules for specific tasks or locations** typically cover the requirements for people entering hazardous areas or undertaking hazardous tasks, and they may be issued to individuals on a card or posted at the entrance to the hazardous area. A set of rules of this type which might be issued to visitors to a location could cover the following:
   - 'if an alarm sounds then...'
   - 'if you find a suspicious object then...'
   - 'if you discover a fire then...'
   - 'do not bring the following articles into the area...'
   - 'in case of an emergency, phone...'

   Some organisations prepare sets of rules of this type in the form of credit card-sized, laminated sheets. These are then issued to everyone likely to carry out the relevant task, or enter the relevant area, to serve as an *aide-mémoire*.
3. **More or less extensive sets of rules** for carrying out tasks which are so variable that a safe work procedure is not appropriate. In the UK, probably the best known of such sets of rules is the Highway Code, which describes the rules to be followed when driving. In the occupational setting, safety in fork-lift truck driving, for example, is usually specified with a set of rules, and typical rules making up such a set would include the following general topics:
   - the competences required to drive fork-lift trucks
   - the checks to be made before each use of the fork-lift truck
   - the rules to be followed when moving large loads
   - a prohibition on carrying passengers or using the fork-lift truck as a work platform except when driving appropriately modified vehicles
   - rules to be followed when refuelling or recharging.

In general, a limited number of clearly specified rules can improve health and safety performance, so long as there are adequate arrangements for enforcement. However, as the number and complexity of the rules in a set increase, the rules begin to lose their effectiveness. This is partly because people are poor at remembering long lists of rules and recalling them at the relevant time, but two other factors may also be involved.

1. **The rules may contradict each other.** It is extremely difficult to work out the implications of various rules and ensure that no two rules conflict. In long-established sets of rules, which have been added to piecemeal over the years, this can be a major problem.

2. **The rules may be seen as protecting the people who issue the rules, rather than the people who have to comply with them.** Organisations have been known to have written rules so that if anything goes wrong they can claim that the people involved were 'not following the rules'. The fact that these rules were known to be impractical, and never enforced, is conveniently ignored.

Where either of these conditions applies, the whole set of rules falls into disrepute and even the non-conflicting rules, and the rules which are not primarily for the protection of the rule makers, are ignored.

To be at their most effective, safety rules should meet the following criteria:

- there should be few of them and there should be no conflicts between them
- each rule should be straightforward and the reasons for it self-evident or clearly stated
- they should be available at the time or place of need, or presented sufficiently often that people remember them
- they should be consistently enforced without, for example, one set of rules for managers and another set for the workforce.

Links between safe work procedures, permits and rules

Apart from the terminological difficulties already described, the topic of 'safe systems of work' is further confused by the fact that organisations often combine two or more of the different concepts dealt with in this chapter in the same procedure. For example:

- a list of safety rules may include a rule which requires a particular safe work procedure to be followed, or require a permit to work to be obtained
- a written safe work procedure may include, at one or more points in the procedure, the need to obtain a permit to work and to follow the procedures associated with it
- permit to work documentation may have a list of safety rules printed on the permit or issued as part of the permit to work procedure.

The particular implementation of a 'safe system of work' must, therefore, be judged by its specific aims and how these aims are to be achieved. However, there are three conceptually different elements of safe systems of work as described in this chapter and they are summarised below.

Summary

This chapter pointed out that the term 'safe system of work' is used as a general description of three different categories of risk control, which are based on attempts to get people to carry out work activities in a particular way. The three categories are:

1. **safe work procedures** – ways of carrying out hazardous tasks in a way which minimises the risks associated with the tasks' hazards, which take the form of written descriptions of how work should be carried out

2. **permit to work procedures** – a method of making hazardous tasks less of a risk by specifying procedures to be followed that will address the hazards associated with the task

3. **safety rules** – lists of dos and don'ts that are mainly used for highly variable tasks such as using ladders or driving fork-lift trucks, where it is impossible to specify, in a safe work procedure, all the possible types of task which may have to be done.

8: Monitoring and measuring losses

Introduction

In Chapter 4, the following topics for monitoring and measuring were briefly outlined:
- incidents, including accidents
- conformity with standards
- spin-offs and side effects.

This chapter deals with incidents, spin-offs and side effects; conformity with standards is dealt with in Chapter 10.

Remember that incidents include accidents and near misses, and that the term 'accident' is defined in various ways. In particular, some definitions of 'accident' include damage to physical assets, while others restrict the term to harm to humans. For the purposes of this chapter, we will use the term 'losses' to cover the adverse consequences of accidents, near misses and side effects.

Losses

This section on losses deals with the following topics:
- **Types of loss and choosing types of loss to monitor and measure.** The most common type of loss monitored and measured in organisations is the loss arising from injury accidents, but other types of loss can be important. We have to decide what types of loss we will monitor and measure.
- **Monitoring losses.** Systems have to be developed to identify those incidents and side effects that result in the losses we wish to monitor, and to collect the information required about the incidents and side effects.
- **Measuring losses.** Losses are measured for a variety of reasons, including assessing the effectiveness of risk control measures and demonstrating continual improvement.

Types of loss

This book is primarily about health and safety, but before looking specifically at losses in these areas, it is worth putting them in context with quality and environmental losses, since these are two other areas of risk with which health and safety management may be linked.

The main types of loss are shown in Table 8.1, together with some examples of each type. To a certain extent, the divisions shown in Table 8.1 are artificial. For example:
- a customer complaint may be about product safety
- a spillage may result in injury, ill health, damage to the environment and asset damage
- an injury accident can also involve asset damage and damage to the environment
- some injuries, such as back injuries, can result in chronic illness.

Although these divisions are artificial, the tradition is to keep them separate, with different specialists dealing with particular types of loss. However, the risk management principles described in this book can be applied equally well to any type of loss, although we will restrict discussion to the following categories:
1. damage to people, including mental and physical damage, whether this occurs instantaneously (mostly injuries) or over a longer period of time (mainly ill health)
2. damage to assets – including assets of the organisation, such as plant and equipment, and assets belonging to other people – caused by the organisation's activities.

Table 8.1

Typical losses

Quality	Environment	Injuries	Health	Asset damage and other losses
Customer complaints Product nonconformities Service nonconformities Damage resulting from unsafe products	Spillages Emissions above consent levels Discharges above consent levels	Injuries to employees at work Injuries to others at work Injuries during travel Injuries at home Injuries resulting from unsafe products Injuries resulting from work activities	Sickness absence Chronic illness* Acute illness Sensitisation	Damage to the organisation's assets, eg plant and equipment Damage to other people's assets Interruptions to production Losses from theft and vandalism Losses from fire

From now on, these will be referred to as key losses to distinguish them from all the other types of loss. The use of this term is not intended to imply that these losses are more important than other losses; it is simply a term used to label the losses which are of current concern.

Typically, an organisation will have a number of key losses but, in order to be used effectively in risk management, each one will have to be clearly defined. There are three important aspects of this definition:
1. **the nature of the incidents to be included** – for example, injury, sickness or damage to assets
2. **the severity of the incidents to be included** – this can be difficult, particularly when minor injuries or minor damage are to be used as a key loss. However, if trend analysis is to be meaningful, there has to be consistency
3. **the population to be covered by the key loss** – for example, will it be restricted to the organisation's personnel and assets, or will contractors and/or members of the public be included?

The key loss will also have to meet a number of other criteria as follows:
1. The key loss must be measurable in a quantifiable manner. Typical measures used for key losses include:
   - numbers of incidents
   - hours or days lost
   - costs of incidents
   - relevant units (for example percentage of rejected products, litres spilled).
2. There must be a way of measuring the key loss. The fact that something is measurable does not mean that an organisation has the facilities to measure it. For example, the cost of injuries is measurable but few organisations have the investigative and accountancy procedures in place to carry out the measurements.
3. There must be something which can be done to influence the key loss, and the organisation must be willing to commit the resources required to do what is necessary.

Choosing key losses

In the majority of organisations there will be a number of possible key losses and a decision will have to be made on which to use. A number of factors will influence this decision and the main ones are:

---

* A chronic illness is one that continues for a period of time, irrespective of its severity.

- **Legal requirements.** Many countries impose a legal requirement to record incidents which are suitable for use as key losses.
- **Third party accreditation.** Where organisations are committed to a health and safety standard such as BS OHSAS 18001, this will impose recording requirements; these requirements can be extended to cover key losses.
- **Cost reduction.** Where a particular type of incident has been identified as costly, setting it up as a key loss will be a first step in reducing these costs.

An important stage in effective risk management for any organisation is a clear definition of the losses to be managed. However, this initial description need not be exhaustive. What is important is to identify some key losses and put in place effective procedures for managing them. The review procedure, which will be described in Chapter 11, will then be the mechanism for judging the appropriateness of this initial list of key losses and identifying whether, and how, the list should be changed.

As with most aspects of risk management, the danger is that 'the best will be the enemy of the good'. In other words, so much time will be spent on trying to decide on the 'best' list of key loss indicators that no work is done on managing a 'good' list which could be decided on quite quickly.

For most organisations making a start on systematic risk management, the following initial list of key loss indicators is likely to be adequate:
- number of injury accidents and/or the number of days lost as a result of injuries
- number of days lost as a result of illness
- number of repair jobs necessitated by damage to assets, and/or the cost of these repair jobs.

Whichever key losses you choose, the same principles of risk management apply. If the examples of key losses used in this book to illustrate these principles do not suit your current list, you should translate the examples as appropriate.

Monitoring key losses

The primary purpose of monitoring key losses is to identify all of those incidents and side effects that give rise to the key loss. If these are not identified, no data concerning them can be collected, and there can be no accurate measurement of the key loss and no corrective action can be taken to deal with the incidents and side effects. This chapter deals with monitoring key losses for the purposes of measurement; monitoring for the purposes of corrective action is dealt with in Chapter 9.

Identifying key losses

The most common way of identifying key losses is some form of procedure for reporting 'accidents', 'incidents' or 'safety concerns'. Typically, this procedure will take the following form:
- There is a reporting form that specifies what type of loss is to be reported and the information to be collected and recorded about each loss.
- Relevant personnel are informed of the need to complete the form when a specified loss occurs.
- When a specified loss occurs, data are collected and recorded for measurement purposes and, where appropriate, corrective action is taken.

This type of data collection procedure has a major failing in that, in general, the less serious an incident is, the less likely it is to be reported. It is extremely difficult in most organisations to 'cover up' a death or major injury, but minor injuries often go unreported. However, there are various things which can be done to improve reporting:

- **Have a 'user-friendly' system.** Reporting and recording systems which try to collect too much information given the severity of the incident will not be used. An example is using forms designed for 'major' accidents to report 'minor' accidents.
- **Emphasise continual improvement and prevention of recurrence.** The reasons for collecting the data (continual improvement and prevention of recurrence) should be clearly stated and repeated often.
- **Avoid a 'blame culture'.** If incident reports are followed by disciplinary action, or more minor forms of 'blame', people will stop reporting.
- **Demonstrate that the data are used.** If people who have to report and record cannot see that their work and effort are being used, they will stop making the effort.
- **Always give feedback.** It is not always possible to take action on a report, but when it is not, there should be feedback on the reasons for the lack of action to the people concerned.

Details of how these actions can be implemented are dealt with at appropriate points in later chapters.

Checking for non-reporting

Where it is important to have an accurate measure of a particular key loss, it is essential to check that all of the relevant incidents are being reported. Three methods of carrying out such checks are:

1. **Interviews with people who are likely to have experience of the relevant losses.** People are usually quite willing to talk about the incidents they did not report if there is no adverse consequence as a result of their revelations. They are even more willing to talk about incidents that other people did not report, unless this is likely to result in adverse consequences for the people being identified. A skilled interviewer who has carried out an appropriate sample of interviews should be able to make a reasonably accurate assessment of the proportion of incidents that is going unreported. Interview technique is described in Chapter 9.
2. **Inspections of locations and people.** The simplest example in this category involves inspecting plant and equipment for damage and comparing the findings from the inspection with the most up-to-date damage records. A similar approach can be used for minor injuries by, for example, inspecting people's hands, checking for dressings, cuts, grazes and burns, and then comparing the inspection findings with the injury records. However, this approach can create resentment and should be used with care. In some organisations, the dressings for minor injuries are a characteristic colour, easily recognised even from a distance. The use of this type of dressing makes inspections for minor injuries much easier.
3. **Cross-checking one set of records with another.** The usefulness of this type of technique will depend on the records available and their accuracy, but possible cross-checking includes the following:
   - where there are records of what is taken from a first aid box, these can be checked against injury records to see whether everyone who has made use of the first aid box has reported an injury
   - where there are records of plant and equipment maintenance, these can be checked against records of accidental damage to plant and equipment to see whether all of the relevant repairs which have had to be carried out have been recorded as accidents
   - where records of 'cradle to grave' or 'mass transfer' are available for particular chemicals or substances, these can be used to check whether all unexplained losses of chemicals or substances have appeared as, for example, accidental spillage records.

Any or all of the above techniques can be used to check the adequacy of loss data. For the remainder of this chapter we will assume that the relevant checks have been made and that 'good' data are available for analysis.

It can be seen from the previous description that there are two aspects to monitoring losses:

- waiting for information on losses to be reported
- going out and collecting information on losses.

In quality management, the first type of monitoring is referred to as 'passive' monitoring and the second as 'active' monitoring. The example quoted in ISO 9004:2004*[24] deals with customer complaints. Passive monitoring in this case is waiting for customers to complain, and active monitoring is asking customers whether they have any complaints. In most organisations, loss information is collected only passively, even though it is known that not all of the relevant losses are being reported. Active and passive monitoring are dealt with in more detail in Chapter 19.

Measuring key losses

It has been said that 'if you can't measure it, you can't manage it' and this is undoubtedly true for key losses. The ultimate aim in measuring key losses is to determine accurately whether performance is improving or deteriorating and, if key losses cannot be measured accurately, this cannot be done.

There are two commonly used measures of key losses – the numbers of incidents and their severity. Table 8.2 illustrates, for a number of key losses, the sorts of measure which fall into these categories.

Numbers of incidents	Severity of incidents
Number of injuries resulting in more than three days away from normal work	Number of days away from work as a result of 'over-three-day' injuries
Number of incidents resulting in damage to production machinery	Number of hours of downtime as a result of accidental damage to production machinery
Number of claims for noise-induced deafness	Compensation paid on claims for noise-induced deafness†
Number of spillages of diesel fuel	Number of litres of diesel fuel spilled

Table 8.2
Numbers of incidents and severity for a range of key losses

Where possible, it is preferable to use severity measures, since these give a more accurate estimate of losses. Yet even when severity measures are used, it is still difficult to compare one type of loss with another. For example, is having a person away from work for a week 'more severe' than spilling 500 litres of diesel? For this reason, the best measure of loss is financial cost, since this enables direct comparisons, not only between incidents of the same type but also between incidents of different types. In addition, it enables the comparison of the cost of preventing the loss and the likely saving from the prevention. However, few organisations have the accounting procedures necessary for these sorts of comparison, and the issue will be considered later (see Chapters 10 and 24) in the context of the financial aspects of risk management.

Whichever type of measure we use, there are certain basic principles we must follow if we are to make an accurate assessment of performance over time. These basic principles are described in the remainder of this chapter.

---

* ISO 9004:2004 was replaced by ISO 9004:2009 which does not include a description of active and passive monitoring. However, this terminology is useful and has been retained for this book.

† Compensation may be paid via an insurance company and in these cases it may be preferable to use the insurance premiums with respect to noise-induced deafness as the loss measure. However, for 'self-insured' companies, compensation payments are a direct measure.

Trend analysis of loss data

If continuous measurements of loss data are being made, the next step is to calculate performance in different time periods and compare one time period with another – that is, to carry out trend analysis over time. However, any such analysis can be influenced by changes other than those in the effectiveness of risk management. For example, if an organisation is reducing the amount of work it does, for whatever reason, it is likely that the number of accidents will decrease, whether or not there are any changes in risk management practices. Similarly, if the amount of work being done is increasing, it is to be expected that the number of accidents will increase.

Because numbers of accidents can be influenced by these sorts of change, trend analysis will be dealt with in two stages. First, we will assume that there is a steady state, with no relevant changes; this will enable us to consider the basic techniques without undue complication. We will then describe the techniques needed to take into account the sorts of change mentioned in the previous paragraph.

Figure 8.1

Quarterly figures for major accidents in 2010

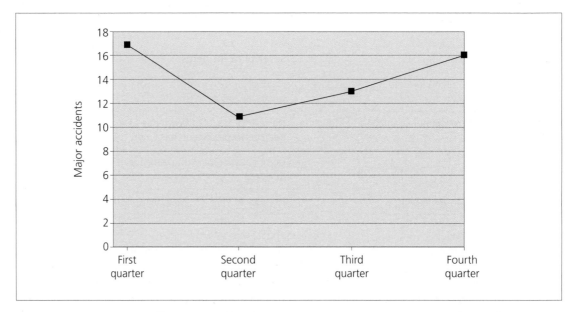

Figure 8.2

Days lost per month through sickness in 2010

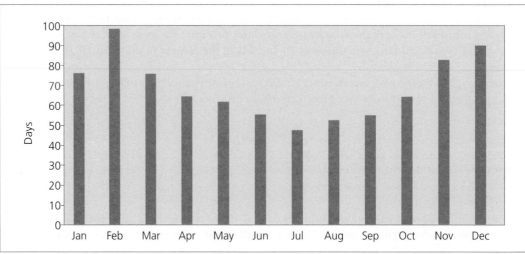

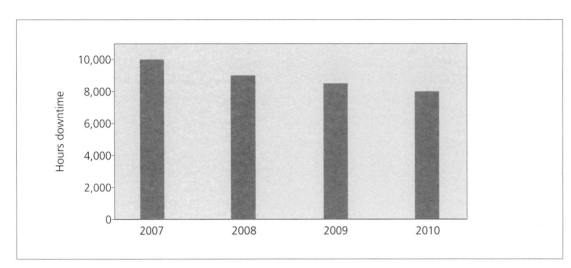

Figure 8.3
Hours of downtime per year
(2007–2010)

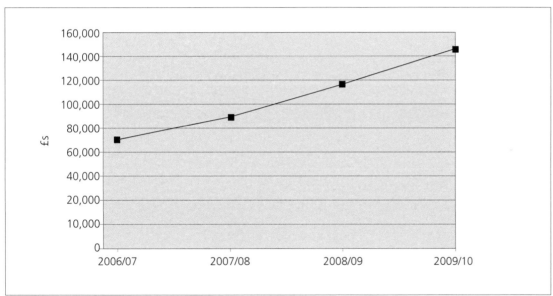

Figure 8.4
Damage costs in financial years
2006/07–2009/10

Trend analysis with a steady state

The most straightforward method of trend analysis is to plot the level of loss against a suitable measure of time. Typical losses for this sort of analysis include the number of major injuries, days lost through sickness, litres of lost fuel and costs of damage repair. Typical time measures include monthly, quarterly and annually. Figures 8.1 to 8.4 show examples of these kinds of plot.

One practical problem with plots is that the more detailed they are, the more difficult it is to judge the trend 'by eye'. Compare the two plots shown in Figures 8.5 and 8.6 overleaf. Although the same data form the basis of both figures, it is easier to see from the quarterly plot that there appears to be a slight downward trend. It is generally the case that grouping data in this way 'smooths out' variations and makes trends easier to identify. One technique for smoothing data is the quarterly moving mean.

The calculation of the quarterly moving mean is quite simple and the steps required can be illustrated by using as an example the data from Figure 8.5.

Figure 8.5
Monthly minor
injury figures for
2010

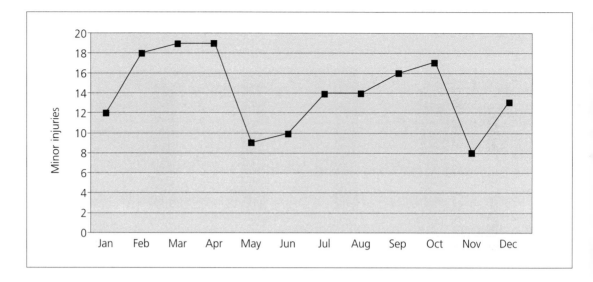

Figure 8.6
Quarterly minor
injury figures for
2010

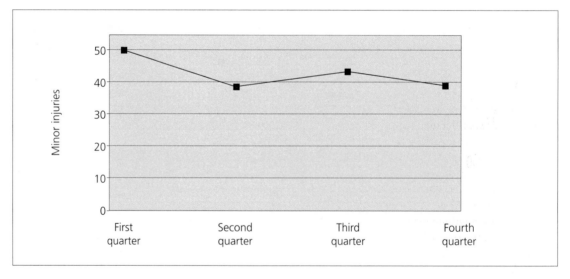

1. For the first two months of the year, the accident numbers are plotted on a month-by-month basis as in Figure 8.5.
2. For the third month, the numbers of accidents for January, February and March are added together and the result divided by three to give the mean number of accidents, and this mean is plotted. Readers may be more familiar with this calculation described as the 'average' number of accidents per month. However, there are different types of average, as will be seen in Chapter 19, one of which is the mean.
3. For the fourth and subsequent months, the quarterly mean is calculated from the current month's and the previous two months' figures, and the results of these calculations plotted.

The quarterly moving mean for the data in Figure 8.5 is illustrated in Figure 8.7. It is usual to plot both the actual monthly figures as well as the moving mean, and this has been done in Figure 8.7.

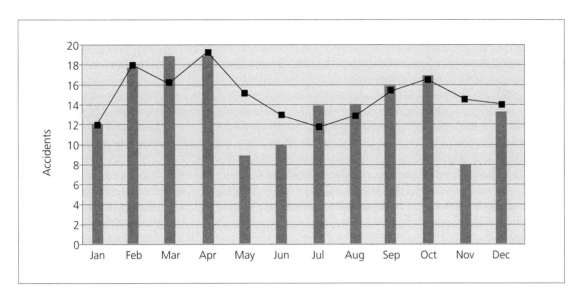

Figure 8.7
Monthly accidents
and quarterly
moving mean
(2010)

There are many ways in which the principle of a moving mean can be used to smooth data, but detailed consideration of statistical techniques will be left to Chapter 19. At this point, it is important to appreciate that the numbers of accidents can fluctuate widely from month to month and that some smoothing technique must be used if trends are to be identified 'by eye' rather than by using one of the statistical techniques described in Chapter 19.

Whichever method of trend analysis you use, it is always necessary to check that any change in direction is more than random fluctuation.

Suppose that in a particular year there were 100 accidents in an organisation and that in the following year the organisation was going to carry out the same amount of work with no changes which would affect risk. In these circumstances, we would expect that there would be around 100 accidents in the year following the one for which records were available. The expectation would not be exactly 100 accidents, but around 100. If there were 99 or 101 accidents, it would be possible to say that this was due to random fluctuation; more generally, anything between say 95 and 105 accidents could also be random.

The difficulty arises when there are 85 or 90 accidents. Are these numbers due to random fluctuation, or is someone doing something which is improving risk control and influencing the accident numbers? Statisticians refer to fluctuations in numbers which cannot reasonably be attributed to random fluctuation as 'significant' and they are able to calculate confidence limits that allow statements such as:

> There is only a 5 per cent chance that the improvement in accident performance is due to random fluctuations.

> This deterioration in accident performance would have occurred by chance in only 1 per cent of cases.

This use of confidence limits, and the calculation of the significance of fluctuations in numbers, has practical importance in the more advanced techniques of loss management since it is only possible to draw valid conclusions when it is known whether or not particular fluctuations in numbers are significant. However, for the purposes of Part 1.1, it is necessary only to know that care should be taken in interpreting the results of trend analyses unless appropriate confidence limits have been calculated. The actual calculation of these limits is left to Part 2.1.

Trend analysis with variable conditions

So far, for the sake of simplicity, we have assumed that everything has remained stable in the organisation. In the real world, however, things rarely remain the same for any length of time and methods of trend analysis which can take this into account are needed.

In an ideal world, it would be possible to measure changes in risk in an organisation since, if this was known, it would be possible to determine how well the risk was being managed. For example, it would ideally be possible to be able to say such things as 'despite a 50 per cent increase in risk due to additional work being done, the accidents increased by only 25 per cent', or 'there was a 10 per cent reduction in risk because of the new machines and work procedures, but accidents increased by 5 per cent'.

Unfortunately, it is rarely possible to measure risk in this sort of way. In practice, we need to find some proxy for risk which can be measured and use this instead. Two such proxy measures are commonly used: numbers of people employed and numbers of hours worked. These are used to calculate two accident rates.

Incidence rate

This is the number of accidents divided by the number of people employed, and the result is usually multiplied by 1,000. It is used to take into account variations in the size of the workforce. The equation is:

$$\text{Incidence rate} = \frac{\text{Number of accidents} \times 1,000}{\text{Number of people employed}}$$

Frequency rate

This is the number of accidents divided by the number of hours worked, and the result is usually multiplied by 100,000. It is used to take into account variations in the amount of work done caused by, for example, overtime and part-time employment. The equation is:

$$\text{Frequency rate} = \frac{\text{Number of accidents} \times 100,000}{\text{Number of hours worked}}$$

Note that the number of hours worked must be properly established. Some 'frequency rates' are calculated by multiplying the number of people employed by 40 hours (for a weekly rate), 160 hours (for a monthly rate) and 150,000 for an annual rate. While this looks like a frequency rate, it is actually an incidence rate, since no accurate information on hours worked is used in the calculation.

There are, however, problems with these accident rates:

- **Terminology.** Although the versions given above are in general use, there is no universal agreement. It is not possible to interpret a rate unless the equation on which it was based is known.
- **Definitions.** There is no general agreement on what constitutes an 'accident' – some organisations base their rates only on major injuries, while others use both major and minor injuries. Similarly, there is no general agreement on what constitutes an employee; incidence rates can be reduced simply by employing more part-time workers! Hours present similar problems, with different types of hours having significantly different types of risk. For example, 'working' time, when the risk is high, may differ from 'waiting' time, when the risk is low.
- **Multipliers.** As there is no general agreement on the multipliers which should be used, it is again necessary to know the equation.

In general, the value of accident rates depends on the quality of the data on which they are based and the competence of the person preparing them. Quoted rates should always be treated with caution until it is possible to be reassured on these points.

Using incidence and frequency rates enables us to carry out sensible trend analyses during periods when organisations are changing the numbers in the workforce, or when there is variability in the amounts of work being done. If used properly, they can be quite accurate and they can enable comparisons between one organisation and another, or between different parts of the same organisation. These sorts of comparative analysis are described next.

Comparisons of accident data

As we have seen, it is possible to make valid comparisons only when there is some measure of the risk being managed. Even when the numbers employed or the hours worked are taken into account, these are only proxies for risk that are used because they can be measured, rather than because they are good indicators of risk.

If there are two organisations, each with a frequency rate of 100, this could be for either of the following reasons:
- the organisations have roughly equal levels of risk and are managing them equally effectively; or
- one organisation has high levels of risk and is managing risk well, while the other organisation has low levels of risk and is managing risk badly.

This should be borne in mind when making, or interpreting, comparisons of accident data, since it is a fundamental weakness of such comparisons. In general, a comparison will be valid only to the extent that the risk levels in the organisations being compared are equal.

Having dealt with this caveat, it is now possible to look at the types of comparison which can be made.
- **Comparisons between parts of an organisation.** In theory, these are the simplest and potentially most accurate comparisons. This is because the measurement of risk, the definition of what has to be reported, reporting procedures, and methods of calculation are all under the organisation's control and can be standardised. However, whether this standardisation is carried out effectively in practice is usually open to debate.
- **Comparisons between one organisation and another.** Industries in the same sector can compare accident data with one another, assuming that they are willing to do so. In the UK, for example, there are national associations for health and safety professionals from particular industry and service sectors and meetings of these associations provide a forum for comparing accident data. More formally, there have been moves recently to include accident data in 'benchmarking' exercises where organisations compare various aspects of their performance with those of their competitors. For further information on benchmarking, see *Health and safety benchmarking.*[25]
- **Comparisons between an organisation and the relevant industry or service sector.** Some trade organisations publish aggregated data on accidents for their industry or service sector, which give, for example, the 'average' frequency and incidence rates for a particular year. In addition, national accident statistics may be available, as they are in the UK. These national figures may be presented in a variety of ways, including a breakdown by industry and service sector. There is a Standard Industry Classification (SIC) which can be used for this purpose and further details of this, and other sources of UK accident data, are given in the *UK box – accident data*, overleaf.
- **Comparisons between countries.** Where appropriate data are available, for example from the European Safety Agency, the ILO or the World Health Organization (WHO), comparisons can be made between accidents in one country and another, either for the country as a whole, or by industry or service sector. However, there are major variations in accident reporting procedures between countries so that comparisons of this type should be made with great care.

A particular problem with all the comparisons just described is that there is no consistency about what constitutes an 'accident'. You will remember that this was one of the problems with comparing incidence and frequency rates. One way of improving comparisons is, therefore, to calculate a rate which takes into account the severity of the accidents which are occurring. This is usually done using the following equation:

$$\text{Severity rate} = \frac{\text{number of days lost as a result of } n \text{ accidents}}{n}$$

Effectively, this gives the mean number of days lost per accident and, indeed, it is sometimes referred to as the mean duration rate.

This severity rate can be used in trend analysis in the same way as other rates but, for the purpose of monitoring losses as a result of injury accidents, it is preferable to use the numbers of days lost. This is because the severity rate can go down while, at the same time, the number of days lost is increasing (more accidents, but less severe ones), so that the severity rate can give a misleading picture of actual losses.

The final point to make on comparisons is that the three rates described above should, when the relevant data are available, be used in conjunction with each other. This is because they need not necessarily give the same result, as Table 8.3 illustrates for four departments (A, B, C and D) using simplified data.

Table 8.3
Comparisons using incidence, frequency and severity rates

Departments	A	B	C	D
Injury accidents	100	80	60	20
Numbers employed	100	40	60	20
Incidence rate	1,000	2,000	1,000	1,000
Hours worked	10,000	8,000	3,000	2,000
Frequency rate	1,000	1,000	2,000	1,000
Days lost	100	80	60	40
Severity rate	1	1	1	2

Key loss data as a measure of risk

Accurate key loss data will provide a measure of what has gone wrong in the past, and enable comparisons over time (trend analyses) and comparisons between one organisation and another. What key loss data will not do, even if they are accurate, is to provide a measure of risk. It is possible to illustrate why this is so by considering the earlier example about rain and missing roofs (see Chapter 3). In that example, the risk was defined as the likelihood of rain multiplied by the severity of the damage to assets if it were to rain. It follows from this that there are three ways in which damage to assets could have been avoided:
1. there was no rain
2. the assets were of a kind that is not damaged by rain
3. both of the above.

If the only data available are for the number of times assets were damaged, then there is no information on the extent of the risk, only that a risk exists. It is not known, for example, how often it rained, how many roofs were missing or what proportion of houses with missing roofs contained assets which would have been damaged by rain.

More generally, data on the number of accidents provide very little information about risk. As was mentioned earlier, two organisations can have the same frequency rate because one is managing high levels of risk very well, while the other is managing low levels of risk very badly. Alternatively, because risk is

UK accident data

National figures – HSE

There are two practical problems associated with obtaining accident data. First, there is always a delay in publication, with some detailed statistics being published two or three years in arrears. Second, changes in legislation and the government departments responsible for particular aspects of health and safety mean that contact points for obtaining data change from time to time. However, despite these practical difficulties, a wide range of information is available from the HSE. Check the statistics webpage of the HSE's website at www.hse.gov.uk/statistics/index.htm (Great Britain only).

National figures – other

Separate statistics are compiled for:

- social security
- domestic accidents
- road traffic accidents.

Details on these are available from the relevant government departments.

Industry and trade figures

In addition to the industry analyses provided by the HSE, certain industry associations provide their members, and others, with accident statistics.

probabilistic, two organisations with the same levels of risk can have widely different numbers of accidents because one was 'lucky' and the other was not.

True levels of risk in an organisation can only be determined accurately using an appropriate risk assessment methodology and such methodologies were considered in Chapter 5.

Spin-offs and side effects

Monitoring and measuring losses enables checks to be made on the effectiveness of risk control measures and may identify the need for further control measures. However, implementing and maintaining risk control measures may have unforeseen effects, which may be beneficial (spin-offs) or detrimental (side effects). For this reason, there should be procedures for monitoring and, where appropriate, measuring these effects. The nature of these procedures is the subject of this section, which begins with a discussion of the types of spin-off and side effect that may arise.

Types of spin-off and side effect

Although the range of spin-offs is potentially wide, a large number of them will fall into one or other of the following categories:

- **Increased productivity.** The required changes identified during a job safety analysis often result in more efficient as well as safer procedures – this can lead to increased productivity.
- **Increased morale.** Attention to health and safety and a reduction in incidents may increase general morale.
- **Fewer breaks from work.** Risk control measures such as those that reduce the physical strain of tasks may mean that fewer breaks are needed.

- **Lower staff turnover.** Lowering the risk of injury and ill health associated with jobs may make it less likely that people will leave those jobs.
- **Increased comfort.** Risk control measures that increase comfort will be beneficial irrespective of whether they influence productivity or morale.

Note that most of these categories are related, in that they can all have a direct or indirect effect on productivity. In general, side effects are the opposite of spin-offs, for example reduced productivity, lower morale and more breaks from work.

Monitoring spin-offs and side effects

As with losses, spin-offs and side effects can be monitored passively or actively, and these two types of monitoring are outlined below.

- **Passive monitoring.** This requires a procedure similar to that for losses, with suitable information being provided on what to report and what data to record.
- **Active monitoring.** This requires a procedure which checks individual risk control measures for spin-offs and side effects. Since risk control measures have to be monitored for other purposes, it is convenient to include active monitoring of spin-offs and side effects with this more general monitoring of risk control measures. This type of monitoring is dealt with in detail in Chapter 10.

Measuring spin-offs and side effects

A number of spin-offs and side effects can be measured in ways which enable a monetary value to be assigned to them. However, for spin-offs and side effects such as morale and comfort allocating a monetary value is more difficult, but they can be measured in other ways. Measuring monetary value is dealt with first, followed by other measures for spin-offs and side effects.

It is important to measure the monetary value of spin-offs and side effects since this will affect the true cost of the risk control measure and will mean that the relevant business case will have to be revised. Business cases and their revision are dealt with in Chapter 24.

Measuring morale, comfort and other spin-offs and side effects which cannot be allocated a monetary value is best done by appropriate survey techniques such as job satisfaction questionnaires. These techniques are dealt with in Chapter 26.

Summary

This chapter described the broad range of losses which might be the subject of risk management activity, and how appropriate losses might be chosen. We also considered how the chosen losses might be measured and how these measures can be used, via trend analysis, to keep track of an organisation's performance. The chapter ended with a brief consideration of spin-offs and side effects.

However, it is also possible to learn from losses by investigating the causes of losses and by analysing all of the losses to determine whether there are patterns in their occurrence. These two topics – accident investigation and epidemiological analysis – are the subjects of the next chapter.

9: Identifying causes and patterns

Introduction

This chapter is concerned with two topics:
- identifying the causes of incidents, nonconformities, spin-offs and side effects by investigating individual examples of each
- identifying patterns in groups of incidents, nonconformities, spin-offs and side effects – that is epidemiological analysis.

The best-known type of investigation is that of accidents, and we will consider this topic first.

Accident investigation

Before starting on accident investigation, we need to consider some problems with terminology and, in particular, the way the phrase 'accident reporting' is used to mean a number of different things.

The chronology of an accident and its aftermath have the following identifiable stages:
1. the person who sustains the injury, or someone else, **reports** that an accident has happened, usually to a supervisor or manager
2. the person to whom the accident is reported makes a written record of the salient points, usually on an **accident report form** or **accident record form**
3. the accident is investigated and, if it is sufficiently serious, it has to be **reported** to the relevant national authority, for example the HSE in the UK
4. the person who investigates the accident writes a **report** on his or her findings, together with any suggestions for remedial action*
5. the person who investigates the accident **reports back** to those involved on the outcome of the investigation and the action to be taken.

Since it is not always obvious from the context which of these uses of 'report' is the relevant one, we will use the following terminology in this chapter.
- **Accident report** – the report made by the person who sustains the injury, or someone else on his or her behalf.
- **Accident record** – the written record of the salient points, usually on an accident record form.
- **Accident notification** – the notification of an accident to the relevant national authority, for example the HSE in the UK. Accident notification will also be used to describe other required notifications, for example to insurance companies.
- **Investigation report** – the written report on the findings of an accident investigation, together with any suggestions for remedial action.
- **Feedback** – reporting back to those involved on the outcome of the investigation and the action to be taken.

Since the collection of the information required for accident notification is an essential part of a complete accident investigation, the requirements for accident notification in the UK will also be dealt with in this chapter.

---

* 'Remedial action' is used here as shorthand for the mix of corrective and preventive actions which typically make up the recommendations following an accident investigation.

Reasons for accident investigations

Accident investigations can be carried out for a variety of reasons, including:

- collecting the information required for accident notification
- obtaining the information required to initiate, or defend, an insurance claim
- obtaining the information necessary to prevent a recurrence
- establishing who, if anyone, was at fault.

In theory, a thorough investigation will result in the collection of the information required to satisfy all of these purposes but, in practice, this is rarely the case. If, for example, the primary purpose is to collect the information required for accident notification, then the investigation is usually stopped when the relevant information has been collected, whether or not this information includes that required for identifying remedial action. When the primary purpose is to establish who is at fault, there is an additional problem in that the investigation may become adversarial – that is the investigators are on one 'side' or the other, for example the employer's 'side' or the injured person's 'side'. This can lead to biases in data collection, such as information which does not support a particular investigator's 'side' being ignored or not recorded.

The ideal investigation is, therefore, one which is neutral with respect to fault and has the primary purpose of obtaining the information necessary to prevent a recurrence.*

Which accidents should be investigated and by whom

It is often the case that only the more serious accidents are considered worthy of investigation. The rationale for this is usually that investigations take time and, therefore, cost money, so that they are only worth doing when there has been a significant loss. However, various researchers have demonstrated that there is no relationship between the causes of accidents and the seriousness of the outcome, and that, for example, minor injuries have the same range of causes as major injuries.†

It follows from this that as much can be learned from investigating individual minor accidents as can be learned from investigating individual major accidents. Since it is also the case that there are many more minor accidents than major ones, investigating minor ones gives many more opportunities to learn from what has gone wrong.

However, there are so many minor accidents that there is the practical problem of the time required for adequate investigation of them all. There are two ways of dealing with this problem:

1. Provide managers with the competences to carry out proper investigations so that the required work is spread among a number of competent personnel.
2. Identify patterns in minor accident occurrence and investigate groups of minor accidents which are likely to have related causes. How this pattern identification is carried out is described later in this chapter in the section on epidemiological analysis.

In general, the first option is preferable. However, the second option can provide an acceptable alternative and, in any case, it should be used as a backup when managerial investigations are in place.

In an ideal world, all accidents would be investigated in detail by competent personnel, either managers who have been provided with the necessary competences or health and safety professionals who should, in theory, have the necessary competences already. In the practical world, 'serious' accidents tend to be investigated in detail by health and safety professionals while 'minor' accidents tend to receive superficial investigation by managers.

---

* This is an oversimplification which suits the purposes of Part 1.1. There is a more detailed discussion of the purposes of accident investigation in Chapter 20.

† The validity of this finding is the subject of debate; this is dealt with in Chapter 20.

However, irrespective of who is carrying out the investigation, there is a general procedure which should be followed, and this is described next.

The accident investigation procedure

Figure 9.1 gives an overview of the accident investigation procedure and the notes which follow deal with various aspects of the implementation of this procedure. Note, however, the following general points about the procedure described in Figure 9.1:

1. Not all of the stages described in the figure will be required in all accident investigations. For example, statutory notifications will normally only be needed for accidents resulting in certain categories of loss.
2. The process is shown as a linear one but in practice it is iterative. For example, findings from off-site interviews may suggest that it is desirable to revisit the site of the accident to collect more detailed information.

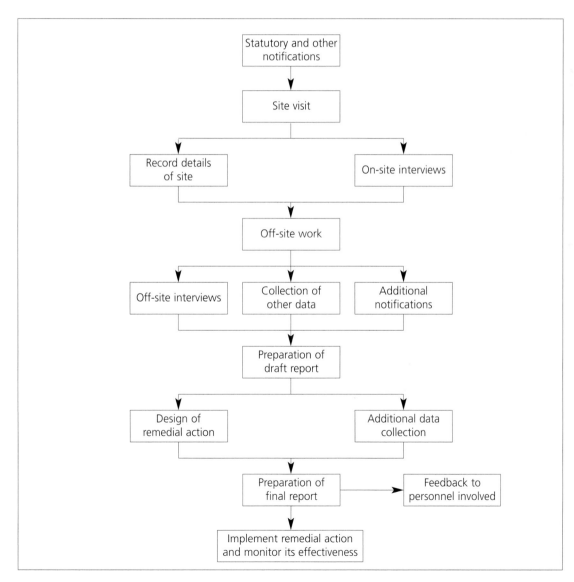

Figure 9.1

Accident investigation procedure

3. The last item in Figure 9.1, 'Implement remedial action and monitor its effectiveness', is not really part of the accident investigation procedure. It is included as a reminder of the real purpose of accident investigation.

Statutory and other notifications

For certain types of accident, there is a legal requirement in the UK to inform an enforcing authority by the 'quickest practicable means', and this should be done even before a detailed investigation has been carried out. Indeed, it may be the case that the enforcing authority wishes to carry out its own investigation and it may issue instructions about how to deal with the site of the accident. Details of the UK procedure for accident notifications are given later in this chapter in the *UK legislation – reportable and recordable incidents* box.

In addition to any statutory notifications, other people and organisations which may have to be notified, depending on the circumstances, will include:
- the next of kin of the injured person
- the supervisor and manager of the injured person
- the owner of any assets damaged
- the organisation's insurers
- the organisation's health and safety department
- the organisation's personnel department.

The timescales for these notifications and the information required will vary from organisation to organisation. It is good practice, therefore, for organisations to provide relevant personnel with an *aide-mémoire*, which can be updated when requirements change, rather than rely on training in accident notification, which may have been received many years before it has to be put into practice.

Site visits

There are various reasons why an investigator should visit the site of the accident and the main ones are listed here.
1. **To obtain first-hand experience of the physical layout of the site, the equipment in use, and so on.** If investigators do not go to the accident site themselves, any conclusions they reach will have to be based on reports from people who were there. This verbal transfer of information provides opportunities for misunderstanding, confusion and misinterpretation. Where a site visit is not possible, investigators can request photographs of the site – this is an improvement on relying on verbal descriptions, but is not ideal.
2. **To obtain first-hand experience of the normal activities at the site, including the personnel involved in these activities and any relevant materials, substances or components in use.** In part, this information can be obtained by observation but this should be supplemented by interviewing those who are engaged in, and familiar with, the site's activities.
3. **To record details of the accident site.** This is usually best done with photographs, supplemented with sketches where it is necessary, for example, to record details of dimensions. This record of the site serves three purposes: first, it is an *aide-mémoire* for the investigator; second, it can be used to enhance the investigation report; and third, it can be a useful tool during off-site interviews (see below). With modern imaging techniques, it is possible to make quite detailed site records by, for example, a combination of video and still images.

However, as will be explained in more detail later in this section, it is important to ensure that the correct site or sites are visited. For example, a distinction may have to be made between the site where the injury occurred and the site where the behaviour leading to that injury occurred. A simple example will be sufficient to illustrate this sort of distinction. If a person is burned by a corrosive liquid in a container labelled as 'hand cleaner', then the location where the container was filled will be relevant, as well as the immediate site of the injury.

Ideally, site visits should be carried out as soon as possible after the accident has occurred. This provides the minimum opportunity for changes at the site and for people to forget the details of what happened. However, even when it cannot be done immediately, a site visit should be undertaken at some point for the reasons listed above.

On-site and off-site interviews

The reasons for conducting interviews, both on and off site, are largely practical and will be dealt with in this section, which also considers some of the main problems which may be encountered with interviewees. However, on-site and off-site interviews use the same types of interview technique; it is therefore convenient to deal with interview technique as a single topic, and this is done in the next section.

Conducting interviews on site has the major advantage that interviewees can point to locations, machines and so on, rather than having to describe them. It is well known that verbal descriptions are difficult to generate and susceptible to misinterpretation – there are numerous word games which rely on this phenomenon for their effectiveness.

Ideally, all of an interview should be conducted on site, but it is often the case that the site is not a suitable venue for an extended interview. For example, the site may be noisy, or in a location which makes interruptions inevitable. Where this is the case, there should be an initial interview on site, to obtain the main benefits of a site interview, and then the interview should be continued in a more suitable location.

Before moving on to interview technique, it is worthwhile dealing with certain practical difficulties which may be encountered with interviewees during both on-site and off-site interviews:

- **Interviewees who are poor at communicating.** This may be a temporary condition brought on by the circumstances of the accident, in which case the interview should be delayed. This is one circumstance in which the general rule of interviewing as soon as possible after the accident should be overridden. However, it may be that the interviewee is always a bad communicator. Obtaining information from people like this requires good interviewing skills – the key features of these skills are described in the next section.

- **Interviewees whose first language is not English.*** Often, the accurate identification of accident causes relies on detailed information about what may be subtle matters of interpretation. Interviewees operating in a second or third language may be at a disadvantage unless they are extremely fluent. Where they are not extremely fluent, a translator should be used.

- **Interviewees who are unwilling to provide information.** Interviewers require two skills in these circumstances. First, they need the skill required to differentiate between unwillingness and inability to provide information. An unwillingness to provide information can take the form of responding with 'I can't remember' to a number of questions, and the interviewer must be able to discriminate between unwillingness to provide information and genuine amnesia. The second skill is to be able to identify the causes of the unwillingness to provide information. These can take a variety of forms, such as a fear of disciplinary action against the interviewee or others, a fear of loss of compensation for the interviewee or others, or a general unwillingness to provide the 'management' with any information at all. While the techniques of interviewing described in the next section can help investigators with problems like these, there is little, in practice, that can be done. The unwillingness to provide information arises from a 'blame culture', a 'compensation culture' or an organisational culture which divides people into 'us' and 'them'.

* More generally, interviewees whose first language is not the same as that of the interviewer.

These problems can be solved (see Chapter 21, which deals with safety culture) but not in the context of a particular accident investigation.

- **Interviewees who are not telling the truth.** Again, the interviewer needs two sets of skills: establishing that someone is not telling the truth and identifying why. Great care has to be taken to discriminate between what interviewees believe happened (which is the truth as far as they are concerned) and what actually happened (the 'objective' truth). There have been many studies that have shown that, if a number of people witness an incident, they all give different descriptions of what happened – in other words, they all have their own version of the 'truth'. It cannot be assumed automatically that if someone makes a statement which is manifestly not the 'objective' truth, or contradicts what someone else has said, that this person is lying. Indeed, one of the few ways of establishing that people are lying is if a group of people who have witnessed or been involved in an accident all give exactly the same story. As with interviewees who are unwilling to provide information, there is little that can be done, in practice, when interviewees are not telling the truth.*

The discussion of interview technique which follows assumes that the problems listed above are not present, since this is likely to be true in the majority of cases. However, at various points in the discussion there are suggestions on how to deal with particular problems.

Interview technique

In the majority of accident investigations, the main source of information will be the people involved. Relevant people to be interviewed may include any or all of:

- those who sustained injuries
- those who caused damage to assets
- witnesses to the accident
- managers and supervisors of personnel involved
- others who may have contributed to causes (remember the example of the person who filled the wrong container with a corrosive liquid).

In order to obtain information, it is usual to talk to the relevant personnel or conduct an interview. Note, however, that there is an important difference between talking to people and interviewing them. Talking to people is usually an informal process, while interviewing is a structured, planned event with specific objectives. Unfortunately, there is a tendency to believe that because a person can talk well, they can also conduct an accident investigation interview well. This is not necessarily the case, since accident investigation interviewing is a skill and, like all skills, someone will possess it only if he or she has the necessary background knowledge and has had a period of practice with feedback from a competent person. Background knowledge can be provided by texts such as this book, and the relevant background knowledge for accident investigation interviewing is dealt with in the remainder of this section. However, skill in accident investigation can only be obtained through supervised practice.

The background knowledge needed for accident investigation can be grouped under three main headings:

1. **information on accident causation at the individual level** – what features of human beings make them more or less likely to have accidents
2. **information on accident causation at the organisational level** – what features of work, working environments and organisations in general make it more or less likely that people will have accidents
3. **information on accident investigation interview techniques.**

---

* There are techniques that can be used to obtain information from unwilling interviewees and to check whether interviewees are telling the truth. However, they should be used only by specialists and only in carefully selected cases.

The next three subsections deal with these three types of information, beginning with the Hale and Hale model,[28] which is a convenient summary of accident causation at the individual level. There is further information on human error in Parts 1.2 and 2.2, which deal with human factors.

The Hale and Hale model

In the Hale and Hale model, an accident is defined as the failure of people to cope with the true situation presented to them. The cause of the failure may lie with the person, with the situation or, more usually, with both.

Whether or not the accident produces an injury is a separate consideration from the causation of the accident. For example, a car driver going across a set of red traffic lights at 3 am when there is no other traffic around would, in terms of the model, be considered to have had an accident, although it produced no injury; whereas the case of someone deliberately jumping from the tenth floor of an office block to commit suicide would not be regarded as an accident, although it produced a fatal injury. Equally, the passengers in a train who are injured when it collides with a heavy lorry at an unmanned level crossing have not had an accident, in the terms of the model. The accident has happened to the train driver and the lorry driver, who may have both escaped injury.

The model is in the form of a closed loop for handling information, with the person as the information channel perceiving, processing and acting on the situation which is thus changed for his or her later perception. The person and the situation, both past and present, interact at all stages. The main elements in the Hale and Hale model and some of the factors involved in the breakdown of function at each of the stages are shown in Figure 9.2, which is followed by notes describing the model's main elements.

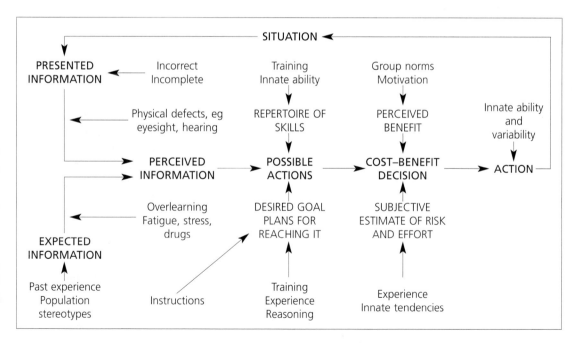

Figure 9.2
The Hale and Hale model[28]

Presented information
Information is presented minute by minute to the person's senses by the situation. If this information is incorrect – for example false instrument readings or the failure of an alarm – an error may result. Equally, if the information is not sufficient to enable the people to formulate a correct solution to the situation, and they

do not seek more information, an error may occur. In general, the identification of failures in presented information are an indication to the investigator that he or she must seek information from people other than the injured person. Two general examples will illustrate this:

1. Where the information is missing or incomplete, we need to establish why this is the case. For example, if warning labels have been removed or obscured, or no instructions (or inadequate instructions) have been given, the investigation will have to move to the sources of these failures.
2. Where the information is incorrect, a similar procedure will be required. This would apply, for example, to substances in wrongly labelled containers, faulty instrumentation and incorrect instructions.

Expected information

It is not only the presented information which gives rise to perception. Everyone in a given situation has their own expectation about the information with which they will be presented. This expectation, which is often referred to as a 'perceptual set', can distort perception by causing people to overlook aspects of the situation which they do not expect to be there or to 'see' other information which is not in fact present. Such reliance on expectation rather than on presented information occurs particularly in habitual or overlearned tasks, under conditions of time pressure, and when the person is fatigued or under the influence of certain drugs. For example, a person who walks along the same corridor every day may fail to 'see' that on a particular day it has been given a high polish and may fall flat on his or her back; or an assembler coming in to work one morning may brush against a soldering iron which is normally cold at that time and sustain a burn because on this particular occasion it has been left switched on by the night shift.

In addition, there are circumstances where a particular item of information is required in order to avoid an accident – for example, what to do if a car skids. Accidents can occur if there is a failure to retrieve the required information, and this can happen either because the information is not there (no training in dealing with skids) or because it cannot be retrieved in time (not enough practice in dealing with skids).

Perceived information

The perceived information comes from two sources: the presented information, which is external to the person; and the expected information, which is internal. It is important to note that decisions (see below) are made on the basis of perceived information and that this is unique to an individual during any one cycle of the model. It is not possible for an investigator to infer what an individual's perceived information was at the time of the accident. This is because the investigator's presented information will be different and the investigator will have his or her own expected information. The only way to identify an individual's perceived information is to ask. (Note that because there is, in theory at least, the possibility of an objective perception of events, deviations from this arising from an individual's interpretation are often described as 'perceptual distortion'.)

Possible actions

The actions people consider during a particular cycle of the model depend on their goals and their plans for achieving these goals.

What people want to achieve, and how they want to achieve it, have a strong influence on the actions considered. These in turn depend on a number of factors, including:

- **Instructions.** What people are told to do can have a range of effects, for example by taking them beyond their competence. They may also be told to act in an unsafe manner.
- **Training and experience.** These influence the range of plans people know about and, critically, whether they know which are safe and which are not.
- **Reasoning.** People can work out plans of action. However, people are notoriously poor at planning, particularly over short timescales. Typically, reasoning stops at the point where a possible plan is devised; rarely do people create a number of alternatives. This is dealt with in detail in Chapter 21.

Cost–benefit decision

At the next stage a decision has to be made on the action to be taken to cope with the situation. Various factors come into play at this time. First, the desired outcome or goal of the process has a bearing, since, if people have been given the wrong instructions about what they should try to achieve, an error is likely. Again, the individual may have formulated, several moves ahead, his or her own plans of action for attaining the goal. If these are inadequate because of faulty reasoning, poor training, or incorrect application of past experience, errors may occur. For example, an engineer may set about dismantling a piece of equipment in the same way as he has done with a dozen others earlier in the week, but without realising that this is a slightly different model with a steel pin held under pressure by the plate he is unscrewing. Consequently, the pin is driven into his face as he bends over to take the last screw out.

Second, a person's repertoire of skills, both physical and mental, acts as a limiting variable at the decision stage. This may be the most important factor which makes the difference between an injury and a near miss, where the person has the skill to avoid the injury in the last few crucial seconds. For example, some drivers may have a reaction time that is short enough for them to stop before colliding with a bus which pulls out in front of them, while others, perhaps older and less fit, run into it.

Third, when the possible actions have been defined, the decision about which one to take can be seen in cost–benefit terms, with the cost in effort required and the subjective estimate of risk being weighed against the benefit in terms of estimated size and type of reward. For example, an operator deciding whether to allow the chuck on a lathe to stop rotating before removing the component weighs the risk of catching a fingernail in the chuck and losing the finger, or of being lacerated on a sharp revolving edge, against the time and bonus he or she would lose by waiting. The degree of risk to be run for a given reward will vary from person to person and from time to time; so will the accuracy with which someone estimates the risk and the amount of knowledge on which he or she can base the estimate.

Action

At the final stage of carrying out the action, there is scope for further error due to innate variability of response. For example, a butcher cutting up some liver may move the knife a little further to one side than usual and cut the end off his or her finger.

During investigations, it is helpful to think of the behaviour which actually occurred as one of a number of possible actions which might have been taken. This thinking process is useful because it leads directly to remedial actions; some examples are given below.

- If a 'safe' action was not chosen because the person did not know there was a 'safe' action, then one remedy is information and training.
- If a 'safe' action was not chosen because the person was instructed not to use it, then the remedy lies at the supervisory or managerial level.
- If a 'safe' action was not chosen because the person deliberately chose the 'unsafe' alternative, then the remedy may be with the person's motivation or it may be with organisational factors such as a piecework payment system.

Monitoring loop

A monitoring loop connects the action back to the situation. In highly practised tasks, the monitoring of actions takes place at much longer intervals than in unpractised tasks, and if the situation has changed unpredictably since it was last monitored, habitual actions may have become inappropriate. Such monitoring is important for the detection of errors before they lead to serious consequences. For example, someone may notice that a steel strip coming out of a machine is about to collide with a trolley incorrectly placed to receive it, and that this will push the trolley onto his or her feet. If monitoring is effective, he or she will step back to avoid injury.

Secondary factors

Hale and Hale point out that a breakdown in any of the major steps in their model could be regarded as a primary 'cause' of accidents. However, they also show that at one remove there is a host of secondary factors, changes which affect the functioning of different parts of the process. These factors range from simple psycho-physical variables such as reaction time and defects in eyesight or hearing, through variables of perceptual motor co-ordination and level of skill, to more complex factors such as the level of subjective risk at which someone prefers to operate. Others are age, experience and social factors concerned with the group in which the person is working. These and many others have formed the basis of most past research on accidents, but almost all of them have an effect at more than one stage of the process represented by the model, and this effect may be extremely complex. For example, a factor reducing the likelihood of breakdown of one part of the process may increase the likelihood of breakdown of another part. A loud noise may counteract to some extent a decrease in performance due to fatigue or boredom in the collection of visual information, but it may also mask auditory information and so increase the likelihood of it being missed. Much confusion over accident causation has arisen from the failure to recognise this complex interaction.

At this stage, it is necessary only to be aware that these secondary factors exist. They are dealt with in detail in the human factors parts of this book.

Accident causation

Appropriate use of the Hale and Hale model enables accident investigators to deal effectively with the causes of accidents at the level of the individuals involved. However, it is rarely, if ever, the case that an accident has a single cause or that the causes are only at the level of the individual. In practice, accidents happen because a number of things have occurred simultaneously and it is the combination of these causes which results in the accident.

It follows from this that removing or changing any one of the causes could prevent similar accidents in the future. On a practical level, since some causes are easier to influence than others, it is important for the investigator to identify all of the relevant causes.

There are two types of causes. The first type is usually referred to as immediate or proximate causes, and they are to do with **what** happened. In investigation terms, they can be identified by obtaining a detailed description of the accident and the events leading up to it.

The second type of accident cause (usually referred to as the underlying or root causes) deals with **why** the accident happened. These causes are more difficult to identify – it is generally easier to establish what occurred than why it occurred. Nevertheless, it is only if these underlying causes are accurately identified that appropriate recommendations on remedial actions can be made. For this reason, we need to consider techniques which can be used to make identification of causes more likely. A number of these techniques are described in Chapter 20 in the context of advanced accident investigation, but for current purposes, Domino Theories, which deal with the multiple underlying causes of accidents, are adequate.

There are several variants of the Domino Theory. Our description begins with a generalised version and there is a note at the end of this subsection on its variants.

Domino Theories

The basic idea behind Domino Theories is that individual errors take place in the context of organisations and that a framework is required for identifying these more general causes of accidents. There are various versions of the Domino Theory and the one illustrated in Figure 9.3 is a generalised version. The notes below explain the main elements of this generalised Domino Theory.

The Domino Theory says that if one of the dominoes to the left of the 'loss' domino falls to the right, it will knock over the other dominoes and a loss will occur. For example, consider the circumstances in which someone slips on a patch of oil, falls, and breaks an arm. The sequence of events will be as follows:

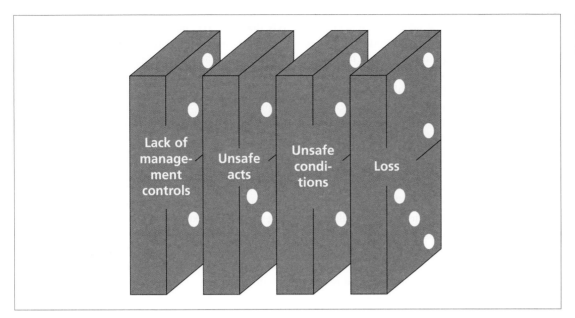

Figure 9.3
Generalised Domino
Theory

- lack of supervision (management control) results in a situation where oil can be spilt and not cleared up
- an unsafe act occurs – spilling oil and not clearing it up
- an unsafe condition results – a pool of oil on the floor
- a loss occurs when someone slips on the oil, falls, and breaks an arm.

When investigating the loss, it is possible to identify unsafe conditions, unsafe acts and lack of management controls and establish causes for these, as well as causes for the loss. For example:
- possible causes of someone slipping on a patch of oil may be not looking where they were going, or not wearing appropriate footwear
- possible causes of not clearing up spilled oil might be lack of time, or not seeing it as part of the job
- possible causes of spilling oil might be working in a hurry, inappropriate implements or a poor method of work
- possible causes of poor management control might be excessive pressure for production (resulting in hurrying), lack of funding for proper implements, or insufficient attention to designing appropriate systems of work.

The further to the left the investigator goes with the dominoes, the greater the implications of the causes identified. For example, lack of appropriate systems of work may apply to a large number of operations, not just to those that can result in oil spillages. It follows that if failures in management controls can be identified and remedied, there is the potential to eliminate large numbers of losses. Thus, the usefulness of the investigation can extend beyond simply preventing a particular type of accident happening again.

One way of doing this is to look systematically at each of the dominoes and determine which ones fell and why.

It is important to remember that there is rarely one domino involved in a given accident and rarely one cause associated with each domino. Rather, there are various causes which contribute to the fall of the domino. Investigators should continue to ask why things happened until they are sure that they have identified all of these contributory causes. We can use the oil spillage example again to illustrate this aspect of investigation.

Possible reasons for the person slipping on the oil were that they were not looking where they were going and that they were wearing inappropriate footwear. Investigators should ask 'why?' about each of these to see whether further useful information can be obtained. For example, the person may not have been wearing appropriate footwear because:

- they did not know they should be wearing special footwear
- they did not know which type of footwear was appropriate
- the appropriate footwear was uncomfortable
- the appropriate footwear was too expensive
  and so on.

The different answers to this 'why?' question will have different implications for remedial action, so it is important to establish the 'why?' before making recommendations. A similar technique should be applied to the other dominoes and again this will be illustrated using the oil spill example.

Unsafe condition

The possible reasons for not clearing up the oil spillage were lack of time and not seeing it as part of the job. Asking 'why?' about the lack of time could produce the following types of answer:

- management pressure
- piecework
- wanted to get home
- understaffing
  and so on.

Again, the remedial action suggested will depend on the answer obtained. There is little point suggesting that people take time to clear spillages if managers are continuing to insist on rapid working.

Unsafe act

Possible reasons for the spillage of oil included using the wrong implements or an inappropriate system of work. Asking 'why?' about the inappropriate system of work might produce the following types of answer:

- no-one has prepared a system of work
- the people who do the work do not know about the system of work
- the recommended system of work is impractical
- the recommended system of work is out of date
  and so on.

As before, whichever reason is identified, it should be followed up so that any remedial action suggested is as relevant and practical as possible.

Lack of management controls

Possible reasons for the lack of management controls were pressure for production, lack of funding and failure to produce written systems of work. Asking 'why?' about written systems of work might produce the following types of answer:

- no-one knows it is necessary
- no-one has the time
- no-one has the skills
- no-one has clear responsibility
  and so on.

It is obvious that, as the investigation moves from the 'loss' domino to the 'lack of management controls' domino, a wider range of people will have to be interviewed. The injured person, for example, is unlikely to have the required information on lack of management controls. He or she can probably give information about the effects of lack of controls but is unlikely to know the reasons why the controls are not in place.

Identifying who should be interviewed in the course of an investigation and knowing which questions to ask are matters of experience and practice and, as with the other skills involved in accident investigation, they should be practised when possible.

Note also that managers investigating an accident within their sphere of control may encounter a conflict of interests. If the manager is the person who should be implementing management controls, there will be a natural tendency to avoid going into detail of weaknesses in management control as thoroughly as may be required. In these circumstances, it may be preferable for the manager concerned to hand the investigation over to, or seek assistance from, a neutral investigator.

Health and safety professionals have a related problem when managers ask to take part in the health and safety professionals' investigations. In general, this is to be encouraged, since it increases management involvement in health and safety matters, but it should be explained to these managers that they may have to be interviewed as part of the investigation if lack of management control is identified as an underlying cause.

As was mentioned earlier, there are various versions of the Domino Theory. Table 9.1 contains a summary of the various dominoes proposed by different theorists.

Generalised	Heinrich[29]	Bird (and Loftus)[69,30]
	Ancestry and social environment (bad character traits and attitudes)	Management (lack of control)
Lack of management controls	Fault of person (bad habits, ignorance)	Origins (basic cause)
Unsafe acts	Unsafe act	Symptom (immediate cause)
Unsafe conditions	Accident	Contact (the accident)
Loss	Injury	Loss (injury, damage)

Table 9.1
Summary of the Domino Theory variants

Heinrich[29] believed that people could be, for example, reckless or stubborn, either because they had inherited these traits or because of their environment, or a combination of both. The bad character traits led to people committing unsafe acts which could cause them to have an accident, or create conditions which made it more likely that other people would have accidents. Some of these accidents would result in an injury.

The Bird and Loftus version[30] of the Domino Theory is similar to the generalised version but it splits causes into basic causes, such as personal factors and factors associated with the job, and immediate causes, such as substandard practices and errors.

As far as accident investigators are concerned, it makes little difference which version of the Domino Theory is used. What is important is to remember to continue asking 'why?' until causes, and the reasons for them, are accurately identified.

An alternative approach to the Domino Theories is a model devised by Reason[93] that is based on barriers and their failure. This model is often referred to as the Swiss Cheese Model and it shows that an accident can happen only if all the barriers intended to prevent it fail simultaneously. Using the Swiss Cheese Model in accident investigation involves identifying the barriers that failed and establishing why they failed. It is also usually necessary to consider whether additional barriers might be required. There is further information on barrier analysis in Chapter 20.

Figure 9.4
The Swiss Cheese
Model[93]

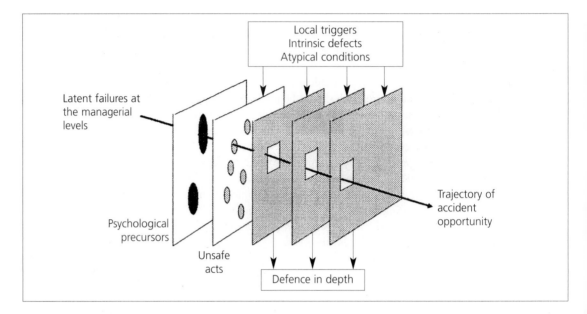

Interviewing for accident investigations

There are three important aspects of interviewing to consider:

1. coverage
2. keeping an open mind
3. getting people to talk.

Coverage

This aspect of interviewing deals with the nature and amount of information which has to be collected. How do interviewers know when they have all the relevant information and how do they avoid collecting information which is of no value?

What is relevant and valuable will, of course, depend on the purpose of the investigation and, as a general guide, interviewers should make sure that their coverage includes all the information necessary to enable them to make a decision about appropriate remedial action. However, in this first stage of the investigation, the purpose is to establish a clear idea of what happened. Information of this type falls into two categories:

1. Information which is common to all types of accident and which is best dealt with using a form containing spaces for the information required. An accident record form can be used for this purpose. The sorts of information which should be included on a form of this type include the following:

    • details of the incident – for example time, date and location
    • details of person injured – for example name, age, sex, occupation and experience
    • details of the injury – for example part of body injured, nature of injury (cut, burn, break and so on), the agent of injury (knife, fall, electricity and so on), and time lost
    • details of asset or environmental damage – for example what was damaged, nature of damage, and the agent of damage.

2. Information which varies widely from incident to incident and has to be recorded as a narrative. For convenience, space for this narrative is usually included on the accident record form used for the first type of information, but in many cases, this brief summary has to be supplemented by a more detailed report.

A thorough knowledge of the Hale and Hale model and the Domino Theory is a significant aid to ensuring adequate coverage, since investigators equipped with such knowledge will at least be aware of the range of issues which may be of relevance in any accident investigation.

Keeping an open mind

One of the main difficulties during an investigation is avoiding assumptions about what has happened. The more experience interviewers have of the type of site involved and the nature of the work and the people, the more likely it is that assumptions will be made. There is always the possibility that investigation will result in a summary of what was thought likely to have happened, rather than what actually happened. Remember the Hale and Hale model and the caveats on inferring perceived information on the basis of the investigator's presented and expected information.

The best way of avoiding assumptions is to make sure that questions are asked about all aspects of what happened, even if the interviewers are sure what the answer will be – perhaps even especially when the interviewers are sure what the answer will be!

Obviously, if making assumptions leads interviewers to form an inaccurate picture of what happened, this is unsatisfactory. However, it has more serious implications in that it can lead to suggestions for remedial action that are inappropriate. If interviewers find that they are identifying remedial actions early in their investigation, this is a warning sign that they may be making too many assumptions.

Getting people to talk

Interviewees will volunteer information more readily if the interviewer can establish and maintain a rapport with them. 'Rapport' is the term used to describe the relationship between people that enables a ready flow of conversation without nervousness or distrust. Interviewers will collect a wider range of information, and more accurate information, if they can establish a rapport with the people they are interviewing. There are no techniques which are guaranteed to establish a rapport, but the guidelines below will make it more likely.

- **Interview only one person at a time.** It is difficult to establish rapport with two or more people simultaneously, since each will require different responses from the interviewer. Interviewing one person at a time may not be possible in some circumstances – for example, if the person interviewed asks for a representative to attend. In these circumstances, the status of any other attendees should be clearly established at the start of the interview, including whether they will be answering questions on the interviewee's behalf and whether they will be entitled to interrupt.
- **Have only one interviewer at a time.** Avoid 'board' or 'panel' interviews since this requires the interviewee to communicate with more than one person and this is rarely successful. Note, however, that there are circumstances where it is preferable to have more than one person involved in conducting the interview. One example is when inexperienced interviewers prefer to have someone there to take notes for them. Another example is when certain aspects of the interview will require specialised technical knowledge. In either of these circumstances, one person should 'lead' the interview and hand over to the second interviewer at an appropriate point for any supplementary questions the second interviewer may wish to ask. If the fact that this procedure is going to be used is explained to the interviewee at the start of the interview, it has a minimal effect on rapport.
- If board or panel interviews have to be used for any reason, they should be run in ways that ensure the interviewee always knows from whom the next question is coming. This requires clear handovers from

one interviewer to another and discipline on the part of the board or panel members, since they are not allowed to 'butt in'.

- **Introductions and explanations of the purpose of the interview.** This should be done even if the interviewer has already been introduced by someone else. The interviewee will gain confidence if he or she knows who the interviewer is and why the interview is taking place. There should be an emphasis on the facts that the primary purpose of the interview is to prevent the accident happening again, and that action will be taken on the results of the investigation. This step can be particularly important when there is reason to suspect that interviewees may be unwilling to provide information, or may wish to be 'economical with the truth'.
- **Check the interviewee's name and the part they played in the incident.** This may sound obvious but checking now can save embarrassment later on. Confusion can arise when, for example, more than one person has been injured, where more than one accident has occurred in the same area, or where interviews are in progress for some other purpose, for example internal audit.
- **Start the interview on the interviewee's home ground.** The idea is to start the interview with things which are familiar to the interviewee and then move on to the details of the accident once rapport has been established. This is helped by beginning the interview at the interviewee's place of work and talking about their normal job before moving on to a discussion of the accident.
- **Use language which is appropriate to the interviewee.** Use vocabulary and sentence structure which matches the intellectual level and verbal competence of the person you are interviewing. This is particularly important when interviewees have poor communication skills since using complex words or phrases may mean that the interviewee does not fully understand the questions.

It is important to establish rapport before moving on to collect detailed information. If this is not done, the interview may degenerate into a series of stilted questions and one-word answers. This can also happen if rapport is not maintained; there are several things which will help maintain rapport.

- **Make sure the interview is not interrupted.** Interruptions can come from other people, and the best way of preventing this is by choosing a suitable place for the interview. However, interviewers can interrupt the interview themselves by stopping the interviewee to ask questions. In general, it is best to let the interviewee talk and ask any questions when he or she gets to a natural break in their story.
- **Use open rather than closed questions.** Open questions are ones which cannot be answered with 'yes' or 'no'. Closed questions are ones which can be answered with a 'yes' or 'no'. For example, 'What was the noise level like?' is an open question; 'Was it noisy?' is a closed question. In general, closed questions should be used only to check on specific points already made by the interviewee.
- **Avoid multiple questions.** For example, a question such as 'Can you tell me what everyone was doing at the time?' is better asked as a series of questions starting with 'Can you tell me who was there at the time?' and then a single question about what each of them was doing. Asking multiple questions is likely to result in only part of the question being answered.
- **Avoid leading questions.** Leading questions are ones which imply that the interviewee has done something which is unacceptable or incorrect. The most famous leading question is 'When did you stop beating your wife?', but this, fortunately, has limited application to accident investigations. Questions to be avoided are those which take the forms 'When did you realise you had made a mistake?' and 'What was the correct way of doing the job?'
- **Keep a positive and uncritical manner.** Interviewees will form an opinion of your manner both from what you say and from your body language. Avoid expressing your views and opinions during the interview, especially if they are critical of what the interviewee has done or not done. Similarly, avoid such obvious signs of lack of interest as not listening, yawning or looking at your watch.

In terms of interview style, establishing and maintaining rapport provides the interviewer with the most favourable circumstances for obtaining information. However, the format of the interview also influences the quantity and quality of the information which can be collected. In order to obtain the maximum information, the following general format is suggested.

- Begin with a general enquiry of the type 'Tell me about what happened' and let the interviewee talk for as long as he or she wishes without interruption. The opening description by the interviewee may be disjointed, not particularly clear, and even confused. However, it will usually contain those items of information the interviewee considers important.
- Only when the interviewee has had the opportunity to describe the circumstances in his or her own words should you try to establish a more coherent story and collect additional information. This is usually best done by trying to get a chronologically ordered description, either from the time of the accident backwards, or forwards from some relevant time before the accident. Using this as a structure, you can ask supplementary questions as appropriate to discover whether the interviewee has additional relevant information.

When asking these supplementary questions, give the interviewee time to consider the answer. Many inexperienced interviewers are afraid of periods of silence during an interview and they do not allow the interviewee sufficient time for reflection. Remember that the more difficult a question is to answer, the more time the interviewee will need to think before answering it. This is particularly noticeable with 'what?' and 'why?' questions – interviewees can normally answer 'what?' questions quite quickly but they usually require some time for thought before they can answer a 'why?' question. If interviewers ask a question, they should wait for an answer; they should not get embarrassed by silence and ask a second question before the first has been answered.

This general format for questions can be used for all of the interviews which are specifically about losses, for example injury or damage. However, when interviewing people about contributory causes (such as filling the wrong container with corrosive liquid), the 'accident' should be redefined for these interviews as the creation of the contributory cause. The opening enquiry in these sorts of interviews would, therefore, be of the type 'Tell me how this container came to be filled with corrosive liquid'. Interviews of this type can be considered as investigations of nonconformities; nonconformity interviews are dealt with later in the chapter.

One final point on interviewing. No matter how skilled an interviewer is at establishing and maintaining rapport and asking questions, little useful information will be collected unless the interviewer listens carefully to what the interviewee says. It can be the case that interviewers are so busy formulating their next question, or working out a hypothesis on why the accident happened, that they simply fail to listen carefully to what is being said. However, listening is a skill and can be learned like other skills. Details of where to find information on positive listening skills are given in Chapter 12 in the section on communication.

Recording the interview
It is essential to learn to record what is said during interviews. There are two reasons for this.

- **So that you do not forget what has been said.** Most people believe that their memory is much better than it really is. Few people can remember all the relevant facts raised during even a short interview.
- **So that you do not confuse what different people have said.** In most investigations, you will have to interview more than one person and, unless you record each interview, it is unlikely that you will remember who said what, especially if there is a delay between the interviews and writing the investigation report.

There are three methods of recording interviews:

- **Audio or video recording.** This gives a very accurate record of what was said and, in the case of video recording, there is also a record of the body language. Although interviewees may initially be ill at ease while being recorded, they soon forget that the recording is taking place and it rarely has any detrimental effect on rapport.

- **Note-taking by a second person.** If detailed notes are being taken by a second person the interviewer has more time to concentrate on questioning and listening. This approach has the additional advantage that the second person can act as a second interviewer, thus increasing the chance that an appropriate range of questions will be asked.
- **Note-taking by the interviewer.** All interviewees should learn to make notes of the supplementary questions they wish to ask so that they do not interrupt the interviewee by asking these questions as soon as they think of them, or forget what they wanted to ask. However, making more detailed notes is a difficult skill to learn and people who interview only infrequently may prefer to use audio or video recording, or have notes taken by a second person.

Where notes are being taken, bear the following points in mind:

- **Timing.** Wait until rapport has been established before starting to take notes. Establishing rapport is difficult enough without the added distraction of note-taking.
- **Agreement.** You should always tell the interviewee that you will be taking notes and should gain the interviewee's agreement to do this.
- **Content.** Record notes on everything that is said. You may think that parts of what the interviewee says are irrelevant but you should record them anyway. This is because you should judge the relevance of particular items of information only when you have collected all the information, including, for example, from other interviews. In addition, if you are selective in what you record you may bias the interview as the interviewees may stop giving information on topics they consider of no interest to the interviewer (because the interviewer is not making notes on these topics).
- **Take time.** Note-taking shows the interviewees that you are interested in what they are saying. They do not consider it an interruption to the interview and are usually happy to wait while you make your notes.
- **Review.** At the end of the interview, go over your notes with the interviewee, checking that what you have written down is an accurate record of what has been said. This may also provide additional information since it may jog the interviewee's memory.

In some circumstances it may be necessary to have a written, signed statement from the interviewee. This may be prepared either at the end of the interview, or as a substitute for the interview. Such statements are usually only necessary when the accident has resulted in, or may result in, legal action or a claim for compensation. Where such accidents are being investigated, advice should be taken from a competent person on the detailed requirements for the written statement, given the particular circumstances that make the statement necessary.

It is important to remember that there is a significant difference between a statement and the transcript of an interview. A statement, written or dictated by a witness or another person involved in an accident, will be a record of what that person considers important. However, this person is unlikely to be an expert in accident investigation and what he or she considers important is not necessarily relevant in preventing recurrence. In contrast, the transcript of an interview contains information on the topics the interviewer considers relevant and, it is assumed, the interviewer is the expert.

A clear distinction should be maintained between statements (what the interviewee thinks is important) and interview transcripts (what the interviewer thinks is important); whether these statements or transcripts are signed as a true and correct record is a separate issue.

Collecting additional information

Although the majority of information collected during accident investigations is usually obtained from interviews, it is rare that interviews supply all the necessary information. For example, if an interviewee claims to have been following a written procedure, it will be necessary to examine the relevant procedure in order to establish whether this was the case. As can be seen from Figure 9.1, it is also necessary to collect additional information when preparing a draft report. This requirement for additional information will also be dealt with in this subsection.

Typical types of additional information which may be required include:

- **records relevant to the type of accident being investigated:** these include accident records, hazard records, risk assessment records, and records of tests and maintenance of machinery and equipment
- **documents relevant to the type of accident being investigated:** these include written procedures, safety rules, manufacturers' instruction manuals and other data available from manufacturers and suppliers, and safety manuals
- **reference material, including Approved Codes of Practice (ACoPs) and guidance:** it may be necessary to seek information on a range of subjects in order to clarify particular issues which arise during an investigation. In addition, reference material is often needed as a source of ideas for remedial actions
- **legislation:** this will normally only have to be consulted where a civil or criminal action may result from the accident
- **various experts:** lawyers, doctors, engineers and so on can provide the sorts of information available in reference material, and where access to such experts is available, it is often a more convenient method of obtaining the required information
- **the internet:** there is a wide range of material on the internet that is relevant to risk management – it can often be found quite quickly with a suitable keyword search.

The list of sources given above is not intended to be comprehensive. Rather, it is intended to make the point that interviews are not the only source of information relevant to an accident investigation. In particular, in making decisions on remedial actions, it is important to obtain as much information as possible about the alternatives which might be available. This point is dealt with in the next subsection on preparing a draft report.

Preparing a draft report

It is preferable to prepare a draft report, since this provides an opportunity to check that nothing has been omitted from the investigation. In particular, investigators should check that they have the information for:

- making any statutory or other notifications (this aspect is dealt with separately later in this chapter)
- making reasoned suggestions on measures for preventing the accident happening again
- any supplementary tasks, such as completing insurance claims.

Some people find that drafting reports is best done with techniques such as system diagrams, system maps and flow charts. These techniques are dealt with in Chapter 21 (in Part 2.1) and it is worth experimenting with these techniques to find out whether they suit you.

Whichever drafting techniques are used, the main work during the draft reporting stage should be the design or selection of suitable remedial actions. In general, there are two main weaknesses which occur during the preparation of a draft report:

- **Insufficient attention is paid to underlying causes.** It is very easy during interviews to become distracted and stop asking 'why?' too soon. However, quiet reflection while preparing a draft report provides an opportunity to spot where this has occurred. Additional data can then be collected by, for example, conducting supplementary interviews.
- **Insufficient creative thought is given to the suggestions for remedial action.** Partly this arises from the lack of attention to underlying causes just described and, in the worst cases, it gives rise to reports in the form

'The man tripped – tell him to be more careful in future'. Unless the investigation discovers why the man tripped, it is not possible to go beyond this trivial sort of recommendation. However, even when accurate data on causes have been collected, this part of the procedure can be conducted inadequately because of lack of creativity. Few people are 'naturally' creative in designing remedial actions; they need some awareness of the techniques which can be used to improve creative thinking, and they need to practise using them. This topic is discussed in more detail in Chapter 21, which deals with systems thinking.

Ideally, the output from the draft reporting stage should be a recommendation for remedial action, or actions, which meet the following criteria:

- **The remedial action does not rely for its effectiveness on human behaviour.** For example, no recommendations in the form 'be more careful', 'pay more attention' or 'always wear personal protective equipment'. Ideally, the remedial actions should eliminate the causes. The whole question of appropriate and inappropriate remedial actions is discussed in detail in Chapter 6, which is about risk control.
- **The remedial action is not restricted to the prevention of a particular accident in specific circumstances.** We have already pointed out that successively asking 'why?' during the accident investigation leads to more fundamental issues with a correspondingly wider range of implications. If, at the draft report stage, it is found that the recommendations being made have a very limited scope, this is an indication that the investigation may have been inadequate – in other words, the 'why?' questions have been discontinued at too early a stage.
- **The remedial actions take into account risk management system issues.** For example, one finding of the investigation might be that there was no risk assessment for the activity which resulted in the accident, or that the risk assessment was inadequate. In these circumstances, one recommendation will be to carry out or redo the relevant risk assessment. However, the draft report should also address why the risk assessment was not carried out, or was inadequate, and suggest ways in which the risk assessment procedure can be improved.

The final step at the draft report preparation stage is to review the complete draft report and, ideally, have it checked by someone who is competent to make such a check but who has not been involved in the investigation. This review should be used as an opportunity to check that all the relevant information is there and, in particular, that the draft material adequately covers the following:

- a coherent description of the events leading up to the accident, the accident itself and any relevant outcomes, such as injuries and damage
- a rational view of what should be done in the future in the way of remedial actions
- that the view of what should be done follows logically from the description of the events leading up to the accident, the accident itself and the outcomes.

Preparing a final report

The first decision to make about a final report is whether or not it is necessary. The answer will be based on whether there is an audience, apart from the investigator, and what this audience needs from the report. In many cases, a detailed draft report is adequate as a record and as a basis for justifying remedial actions.

Where a final report is required, the techniques of good report writing should be adopted and, in particular, authors should make sure of the following:

- **Good signposting.** Any report, but especially a long one, will be difficult to read and act on if the various sections are not clearly distinguishable. If the report is intended for more than one audience, the sections relevant to particular audiences should be clearly identified.
- **Separate fact and opinion.** Facts should be unarguable; the investigator's opinions can, and should, be debatable. It is good practice to keep the two separate.

- **Base opinions on the facts.** Reports should not arrive at conclusions which cannot clearly be supported by the facts presented, nor should they arrive at conclusions which do not take all of the relevant facts into account.

Feedback of investigation results

The relevant results of investigations, including any recommendations for remedial action, should be fed back to everyone involved in the investigation. If this is not done, there will be the following detrimental effects:

- subsequent investigations will be more difficult – less information will be given because people will see no results from helping with investigations
- investigators' credibility will be damaged, since they will have told people that the purpose of the investigation is to prevent recurrence.

Even when no action can be taken on the basis of the results from a particular investigation, the results, and the reasons why no action can be taken, should be fed back to those who were involved.

Accident notification

Accident notification requirements are specific to a particular country and readers outside the UK should make sure they understand the requirements of their local legislation, or requirements imposed through other means. The remainder of this section summarises the UK requirements for accident notification.

UK legislation – reportable and recordable incidents

The Reporting of Injuries, Diseases and Dangerous Occurrences Regulations 1995 (as amended) (RIDDOR)[31] place duties on employers and the self-employed to report certain incidents that occur in the course of work. These reports are used by the enforcing authorities to identify national trends in incident occurrence. The reports also bring serious incidents to the attention of the authorities, so that they can investigate them if they wish.

Reports must be made by the 'responsible person', who, depending on the circumstances, may be an employer, a self-employed person, or the person in control of the premises where the work was being carried out. The method of reporting depends on the type of incident. If an incident results in any of the following outcomes, the relevant enforcing authority must be notified by the quickest practicable means (this is usually by telephone or email):

- fatality as a result of an accident
- major injury to a person at work as a result of an accident
- an accident which results in someone not at work being taken to a hospital
- a 'dangerous occurrence' (see below).

This notification must be followed by a written report within 10 days; this is usually done using form F2508 (see below). If they wish, the enforcing authorities can ask for more information about any incident.

Continued...

... continued

The tables below list the incidents which must be notified by the quickest practicable means. Note that dangerous occurrences are, in general, specific to particular types of machinery, equipment, occupations or processes, so it is necessary to find out about the detailed reporting requirements only if they are relevant. For guidance, we include some examples below.

Incidents to be reported by the quickest practicable means

- Fatalities
- Major injuries:
  - any fracture (other than to the fingers, thumbs or toes)
  - any amputation
  - dislocation of the shoulder, hip, knee or spine
  - loss of sight (whether temporary or permanent)
  - a chemical or hot metal burn to the eye or any penetrating injury to the eye
  - any injury resulting from an electric shock or electric burn (including any electric burn caused by arcing or arcing products) leading to unconsciousness or requiring resuscitation or admittance to hospital for more than 24 hours
  - any other injury:
    - leading to hypothermia, heat-induced illness or unconsciousness
    - requiring resuscitation
    - requiring admittance to hospital for more than 24 hours
  - loss of consciousness caused by asphyxia or by exposure to a harmful substance or biological agent
  - either of the following conditions which results from the absorption of any substance by inhalation or ingestion, or through the skin:
    - acute illness requiring medical treatment
    - loss of consciousness
  - acute illness which requires medical treatment where there is reason to believe that this resulted from exposure to a biological agent or its toxins or infected material.
- Dangerous occurrences – specified incidents involving:
  - lifting machinery (including fork-lift trucks) – collapse, overturning or failure of any load-bearing part
  - pressure systems – failure of any closed vessel or associated pipework where the failure has the potential to cause death
  - freight containers – failure of container or its load-bearing parts while it is being raised, lowered or suspended
  - overhead electric lines – unintentional incidents involving plant or equipment
  - electrical short circuit attended by fire or explosion resulting in plant stoppage for more than 24 hours, or with the potential to cause death
  - explosives
  - biological agents
  - breathing apparatus – malfunctions
  - malfunction of radiation generators
  - diving operations
  - collapse of scaffolding
  - train collisions

Continued...

... continued

- ○ wells (not water wells)
- ○ pipelines
- ○ fairground equipment
- ○ carrying of dangerous substances by road
- ○ collapse of building or structure
- ○ explosion or fire due to the ignition of any material which results in stoppage or suspension of normal work for more than 24 hours
- ○ escape of flammable substance – the sudden, uncontrolled release inside a building of for example 100kg or more of a flammable liquid or 10kg or more of a flammable gas; if in the open air, 500kg or more of flammable liquid or gas
- ○ escape of any substance in a quantity sufficient to cause death, major injury or damage to the health of any person.

If an accident results in a person being 'incapacitated for work of a kind which he might reasonably be expected to do, ... for more than seven* consecutive days (excluding the day of the accident but including any days which would not have been working days)', the responsible person shall 'as soon as practicable and, in any event, within 15 days of the accident send a report thereof to the relevant enforcing authority' (regulation 3(2)).

Fatalities, major injuries and dangerous occurrences are usually reported on form F2508. The main requirements for information on this form are:

- the date and time of the accident or dangerous occurrence
- for a person injured at work, his or her full name and occupation, and the nature of the injury
- for a person not at work, his or her full name and status (for example visitor, passenger), and the nature of the injury
- the place where the incident happened, a brief description of the circumstances, the date of the first report to the enforcing authority and the method of that report.

If an employee dies as a result of a reportable accident within one year of the accident, the employer has to inform the enforcing authority, whether or not the accident was originally reported.

If a person at work suffers from an occupational disease (see the table overleaf) and his or her work involves one of a specified list of activities (see the same table), the responsible person must send a report to the enforcing authority immediately. This is normally done using form F2508A, the main requirements of which are:

- the date of diagnosis of the disease
- the name and occupation of the person affected
- the name and nature of the disease
- the date of the first report to the enforcing authority and the method of that report.

The table overleaf gives a sample of the diseases and activities listed in the Regulations (Schedule 3).

Records of reportable incidents must be kept (regulation 7) and retained for at least three years. There are various ways of doing this:

- highlight reportable incidents in the accident book (the requirement for accident books is dealt with later)

Continued...

---

* Prior to April 2012, accidents resulting in more than three days away from normal work had to be reported. These are the 'three-day accidents' referred to in many health and safety publications and they still have to be recorded, but not reported.

... continued

Occupational diseases (sample only)	
Disease	Activities
Cataract due to electromagnetic radiation	Work involving exposure to electromagnetic radiation (including heat)
Cramp of the hand or forearm due to repetitive movements	Work involving prolonged periods of handwriting, typing or other repetitive movements of the fingers, hand or arm
Beat hand, beat elbow and beat knee	Physically demanding work causing severe or prolonged friction or pressure on the hand, or around the elbow or knee
Hand–arm vibration syndrome	Work involving a specified range of tools or activities creating vibration
Hepatitis	Work involving contact with human blood or blood products, or any source of viral hepatitis
Legionellosis	Work on or near cooling systems which are located in the workplace and use water; or work on hot water service systems located in the workplace which are likely to be a source of contamination
Rabies	Work involving handling or contact with infected animals
Tetanus	Work involving soil likely to be contaminated by animals
Tuberculosis	Work with persons, animals, human or animal remains, or any other material which may be a source of infection
Poisoning by specified substances, including mercury and oxides of nitrogen	Any activity
Various cancers	Various activities
Occupational dermatitis	Work involving exposure to a range of substances
Occupational asthma	Work involving exposure to a range of agents

- photocopy forms sent to the enforcing authorities
- keep records on computer – this will require registration under the Data Protection Act.

The Regulations cover employees, the self-employed, those receiving training for work, members of the public, pupils and students, and other people who suffer injuries or diseases as a result of work activities (regulation 2). However, the Regulations do not cover:
- patients who die or are injured while undergoing medical or dental treatment
- some incidents aboard merchant ships

- death or injury where the Explosives Act 1875 applies
- death or injury as a result of escapes of radioactive gas
- incidents that are reportable under the Road Traffic Act.

Other requirements

As we have seen, serious incidents as defined earlier have to be recorded and reported to the relevant authority. But there is also a requirement to record details of incidents, such as minor injuries, where these may result in a claim for benefit. This requirement is imposed by the Social Security (Claims and Payments) Regulations 1979.[32] These less serious incidents can be recorded either in an accident book (BI510) or on an organisation's own form or forms (paper or electronic). In either case, the minimum information which must be recorded is:

- the full name, address and occupation of the injured person
- the date and time of the accident
- the place where the accident happened
- the cause and nature of the injury
- the name, address and occupation of the person reporting the injury, if not the injured person.

All accidents recorded must be investigated to confirm the recorded information. Where an organisation uses its own form, additional information can be recorded. As with the serious incidents, if minor injury records are kept on computer, registration under the Data Protection Act will be necessary.

Other types of investigation

Investigations of near misses, nonconformities, spin-offs and side effects should be carried out using the techniques just described for accident investigations. However, the extent to which these techniques are applied to particular occurrences should be commensurate with the importance of what is being investigated. For example, a near miss or nonconformity that has the potential for serious injury or death should be investigated with as much rigour as an actual serious injury or fatality, while an informal interview and a brief record of the action taken is likely to be adequate for a trivial side effect.

The major practical distinction between accident investigations and other types of investigation is the numbers involved. In typical organisations, few managers have to conduct accident investigations because accidents are rare. However, all managers who are monitoring effectively are likely to have to investigate nonconformities, and there are many more near misses than there are accidents. The logical follow-on from this is that it would be more beneficial in risk management terms to train managers to investigate nonconformities rather than to investigate accidents.

Epidemiological analysis

The statistical techniques of epidemiological analysis were first applied to the study of disease epidemics, hence their being described as epidemiological. However, the techniques used are statistical, not medical, and although they were first applied to medical problems, and still are, these statistical techniques are also used in a wide variety of circumstances. In general, the statistical techniques associated with epidemiological analysis can be used in any circumstances where it is necessary to analyse unwanted events, including accidents.

Since not everyone is familiar with the techniques of epidemiological analysis, a historical example will be used first as an illustration. We will then show how epidemiological techniques can be applied to accidents.

Cholera was a major cause of death in cities for many years. No-one knew what caused the disease, but many doctors looked for patterns in how the epidemics occurred. This was done by trial and error, with different people looking at where cholera victims lived, what they ate and the work they did. Eventually it was discovered that cholera epidemics were centred around certain wells from which the city dwellers of those days obtained their drinking water. It was also found that closing these wells stopped the spread of the disease in those areas. Although no-one knew why the wells, or the water from them, caused the disease, they had found an effective way of stopping it spreading. In fact, it was not until many years later that the waterborne organisms responsible for cholera were identified.

This example illustrates the essential elements of epidemiological analysis. It is the identification, usually by trial and error, of patterns in the occurrence of a problem of interest to the analyst. These patterns can then be investigated to see whether causal factors can be identified and remedial action taken.

Epidemiology is used to identify problems which would not be apparent from single incidents, such as accidents occurring more frequently at particular locations. The results provide a guide as to where investigation will be most cost-effective, although they provide no information on causes.

The notes below are an introduction to epidemiological analysis; a more detailed treatment is provided in Chapter 19.

- **Data dimensions.** Epidemiological analysis is only possible when the same type of information (data dimension) is available for all (or a substantial portion) of the accidents being analysed. Typical data dimensions are location, time, day of the week and part of body injured.
- **Single dimension analysis.** This is the simplest form of epidemiological analysis. The incidents in the population are compared on a single data dimension, such as time of occurrence or type of product. The analyst is looking for any deviation from what would reasonably be expected. For example, if work is spread evenly over the working day, it would be expected that times of injuries would also be spread evenly. Where peaks and troughs are found in accident occurrences, these should be investigated. The analysis is slightly more complicated when an even spread is not expected, as the analyst has to carry out preliminary work to determine the expected spread.

    Note that the analyst will look for both over-representation and under-representation when carrying out the analysis. Both should be investigated: over-representation because it suggests that there are risks which are being managed poorly; under-representation since it suggests either a degradation in the reporting and recording system, or particularly effective management of risk from which others might learn.
- **Multiple dimension analysis.** The principles and practices described above for single dimension analysis can also be applied to two or more dimensions analysed simultaneously. This can identify patterns which would not be apparent from analysing the dimensions separately. Examples include part of body injured analysed with department, and time of day analysed with nature of injury.
- **Full-scale epidemiological analysis** of a set of data will involve analysis of all the single data dimensions separately and analysis of all of the possible combinations of these single dimensions. For this reason, full-scale epidemiological analysis is a very time-consuming process and where more than a trivial number of data are involved, the only practical approach is to use a computer.
- **Follow-up on epidemiological analysis.** Epidemiological analysis merely identifies patterns in data distribution. It does not give information on why these patterns are occurring. This can be determined only by appropriate follow-up investigations using the investigation techniques described earlier in the chapter.

Epidemiological analysis with limited data

The detailed data described earlier as necessary for full-scale epidemiological analysis may not always be available. However, this does not prevent the application of the techniques to limited amounts of information.

Valuable results can often be obtained simply by tabulating accident data for the past two or three years and looking for patterns in their occurrence. It is also worth trying to discover whether there are no accidents for particular places, times, people and so on, since this can provide clues on non-reporting or effective risk control measures.

Loss monitoring systems summary

This and the previous chapter have dealt with all of the requirements for an effective loss monitoring system. The key requirements for an organisation's internal loss monitoring system are summarised below:
- identification of key losses
- ensuring all key loss occurrences are reported and adequately recorded
- ensuring that incidents are investigated by competent personnel and that action is taken on the results of investigations
- trend analysis of key loss data to check overall performance
- making relevant comparisons of key loss data
- epidemiological analysis to identify patterns in key loss occurrences and investigation of any patterns identified.

Only if all of these requirements are in place will an organisation have an accurate measure of its losses and be making the best use of loss data as a means of learning from what has gone wrong in the past.

However, even good loss monitoring systems have limitations, which is why risk assessment and monitoring of conformity are required. It is worth ending this chapter, therefore, with a summary of the main limitations of loss monitoring.
- **It is too late.** By definition, something has happened which was undesirable and it would be preferable to have prevented it.
- **There are generally too few data for adequate analysis.** Except in exceptional circumstances, or where there is good collection of minor accident data, there are too few data for sound statistical analyses to be carried out.
- **Loss data are too random.** While some patterns can be found using epidemiological analyses, the majority of accidents are random events and it is difficult to learn from them.
- **Loss data give little information on risk.** Accidents mean that there were risks and that these risks were inadequately controlled. However, because accidents are random events, they give little information about the underlying risk in an organisation.

10: Monitoring and measuring conformity

Management gets what management inspects, not what management expects. (Anon.)

Introduction

Monitoring and measuring conformity are two of the main elements in the risk management model introduced in Chapter 4. Figure 4.1 from that chapter is reproduced below as Figure 10.1 as a reminder of the role of conformity monitoring and how it fits with the other elements in the risk management model.

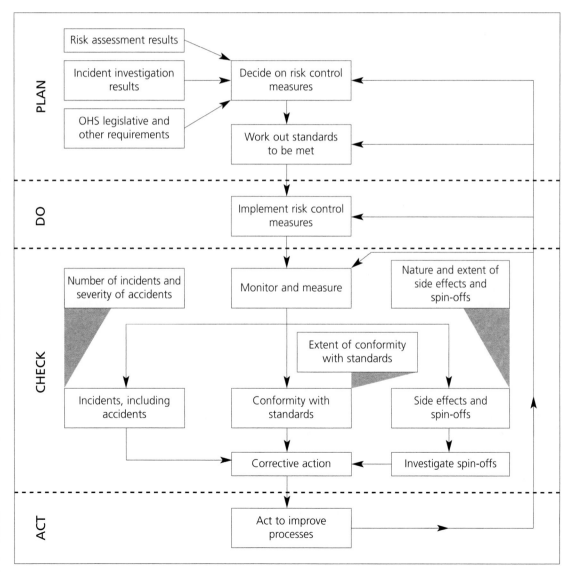

Figure 10.1
Core elements in the risk management model

We know from experience that the vast majority of risk control measures deteriorate over time. For this reason, we need to monitor risk control measures, establish the extent to which they are still effective and take corrective action where necessary. This process is monitoring conformity with standards, and is the subject of this chapter. However, it is usual to combine the monitoring of risk control measures with that of conformity with other health and safety standards, such as document availability and record-keeping. For this reason, for the rest of this chapter, 'monitoring' will be used to mean monitoring of conformity with health and safety standards in general.

The chapter begins with a more detailed look at the purposes of monitoring. We will then describe the techniques that can be used for monitoring. However, as with many other aspects of risk management, there are terminological difficulties in describing monitoring and its techniques. In an attempt to minimise these, the following conventions are used in this chapter:

- 'Check' or 'checking' is used to mean establishing whether or not a particular standard is being met – for example, checking whether the required PPE is being worn, whether risk assessment forms are being completed correctly, or whether tasks are being carried out according to the written procedures.
- The mechanics used to carry out checks are variously described as inspections, examinations and observations. The choice of word used for these mechanics is largely arbitrary – documents can be inspected or examined and work activities can be observed, inspected or examined. In this chapter, these words are used interchangeably but 'check' is always used as described above.

The last section of the chapter deals with measuring conformity with standards.

Purposes of monitoring

The main purposes of monitoring are to enable organisations to check that they are doing what they said they would do, and to take corrective action when they find that they are not doing what they said they would. However, these basically straightforward purposes are only possible if the following criteria are met:

- Everything which needs to be done on health and safety grounds has been clearly identified – in other words, the organisation's decisions about what it says it will do are both necessary and sufficient. If this criterion is to be satisfied, the planning process must be functioning effectively. (Refer to Figure 10.1 if it is not clear why the effective functioning of this process is necessary.)
- Responsibilities for everything which has to be monitored have been clearly allocated so that, for example, no risk control measures are left unmonitored.
- Those responsible for monitoring know what should be in place and have the competence to tell when it is not.

In theory, if these criteria are met, monitoring can be a mechanistic process of checking what is actually happening against what should be happening. In practice, however, this is rarely possible; the primary reason for this is that in most organisations, things change so fast that the setting and recording of standards is not always complete. Obviously, this is not a desirable state of affairs, but in health and safety terms it is better to recognise that it exists, and deal with it, than to assume that the HSMS is functioning more effectively than it actually is.

In order to overcome this potential weakness in the HSMS, monitoring can be extended in a number of ways. These are dealt with in the section on monitoring techniques.

The scope of monitoring

There are four topics which should be covered by monitoring:

- **Documents.** Examples of documents which should be checked include health and safety policies, health and safety manuals, and safe working procedures. These documents should be examined periodically to check that they are still up to date, accurate and relevant.
- **Records.*** Examples of records which should be checked include accident records, permits to work and records of risk assessments. These records should be inspected to check that all the required records are being completed, that they are being filled in accurately and that they are being stored appropriately.
- **Locations.** All locations for which an organisation is responsible should be checked at appropriate intervals. Location inspections should be used as an opportunity to check that the physical conditions are adequate and that the required risk control measures are in place and effective.
- **Activities.** All the activities for which an organisation is responsible should be checked at appropriate intervals. These observations should include not only core production or service activities but also activities such as maintenance and cleaning, and they should be used as an opportunity to check that all activities are being carried out in a safe manner.

Documents, records, locations and activities should be monitored, as appropriate, by the person responsible for them. This means that everyone in an organisation will have to carry out some monitoring, but the nature and extent of this will depend on a number of factors. Three of the more important factors influencing the requirements for monitoring are:

- **the level of risk being controlled** – the higher the risk, the more detailed and the more frequent the monitoring should be
- **the degree of reliability of the risk control measures** – the lower the reliability of the risk control measures, the more detailed and the more frequent the monitoring should be
- **the rate of change** – as change is normally associated with increased risk, monitoring should be increased during times of change.

Another important factor that influences the nature of the monitoring required is the level of management. At lower management levels, the detailed examinations, inspections and observations described earlier are needed, but these would not be appropriate for higher levels of management. At successively higher levels of management, the emphasis changes from monitoring the workplace to monitoring the monitors. In other words, the main function at higher management levels is to check that monitoring at lower levels is being done effectively.

Figure 10.2 shows part of an organisation chart with levels of management below board level identified as Level 1 to Level *n*. If monitoring in the organisation as a whole is to be effective, there must be a 'cascade' of monitoring, with each level checking that the level below is discharging its monitoring responsibilities effectively. Figure 10.2 also illustrates the role played by audit, which is an independent check on how well an HSMS is functioning. This independence can be achieved in either of two ways: by different management functions auditing each other, as indicated by the 'Audit' boxes included in the organisation chart; or by external auditors carrying out an audit of any or all of the management levels, as indicated by the 'Audit' box on the right of the diagram.

In Figure 10.2, note that the monitoring follows the organisation's line hierarchy, whereas audit goes across the hierarchy, either from outside it, as indicated by the box on the right, or from one department to another, as indicated by the 'audit' boxes inside the diagram. We will deal with the subject of audit in Chapter 11.

There is an additional type of monitoring which can be used; this is illustrated in Figure 10.3.

---

* Records are a subset of documents, but it is convenient to deal with them separately because they require different checks.

Figure 10.2
Monitoring
'cascade' and audit

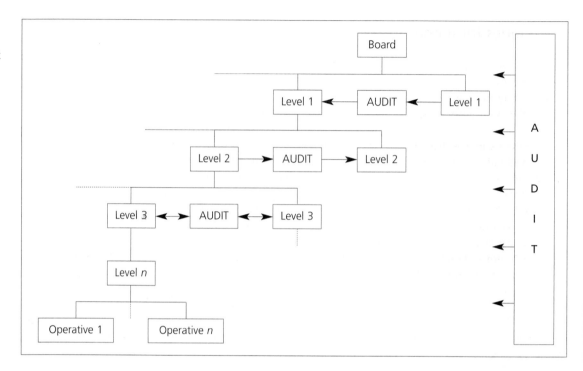

Figure 10.3
Upward monitoring

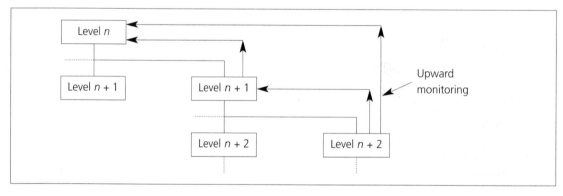

In upward monitoring, lower levels in the organisation check the adequacy of the health and safety work done at higher levels. For most organisations, upward monitoring is a fairly radical departure from normal practices, and it may not be an acceptable approach. However, it is the case that failures in performance at higher management levels can put people at lower levels, and particularly operatives, at higher risk.* It is, therefore, difficult to argue that lower levels do not have a good reason for checking that the higher-level health and safety functions are being carried out adequately. Where, for whatever reason, upward monitoring as illustrated in Figure 10.3 is not in place, alternative means of monitoring should be used and, in this context, health and safety representatives and health and safety committees can play a valuable role.

---

* These management failures do not normally have an immediate and obvious effect. Rather, the effects of decisions made remain in the system until circumstances occur which lead to an accident. These have been described as 'latent' failures and are dealt with in more detail in Chapter 27.

Techniques for monitoring

The fundamental monitoring techniques are:
- examining documents and records
- inspecting work locations and observing work activities
- questioning relevant individuals.

All of these techniques have been dealt with in previous chapters in the context of investigation, and this chapter will only consider how they are applied to monitoring.

The statement at the start of this chapter encapsulates what monitoring has to achieve. We need to make sure that everything that is important in health and safety terms is checked in the course of monitoring. In order to do this effectively, we need to have some way of remembering all the things that must be checked and the standards that have to be achieved. This is not necessarily easy, because:
- some inspections are only required at relatively long intervals, such as every 12 or 14 months
- many health and safety activities are peripheral to 'normal' work and can easily be forgotten
- managers may not have the skills necessary to work out what is needed in health and safety terms and may require guidance on what should be in place.

These problems have long been recognised and many organisations have produced documentation to help with monitoring. This typically takes the form of checklists which remind managers what they need to check and what standards they need to meet. Some examples of questions that could be included in such a checklist are given in Table 10.1.

Document checks
Are the contents of the health and safety policy up to date? Are the contents of all written work procedures up to date? Are any work procedures that should be in writing not yet written down?
Record checks
Are there risk assessment records for all tasks and equipment? Have all risk assessments been reviewed by the required date? Have all scheduled location inspections been carried out? Have the results of all location inspections been recorded?
Location checks
Are all the emergency evacuation routes free from obstruction? Are all electrical cables and casings free from wear and other damage? Are all containers for chemicals clearly labelled? Are all required machine and equipment guards in place and used?
Activity checks
Are all computer users maintaining appropriate postures during computer work? Is all manual handling being carried out in the correct manner? Where written work procedures are available, are they being followed?

Table 10.1

Possible questions for a monitoring checklist

The wording of questions in checklists can be varied to suit the organisation concerned and the type of personnel who will be doing the monitoring. What is important is that the questions accurately represent the standards to be met and that there is a sufficient range of questions to cover all the health and safety standards to be monitored. There is more information on preparing sets of questions in Chapter 22.

Checklists have two other possible uses:

- **They can be used to record the results of particular monitoring exercises.** These records are useful in a number of ways, in that they provide a verifiable record of what monitoring has been carried out, and the data they contain can be analysed to determine, for example, whether performance is improving or deteriorating.
- **Where the questions are phrased appropriately, the results of particular monitoring exercises can be given as a numerical measure.** These results can then be used in a variety of ways, including making comparisons between departments or organisations and setting targets for individual managers. This topic is dealt with in more detail in the section on measuring conformity later in this chapter.

While checklists can be a valuable aid to monitoring, they have a major weakness in that people may check only those items on the checklist and ignore hazards which exist on site but are not included on the checklist. Obviously, this could have serious consequences if the checklists being used are not sufficiently comprehensive and up to date. In order to overcome this problem, some organisations adopt an alternative approach and ensure that managers have the competences to carry out effective monitoring of conformity. With this approach, if a checklist is used at all, it is used as an *aide-mémoire*, not as a definitive list of items to be checked.

In practice, the combination of competence and an *aide-mémoire* checklist is likely to be the most effective, but not all organisations have the in-house resources to train managers for monitoring and to develop appropriate checklists. This has been recognised by a number of commercial health and safety organisations and several monitoring packages are available under a variety of trade names. These packages vary in content but they are all essentially concerned with delivering competences, backed up by checklists. The checklists can be 'fixed' – in other words, purchasers of the system are not allowed to alter them – or they can be 'tailored' to suit more accurately the needs of a particular organisation. If you want to use one of these packages, the most important task is to identify accurately your organisation's needs and then select, or tailor, a package which will meet these needs. In some cases, it may be possible to select a package and then alter your organisation's monitoring practices to meet the requirements of the package. This approach can be suitable, particularly where there are few or no monitoring practices already in place.

Frequency of monitoring

There is a strong argument for saying that monitoring should be a continuous activity. That is, anyone with monitoring responsibilities (effectively everyone) should, in the course of their normal work, check that things are being done to the relevant standard and take corrective action if they are not. Indeed, it is only if this continuous monitoring takes place that the continuing effectiveness of risk control measures can be ensured.

However, superimposed on this continuous activity there should be a formal structure of systematic checks, the results of which are recorded and analysed. These are sometimes referred to as 'self-audits' but self-audit in health and safety terminology is a contradiction in terms, since auditors should be independent of what is being audited.

As we have seen, the appropriate frequency for formal monitoring depends on three factors: the absolute level of the risk; the reliability of the risk control measures; and the rate of change at the location being monitored. It is not possible, therefore, to give detailed recommendations on how frequent monitoring should be, although anything less frequent than once per quarter is unlikely to be adequate even for low-risk, stable environments with reliable risk control measures. At the other extreme, high-risk phases of construction work may require daily monitoring.

Failures in monitoring

It can be argued that, in the majority of organisations, monitoring is by far the weakest part of the HSMS. Given the importance of monitoring, it is worth considering some of the reasons why this is the case so that they can be taken into account when reviewing or setting up monitoring systems.

The first two reasons are fairly straightforward: either managers do not know that health and safety management requires monitoring; or they know that it is required but do not know how to carry it out effectively. In either case, the solution is to provide the necessary knowledge and skills.

The next reason for weaknesses arises from the fact that failure to carry out monitoring rarely has any short-term or obvious consequences. Thus, if busy managers choose not to do the required monitoring, there are often no adverse effects unless the manager is taken to task by his or her superior. Since this immediate superior is probably just as busy, this does not happen and the monitoring system gradually degrades. (If the manager gets away with not doing it once, he or she is likely to continue not to do it!)

Another cause of gradual degradation of the monitoring system arises when repeated monitoring exercises fail to find anything wrong. In these circumstances, managers argue that there is no reason to continue what they see as a pointless exercise.

The first thing to check when this happens is that the monitoring is being carried out properly. Where busy managers will have to spend time and resources sorting out problems identified during monitoring, there is always a temptation, even if it is subconscious, to turn a 'blind eye' to problems. The perceptual mechanisms which explain this 'blind eye' phenomenon are described in the human factors section of this book (Chapters 14 and 28).

However, assuming that the monitoring is being carried out correctly and no problems are being found, a good compromise is to reduce the monitoring frequency. This will reduce the workload created by the monitoring while maintaining its essential function.

For some managers, another reason for failing to carry out adequate monitoring is that they are reluctant to become involved in the confrontations which arise when people are not behaving in a healthy and safe manner. This reluctance may be a general fault with the manager concerned, in that they avoid all confrontation and not just disputes over health and safety matters. Where this is the case, training in the relevant managerial skills will be needed, together with training in skills such as assertiveness where appropriate. Where managers avoid confrontation only on health and safety issues, the solution is rather different: it will require enforcement by a more senior manager, who must, of course, have the necessary skills and motivation.

The final group of reasons for the failure of a monitoring system is a category labelled 'There is no point, because...'. This category arises because various failures in other parts of the HSMS mean that monitoring cannot be carried out successfully. The commonest of these failures are that:

- there are no sanctions available to use against people who are behaving unsafely; there is therefore no point in identifying the unsafe practices, and consequently no point in monitoring
- the rules people have to follow are stupid (impracticable, designed to protect managers and not workers, contradictory and so on); therefore there is no point in enforcing them, no point in identifying contraventions, and in the end no point in monitoring
- senior managers do not set a good example by following safety rules themselves, so trying to get junior staff to follow them is a waste of time.

The solutions to these problems are self-evident and have to be addressed at other points in the HSMS. This has been done in the relevant parts of other chapters.

Links with other inspection activities

As we have seen in other chapters, there is a wide range of terms used for health and safety monitoring, including safety inspections, safety tours and safety walkabouts. Each of these inspection activities has its place in the HSMS, and in some circumstances they may take the place of monitoring. However, there is a critical difference between true monitoring and these other techniques. Monitoring involves checking that what should be in place is in fact present; the other activities involve simply finding out what is in place. These other inspection activities do, however, have a role to play, which we discussed in Chapter 5, under 'Other uses for risk assessment techniques'.

Dealing with nonconformities

Two things are usually required when a nonconformity is identified:
- correction – putting the nonconformity right
- corrective action – putting the causes of the nonconformity right.

Usually, the latter will require an investigation and the techniques described in the context of accident investigation in the previous chapter can also be used for nonconformity investigations.

Measuring conformity

Measuring conformity is possible if checklists are available of the type described earlier in this chapter. Where they are, the proportion of correct answers gives a score and this score is a measure of conformity. This approach can be elaborated in a number of ways:
- The questions can have percentage answers as well as 'yes/no' answers. For example, the second question in Table 10.1 is 'Are the contents of all written procedures up to date?' This could have a percentage answer so that a distinction can be made between areas with low or high percentages of up-to-date procedures.
- The questions can have 'no' as the answer indicating conformity, as is the case with the third question in Table 10.1: 'Are any work procedures which should be in writing not yet written down?' This form of question is used to prevent people using the checklist simply answering 'yes' to all questions requiring a 'yes/no' answer.
- Questions can be given weightings to reflect their importance. For the questions in Table 10.1, for example, the questions on document and record checks might be given a weighting of 1, while the questions on location and activity checks might be given a weighting of 5.

The monitoring packages described earlier in the chapter typically use all of the elaborations just described. However, this can make the calculation of conformity scores a time-consuming job and for this reason many of the more detailed monitoring packages are software-based. This type of software allows for automatic calculation of scores and, depending on the particular software used, a range of other facilities such as the graphic display of results.

The questions in these monitoring packages are usually based on one or both of:
- the requirements of an HSMS, such as BS OHSAS 18001[3] or HSG65[1]
- UK legal requirements.

Where a suitable subset of UK legal requirements is included in the package, it can be used for measuring conformity with legal requirements, referred to in BS OHSAS 18001 as 'evaluation of compliance'.

Although these detailed packages are available, it is possible for organisations to draw up their own sets of questions, and guidance on how to do this is given in Chapter 22.

Conformity data should be the subject of trend and epidemiological analysis just as loss data are. The trend analysis will enable the identification of, for example, deterioration in the implementation of risk control measures over time, and epidemiological analysis will identify any patterns in this deterioration which might indicate common causes.

However, just as there are problems with the collection and analysis of loss data, there are also problems with conformity data. The main ones are:
- There has to be some method of ensuring that the monitoring covers all relevant risks and their associated risk control measures. If this is not done, the conformity data collected will not give a true measure of performance and comparisons of, for example, departments or organisations will not be valid.
- Where weightings are used, it is necessary to ensure that the individual items are weighted correctly. If this is not done, there may be a tendency to deal with issues with a high numerical weighting, rather than issues with a high risk.

The statistical techniques needed for the basic analysis of monitoring data have already been dealt with in Chapter 8. More advanced statistical techniques are dealt with in Part 2.1 (Chapter 19).

Summary

This chapter described what we mean by monitoring conformity and how to carry it out. We also discussed some of the reasons why monitoring is not carried out as effectively as it could be. The chapter ended with a description of how conformity can be measured and how these data should be used.

11: Other elements of occupational health and safety management systems

Introduction

This chapter deals with those elements of BS OHSAS 18001[3] and HSG65[1] which have not been dealt with in detail earlier in Part 1.1. These elements are:

- policy
- planning
- documentation and control of documents and records
- operational control
- audit
- review.

Policy

It should be noted that the Policy element of the HSMS is the only element which is not a process. However, since there are requirements to communicate and review the contents of the policy, there are processes involved in this element.

The main material on policy from BS OHSAS 18001 and HSG65 is summarised in the next two subsections.

BS OHSAS 18001

BS OHSAS 18001 requires that top management 'shall define and authorize the organization's OH&S policy' and specifies various requirements with regard to this policy, including the following:

> The policy must:
> - be appropriate to the nature and scale of the organisation's OH&S risks;
> - include commitments to:
>   - prevention of injury and ill health
>   - continual improvement in OH&S management and OH&S performance
>   - comply with applicable legal requirements
> - be communicated to all persons working under the control of the organisation; and
> - be reviewed periodically.

HSG65

HSG65 contains a list of typical statements of 'safety philosophy', and a selection from this list is given here:

> We believe that an excellent company is by definition a safe company. Since we are committed to excellence, it follows that minimising risk to people, plant and products is inseparable from all other company objectives.
>
> Prevention is not only better, but cheaper than cure. There is no necessary conflict between humanitarianism and commercial considerations. Profits and safety are not in competition. On the contrary, safety is good business.

Health and safety is a management responsibility of equal importance to production and quality.

In the field of health and safety [we] seek to achieve the highest standards. We do not pursue this aim simply to achieve compliance with current legislation, but because it is in our best interests. The effective management of health and safety, leading to fewer accidents involving injury and time taken off work, is an investment which helps to achieve our purposes.

The identification, assessment and control of health and safety and other risks is a managerial responsibility and of equal importance to production and quality.

HSG65 also lists the issues which could form the basis of a policy statement:

1. Set the **direction** for the organisation by:
   - demonstrating senior management commitment;
   - setting health and safety in context with other business objectives;
   - making a commitment to continuous improvement in health and safety performance.
2. Outline the **details** of the policy framework, showing how implementation will take place by:
   - identifying the Director or key Senior Manager with overall responsibility for formulating and implementing the policy;
   - having the document signed and dated by the Director or Chief Executive;
   - explaining the responsibilities of managers and staff;
   - recognising and encouraging the involvement of employees and safety representatives;*
   - outlining the basis for effective communication;
   - showing how adequate resources will be allocated;
   - committing the leaders to planning and regularly reviewing and developing the policy;
   - securing the competence of all employees and the provision of any necessary specialist advice.

It is good practice for an organisation to establish and record its policy on risk management. There are two main reasons for this:
- The work required to produce and record a clear statement of policy gives senior managers an opportunity to think about risk management and what they want their organisation to achieve in risk management terms. This can be reinforced by the senior managers stating their personal commitment to the policy and signing the relevant document.
- The policy can be promulgated throughout the organisation and, if necessary, outside the organisation as a clear statement of what has to be achieved.

As with any other documentation, the policy should be reviewed from time to time, amended as necessary, and any amendments promulgated to relevant personnel and organisations.

---

* This should now include representatives of employee safety – see Chapter 12.

Planning

The BS OHSAS 18001 requirements for planning can be summarised in the form of a diagram of the type shown in Figure 11.1. The figure is followed by explanatory notes.

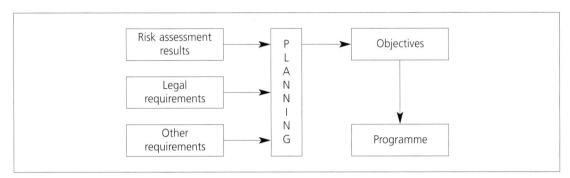

Figure 11.1
BS OHSAS 18001
planning
requirements

There are three main inputs to the planning process – risk assessment results, legal requirements and other requirements. The last of these includes health and safety requirements such as:
- requirements to which the organisation voluntarily subscribes – for example, trade or industry sector health and safety schemes and the requirements of BS OHSAS 18001
- requirements which are imposed on the organisation other than by legislation, such as by parent companies or customers.

The outputs of the planning process are objectives and the programme which will have to be implemented and maintained in order to achieve those objectives. The term 'programme' is carried over from quality management and it is what is referred to in HSG65 as a 'plan'. The terms 'programme' and 'plan' are used interchangeably for the rest of this book. There is more information on planning in Chapter 22.

Documentation and control of documents and records

Documentation

The requirements for documentation and the control of documents and records are not well specified in HSG65 but in BS OHSAS 18001 they are similar to the requirements in ISO 9001[5] and ISO 14001.[6]

The text from the BS OHSAS 18001 Documentation subclause is reproduced below, followed by explanatory notes.

The OH&S management system documentation shall include:
a) the OH&S policy and objectives;
b) description of the scope of the OH&S management system;
c) description of the main elements of the OH&S management system and their interaction, and reference to related documents;
d) documents, including records, required by this OHSAS Standard; and
e) documents, including records, determined by the organization to be necessary to ensure the effective planning, operation and control of processes that relate to the management of its OH&S risks.

BS OHSAS 18001 is intended for external accreditation, as are ISO 9001 and ISO 14001, and this is one of the main reasons why the documentation above is required. For example:

- The scope of the OH&SMS ((b) above) is required so that external auditors know what to audit and what samples to take.
- The description of the main elements of the OH&SMS ((c) above) is required because there are many ways of satisfying the BS OHSAS 18001 requirements and auditors need to know how the organisation being audited will do this. This is also part of the reason for the more detailed documents required by (d) and (e).
- The records required by (d) and (e) are needed to prove to auditors that necessary actions have been carried out and, where appropriate, provide information on the results of these actions.

Because these requirements for documentation are so similar to those for quality management, those with responsibility for health and safety documentation are advised to study ISO/TR 10013:2001 *Guidelines for quality management system documentation*.[33]

Control of documents and records

As with the requirements for documentation, the requirements for control of documents and records are similar to their equivalents in quality management.

Key points on control of documents are that there must be procedures for the following:

- approval, review, update and re-approval of documents
- identifying changes to documents and their current revision status
- ensuring relevant versions of documents are available at the point of use and preventing the unintended use of obsolete documents
- ensuring that documents remain legible and easily identifiable.

For the control of records there must be procedures for the identification, storage, protection, retrieval, retention and disposal of records, which 'shall be and remain legible, identifiable and traceable'.

Operational control

In BS OHSAS 18001 the 'Operational control' subclause is rather confusing but study of its equivalent in ISO 14001 allows the following interpretation:

- Where an operation requires risk control measures, this operation must be included in the OH&SMS.
- For each of these operations the following are required:
  - operational controls must be implemented and maintained; these can be thought of as the safe systems of work described in Chapter 7
  - where necessary, procedures should be documented – for example, a written safe system of work would be required for tasks that were carried out only infrequently, so that operators were likely to forget the details of the safe system of work
  - where necessary, operating criteria must be specified; these can be defined as parameters of the operation that can be used to control health and safety risks, such as speed of operation of machines, electrical voltage for hand tools, temperature of water supply to wash basins, and tyre pressures.

The other requirements under 'operational control' relate to change and imported risks which would be more appropriately dealt with under 'hazard identification, risk assessment and determining controls', where the majority of them have already been mentioned.

Audit

The term 'audit', like most terms used in risk management, can mean a number of different things. The HSG65 and BS OHSAS 18001 definitions of audit are reproduced below:

> HSG65: The structured process of collecting independent information on the efficiency, effectiveness and reliability of the total health and safety management system and drawing up plans for corrective action.

> BS OHSAS 18001: systematic, independent and documented process for obtaining 'audit evidence' and evaluating it objectively to determine the extent to which 'audit criteria' are fulfilled.

There is a note to the BS OHSAS 18001 definition that reads: 'For further guidance on "audit evidence" and "audit criteria" see ISO 19011.' The document referred to is *Guidelines for quality and/or environmental management systems auditing*. As the title suggests, it does not cover the auditing of HSMSs. However, this Standard was replaced in 2011 by BS EN ISO 19011, *Guidance for auditing management systems*, which does cover HSMS auditing.

Health and safety professionals should follow ISO 19011 when auditing – this Standard is dealt with in detail in Chapter 23. What follows is a less formal consideration of auditing.

In this chapter, the term 'safety audit' is used to describe a process which meets the following criteria:

- the audit measures an organisation's performance in relation to pre-defined written standards (these standards are the 'audit criteria')
- the output from the audit is praise for strengths and suggestions for remedying identified weaknesses
- the audit takes a structured approach to data collection, with the structure being decided before the audit begins
- the audit is carried out by people who are independent of the line management being audited
- auditors have the necessary competences.

In the remainder of this section, each of these criteria is considered in more detail.

Pre-defined standards

It is important to realise that audit is not a straightforward fact-finding exercise. The auditors do not ask 'What is this organisation doing?'; rather, they ask 'Is this organisation meeting pre-defined standards?' It follows from this that auditors need to know what these standards are before they can conduct an audit. There are three main sources for these standards:

- **Legislation.** Organisations should be satisfying all relevant legal requirements. Auditors who undertake the legal compliance parts of audits will have to have a detailed knowledge of these legal requirements.
- **Good practice.** There are good practice standards set out in such documents as HSG65 and BS OHSAS 18001, and organisations can be audited against these standards during a good practice audit.
- **The organisation's own standards.** Organisations should have written standards for all relevant aspects of their HSMS. Auditors can conduct an internal standards audit to check whether these standards are being met.

In some audits, for example most quality audits, it is only necessary to use the organisation's own standards. However, in health and safety auditing this may not be adequate, since, even though the organisation may be meeting its own standards, those standards themselves may not satisfy legal requirements. This has implications for the competences required of auditors, and these are discussed later in the section.

There are various ways in which auditors record details of the standards which have to be met in order to make the audit process more manageable. A commonly used method is to record the standards in the form of 'audit questions'; this technique is similar to that used for preparing questions for measuring conformity as described in Chapter 10.

Output from audits

The output from audits should be recommendations for corrective actions with timescales for their completion. However, since audits may not be conducted by personnel from within the organisation being audited, it may not be appropriate for the auditors to allocate responsibilities for the implementation of corrective actions.

However, as HSG65 points out, 'it is important that auditing is not perceived as a fault-finding activity but as a valuable contribution to the health and safety management system and learning.' Given the average manager's response to being audited, this is more of a 'fond hope' than a likely achievement, but it may be helped by the advice on outputs in HSG65 that 'auditing should recognise positive achievements as well as areas for improvement'.

Structured approach

A full audit should cover all the aspects of a HSMS described earlier in this book. There are various ways of ensuring a structured approach, including the use of appropriate sets of audit questions, as mentioned above in the section on pre-defined standards. We will discuss how to develop and use these sets of questions in Part 2.1 (Chapter 22).

A full audit of an organisation's HSMS would be a prohibitively time-consuming task, and it is more realistic to conduct a sample audit and draw inferences about the whole of the organisation from the sample results. Where sample audits are used, it is important that they have an appropriate structure. Typical methodologies for ensuring appropriate structures in sample auditing are described in Chapter 23.

Sample audits provide valuable data on the actual circumstances in the sample which has been audited and, if the sample has been chosen to be properly representative, these findings can be generalised to the remainder of the organisation.

Independence

We noted earlier in this chapter that there are health and safety functions such as risk assessment and monitoring which are the responsibility of line managers. However, even the best managers, carrying out all the required functions, are prone to overfamiliarity, and this can lead to a failure to perceive weaknesses in their own areas. Audit provides the opportunity for 'a fresh pair of eyes', which are not suffering from overfamiliarity, to look over an HSMS and identify its strengths and weaknesses.

As we have already seen, in the definition of audit used in this book, the auditors have to be independent of the line management they are auditing, so the phrase 'self-audit' is a contradiction in terms. However, this is purely a terminological difficulty. The function referred to as self-audit elsewhere is referred to in this book as monitoring conformity.

To achieve independent audits, it is not necessarily essential for someone from outside the organisation to carry out the audits. As we noted in Chapter 10 on monitoring conformity, one branch of an organisation's line management can audit another branch and, so long as the process is managed carefully, this can provide an adequate degree of independence. A useful rule of thumb to follow is that auditors should never be in a position where they have to rectify any of the problems they discover. It is a recognised characteristic that humans are poor at perceiving problems when solving these problems will require time and effort on their part. This characteristic is so well known that it has a name, 'perceptual defence', and ensuring that auditors do not have to solve the problems they identify helps to avoid this difficulty.

Although sufficient independence can be achieved with one group of managers auditing another, there are other options. The complete range of options is listed below, in descending order of independence:

1. **Audit by an external agency with no vested interest in the outcome.** For example, a health and safety audit by the HSE or one of the accreditation organisations.
2. **Audit by insurance companies.** These audits are conducted for the benefit of the insurance company, although the results may be of value to the organisation being audited.
3. **Audit by commissioned consultants.** These audits can be totally independent but there is always the possibility that the consultants will be less critical than is necessary since, if they are too critical, the client may go elsewhere for the next audit.*
4. **Audit by in-house audit teams.** These teams can be totally independent but, ultimately, there will be one person at the top of the organisation who controls both the auditors and the line management being audited. There is always the possibility, therefore, that influence can be brought to bear on auditors to be less critical than the situation warrants.
5. **Managers auditing other managers.** These audits can be successful but there is always the possibility of collusion of the 'you go easy on me and I'll go easy on you' type.

In most larger organisations, over a period of time, it is likely that all of these types of audit will be used and this is to be encouraged. The more people looking at the operation of an HSMS the better, especially if they are all looking at it from slightly different points of view.

Competence

Irrespective of who carries out an audit, it will be successful only if the auditors have certain competences. Two broad categories of competence are required:

- **Competences in audit.** These include inspection and investigation skills, competence in document examination techniques, report-writing skills, skills in chairing meetings, and presentation skills for the feedback of the audit results. All of these skills have already been discussed or are covered in the next chapter, which deals with communication and training.
- **Competences in health and safety.** These include hazard identification and risk rating skills, and the skills required to assess the effectiveness of existing risk control measures and to suggest, where necessary, improved risk control measures. Where legal compliance or good practice audits are being carried out, the auditors will also need the relevant range of underpinning knowledge.

Audit competences are, in general, transferable across a number of different types of audit, including health, safety, environment and quality. However, someone carrying out a health and safety audit does not need to have environment or quality competences.

Review

In risk management, 'review' is used to mean a number of different things, but they can be grouped into two broad categories:

- Those reviews where the word 'review' is used in the same sense as it is with book and film reviews – in other words, assessing what is there and preparing some form of critique. In risk management, the 'initial status review' as described in BS 18004 falls into this category, as do the reviews of organisations' HSMSs conducted by external consultants.

---

* This may also apply to accreditation organisations and insurance companies.

- Those reviews where the word is used in the sense of 're-view' – that is, looking again at some aspect of the organisation and suggesting corrections or improvements. The 'status review' described in BS 18004, and defined as 'formal evaluation' of the OH&SMS, falls into this category as does the management review in BS OHSAS 18001.

It is the second category of review that will be considered in this section, beginning with why this type of review is important.

The importance of review

There are two main reasons why review is important: first, it closes the feedback loop in the HSMS; and second, it is the primary driver for continual improvement. Each of these reasons will now be considered in more detail.

Closing the feedback loop

In any system like an HSMS, control is only possible if there is a feedback loop. In essence, the output of the system is measured and the level of the output used to determine what the system does next. A very simple electromechanical system like a refrigerator illustrates the basic principles very well. The main part of the refrigerator system is a heat exchanger, the designed output of which is to cool the air inside the refrigerator. This output is measured by a thermostat and the reading on the thermostat is used as the basis for switching the heat exchanger on and off. This is illustrated in Figure 11.2.

Without the thermostat and the feedback loop, the interior of the refrigerator would either get colder and colder (heat exchanger permanently on) or warm up to room temperature (heat exchanger permanently off).

An HSMS is not, of course, a simple electromechanical system and cannot, therefore, rely on an electromechanical feedback loop. In an HSMS, the feedback loop is created by the actions of people during the review process and if these actions are not carried out, the HSMS is out of control. Because the HSMS is complex, it is difficult to predict the consequences of this lack of control, but experience from major disasters has shown that they can be very severe.

Figure 11.2
A simple electromechanical system

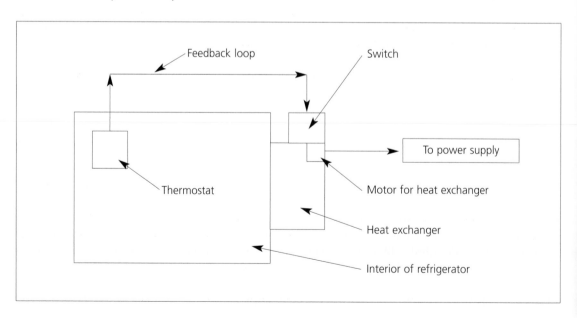

If you have difficulty with the idea of people forming a feedback loop, go back to the refrigerator illustration and imagine that the thermostat has broken. In these circumstances, a thermometer could be put inside the refrigerator and a person could check its reading every hour or so, and switch the power supply to the refrigerator on or off depending on the observed reading on the thermometer. This would be tedious (which is one of the reasons why thermostats were invented) but it would work because the people doing the measuring and the switching on and off are closing the feedback loop.

In the HSMS, the actions required to close the feedback loop are rather more complex and they will be dealt with later in the chapter.

Continual improvement

Systems for use in organisations are developed using currently available knowledge and technology and to meet an individual's or an organisation's perceived needs at the time of development. However, all of these things (knowledge, technology and needs) change as time goes on and, unless the relevant system changes to match, it will gradually become less and less appropriate. If a refrigerator is thought of as a system for keeping food and drinks cool, then this can be illustrated quite simply.

Older readers will remember the days when refrigerators were a rarity and there was a variety of ways of keeping food and drinks cool, including cellars, ice boxes and water evaporation. They will also remember the early refrigerators where the ratio of storage space to insulation and electromechanical equipment was low, and defrosting was a frequent (and major) requirement. Continual improvement has, however, produced the current generation of refrigerators and it is possible to look forward to self-cleaning refrigerators.

It is important to remember that systems either deteriorate or improve; they do not remain the same. This is the case for electromechanical systems like refrigerators, which deteriorate until they are repaired or replaced, and living systems which, after an initial growth phase, deteriorate with age until they die. It is also the case for management systems, which deteriorate unless they have some mechanism for ensuring continual improvement.

This process of continual improvement also applies to risk management and, like refrigerators, risk management has changed over the years. This is not the place for a detailed history of the evolution of risk management but one example will be useful. Thirty years ago there was very little formal risk assessment and what did exist was almost all concerned with low likelihood, high severity risks. Over the years, techniques of risk assessment have been developed for other categories of risk and these techniques are gradually being introduced in more and more organisations. In future, it is to be hoped that this process will continue with better risk assessment techniques being developed and used ever more widely.

If an organisation is to maintain an effective and efficient HSMS, it must, therefore, have an in-built procedure for continual improvement. As a minimum, this procedure should ensure that the HSMS adapts to meet internal and external changes and so stays up to date. However, the preferred type of continual improvement procedure will go beyond this and result in absolute improvements. The possible nature of these absolute improvements will be discussed in Part 2.1 (Chapter 23); this section continues with a discussion of the functions of review.

The functions of review

Although review is often referred to as a single process, it is possible to identify a number of different functions of review:

- checks that the individual elements of the HSMS (for example incident investigation and risk assessment) are functioning as intended
- checks that the HSMS as a whole is functioning as intended (for example that data are being passed from one element to another)
- data analysis that takes into account the output information from all elements of the HSMS, such as

comparing loss data and conformity monitoring data to check that risk control measures are having the intended effect
- looking for ways in which what is being done now can be done more efficiently or effectively*
- checking that what is being done now is still appropriate (that is nothing is being done which is not required) and adequate (that is everything that is required is being done).

Each of these functions is dealt with in detail in Part 2.1 (Chapter 23), but one example is given below to illustrate how review goes beyond what can be done within the other elements of the HSMS. The example used is the analysis of combined data.

Analysis of combined data

As we pointed out in earlier chapters, much can be learned by analysing the output data from each of the main elements in the HSMS, and this is one of the possible activities which can be carried out during review. However, there are opportunities to learn more if loss and conformity monitoring data are combined for the purposes of analysis.

Put simply, loss data are a measure of the number and severity of losses or failures, and conformity data are a measure of how well risk control measures are being maintained. It would seem reasonable, therefore, that if conformity monitoring shows that risk control measures are being maintained at the desired level, then the number and severity of losses should remain stable or decrease. However, this does not necessarily occur. Table 11.1 shows four of the possible results from combined analysis of these two types of data, together with the actions required if these results are obtained.

The top left and bottom right cells in Table 11.1 are fairly straightforward, in that they are both what we would expect. When the results fall into one of the other two cells, rather more interpretation is required and this usually involves further analysis. However, details of how these analyses are carried out are beyond the scope of Part 1.1 and they are dealt with in Part 2.1 (Chapter 23). There are two final points on the data comparisons in Table 11.1:
- The analysis shown in the table should be carried out separately for each key loss, since each of these will have its own risk control measures.

Table 11.1
Combining loss data and conformity data

Loss data	Conformity data	
	Risk controls well maintained	Risk controls not well maintained
Losses stable or decreasing	Desired situation; no further action required, continue to monitor	Either you have been lucky or the risk controls are not relevant. Find out which is the case and take appropriate action. Continue to monitor
Losses increasing	The risk control measures are inadequate. Carry out additional risk assessments and implement improved control measures. Continue to monitor	To be expected. Maintain risk control measures properly and continue to monitor

* These terms are formally defined in quality management – see the review section in Chapter 23.

- The table shows a binary choice of 'well maintained' and 'not well maintained'. In practice, there will be a continuum between very poor and very good which complicates the analyses. However, the statistical techniques required for these more complex analyses are dealt with in Part 2.1 (Chapter 19).

Other points on review

Since review is a risk management function, every organisation should have its own written standards for review. These standards should cover such issues as the following:

- **Who will carry out the reviews and with what frequency.** HSG65 recommends relatively frequent reviews (for example monthly) for small units such as sections or teams, with decreasing frequency for progressively larger units. It recommends an annual review of the whole organisation. BS OHSAS 18001 requires reviews only by top management 'at planned intervals'.
- **The procedures to be used for review.** These will include such things as the data which will be included in the review (for example loss data and audit results), and the sorts of analyses to be carried out on these data.
- **The output from the review.** It is important that when the review identifies the need for action, responsibility for this action is clearly allocated and that there is a timescale or deadline for its completion.

Even at a basic level, review can be an extremely powerful management technique and more advanced techniques are available to extend its usefulness. Details of these more advanced uses of review are given in Part 2.1 (Chapter 23).

Summary

This chapter described those elements of the HSMS that do not deal with specific risks or go beyond the scope of individual managers.

12: Communication and training

Introduction

For the majority of health and safety professionals, their primary role is advisory. That is, they advise line managers on the sorts of risk their activities will create and how these risks can best be controlled. Obviously, this advisory role cannot be carried out effectively unless the health and safety professional has good communication skills. Similarly, it could be argued that a large part of management depends on effective communication. The first part of this final chapter in Part 1.1 is, therefore, devoted to communication skills.

While it is possible to think of training as an exercise in communication, training has certain features which make it convenient to deal with it as a separate subject. The second part of this chapter is, therefore, directed at training skills.

Communication skills

Introduction
Communication serves many purposes. It can be an end in itself (people like chatting), or it can be a means of passing on or obtaining information. However, the important function of communication, as far as risk management is concerned, is to change people's behaviour. Information is communicated to people in the hope that they will do something they do not do at present, or stop doing something they do at present, or do something differently. This is an important point, since it can be argued that a large amount of communication effort is wasted because of a failure to concentrate on behaviour. This section will, therefore, begin with a description of the steps required for effective communication.

Effective communication
It is possible to identify the following steps as being required if health and safety professionals are to maintain effective communication:
1. **Identify the target audience.** The target audience are those people whose behaviour is to be changed. There is a tendency to try to communicate everything to everyone 'just in case'. While this might look like diligence, in fact it is a cover for ineffective communication. If you consistently send everything to everyone, whether it is relevant to them or not, then eventually (and usually fairly quickly) everyone will ignore everything you send them. Memos put out by health and safety departments with the heading 'To all managers' are usually a sign that the health and safety professionals are not identifying their target audiences.
2. **Identify what the target audience has to do.** If it is not necessary for the target audience to do anything differently from the way they are doing it now, why is the communication necessary? For example, suppose that a new piece of health and safety legislation has been enacted and your job is to communicate the implications of this legislation to the managers in your organisation. The ineffective response is to make a summary of the legislation and send it to all managers since they will, typically, think 'so what?' and file the summary (at best) or bin it (at worst). The effective response is to translate the legislative requirements into the actions required and send each manager a tailored list of the things that he or she has to do in order to meet the requirements of the legislation. This may take more work and more time, but effective communication does require more work and more time.
3. **If necessary, identify the resources required to implement the actions.** Managers are not, in general, health and safety experts and they may need information or training before they can carry out the required

actions. They may also need more tangible resources, such as machine guards or PPE. Where possible, you should identify and suggest appropriate sources of supply.

4. **If possible, predict likely reactions and take pre-emptive action.** A typical, some might say inevitable, reaction is that the required actions are not possible because of lack of resources. While you as a health and safety professional may well not be in a position to provide these resources, you could already have raised the matter with relevant senior management, so that any resources issues have been addressed before you send out the communication. Remember: if you keep sending out what managers perceive as being impossible or unrealistic demands, the managers will sooner or later ignore anything you send.

5. **Decide on an appropriate medium for the communication.** Various media are available, including both written and oral means. The various ways of communicating will be dealt with later in this section.

6. **Communicate using the chosen medium or media.** This will require competence in the use of the chosen media – this is covered later in this section.

7. **Monitor the effects of the communication.** If you do not do this, you are unlikely to find out, in any systematic way, whether you are communicating effectively. However, the monitoring procedure need not be complex; it is only necessary to follow up a sample of communications and check whether they have produced the desired behavioural change. The follow-up procedure will require a range of interpersonal skills, including effective listening, and these will be described later in the section.

Having looked, in overview, at the main steps in effective communication, it is now appropriate to consider the main communication media.

Communication media

These days there is a wide range of media for communication, from the old-fashioned face-to-face discussion to the internet. However, for the purposes of this discussion, it is useful to identify two main dimensions on which they differ:

- **Written *versus* oral communication.** Written communications, for the purposes of this chapter, include memos, letters and reports, and it is also necessary to differentiate between different delivery methods such as email, fax and post. Oral communications include dialogues, meetings and presentations, and again it is necessary to differentiate between delivery methods such as face-to-face discussions, video conferencing, live presentations and video presentations.

- **Interactive and non-interactive communication.** This dimension concerns whether or not those communicating can interact in the course of the communication. In the old days, interactive communication was confined to face-to-face conversations, but now that we have the internet, written communication can also be interactive.

These two dimensions have different ranges of competences associated with them, and these are described in the next two subsections.

Written communications

Written communications include memos, letters, reports and articles in, for example, organisations' magazines. The skills required include the following:

- **The ability to follow rules of spelling, grammar, sentence construction and punctuation.**
- **The ability to write in plain English without being verbose, or using unnecessary jargon or inappropriate slang.**
- **The ability to write in ways which suit the target audience.** This will involve, for example, choosing suitable vocabulary, including a suitable proportion of active words, and average length of sentence to suit the intellectual level of the target audience.

- **The ability to subdivide large documents and tailor sections to particular target audiences.** For example, an abstract or executive summary could be prepared for senior managers, while the body of the text could be aimed at middle managers, and all detailed information and technical information could be put into appendices.
- **Laying out the material for ease of assimilation and impact.** Although these two matters are related, they serve different purposes. Layout for ease of assimilation includes techniques such as isolating key points on the page by using, for example, bullet points, a different typeface, or graphics or pictures which can replace text. Impact is more to do with making the document attractive to look at and so increasing its chance of being read. Techniques here include layout of text on the page, not having too much text on each page, and using graphics and photographs as 'decoration'.

The wide availability of word processors and laser and inkjet printers means that the preparation of suitable documents is technically easier than it has ever been. However, it also means that it is easier to produce poor documents with, for example, too many different fonts and overuse of readily available clip-art graphics.

Oral communications

Oral communications can be divided into two main categories:

- **Oral communication without interaction or with minimal interaction.** These include video presentations, as well as talks, presentations and lectures where there is limited scope for interaction with the audience. The key skills here are identifying the target audience, its requirements, and the sorts of presentation material, vocabulary and so on that will best meet its requirements. Presentation skills such as voice projection, body language and the use of visual aids will also be required.
- **Interactive communications.** In addition to the skills listed for oral communications without interaction, a range of interpersonal skills will be required here, including effective listening, dealing with disagreement and conflict, and using and interpreting body language. It is also necessary to be able to apply these competences in a range of circumstances, including dialogues, informal groups and formal meetings.

More information

There are numerous books on effective communication, including listening and presentation skills and the skills required to run effective meetings. The management section of bookshops and libraries should contain a variety of books and you can select ones that match your current skill and knowledge levels.

If you need more basic information on communication skills, there are texts available that meet the requirements of the General National Vocational Qualification (GNVQ) core skills, including communication. These texts are excellent introductions to the requirements of effective communication; examples are given in the references at the end of the book.[35,36]

Remember that good communication is a skill – you can learn and improve it only if you get feedback on your performance. You can provide your own feedback by monitoring the effectiveness of your communications but, in the early stages of learning any new skill, it is preferable to get more detailed feedback by attending a relevant training course.

Legal and other communication requirements

In the UK there are legal requirements for various types of communication and details are given in the *UK legislation – communication* box overleaf. Both HSG65 and BS OHSAS 18001 have requirements on communication; these were summarised in Chapter 4.

UK legislation – communication

There are two sets of regulations which deal with consultation:
- the Safety Representatives and Safety Committees Regulations 1977,[37] which apply only to members of trade unions
- the Health and Safety (Consultation with Employees) Regulations 1996,[38] which serve a similar but more restricted function for non-unionised workforces.

There is also a set of regulations dealing specifically with information provision, namely the Health and Safety (Information for Employees) Regulations 1989.[39]

In addition, most legislation dealing with specific sources of risk has a clause requiring that information be provided on the nature of the risk and the necessary risk control measures. These clauses are backed up by the more general requirement for information provision in regulation 10 of the Management of Health and Safety at Work Regulations 1999.[13]

Training

Introduction
In Chapter 3, we pointed out that, on ethical and financial grounds, line managers have health and safety responsibilities and, in many countries, they also have legal responsibilities for health and safety. However, not all managers have received formal management training and, even if they have, few management training courses contain sufficient coverage of health and safety issues. For these reasons, many managers rely on the organisation's health and safety professional for their training. Since this is the case, health and safety professionals need to be effective trainers. This section discusses the issues involved in effective health and safety training.

Basic principles
There is a story which circulates in the advertising world to the effect that 50 per cent of the money spent on advertising is wasted, in that it does not get people to buy the product being advertised. The problem is that no-one knows which bits make up the 50 per cent that is wasted.

The same could be said of health and safety training, in that a certain percentage of this training produces no measurable improvement in health and safety performance. However, unlike in advertising, it is possible to remedy the situation by employing a number of basic principles. This section deals with these basic principles and how they can be implemented.

The first basic principle is that the health and safety performance of an organisation will improve only if people behave differently. A good deal of time and effort is expended on 'health and safety awareness' training without any clear idea of what behavioural changes are supposed to result from this increased awareness. There is little point, for example, in making managers aware of their health and safety responsibilities without also telling them what they have to do in order to meet these responsibilities.

The second basic principle is that different people have to do different things if an organisation's health and safety performance is to improve. Senior managers will have to prepare and review health and safety policy, middle managers will have to plan, and monitor compliance with the plan, first-line managers will have to set and ensure compliance with health and safety standards, and operatives will have to work in ways which avoid risks to themselves or others.

The third basic principle is that different industries have different risks to manage. The requirements for risk control on a construction site will be different from those in an office, and the emergency plan for an offshore installation will be different from one for an onshore installation.

The fourth basic principle is that everyone already has the ability to do certain things and that existing abilities differ from person to person. These abilities may be inherent, or obtained by experience or training, but however they have been obtained, they influence the extent to which training will be useful for an individual.

The final basic principle is that everyone has limits on what they are able to do. These limits may be physical, intellectual or motivational, but no training can take people beyond their individual limits. It is true that very good training may take people right up to their limits, beyond what they initially believed they could do, but there are real limits and these have to be recognised.

If these basic principles are applied to a training programme, not only does it improve the programme, but it also helps to demystify training jargon. We will now translate the principles listed above into training terminology and then consider the implications for designing effective training programmes.

Behavioural objectives

The first principle, that people have to behave differently after training, results in the specification of a training course in terms of 'behavioural objectives'. The most straightforward way of doing this is to prepare a list similar to the one below, which, as an example, uses behavioural objectives for interview training. By the end of the course, course members will be able to:

- establish rapport
- use open and closed questions appropriately
- make an adequate record of an interview.

Courses based on behavioural objectives are often referred to as skills courses, or skills-based courses, to differentiate them from knowledge or awareness courses which have objectives in the following form. By the end of the course, course members will be able to describe:

- what is meant by rapport
- what is meant by open and closed questions
- what constitutes an adequate record of an interview.

Usually, a behavioural objective implies a knowledge-based objective. For example, it is difficult to train people to establish rapport without first explaining what is meant by rapport. The behavioural objectives for a skills-based course are, therefore, usually supplemented with what is known as 'underpinning knowledge'.

For the fragment of interview training discussed above, the objectives and underpinning knowledge might be set out as shown in Table 12.1.

	Behavioural objective	Underpinning knowledge
1	Establish rapport	What is meant by rapport Techniques for establishing rapport
2	Use open and closed questions appropriately	What is meant by open and closed questions, and their appropriate use
3	Make an adequate record of an interview	What constitutes an adequate record of an interview

Table 12.1 Behavioural objectives and underpinning knowledge

In summary, therefore, the first principle, that the health and safety performance of an organisation will only improve if people behave differently, requires training courses to be based on behavioural objectives and the underpinning knowledge necessary to meet these objectives.

Performance criteria

The second principle, that different people in an organisation have to do different things, can be translated into training terminology as some form of job or role analysis. Essentially, the organisation's requirements from a particular job are defined as 'performance criteria' – what the job holder must do or achieve in order to satisfy the organisation's requirements. For example, one criterion for managers might be 'carry out accident and incident investigations'. These performance criteria are then supplemented by a more or less detailed description of the standard of performance which is appropriate for the job or role being analysed.

This sort of job or role analysis is very time-consuming and this is one of the reasons why NVQs and SVQs are valuable. Where it is possible to identify a job or role which occurs widely, it is cost-effective to carry out a single detailed analysis and make the results generally available.

Range statements

The third basic principle, that different industries have different risks, is taken into account in training terms by using 'range statements' in conjunction with the performance criteria. For example, the monitoring performance criteria for managers of office-based staff and managers of staff working on a construction site might be very similar, but the range statements would be very different.* In essence, the range statements could specify the risk control measures which a particular individual should be capable of monitoring.

Training needs analysis

The fourth basic principle, that people already have skills and knowledge and that these will vary from person to person, is partly what leads to the requirement for 'training needs analysis'. That is, identifying what people are required to do, what they can already do, and what training they need to bridge the gap.

There is, however, a problem with training needs analysis as a concept because it implies that training is the only way of providing people with skills and knowledge. Since this is manifestly not the case, a more appropriate term could be found and this point will be dealt with later.

Pre-course assessment

The implication of the final principle set out above, that everyone has a limit on what they are able to do, is that not everyone can benefit from certain types of training, and that this limit should be identified before the training has taken place and the individual concerned has 'failed'. This has benefits for the organisation in that training resources are not wasted, but it also benefits the individual since 'failing' a training course can have detrimental effects on self-esteem and motivation.

Having looked at some basic principles and training terminology, it is now appropriate to move on to the concept of competence and consider vocational standards in more detail.

Competence and vocational standards

There are several possible definitions of competence but, for the purposes of this chapter, we will assume that people are competent in a defined task when they have demonstrated that they can carry out that task, to acceptable standards, in 'real life'. It follows from this definition that a judgment on competence can only be made if:

---

* Range statements have other uses but, for the present purpose, we need only consider their use as a way of dealing with different risks.

- the task is clearly defined
- the required underpinning knowledge is clearly defined
- the required standards are clearly defined
- the person has been observed carrying out the task at a place of work.

While it may be the case that a 'trained' person is also a 'competent' person, this is not necessarily so, and a person can be competent without having been formally trained. If it is not clear why this is so, try to think of the things at which you consider yourself competent. For how many of these did you receive formal training? Similarly, think of all the things for which you have received training and ask yourself whether you consider yourself competent in all of them.

When considering the difference between 'competent' and 'trained' in the work context, it is fairly obvious that competent people, not trained people, are needed, and this has been a major driving force behind the idea of vocational standards.

Essentially, vocational standards define tasks, underpinning knowledge and skills, and people can be awarded a qualification (NVQ or SVQ) only when they have performed a task adequately and been assessed as doing so in a work environment. In this system, training is used as one means of imparting skills and underpinning knowledge, but there is no assumption that because people have received training, they are competent.

Obviously, from the health and safety point of view, competent people are needed, and this has been emphasised at appropriate points in earlier chapters. For this reason, the principles underlying the vocational standards and associated qualifications should be applied to health and safety competence at work, and the final part of this section considers how this might be done.

Health and safety competences

In any organisation, it is possible to identify the following categories of required health and safety competences:

- **Organisational competences.** These are competences required by the organisation as a whole but which might be discharged by a single individual or a small group. Examples of such competences include notification of accidents, trend and epidemiological analysis of loss data, and providing specialised advice on health and safety matters. It is these organisational competences which are usually supplied by the health and safety professional. In the UK, there is a legal requirement to have, or have access to, these organisational health and safety competences under the Management of Health and Safety at Work Regulations 1999.[13]
- **Managerial competences.** These are the competences required by all managers, irrespective of what they are managing. Examples of managerial competences include accident investigation, risk assessment and monitoring.
- **Operational competences.** These are the competences required to work safely in a given environment, with specific risks. Everyone in an organisation will require operational competences since, although they may not be operators, they still have to work safely in their own environment. For this reason, it is preferable to subdivide operational competences according to the risks involved. For practical reasons, these subdivisions are usually expressed in terms of job categories, such as electricians, fork-lift truck drivers and office workers.

In order to determine whether an organisation has the required health and safety competences, it is necessary to carry out a competence needs analysis.* The overall procedure for a full competence needs analysis is:

---

* This type of analysis is more commonly referred to as 'training needs analysis'. However, the term 'competence needs analysis' is used here to reinforce the point made earlier that training is not the only method of delivering competence.

1. specify the required organisational competences
2. identify the individual or individuals supplying the competences on behalf of the organisation
3. assess whether the individual or individuals have the required competences
4. specify the required managerial competences
5. identify the individuals in the organisation with managerial responsibilities
6. assess whether the individuals with managerial responsibilities have the required competences
7. specify the required operational competences
8. identify which individuals require which operational competences
9. assess whether each individual has the required operational competences.

Where a required competence is found to be absent, there are various options for putting it in place:

- **Buy in the competence on an 'as required' basis.** This is appropriate for a number of health and safety competences, especially those which are highly specialised, or only required infrequently. Examples include auditing, advanced risk assessment and health and safety training competences. However, it is also appropriate for high-risk tasks which are 'one off' or infrequent and which require high levels of competence if they are to be carried out safely. Examples of these include window cleaning for high buildings, asbestos removal and maintenance of specialised equipment.
- **Recruitment.** This is similar to buying in competence but is more appropriate when there will be a continuing requirement for the competence being considered. Typical examples include the recruitment of health and safety professionals to provide organisational competences and of competent operators for high-risk tasks.
- **Placement.** This is similar to recruitment but involves moving an individual within the organisation rather than bringing someone in from outside the organisation.
- **Self-development.** Individuals are informed of the competences they require but do not have and are expected to obtain them themselves. This may be supplemented with various forms of assistance, including guidance on sources of help in achieving competence, and mentoring. Note that self-development is not an acceptable option for operational competences where the individual is currently exposed to risk and requires the competences to deal with this risk. This has been recognised in UK legislation and there are legal requirements to assist in the provision of competence. These requirements are summarised later in this chapter in the *UK legislation – information, instruction and training* box.
- **On-the-job training.** In this procedure, individuals are told or shown what is required, observed carrying out the task, and given feedback on their performance. In theory, this should be the most effective form of training, but it depends critically on the competence of the trainer. First, it works only if the trainer is competent in the relevant task, since, if he or she is not, poor practices and unsafe methods of work will be passed on. Second, the people giving the training need the competence to train, including the ability to provide adequate feedback on performance.
- **Off-the-job training.** This covers a very wide range of activities which can be categorised in order of general effectiveness as follows (note that each level assumes that the content of the previous levels is included):
  - People are told what is to be achieved but given no information on how it is to be achieved. Much 'awareness' training falls into this category.
  - People are told or shown what is to be done but given no opportunity to practise the skills involved. Much 'chalk and talk' training falls into this category.
  - People are given the opportunity to practise the required skills, with feedback on their performance, in more or less realistic exercises, for more or less extended periods of time. Much 'competence-based' or 'skills-based' training falls into this category.

It can be seen from this list of options that training is only one of the ways in which an organisation can achieve the competences it requires and that this is also true for individuals. Despite this, training is given great emphasis and it is interesting to consider why this is so.

Two extreme reasons can be identified – the charitable and the uncharitable.

- **The charitable view is that people believe that training provides competences** – so when competences have to be provided, training is the answer. This view is held by large numbers of people despite the evidence that a lot of training has done very little to improve competence.
- **The uncharitable view is that training is emphasised because it is the easy option.** It is time-consuming and, therefore, expensive to assess a person's competence adequately, since it involves observation of their performance at work. Dealing with people who have been assessed as not being adequately competent also creates a lot of management problems. It is much easier to send people on a training course, thus demonstrating the organisation's commitment to health and safety, and ignore whether or not this training has any effect on their performance at work. If the training course ends with an examination of underpinning knowledge, preferably in a tick-box format which everyone can 'pass', then it is also possible to claim that the training has an assessment procedure.

The truth, as usual, probably lies somewhere between these two extremes. Most organisations genuinely want their personnel to have the relevant health and safety competences but they have limited resources for providing them. They know, or are soon told, that providing properly assessed competences is a much more expensive option than providing assessed training.* They opt, therefore, for the cheaper option, recognising that it may not be the most effective.

It is difficult to criticise organisations that operate in this way since, realistically, there will always be limited resources. However, organisations can be criticised for not spending what resources they have in the most cost-effective ways. Typical examples of inadequate resource use include the following:

- **Failure to spend resources on adequate competence needs analysis, resulting in the training of people who do not need training, or providing people with inappropriate training.** The lack of adequate competence needs analysis can also result in key health and safety competences not being provided while extensive resources are devoted to 'awareness training'.
- **Failure to train senior managers in their key health and safety skills, including setting policy and review.** The arguments appear to be that, first, senior managers should have these competences already and that, second, senior managers do not have the time for health and safety training. Neither of these arguments is very strong. The general experience, and one of the main lessons from inquiries into disasters, is that senior managers do not have the required competences, and for senior managers to say that they are committed to health and safety but do not have the time for training does not create a good impression with anyone.
- **Failure to tailor courses to meet the requirements of the population to be trained.** The tendency is to design general courses which are then delivered to a large, heterogeneous population, what is sometimes referred to as 'sheep-dip' training. A better approach is to subdivide the population according to their competence requirements and provide more focused courses, each designed for a homogeneous population.
- **Failure to provide skills-based courses.** The tendency is to provide courses which are cheap to deliver and assess. It could be argued that resources would be better employed in doing less but more effective training.

---

* Assessing competence can be very much more expensive. Consider the following example: 20 people attend a training course and complete a 30-minute tick-box examination which requires 5 minutes per person to mark. Total assessment time: 1 hour and 40 minutes. Twenty people have one aspect of their job competence assessed after a training course. This will require a minimum of one hour's observation per person and, say, 15 minutes to write an assessment. Total assessment time: 25 hours. However, this does not take into account the assessor's travel time or travel and subsistence expenses.

Many of the current problems with health and safety training would be resolved if a clear distinction were made between 'competent' and 'trained'. Unfortunately, there are still senior managers (and politicians) making statements along the lines of 'What we need is a trained workforce', when what they mean is a 'competent workforce'. This is an important distinction.

- A **competent person** is someone who has demonstrated to the satisfaction of a competent assessor that he or she can carry out tasks to the required standard in the workplace.
- A **trained person** is someone who has been on a training course and, optionally, may have demonstrated to the satisfaction of a tutor, that he or she can tick the correct boxes on an examination paper, or write short essays, or complete a short practical exercise.

While there may be a correlation between trained and competent, it is not a perfect correlation; and the fact that someone is trained does not mean that he or she is competent. Conversely, an individual can be competent without ever having been trained. As a way of summarising this section on training, Figure 12.1 illustrates a systematic approach to training provision. Note that one box has a dotted outline to indicate that it is a decision point.

Figure 12.1
Systematic training model

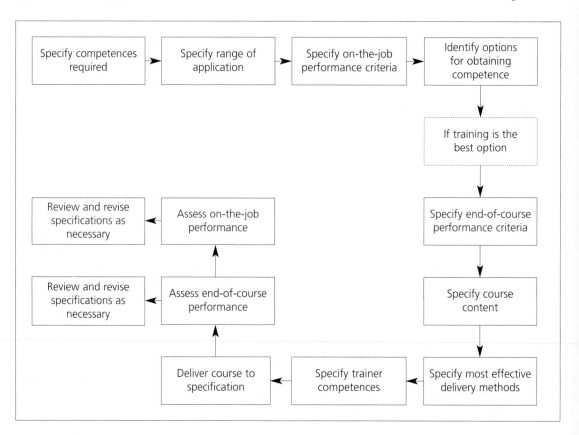

The HSE has produced guidance on health and safety competence issues,[41] which is a useful source of more detailed information on managing health and safety competences.

This guidance identifies the 'competency stages for the individual' as shown in Figure 12.2. It then goes on to describe how a competence management system (CMS) should be devised, implemented and maintained so that all of the competence stages can be dealt with appropriately.

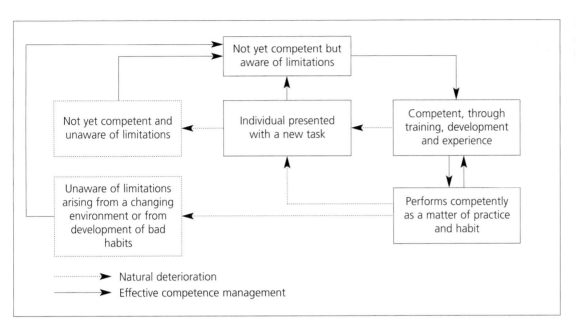

Figure 12.2
Competency stages
for the individual[41]

Legal and other requirements for training

The UK legal requirements for training are summarised below in the *UK legislation – information, instruction and training* box.

Both HSG65 and BS OHSAS 18001 have requirements for training, and these were summarised in Chapter 4.

UK legislation – information, instruction and training

Many pieces of UK health and safety legislation contain duties to provide information, instruction or training. However, historically there have been few legal requirements for competence in the UK, partly because of the burdens this places on organisations. One such area is the railway industry, where certain activities, including train driving and signal operation, which have obvious safety implications, have to be carried out by competent persons. However, it is noteworthy that regulation 5 of the Work at Height Regulations 2005[26] deal with competence as follows:

> Every employer shall ensure that no person engages in any activity, including organisation, planning and supervision, in relation to work at height or work equipment for use in such work unless he is competent to do so or, if being trained, is being supervised by a competent person.

This could be an indication that requirements for competence will be substituted for requirements for information, instruction and training.

Summary

This chapter dealt with two related issues: communication and training. There was also a discussion on competence and the difference between 'trained' and 'competent'.

Part 1.2:
Human factors –
introduction

13: Part 1.2 – Common themes and overview

Introduction

Part 1.2 covers the basic human factors material required for the NEBOSH Diploma and certain aspects of the NVQ/SVQ (see Appendices 1 and 2). It also addresses topics covered in some university health and safety syllabuses and, in particular, those topics required by IOSH for higher level qualification accreditation (see Appendix 3). This chapter provides a route map through the human factors material by outlining what is contained in this part.

However, certain aspects of human factors are common themes across all of the chapters and it is useful to deal with these aspects before dealing with specific human factors topics. We will therefore deal with the following topics before moving on to an overview of the rest of the part.

- **Specialisms.** Disciplines such as psychology, medicine and ergonomics are all involved in human factors work; it is worthwhile reviewing these specialisms and the approach they take to human factors.
- **Individual differences.** Individuals differ from each other in a wide variety of ways and no-one can have a real appreciation of human factors without knowing about the range and complexity of individual differences.
- **Individual variability.** People's attitudes, motivation and behaviours change over time and it is essential to appreciate the possible effects this variability can have.
- **The mutuality of effects.** Although this part is primarily concerned with the effect on individuals of their physical and social environments, it must also be remembered that individuals affect their environments as well as being affected by them.

Specialisms

The subject of human factors covers a wide range of issues and is studied from a variety of viewpoints. This section of the chapter will look at the main scientific disciplines which get involved in human factors, and the approaches they take to the study of mental and physical behaviour.

It is extremely difficult to identify an appropriate taxonomy for the various disciplines since they overlap in many areas. However, for the purposes of this chapter, the following main scientific disciplines are described:

- psychology
- medicine
- occupational hygiene
- sociology
- anthropology
- ergonomics
- physiology.

Within each discipline, only those aspects of the subject that are particularly relevant to health and safety are dealt with, together with any major areas of overlap.

Psychology

Psychology is broadly described as the study of physical and mental behaviour in humans and animals. However, this is such a wide field that it is subdivided in a variety of ways and some of the main subdivisions are described below.

One major division is between human psychology and animal psychology. Animal psychology overlaps with zoology but many aspects of human psychology are studied by using animals as experimental subjects. This use is analogous with the use of animals in medical research; they are used for procedures deemed unacceptable to carry out on humans.

Within human psychology, it is usual to subdivide the subject according to the topics being studied and individual psychologists tend to specialise in the sorts of subject listed below:

- **perception** – how people see, hear, taste, feel and smell
- **motivation** – why people do the things they do
- **emotion** – what emotions are and why particular emotions are evoked
- **learning** – this includes the learning of both skills and knowledge
- **intelligence and aptitudes** – intelligence is usually thought of as a general ability which can be applied to anything, while aptitudes are specific to a subject, such as music or mathematics
- **thinking and reasoning** – how people make a mental model of the world and how they use this model to help solve problems
- **instinctive behaviour** – there is a continuing debate about the extent to which humans behave instinctively, as animals do, rather than on the basis of what they have learned
- **personality** – there have been various attempts to classify people according to their personality; one of the more famous is the introvert/extrovert classification
- **group dynamics** – the study of how being a member of a group changes the behaviour of the people in that group
- **stress** – there are many definitions of stress but, in general, stress is used to describe a range of physical and mental symptoms arising from the effects of events in a person's life. Work-related stress is becoming an increasingly important issue since more and more people appear to be suffering from its effects.

In the academic study of psychology, it is usual to divide the discipline into the sorts of area listed above. However, when applying what has been learned from these studies, it is more useful to take from each area what is valuable for a particular application. This has resulted in practising psychologists being classified according to their area of application, rather than by a subject topic. Typical areas of application include:

- **educational psychologists**, who apply psychological principles to all aspects of education
- **clinical psychologists**, who apply psychology to the treatment of mental ill health. There are major overlaps in this area with abnormal psychologists, who study mental ill health *per se*, rather than its treatment. There is also an overlap with psychiatrists, who, in the UK, are medical doctors who are also qualified in clinical psychology
- **sports psychologists**, who apply psychological principles in order to improve the performance of sportsmen and sportswomen
- **occupational psychologists**, who apply psychological principles to all aspects of work including selection, training, motivation and health and safety
- **forensic psychologists**, who apply psychological principles to a range of activities associated with prisons and prisoners.

Much of the material in this part of the book, and the material in Part 2.2, is based on the work undertaken by occupational psychologists; this material illustrates the nature and range of occupational psychology.

Medicine

Medicine is concerned with disorders of the body and how they are treated. As such, medical practitioners have a number of roles in occupational health, including:

- treating those who are injured or made ill as a result of work activities
- identifying the causal relationships between work activities and ill health
- identifying and implementing methods by which early symptoms of work-related ill health can be detected so that exposure can be stopped before irreversible damage is done.

As with psychologists, medical practitioners are classified according to the field in which they work. In health and safety, the primary practitioners are occupational physicians and occupational health nurses; other specialist branches of medical practice include gynaecology and paediatrics.

As we mentioned above, psychology and medicine overlap in the area of psychiatry. Psychiatrists, like clinical psychologists, aim to remove or alleviate mental distress which may be resulting in pain, discomfort or conditions such as depression or anxiety. The treatments used can be divided into two groups:

- **Treatment by transformation.** These are treatments which involve some physical interaction with the patient. They include drug treatments, electroconvulsive therapy and brain surgery.
- **Treatment by communication.** These treatments include cognitive behaviour therapy (CBT), psychoanalysis and various forms of psychotherapy, including psychodrama. There is a wide range of such treatments reflecting the wide range of theories about what causes mental illness. Probably the best known variants are those based on the theories of Freud (Freudian psychoanalysis) and Jung (Jungian psychoanalysis). Freud's theory, which was the first and arguably the most influential, is dealt with in Part 2.2 (chapter 26).

The work of psychiatrists and clinical psychologists can be very similar but, in the UK, the administration of drugs is strictly controlled and their use by clinical psychologists is restricted.

Occupational hygiene

Occupational hygiene is concerned with the effects of the work environment on the health of individuals and how negative effects can be eliminated or reduced. Typical activities of occupational hygienists include:

- measuring the work environment, for example the nature and amounts of chemicals in the atmosphere, or the levels of noise in the workplace
- measuring the doses of harmful agents individuals receive – for example, how much of a chemical in the atmosphere gets into an individual's body, and by what routes, or the extent to which noise in the environment reaches an individual's ears
- measuring the effects of doses of harmful agents and noise
- selecting or designing risk control measures which will, for example, prevent chemicals getting into the atmosphere, or getting into an individual's body if the chemicals are in the atmosphere.

The work of occupational hygienists is primarily concerned with health effects, particularly long-term effects, and naturally complements the work of safety specialists, whose primary concerns are usually injuries and damage to plant and equipment.

Sociology

Psychology and sociology are both concerned with the study of people and their behaviour, but while psychology is primarily concerned with the individual, sociology is primarily concerned with groups of people. However, there is a major overlap between social psychology and sociology.

In more detail, sociology is concerned with the interactions between people, the groups and institutions which arise out of these interactions, what it is that holds these groups together and makes them break up, and, to complete the cycle, the effects that groups and institutions have on the behaviour and personality of the people who join them.

Sociology is also concerned with the basic nature of human societies, and what it is that creates them and makes them change.

Anthropology

The aim of anthropology is to describe humans and explain their behaviour in terms of biological and cultural characteristics, and particularly in terms of differences between cultures. Originally, anthropology was concerned with 'primitive' cultures but it now involves work in such areas as modern villages, cities and particular industries. This form of anthropology is, however, almost indistinguishable to the lay person from sociology and social psychology.

It is usual to divide anthropology into physical anthropology, which is primarily concerned with the people themselves, and cultural anthropology, which is primarily concerned with the people's customs, rites and social rituals.

The main area in which anthropology impinges on safety is the concept of 'safety culture', which is dealt with in Chapter 21.

Ergonomics

Ergonomics, also known as human factors or human factors engineering, is the collection of data on, and the formulation of principles of, human characteristics, capabilities and limitations in relation to machines, jobs and environments. This aspect of ergonomics is usually referred to as research ergonomics.

The practical application of research findings is concerned with designing machines, machine systems, work methods and environments to take into account human limitations. Thus, 'ergonomically designed' work methods should, in theory at least, be safer, healthier, more comfortable and more productive than equivalent systems which do not adopt ergonomically sound principles.

Ergonomics is often seen as the interaction between man, machine and the environment, and this is a useful lay definition of what ergonomics is about.

One particular branch of ergonomics, anthropometry, has particular importance in health and safety. Anthropometry is the measurement of particular human dimensions or capabilities. The data collected are used to determine, for example, appropriate dimensions for machinery guarding and the weight limits for loads which have to be moved in various ways. Ergonomics is dealt with in more detail in Chapter 30.

Physiology

Physiology is the study of how the human body works as a mechanical and chemical system. Physiologists study, for example, the effects of physical effort on the muscles in the body. These effects can be mechanical – resulting in, for example, torn ligaments – or chemical – resulting in the build-up of the chemicals which result in 'muscular fatigue'. The work of those who study physiology is put into practical use by a number of different categories of people, including physiotherapists.

Individual differences

Everyone is used to the idea that individuals are unique and that, with the exception of identical twins, they can be identified by their physical appearance. However, people may be less familiar with the idea that individuals also differ in their physiological and psychological make-up. Nevertheless, these physiological and psychological differences are just as varied as physical differences; just as individuals are physically unique, they are also physiologically and psychologically unique.

You could, no doubt, make an extensive list of the ways in which individuals differ from each other physically. You would probably start with such obvious measures as height, weight, colour of eyes, and colour and amount of hair, and then move on to more subtle features such as 'build', size of feet and hands, skin colour and birthmarks. You could also include such things as fingerprints, retinal patterns and tone of voice.

Everyone is used to these sorts of physical differences and the idea that they make each individual physically unique. However, few people appreciate that, because of the variation in physiological and psychological factors, individuals are also physiologically and psychologically unique. The sorts of physiological factors which make people different include:

- **general metabolism**, including such things as heart rate, blood pressure and the speed at which individuals digest their food. It also includes individuals' responses to chemicals – for example, some people suffer from hay fever while others do not
- **fitness**, including such things as strength and stamina. However, these have to be subdivided since, for example, different muscle groups are involved in different tasks and someone who can run for hours may not be able to lift heavy weights
- **general health**, including such things as existing medical conditions and predispositions to medical conditions
- **acuity of senses**, which includes extreme conditions such as blindness and deafness but also covers, in general, the fact that some people have more sensitive hearing than others or more sensitive smell receptors
- **speed of reaction** – some people have much faster reaction times than others. This is sometimes referred to as having 'fast reflexes', but not all of the actions involved are true reflexes; rather, many of them are learned activities.

When an individual's state with respect to all the dimensions listed above (and many more) is taken into account, it results in a 'physiological profile' that makes each individual unique.

There are just as many psychological factors on which individuals differ and, taken in combination, these make each individual psychologically unique. Since these psychological factors will be considered in detail later (in Chapter 15), it will be sufficient for now just to list some typical examples:

- intelligence
- beliefs
- attitudes
- motivation
- personality
- emotional stability.

The problem with psychological differences is that they cannot be 'seen' in the same way that physical differences can be seen and, because they are not brought to people's attention, people tend to work on the assumption that everyone has the same psychological profile as themselves. No-one would think of saying 'I have black hair, therefore everyone has black hair'. However, people frequently do say, imply or behave as though they believed 'I think x, therefore everyone thinks x' and 'I would react in a certain way, therefore everyone will react in that way'. There are also problems with physiological differences, but these will be discussed later in this part.

Everyone who wants to make sensible decisions on human factors issues must always remember the variety and extent of individual differences in physiology and psychology. If they do not, they will arrive at solutions to problems which would be appropriate for themselves but not necessarily appropriate for the people for whom the solution is intended.

Individual variability

So far, the emphasis has been on variations from individual to individual but, over a person's lifetime, there is a wide variation within the same person. This variability affects all three categories of individual differences:

- **Physical.** There are continuous changes in physical make-up over time. For example, hair and nails grow longer and, over a more extended period, people can put on or lose weight. However, the major long-term changes in physical appearance are due to ageing and readers can list these for themselves.
- **Physiological.** Again there are various short-term changes, many of them cyclical, such as hunger, thirst and the need for sleep. On a longer timescale, people go through periods of being more or less fit and more or less healthy. And always, and inevitably, there are the physiological effects of ageing. In the last category, deterioration in the performance of the senses is particularly noticeable with all senses becoming less acute ('sans eyes, sans ears...').
- **Psychological.** Although certain psychological differences, such as intelligence, are thought to be fairly stable over time, people's attitudes, beliefs and motivation do change. These changes may be brought on by what happens to them during their life with, for example, exposure to new experiences resulting in changes in attitude. However, there is also a natural progression with age; what motivates people at 15 does not usually motivate them at 50!

Although we have described three categories of individual differences, it is important to note that they all interact and that an individual's state at any particular time will depend on the current balance between all of the factors. Some of these interactions are obvious – for example, state of health and level of fatigue affect motivation. However, others are more subtle: for example, hormonal balance can affect emotional state and level of arousal can affect perceptual acuity.

The key point to remember is that even if an individual has been identified as being in a particular category, such as 'very intelligent', this does not mean that that person will behave in a very intelligent manner at all times and in all circumstances. Similarly, if a person is identified as 'highly motivated', this does not mean highly motivated at all times and for all tasks.

Mutuality of effects

The primary concern in this part is with the individual and the effects on individuals of factors in their environment. However, bear in mind throughout that, although the environment affects individuals, individuals also affect the environment. If a person works in a room, the physical environment will be different at the end of the period of work. The room may be tidier or less tidy, it may be warmer or colder, have more or less oxygen, and be more or less polluted with things like cigarette smoke.

Similarly, if a person joins a group of three or four people, the group is now different. The topic of conversation may change, the members of the group may change their positions and the language the group is using may change.

If a person joins an organisation, the organisation will be changed although, except in special circumstances, the change to the whole organisation is likely to be small. Similarly, people influence society, albeit only in minor ways for the majority of individuals.

One of the effects of any changes brought about by individuals is that, in general, the more they have been able to change the world around them, the more they will be committed to, and influenced by, that aspect of their surroundings. For example, rooms, groups and organisations on which individuals have had some influence will be more attractive to those individuals than strange rooms, groups or organisations. Since they are already more attractive, people are willing to spend time and effort maintaining them, which changes them further, which makes them even more attractive. Thus, the interaction between an individual and his or her environment, groups and organisations tends to be self-reinforcing.

This mutuality of effects can have implications for human reliability, and two examples are given below.

- **People are less likely to leave litter in places which are already clean and tidy, and are less likely to park illegally if no one else is already parked illegally.** This has obvious implications for such things as 'good housekeeping' schemes at work and the visible effects of compliance with safety procedures.
- **Once a person is committed to a group or organisation and wishes to remain part of it, his or her behaviour will change to ensure continued acceptance by the group or organisation.** Over time, people will have some influence on their groups and organisations but if they try to make changes too forcibly, they will be rejected and lose any influence they would have had. For example, if a work group is primarily interested in productivity, irrespective of safety considerations, anyone wishing to remain in that group will have to adopt this approach to working. Only when they have been accepted by the group, and have some influence on it, can they begin to persuade group members towards safer working practices.

In studying human factors, although relevant features of the individual and the effects of the physical and social environment are dealt with separately, it is important to remember that there is a continuous interaction between these factors.

Overview of Part 1.2

Part 1.2 is divided into two main subjects: first, how human beings function; and second, the various factors that influence how human beings function.

Since individual human beings are complex and there is a wide variation between them, the first subject is a substantial one and, for ease of presentation of the required material, it is dealt with in two chapters:

- **Chapter 14: The individual – sensory and perceptual processes.** This chapter describes how the human senses function. The five well-known senses – sight, hearing, taste, smell and touch – are dealt with first, followed by information on two lesser-known senses, balance and proprioception. These last two senses are the ones which enable humans to move around in a co-ordinated manner.
- **Chapter 15: The individual – psychology.** This chapter deals with the 'mental functioning' of human beings. It begins with a discussion of the main sources of mental differences between individuals and within the same individual from time to time, and moves on to describe two groups of related psychological topics:
  - fatigue, boredom and stress
  - attitudes, motivation, personality and intelligence.

So far as is possible, Chapters 14 and 15 deal with the functioning of human beings as though they were independent of the environment in which it takes place. However, this is done solely for ease of description, since human functions are manifestly influenced by the environment in which they take place. In **Chapter 16: The human factors environment**, there is a description of the sorts of factors which influence human functioning. To provide a structure for Chapter 16, it is set out in four main sections:

1. machines and work
2. groups
3. organisations
4. society.

Within each of these sections, the main features relevant to human functioning in the context of risk management are described and discussed.

14: The individual – sensory and perceptual processes

Introduction

This chapter deals with the human senses: how they function, how they can go wrong and how defects in them can be detected. We will consider each sense in turn, in sections on vision, hearing, smell and taste, touch, balance and proprioception. In the course of the chapter, we will also consider the implications of how the human senses work for hazard identification and human reliability.

Vision – the eye

The eyes are the human sense organs which respond to light, which can be thought of as particles of energy (known as photons) travelling through space at, not surprisingly, the speed of light. In the eye, the energy in these particles starts an electrochemical process which the brain eventually interprets as 'seeing'.

Figure 14.1 shows a cross-section through the eye and illustrates its main features. In the notes following the diagram, the function of each of these main features is described, together with their most common defects and methods for screening for these defects.

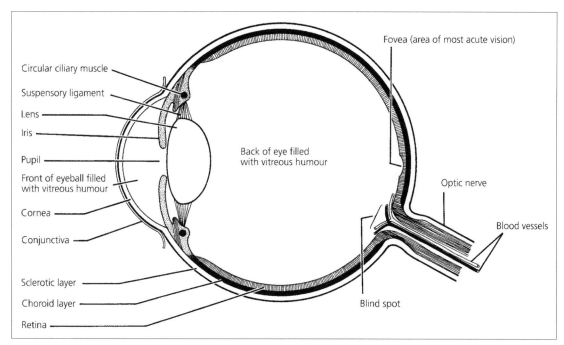

Figure 14.1
Cross-section of the human eye

The cornea and conjunctiva

The cornea is the transparent covering at the front of the eye and forms part of the tough outer layer of the eye known as the sclerotic layer. It is the area through which light enters the eye and its curvature makes it act like a lens, focusing the light which reaches it. The most common problem with the cornea is imperfect curvature, that is distortions which prevent proper focusing. These distortions, known as astigmatisms, can produce blurred or distorted vision. However, glasses or contact lenses can be worn to compensate for the distortions in the cornea and so restore normal vision.

Astigmatisms can be identified during routine eye tests by an ophthalmic optician and this is the usual screening method. These tests are also used to screen for a number of other eye defects; where eye tests are referred to in the rest of this chapter, this should be taken to mean tests by an ophthalmic optician or other suitably qualified individual.

There are many less common but more serious problems that can affect the cornea, which result in it becoming opaque (it will not let light through at all) or translucent (it lets light through but blurs it, like frosted glass). These conditions mainly occur in the elderly, although similar symptoms can be produced by certain diseases at any age. Similarly, certain radiations can damage the cornea and produce opacity or translucency. These problems can be detected during routine eye tests and, with modern surgery, a complete cure is usually possible.

The cornea is protected by a transparent layer of tissue, the conjunctiva, which can become inflamed, a condition known as conjunctivitis. This is caused by infection or an allergic reaction and can result in temporary impairment of vision.

The iris and pupil

In lay terms, the iris is the coloured part in the front of the eye and the pupil is the black circle in the middle. More technically, the iris is a circular muscle which can contract and relax to alter the size of the hole in its centre, the pupil. The size of the pupil determines how much light gets into the eye in the same way that the iris diaphragm on a camera determines how much light gets into the camera. When there is a lot of light, the iris contracts to reduce the amount of light getting into the eyes, and in low light levels, the iris relaxes, the pupil increases in size (dilates), and more light gets into the eye. Note that in extremely high levels of light, people may also 'screw up their eyes' as an additional way of reducing the amount of light reaching the eye.

This contraction and relaxation of the iris is a relatively slow process which explains why people are 'dazzled' when they move from low to high light levels; the iris simply cannot operate quickly enough. The opposite problem occurs when people move from high to low light levels; the iris cannot relax quickly enough and people are, effectively, blind until it adjusts. This has obvious safety implications for tasks which involve quickly alternating between high and low light levels, for example fork-lift truck drivers moving between a poorly lit warehouse and a sunlit yard.

The speed of adaptation of the iris decreases with age since the iris, like all muscles, deteriorates with age. There is little which can be done to remedy this but, fortunately, it rarely causes additional practical problems.

However, anything which causes the iris to set the pupil to a size inappropriate to the light levels can cause problems. Certain drugs can do this, for example atropine, which is used by opticians to dilate the pupil so that they can have a clearer view into the eye.* Other drugs, including some of those abused by drug addicts, have the opposite effect, making the pupil smaller. These possible effects of drugs should be taken into account in tasks which require effective adaptation of the eye for them to be performed safely.

Humours (aqueous and vitreous)

These humours fill the cavities of the eye. The aqueous humour, which fills the space between the cornea and the lens, is a watery liquid, but the vitreous humour, between the lens and the retina, is more like a clear jelly. It is thought that the humours are responsible for maintaining the shape of the eye by exerting a pressure, in the same way that the shape of a balloon is maintained by the pressure of the air inside it. This pressure maintenance

---

* Atropine occurs naturally in deadly nightshade, and women used to smear extracts on their eyes to make their pupils dilate and so, they believed, make themselves more beautiful. Hence the botanical Latin name for deadly nightshade is *Belladonna*, from the Italian for 'beautiful woman'.

system can fail so that the pressure increases to a level which damages the eye. This condition, called glaucoma, can, if untreated, result in blindness. It is checked for during eye tests using a special instrument which blows a puff of air onto the cornea and has electronics for translating the eye's response into a measure of pressure within the eye. If detected in its early stages, glaucoma can be controlled with suitable medication.

A more common problem with the humours is the occurrence of floaters and 'flying flies' (*muscae volitantes*). Flying flies are small pieces of dead tissue which have become dislodged at some time and continue to float or move around in the eye. You may have noticed that after a fit of violent coughing you 'see' a number of black specks. These are pieces of debris which normally lie at the bottom of the eye where they cannot be seen, but the coughing fit stirs them up so that they can be seen as they float slowly down again out of harm's way. These are the flying flies and they rarely cause problems.

Floaters, on the other hand, are permanently in vision, usually because they are lodged in the vitreous humour, and they can cause problems because they block the person's vision. This effect can cause either an overall degradation in visual performance, or the floaters can be confined to one part of the visual field, blocking sight in a particular area. Note that many people with floaters remain unaware of them because they rarely occur in the same place in each eye. These people are using the vision from one eye to fill the gaps in vision in the other eye.

The incidence of floaters and flying flies increases with age and they can be detected during eye tests even if the person involved has not noticed them. The presence of floaters may prevent people from carrying out specialised visual tasks, but they rarely have direct safety implications unless there are so many that they produce a significant degradation in general visual performance.

The lens

The lens is a disc of flexible, transparent tissue attached to a circular muscle (the circular ciliary muscle) by the suspensory ligament. The lens is responsible for the more detailed focusing of the light which has already been roughly focused by the cornea so that people are able to see objects clearly at a distance as well close to the eye. When people look at some distant object, the lens muscles contract, stretching the lens and making it thinner. In optical terms, this makes it weaker so that it focuses at a distance. When the focus has to change to an object close to the eye, the lens muscles relax, the lens becomes fatter and, in optical terms, more powerful. This changing of the shape of the lens to focus objects at different distances is known as accommodation.

Everyone has a limit to their accommodation, which in normal eyes extends from the far distance to about 75–100 mm from the eye. You can readily determine your own accommodation limits by focusing on a finger and gradually bringing it towards your eye. At some point the finger will appear blurred and no action, apart from the use of an artificial lens, will enable the finger to be brought into focus. This can be done for each eye separately and for both eyes together, but note that although the eyeballs rotate inwards when focusing on a close object, it is not this rotation which imposes the limit on close focusing, since the limit exists for each eye separately. Indeed, you may have noticed that you have different accommodation limits in each eye. This is quite usual and in binocular vision (using both eyes), the brain normally uses only the image from the 'better' eye.

Short- and long-sightedness are deviations from the normal range of accommodation. Short-sighted people are able to see objects closer than 75–100 mm but cannot focus on distant objects. Long-sighted people cannot focus properly on objects close to the eye and the limit may be up to several metres from the eye in extreme cases. As people get older, the lens muscles, like most other muscles, get weaker and lose some of their flexibility. Thus, for most people, even those born with normal vision, glasses will be required sooner or later.

Short- and long-sightedness are readily detected during an eye test and, except in very extreme cases, can be corrected with glasses or contact lenses. For some conditions, it is also now possible to correct accommodation problems with eye surgery.

Another problem with accommodation arises because it takes a certain amount of time for the eye to refocus. Try the following experiment. With one eye closed, put a hand as near to the open eye as you can while still keeping it in focus. Look at the hand for a few seconds and then look away at a distant object. Notice that it takes a fraction of a second for the distant object to come into focus. A similar amount of time is required to refocus on near objects after focusing on a distant object.

These accommodation delays mean that, effectively, people are without clear vision during these fractions of a second. If an individual is driving at speed along a motorway and he or she glances down at the speedometer, the times required for accommodation will have to be added to the time it takes to read the speedometer. With increasing age, the time required for accommodation gets longer, and may be two or three seconds in extreme cases. It could, therefore, take someone with very slow accommodation up to nine seconds to 'glance' at the speedometer (three seconds' accommodation, three seconds to register the speed reading, and three seconds' re-accommodation for distance). Such delays can obviously be very dangerous.

Another task which requires rapid accommodation occurs where control operators have a number of controls close to them on a desk which they operate according to the information displayed on large mimic displays on a wall some distance away. In emergencies, it is possible that the required response time is shorter than the time required for changes in accommodation.

However, speed of accommodation can be measured and should be included in eye tests for any tasks where too slow an accommodation rate may increase risk to an unacceptable level.

The lens, like the cornea, may become opaque or translucent with age or as a result of disease or damage. This condition is known as a cataract and it can usually be repaired with surgical techniques, such as the replacement of the affected lens with an artificial one.

The retina

The retina is the part of the eye responsible for converting light into electrochemical signals for transmission to the brain via the optic nerve. In structure and function, the retina is extremely complex but, for our purposes, it is necessary only to consider those aspects of the retina's function which have implications for safety.

Part of the reason for the complexity of the retina is that it contains two separate visual systems: one for colour vision and one for monochrome (black and white) vision. How these two systems operate is considered first. There is then a discussion of three other topics – central and peripheral vision, adaptation and after images – which concern both colour and monochrome vision.

Colour and monochrome vision

As we mentioned above, the retina contains two visual systems: one for colour vision and one for monochrome vision. Each of these systems consists of receptors, which convert light into electrochemical impulses, and nerves, which conduct these impulses to the brain. It is not necessary to go into details of how this works but it is helpful to think of the analogy of a closed-circuit television camera which contains 'receptors' which convert light into electrical impulses and 'nerves', in the form of wiring, which conduct the output of the receptors to a television monitor. Figure 14.2 shows, in a much simplified form, small sections of the retina containing colour and monochrome receptors and their associated nerves.

Several of the features in Figure 14.2 are identical and these will be dealt with first, before moving on to the features which make them different.

The retina (see Figure 14.1) occupies the back wall of the eye, the interior of which is covered with a matt black substance to eliminate extraneous light reflection (the choroid layer). Camera interiors are usually matt black for the same reason. However, the retina is liberally supplied with blood vessels and the 'red eye' often seen in flash photographs is caused by light reflecting from the red blood in these vessels.

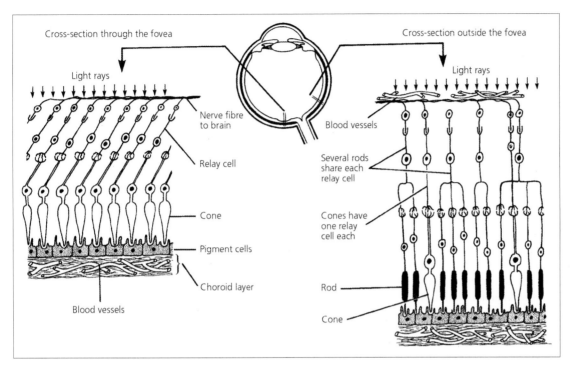

Figure 14.2
Small sections of
retina

As Figure 14.2 shows, the main differences between the colour and monochrome systems concern the shape of the receptors and the arrangement of the nerves. The cone-shaped receptors are responsible for colour vision and there are three types of cone each responding to a restricted range of colours. For example, some cones only respond to red light, while others only respond to green. The brain 'creates' the full range of colours by interpreting the proportion of cones which fire, in the same way that artists create different colours of paint by mixing different proportions of the primary colours. For example, if it is red light which is entering the eye, only the red cones will fire, whereas purple light will cause red cones to fire and other cones, too. The proportion of red cones firing, in relation to the other cones which fire, determines the shade of purple seen.

In order to do this, the brain has to 'know' which cones are firing and, for this reason, each receptor has its own nerve connecting it to the brain via a relay cell. This one-to-one arrangement has the further advantage in that it enables very detailed visual discrimination since the brain 'knows' not only which type of colour receptor is firing but also exactly where it is on the retina.

Obviously, the degree of visual discrimination will depend on how densely the cones are packed together. There is one area of the retina, the fovea, where they are particularly close together. The cones in the fovea are, therefore, responsible for detailed colour vision and it is this part of the retina, and this visual system, which is used most of the time. However, it requires relatively high levels of light to operate effectively and when light levels are low people have to rely on the monochrome visual system.

The receptors for the monochrome system are rod-shaped and not so densely packed as the cones. Rods also share relay cells. This means that they are not particularly good at detailed discrimination but can work in low light levels. The majority of the rods are located away from the fovea and it is, therefore, the rods which are mainly responsible for peripheral vision, that is 'seeing out of the corner of the eye'.

Although both the colour and monochrome systems work together most of the time, people are usually only aware of the colour system. However, in low levels of light, for example moonlight, only the monochrome system has sufficient light to operate.

Defects in the visual system

As might be expected with a complex system like vision, there are several ways in which it can go wrong. The main ones are described in this section.

Adaptation

The eye works best within certain light levels and it is difficult to see when there is either too little or too much light. As mentioned above, the iris, by controlling the size of the pupil, determines how much light gets into the eye. However, the retina itself can adapt to changes in light, although it does so much more slowly than the iris. You may have noticed that when you first go into a cinema auditorium you are practically blind, hence the need for ushers with torches. When you have been in there for 10 minutes, however, you can see quite well. The slow change in the retina which allows this is known as adaptation.

The practical importance of adaptation is recognised in many night-time activities. On the bridge of a ship at night, only dim lights are allowed because bright ones would destroy what is referred to as 'night vision'. This is simply reconfirming the fact that, within certain limits, the longer people stay in poor light, the better they are able to see in that light. Even brief exposure to brighter light destroys the effects of adaptation.

Adaptation does, of course, work in the opposite direction, too. If the eye has adapted to a low level of light and is then exposed to a brighter light, it will take some time for the retina to adapt to the higher level. However, adaptation in this direction is usually more rapid than adaptation from high to low levels of light.

Some occupations can involve people moving between bright and dim environments – for example, fork-lift truck drivers driving in and out of a poorly lit warehouse on a sunny day. In these circumstances, the lighting in the warehouse has to be at its best when the sun is shining. An apparently obvious solution to this type of problem would be the wearing of photochromic lenses, that is ones where the lenses darken in bright light (becoming, in effect, sunglasses) and clear again in dimmer light. In practice, however, the response times of these lenses is too slow, much slower than that of the iris and the retina.

After-images

Given that the retina has adapted to a certain level of light, anything which appears in the visual field which has a much higher light intensity will produce an 'after-image'. If you want to see how this works, look at a light bulb for a few seconds and then close your eyes. You should then see an after-image of the light bulb which, if the eyes are kept closed, will gradually fade, usually changing colour as it does so. Even when the eye is adapted to very high levels of illumination, as in bright sunlight, an after-image can be caused by such things as reflections from glass or water.

The practical importance of after-images is that while the after-image exists on the retina, that part of the retina cannot 'see' anything else. Unlike glare, which only lasts as long as the source of the glare lasts, a strong after-image, such as that produced by a photographic flash gun, can last for quite a long time. If you looked at the light bulb earlier, you will probably still be able to see the after-image if you close your eyes now. Care is needed, therefore, to ensure that after-images cannot be created during tasks where visual performance is important.

Defects in the monochrome visual system

In these days of widespread artificial lighting, it is rare that people make extensive use of their monochrome visual system, but how it operates can be observed on a moonlit night away from street lamps. Many people do not realise that they are only seeing in black and white because the brain adds colours on the basis of past experience. People 'see' the leaves on trees as green because they know that leaves are green. They only realise that they are seeing in black and white if they are asked to describe the colour of an unfamiliar object.

This has important safety implications for tasks that have to be carried out in low light levels, since people will attribute colours on the basis of experience and assume that these attributions are correct. Unless people are trained to use their monochrome vision in low light levels, serious errors can occur.

As we mentioned earlier, the monochrome visual system is also referred to as night vision and people vary in the extent to which their night vision system is effective. In extreme cases, the system does not work at all and people are said to be 'night blind' or to suffer from 'night blindness'. Some people cannot, or refuse to, drive at night because they do not think that they can see sufficiently well. Often they attribute this to the obvious fact that levels of illumination are poorer at night; less often do they identify a degree of night blindness as the cause of their difficulty.

This has obvious safety implications for jobs which have to be carried out in low light levels – personnel who have to carry out tasks in these circumstances should be screened to check that they have adequate night vision.

Defects in the colour visual system

Most people are familiar with the term 'colour blindness', but complete colour blindness is rare. In this condition, only the monochrome visual system works and the affected person sees no colour, even in bright illumination. Much more common are various forms of partial colour blindness where the affected person is unable to distinguish between certain colours. The most frequently occurring variety of partial colour blindness is red–green colour blindness, in which the colours green and red look similar to the affected person.

This is unfortunate given that many important safety indicators are red and green, such as stop and go for traffic lights and port and starboard for ship navigation. This, of course, is partly the reason why colours of warning lights should be backed up by positional indications, such as the red (stop) light always being at the top of the traffic light array in the UK.

For genetic reasons, colour blindness is almost completely confined to males, in the same way that haemophilia occurs only in males. However, where a task requires colour discrimination, those who have to carry it out should be screened, irrespective of their sex. Fortunately there is a very simple test for the various forms of colour blindness and a number of different versions are available. The most frequently used test is in the form of diagrams composed of differently coloured dots. The colours of the dots are chosen such that people with normal colour vision will see, for example, the number '5' when they look at a diagram while people with defective colour vision will see a different number, or no number at all, depending on the nature of their colour vision defect. One of the best known of this type of test is the Ishihara test for colour blindness and samples are freely available on the internet.

Finally, remember that colour vision is only possible in good lighting conditions. While it is easy, for example, to discriminate between different electrical wires in good light, this may be impossible when rewiring inside a machine casing where the light levels are low. Thus, the lighting levels for tasks requiring colour discrimination should always be taken into account in deciding whether or not the task can be done safely.

Defects in peripheral vision

As mentioned above, there is a small area of the retina – the fovea – which is particularly sensitive. Outside this foveal area, the retina is less sensitive, although there is still some vision, known as peripheral vision. You can clearly see the distinction between these two areas while you are reading. The word or phrase you are reading at the moment will be clear and sharp but the words on the rest of the page, although still visible, will not be readable. You can see these words well enough to read them only if you move your eyes so that the words become focused on the fovea.

The main function of peripheral vision is to enable people to remain aware of what is going on around them, even when they are focusing on a particular object. Peripheral vision is poor at discriminating detail but very good at picking up changes in the visual field, particularly sudden movements. People who have been driving for a number of years often have the experience of braking because they saw something move into the road at the side of the car. This braking is carried out before they know what the 'something' is; people react first to the movement identified by the peripheral vision before turning their head or eyes to focus the object on the fovea.

Some people have no peripheral vision at all because the retina, apart from the fovea, does not work. This condition is known as 'tunnel vision' because for the people who have this condition, it seems as though they are looking down a tunnel. It is possible to get some idea of the nature of tunnel vision by holding a cardboard tube (about 125 mm long) to one eye and closing the other eye. Tunnel vision makes walking and moving about difficult for the people affected.

Even in people with normal vision, the quality of peripheral vision is not uniform over the whole of the retina. In general, it tends to get worse with increasing distance from the fovea and the rate of fall-off varies from quadrant to quadrant of the visual field.

Peripheral vision can be important in certain monitoring tasks which involve a number of displays over a wide visual field. There are straightforward tests of the effectiveness of peripheral vision which can be used to screen people who have to carry out this sort of task.

Binocular vision

Remember that the visual system usually operates with two eyes. This binocular vision makes it easier to judge distances and, more importantly, increases our normal angle of vision. If you want to check this, fix your eyes on a point straight ahead and then, keeping your head still, look as far right and as far left as you can. Then close one eye, repeat the process, and notice how much vision is lost on the side with the closed eye.

People with monocular vision, or poor eyesight in one eye, can compensate for this in a variety of ways and can, for example, be as good at judging distances as a person with vision in both eyes. However, the loss of the usual wide angle of vision may pose insurmountable problems for carrying out certain tasks.

General checks on the eyes

The eyes are extremely robust instruments, considering how sensitive they are to light. Most of the things which interfere with their normal operation (glare, dust, poor light levels and focusing at a fixed distance for long periods, for example) rarely cause long-term damage. There has been extensive debate about the possibility that the last of these examples, in the form of using computer screens, does have long-term effects. However, it seems more likely that computer use, because it requires different visual performance from, for example, typing, simply uncovers existing problems. These have not been noticed because they did not interfere with the performance of previous tasks. In a similar way, children often have poor eyesight diagnosed only once they have started to go to school and had difficulty reading the blackboard or display. This weakness in the visual system remained undetected because it was not interfering with earlier visual tasks.

However, the eyes do deteriorate with age and, because it happens so slowly, few people realise that it is happening. As their eyesight deteriorates, they find ways to compensate and it is only when they have an eyesight test that they realise the extent of the deterioration. Visually demanding jobs, such as constant work in low light levels, can speed up this deterioration but, whatever the cause of the deterioration, the sooner it is diagnosed and remedial action taken, the better it will be for the person concerned. There are two reasons for this: first, wearing appropriate spectacles tends to slow down the rate of deterioration; and second, people operating with poor vision are likely to be a danger to themselves and others.

A routine eye test will detect most forms of deterioration but, as has been seen for people doing jobs with particular visual requirements, additional tests are required.

Finally, all of the foregoing discussion assumes that the environment in which people are working is reasonable, both in visual terms (adequate levels of light, absence of reflections and glare, and so on) and in physical terms (absence of dust particles, irritant vapours and foreign bodies which may injure the eyes). The eye is not 'designed' to work for long periods under poor conditions, nor is it designed to withstand repeated physical shock.

The brain and vision

When a person looks at an object, an image forms on the retina. This image can be thought of as a moving photograph of the outside world, but there is no need to consider the nature of this image in detail. With binocular vision, there will be two images of the same object, one from each eye. Look straight ahead and close first one eye and then the other and notice how the images from your left and right eyes are slightly different.

What a person actually sees is only in part a function of the images on the retina. Information about what is on the retina is passed, via the optic nerves of the eyes, to the brain and it is the brain images which are 'seen'. If this sounds odd, just think for a moment how it is that people 'see' only one image when they look at an object with both eyes. A moment ago we demonstrated to ourselves that we are working with two slightly different images, one from each eye. Why do people not see both of them?

Combining these slightly different images from the two eyes is only a minor part of the role the brain plays in vision. There is no vision at all if the parts of the brain responsible for vision are severely damaged, and this is true even if the eyes are undamaged. Perhaps an analogy will help to illustrate the brain's importance in vision. When using film in a camera, the film has to be processed before any photographs are available. Without a darkroom or a laboratory, the camera and its film would be useless. By analogy, the brain is the eyes' equivalent of the laboratory and without the appropriate parts of the brain, the eyes are useless.

This is of practical importance because the fact that an object appears as a retinal image does not guarantee that it will be part of the brain image. Readers must at some time have 'stared blankly' or been 'day dreaming'. In these circumstances, the retinal images are normal but there is no corresponding brain image since the brain is busy with something else. In order to distinguish between these two images, the image on the retina is referred to as the presented information while the image in the brain is referred to as the perceived information (see the discussion of the Hale and Hale model[28] in Chapter 9).

It is this distinction between presented and perceived information which explains how things can be missed or overlooked in visual searches. There are many cases when you know that an image of what you were looking for must have been received by the retina (that is, the necessary information was presented) but you did not perceive the information (that is, it did not reach the brain). How many times have you asked yourself: 'How did I miss that?'*

There are various possible reasons why something can be missed, even though there is an image on the retina, and some of these are considered next.

Lack of attention

Usually it is necessary to attend to an object before it can be perceived. The parts of the brain concerned with vision are linked to other parts of the brain dealing with the other senses and with other functions like attention, memory, motivation and fatigue. All of these functions will be discussed at appropriate points later in the book.

Habituation

Anything which does not change in the visual environment is eventually not perceived and this phenomenon is known as habituation. One practical application of this phenomenon concerns the use of safety posters. When a new poster is put up, people perceive it and may take note of its contents. If the same poster remains on display for any length of time, it will no longer be perceived. Thus, if poster campaigns are to be effective, the posters must be changed frequently, even if the different posters are conveying the same type of message.

---

* For people with good visual imagery, the converse is also true: they can create brain images which have no correspondence with anything currently in the retinal image.

Lack of competence

It is not usual to think in terms of people being competent in perception but this is what occurs in practice. The classic example of this is the shepherd and his or her sheep. If shepherds look at their own flocks of sheep, they can tell one sheep from another whereas to someone with no competence in this type of perception, all the sheep look the same. It is experience or learning or training which gives shepherds this competence to discriminate between the different animals. Note, however, that this is a function of the brain, not of the eye. For the shepherd and the non-shepherd, the images on the retina are identical; it is the changes in the brain as a result of learning which enable the shepherd to tell one sheep from another.

Let us take a less rural example. For most learner drivers, the amount of visual information which has to be taken in and processed while driving along a road is far too much for them to cope with. They attempt, therefore, to reduce this visual overload in three main ways.

- They create for themselves a mental tunnel vision so that they only look straight ahead. Anything happening in the periphery of their vision is simply not perceived, although its image will be on the retina.
- They drive slowly to reduce the rate at which the visual input is changing and give themselves more time to process the images they are receiving.
- They perceive only obvious things. For example, they will perceive what the car in front of them is doing but they will not look beyond that car to see what it might do next, or what cars further up the road are doing.

It is worth noting that, in all of these cases, what the learner driver perceives is not necessarily what the experienced driver would consider important to maintaining safe and economical driving.

As drivers gain experience, one of the most important things they learn is what it is not necessary to perceive. The images of these irrelevant items get onto the retina but nothing further is done about them. This reduces the amount of information which has to be processed and the processing capacity thus freed is used to widen the tunnel vision, increase the speed of driving and perceive less obvious, but more important, visual features.

This idea of learning to perceive can be a very fruitful one when applied to safety. A great many tasks depend for their safe performance on fine visual discriminations which have to be learned as part of the job training. There is even an expression, 'the untrained eye', which encapsulates this. There are people who collect and eat wild fungi. Some of these are highly poisonous, so it is a good idea to be able to tell one species from another before eating them. The 'trained eye' can do this, but it is not just a matter of knowing that one species is edible and another is not. To the untrained eye, the edible fungus and the poisonous one look identical. The only safe way to learn to discriminate between them is via extensive, supervised practice.

It is not generally appreciated that there are almost as many nerves running from the brain to the eye as there are running from the eye to the brain, and that the function of these nerves is to 'tell the eye what to see'. This can be illustrated with a number of simple optical illusions, three of which are given opposite, together with explanatory notes.

The important thing about these figures, and many others like them, is that the visual image on the retina does not change. The Necker cube is a single outline on the retina but the brain tells the eye how to see it. This is extremely important for human factors since it demonstrates that vision is not a passive process. People do not see what there is to see; they see what their brain tells them to see.

The other point about these illusions is that people have to know that they are illusions, otherwise they simply see one alternative, which is why instructions were given to accompany the illusions. This reinforces the point made earlier – that people have to learn to perceive.

Note that these points apply to all the other senses – hearing, touch and so on – but it is more difficult to illustrate for these senses in a text. However, at the moment you are reading this text and totally unaware of the clothes touching your body – except that now you are! This is a function of changes in attention but it illustrates, for another sense, how your brain tells your senses what to perceive.

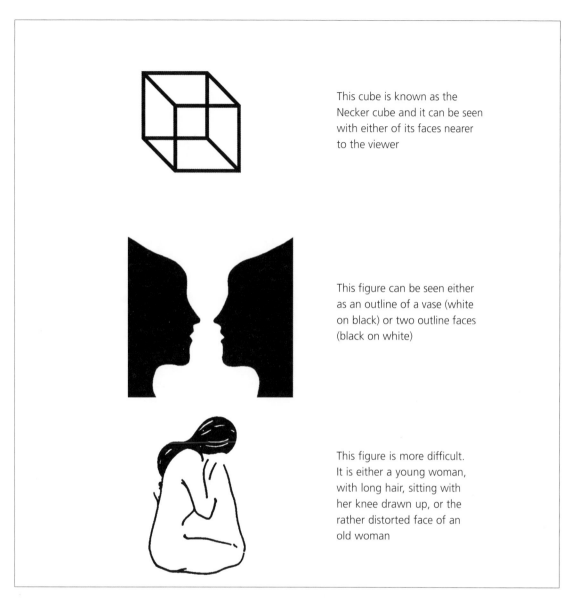

Figure 14.3
Reversible figures

This cube is known as the Necker cube and it can be seen with either of its faces nearer to the viewer

This figure can be seen either as an outline of a vase (white on black) or two outline faces (black on white)

This figure is more difficult. It is either a young woman, with long hair, sitting with her knee drawn up, or the rather distorted face of an old woman

Hearing – the ear

The ears are the human sense organs for sound, which can be thought of as pressure waves travelling through the air at, not surprisingly, the speed of sound. The usual analogy is with ripples on a pond, which are pressure waves travelling through water. In the ear, the energy in the pressure waves starts an electrochemical process which the brain interprets as 'hearing'.

Like the visual system, the system for hearing is in two parts: the ears and the areas of the brain responsible for what we actually hear.

Figure 14.4 shows a simplified cross-section through the human ear and the notes which follow the figure describe each of the ear's main features and the ways in which they can go wrong.

Figure 14.4
Cross-section
through the human
ear

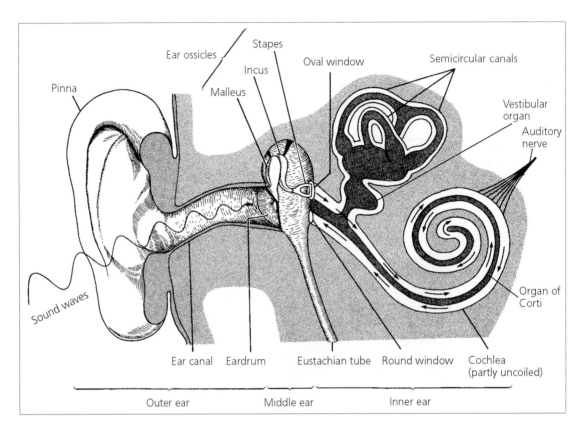

The pinna

The pinna is the visible part of the ear and its primary function is to focus the sound waves into the ear canal (also known as the external auditory meatus). The pinna is fairly rudimentary in humans but in animals that rely heavily on hearing for their safety, or to catch food, the pinna is much larger and can be rotated to pick up sounds coming from different directions. Better focusing of the sound waves can improve hearing and this was the principle behind the old ear trumpets which were used before electrical and electronic hearing aids were available.

Ear canal (external auditory meatus)

This is simply a physical channel for conducting sound into the ear. Blocking of this canal causes loss of hearing, the extent of loss depending on the extent of the blockage. This blocking can be done deliberately by, for example, using ear plugs to reduce the amount of sound entering the ear. However, it does also occur naturally owing to the build up of ear wax. This natural blocking usually occurs slowly and the people affected may not notice that their hearing is deteriorating. Blockages of the external auditory meatus caused by wax or other material can readily be detected by visual inspection using an instrument specifically designed for this task, and it is usually relatively straightforward to remove them.

The eardrum (tympanum)

This is a membrane which closes off the end of the external auditory meatus and prevents dust, bacteria and other foreign bodies getting into the middle ear. It operates rather like a drum skin (hence its name).* When

---

* A tympanum is the 'kettle drum' used in orchestras.

sound gets into the ear, it makes the tympanum vibrate and the nature of these vibrations depends on the type of sound. If this is an unfamiliar idea, think of how a microphone works. The effective part of a microphone is one or more discs of material which vibrate when exposed to sound. These vibrations are then translated into patterns of electrical impulses.

Problems occur when the tympanum is perforated or torn. This can be caused by very loud noises which cause the tympanum to vibrate beyond its natural limits or by physical damage such as might occur if something enters the ear. Just as a torn drum skin will not vibrate properly, a perforated tympanum will not vibrate properly and hearing loss results. The extent of the hearing loss depends on the degree of damage and, usually, this damage will not heal by itself.

Damage to the tympanum can be detected by visual inspection and there are surgical procedures for repairing the damage and, in extreme cases, inserting an artificial tympanum.

Ear ossicles – Hammer, anvil and stirrup (malleus, incus and stapes)

The function of these three small bones (ossicles) is to transmit the vibrations set up in the tympanum to the next part of the ear's mechanism, which is described below. Simple mechanical transmission of sound is not much used in modern artificial sound reproduction systems, but old-style record players used a needle to transmit the vibration created by moving over the grooves in the record to the electronics in the arm which held the needle. This transmission is purely mechanical, in the same way that the transmission of sound by the ear bones is purely mechanical.

In order to transmit sound effectively, the three bones have to be able to move freely, but there is a condition known as ossification which results in two or more of the bones fusing together. When this happens, the bones can no longer transmit vibrations and hearing is lost. However, there are surgical techniques for 'freeing up' the ossified bones or replacing them with artificial equivalents.

The cochlea

This is a snail-shaped, coiled structure made of bone (apart from two membranous windows) and it is hollow and filled with fluid. In Figure 14.4, it is shown partly uncoiled to illustrate its inner arrangement. The last of the three small bones in the middle ear, the stirrup, is in contact with one of the membranous windows, the oval window, at the base of the cochlea. The vibrations of the stirrup cause the oval window to vibrate and these vibrations are transmitted to the fluid in the cochlea. The second membranous window, the round window, acts as a 'damper' so that the vibrations in the cochlear fluid do not persist for too long.

Notice that, so far, the transmission of vibration from the tympanum to the fluid in the cochlea has been purely mechanical. As yet, the vibration has not been transformed into the electrochemical signals for onward transmission to the brain via the auditory nerve. This translation is carried out by the Organ of Corti, which serves the equivalent function for hearing as the retina does for vision.

Like the retina, the Organ of Corti is a complex structure but, for the purposes of this chapter, it is necessary to know only in broad outline how it works. Along its length there are microscopic hair cells. They look like small hairs, hence their name, but they are not made of the same material as external hair. Each of these hair cells responds to a sound of a different pitch so that when its particular sound reaches the Organ of Corti, it fires and sends a signal to the brain. Since different sounds are made up from different mixtures of pitches, each different sound sets up a unique pattern of firing in the hair cells. The brain then interprets these different patterns of hair cell firing as different sounds.

If this is a difficult concept to grasp, think of the Organ of Corti as a piano keyboard in reverse. If a particular key on a piano keyboard is struck, it produces a note of a particular pitch, and if a number of keys are struck simultaneously, it produces a pattern of sound. The Organ of Corti operates the other way round; a note of a particular pitch causes particular hair cells to fire, while a complex sound causes a whole range of hair cells in the Organ of Corti's 'keyboard' to fire.

An important feature of these hair cells is that, if they die, they are not replaced. Most of the cells in the body die and are replaced in a constant cycle, as can easily be seen on the skin where the top layer is constantly being replaced. Similar replacement of cells occurs after most damage to the body and the way that cuts, for example, heal over a period of time is familiar to everyone.

However, many of the cells making up the nervous system are not included in this repair and replacement cycle. For example, once a person has grown his or her full complement of brain cells, usually at about 14 or 15 years of age, no more brain cells are produced. If a brain cell dies, then whatever function it was carrying out is lost, unless that function is being duplicated elsewhere in the brain. Large numbers of a person's brain cells die each day but since there are so many of them to start with, and there is so much duplication of function among brain cells, this rarely causes any problems.

Unfortunately, there is no duplication of hair cell functions apart from there being an Organ of Corti in each ear. This means that if particular hair cells are damaged or die of old age, the person concerned can no longer hear those sounds to which the dead or damaged hair cells used to respond. If the damage is confined to one ear, however, the person will be able to hear the sounds in the undamaged ear.

Deafness arising from damage to the Organ of Corti is usually referred to as 'nerve deafness' and it is rarely reversible. Deafness arising from failures in the mechanical transmission of sound, for example, perforated eardrums and ossification, are usually curable.

There are certain diseases which kill hair cells in the ear, producing nerve deafness, but the most common causes of nerve deafness are exposure to loud sounds and ageing.

Exposure to loud sounds

Extremely loud sounds reaching the unprotected ear can kill hair cells during a single, split-second exposure. Noises of this level would also cause severe pain. However, deafness is more commonly caused by long-term exposure to loud sounds well below the threshold of pain. This is because each time a hair cell sends a signal to the brain, it requires a short period of time to recover before it can send a second signal. However, the mechanism used by the hair cells to transmit information to the brain about how loud a sound is involves the rate at which the hair cells fire. For example, a steady, quiet sound might involve the hair cells firing at a rate of once per second. Doubling the sound level will cause them to fire at twice per second and so on.

There is no long-term damage to the hair cells so long as the sound level does not require them to fire faster than it takes them to recover between firings. However, forcing the hair cells to keep firing by continuing the presence of a loud sound causes them to become fatigued so that they require more and more time to recover their normal state. Anyone who has spent time in a very noisy environment, either at work or in, for example, a disco, will probably have noticed that immediately after the exposure it is not possible to hear quiet sounds, but that hearing gradually recovers. This is known as temporary threshold shift since the threshold of hearing has been changed but the effect is not permanent.

After, say, eight hours in a noisy environment, it can take the hair cells more than 16 hours to recover fully. This means that if the eight-hour exposure is repeated the next day, it will start with the hair cells slightly fatigued so that the effect of this second eight hours of exposure will be more severe than the first. If exposures like this are repeated over months and years, the hair cells eventually fail to recover and die. This is the mechanism underlying noise-induced hearing loss which is a major risk in noisy environments. Since the effect is irreversible, it is known as permanent threshold shift.

Ageing

Most people's hearing deteriorates with age, the condition being known as presbyacusis. Typically, it is the hair cells that respond to higher pitches which die first so that, for example, children can hear the high-pitched noises made by bats while adults cannot. The rate at which hearing deteriorates with age varies widely from individual to individual, but for everyone it is accelerated by exposure to loud sounds.

The Eustachian tube

The Eustachian tube connects the inner ear to the back of the nasal passages but it is normally closed. Its primary function is to provide a mechanism for equalising pressure in the inner ear with the ambient pressure. When going up in a lift in a tall building, or taking off in an aeroplane, the ambient pressure decreases so that the pressure in the sealed inner ear is higher than the ambient pressure. In these cases, the Eustachian tube has to open to let out a small amount of air so that the pressures are equalised. The opposite is required when descending in a lift, landing in an aeroplane or diving under water. Here the ambient pressure is greater than the pressure in the inner ear and the Eustachian tube has to open to let a small amount of air into the inner ear.

Any differences between the inner ear and the ambient pressures cause the tympanum to be pushed out (air pressure in the inner ear too high) or pulled in (air pressure in the inner ear too low). This forced movement of the tympanum causes severe pain which can only be relieved by forcibly opening the Eustachian tube. This can be done by holding the nose and then trying to blow through it. If you do this, you will probably be able to feel your ears 'pop'. This process is forcing the tympana outwards with the pressure of your blowing. It is a good idea to learn how to make your ears pop, since it will save a lot of discomfort if you intend to fly, dive or indulge in any other activity which involves changes in atmospheric pressure.

There are certain conditions in which the Eustachian tube gets blocked, the most common being blockages by catarrh during a cold. If this happens, it may not be possible to equalise the inner ear pressure in the usual way and it is advisable to avoid large changes in atmospheric pressure at these times.

Deafness and hearing tests

From what has been described so far about the hearing system, it is possible to identify two broad categories of deafness:

- Deafness due to failure of some mechanical part of the hearing system such as wax in the ear or ossification of the bones in the middle ear. The causes of this type of deafness can usually be remedied.
- Deafness due to the death of hair cells in the Organ of Corti, either from exposure to excessive noise or through ageing. This type of deafness cannot be cured.

However, there is a third type of deafness which is caused by damage to those parts of the brain responsible for hearing and these will be described later in this section.

Irrespective of the type of deafness, the same tests can be used to detect hearing loss. The procedure is known as audiometry and it is used to measure both the sensitivity of hearing – that is, how quiet a sound a person can hear – and the range of pitches a person can hear. Sensitivity is measured separately for different pitches since, depending on the pattern of deaths among hair cells, people may be able to hear quiet noises at a low pitch but not at a high pitch, or *vice versa*.

The sooner hearing loss is detected, the sooner the person concerned can take action to slow the decline by, for example, avoiding environments with high noise levels. Ideally, all employees should have their hearing tested when they start working in an organisation and then periodically thereafter. Once every two or three years would be adequate; for older people or people working in a high noise environment, the time between tests should be reduced.

It should also be remembered that what people do outside their work will influence their hearing. The noise which can be produced by music amplification systems is so high that hearing loss can result from this cause alone. Long-term use of music players may also cause problems because of the noise levels they can generate close to the ears.

The brain and hearing

As we noted earlier, what is seen is only partly a function of what appears on the retina, the final image being determined by what happens in the brain. There is a similar situation with hearing, in that what is heard is only partly a function of the pattern of the signals from the Organ of Corti. The rest is supplied by the parts of the brain responsible for hearing.

The most dramatic example of this occurs among people who work in noisy environments. Despite the levels of background noise, these people can talk to each other because they learn to filter out the background noise and attend only to what is being said. If someone who is not used to their environment tries to understand what someone is saying in the same circumstances, they will not be able to do so. Note the implication of this: people can learn to communicate in environments which are so noisy that they are damaging their hearing!*

Learning to filter out background noise is only one of the ways that people 'learn to hear'. There is even the expression 'the trained ear', which is used to describe the difference between the way musicians and non-musicians hear music. The presented information for music is the pattern of signals from the Organ of Corti, and it is the same for everyone who has a fully functioning Organ of Corti. The perceived information, what is actually heard, is a function of this presented information and the brain's interpretation of this in the light of past experience and training.

A trained musician will be able to tell the difference between a violin and a viola even when both are playing the same note. For a non-musician, the presented information, the pattern of signals from the Organ of Corti, will be the same but they will not be able to tell the two instruments apart. This is not simply that the musicians know there are two instruments, a violin and a viola, and so can distinguish one from another; informing a non-musician that there are two instruments does not make them capable of telling which is which.

If musicians seem like too special a case, think of the noises heard from the radios of taxis and police cars. For most people who hear these broadcasts for the first time, they are incomprehensible. Nevertheless, the taxi driver or police driver understands them perfectly well since he or she has learned to discriminate what is being said. Anyone who had to spend time interpreting the broadcasts would also 'learn to hear'.

This learning to hear can be important since, for some activities, hazards are signalled by audible information. The example of this with which most people are familiar is the noise generated by a car engine when driving. The experienced driver knows, from the sound of the engine, when to change gear and does so 'without thinking'.

On a more technical level, experienced chain-saw operators use the noise made by the saw blade as it goes through the tree as an indication that they are, for example, encountering a knot which might cause the saw to 'kick back' and injure the operator.

In order, therefore, to maintain the effectiveness of auditory warnings of hazards, it is necessary to:
* teach those concerned how to discriminate between the sounds
* teach those concerned the significance of different sounds
* ensure the warning sounds are not masked by louder noises
* monitor the effectiveness of people's hearing with audiometric tests.

Links between vision and hearing

When you watch a film in the cinema or on television, it appears that each actor's voice comes from that actor's mouth. We all know that this is not the case, since all the voices come from the same speakers, usually

---

* People may also communicate in noisy environments by lip-reading, but this is not the same thing as learning to filter out background noises.

on each side of the cinema or television screen. The whole of cinema and television sound is based on an illusion. The brain is used to associating moving lips with the sound of voices coming from those lips and this learned association overrides the true sources of cinema and television sound.* Ventriloquists exploit the same illusion.

However, not all of the links between vision and hearing are illusions. There are important categories which can be described as the 'inner ear' and the 'inner eye'. If one person describes something to another person, the person hearing the description can use his or her inner eye to create a mental image of what is described. Similarly, an individual might look at a photograph of someone and be able to 'hear' that person's voice. These links between vision and hearing are associated with other links with memory and past experience, and they will be dealt with in more detail later.

Smell and taste

The sense of smell (the olfactory sense) and the sense of taste (the gustatory sense) are quite separate but they operate in a very similar way and, in humans, they normally operate simultaneously. For these reasons they are dealt with together.

It has already been seen that in the retina, light is converted into electrochemical signals for transmission to the brain, and that in the Organ of Corti, vibration is converted into signals, also transmitted to the brain. In the senses of smell and taste, it is the presence of certain molecules which is converted into electrochemical signals and this conversion is carried out by special receptors located in the taste buds on the tongue (for taste) and high in the nasal cavities (for smell). The receptors for smell are located in the olfactory bulb, whose location is shown in Figure 14.5.

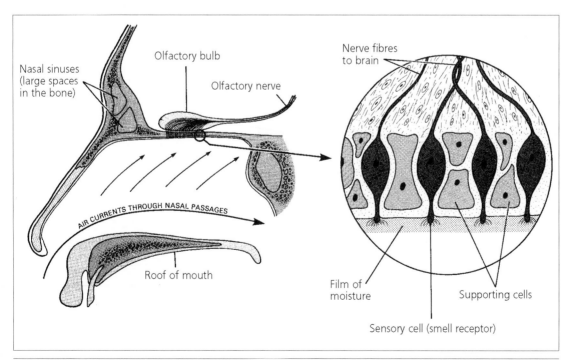

Figure 14.5
Section through the nasal cavities showing the location of the receptors for smell

Nasal sinuses (large spaces in the bone)

Olfactory bulb

Olfactory nerve

Nerve fibres to brain

AIR CURRENTS THROUGH NASAL PASSAGES

Roof of mouth

Film of moisture

Supporting cells

Sensory cell (smell receptor)

---

* If this seems odd, remember that there are no 'moving pictures'. Cinema film is a sequence of 'stills' shown so quickly that people think things are moving. Television pictures operate on a similar principle.

The receptors for smell can be thought of as specially shaped templates, each with a shape corresponding to the shape of a particular type of chemical group, rather like a lock and key. The chemical group (the key) will cause only a certain type of receptor (the lock) to fire. The receptor is inactive until a particular molecule or part of a molecule, of the appropriate shape, comes into contact with the receptor. When this happens, the receptor sends a signal to the brain via the olfactory nerve and the brain interprets this as a particular smell.

In fact, there are not different receptors for every different type of molecule. Instead, particular receptors respond to molecules that contain groupings of atoms which have a shape that fits the receptor. For example, many ketones (containing the keto group of atoms) have a similar smell and many esters have a 'fruity' smell. However, while such rules of thumb are useful, they cannot be used to predict a smell with any accuracy. Not all esters smell fruity. One reason is that other parts of the molecule may be relatively bulky and get in the way of the relevant atomic group, preventing it from fitting into the 'lock' in the receptor.

Note that any substance that gives off some of its molecules so that they can get into a person's nose has a 'smell' in one sense and substances which readily give off molecules are said to be volatile. However, whether a person can respond to this smell is a function of the receptors so that, for example, some animal species can smell water but very few humans can. Similarly, because of individual differences in the sense of smell, some of which are genetically determined, some substances will smell more strongly for some people than for others. An extreme case of this is phenylthiocarbamide (PTC), which to about 75 per cent of the population tastes extremely bitter but has no taste at all, or is very slightly bitter, for the other 25 per cent. In 2003 it was discovered that this difference is due to variations in a single gene.

A certain minimum number of molecules of the same type are required to trigger a perception of smell. This means that circumstances can arise where the number of molecules required to trigger perception is larger than the number of molecules which will cause harm. For these substances, as with substances which humans cannot smell at all, special measures have to be taken to protect people who may be exposed to the substances.

In everyday life, the smells people encounter are usually complex, being a mixture of many volatile substances. These mixtures trigger a number of different receptors and, with time, we learn to associate particular patterns of firings with particular sources of smells. For example, freshly-baked bread and percolating coffee both have their own complex patterns.

Figure 14.6

The upper surface of the tongue showing the location of the receptors for taste

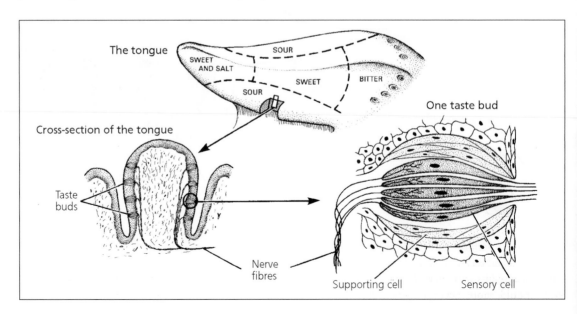

The sense of taste operates in a similar way to the sense of smell, but the taste receptors, which are located in the taste buds on the tongue (see Figure 14.6), have a very limited range of responses. When stimulated, particular taste buds signal sweet, sour, bitter or salt – and a 'sweet' taste bud can only signal sweet.

People vary in the range and number of their taste and smell receptors and this is genetically determined. This means that some people are born with a poor sense of smell while others are born with a very high level of olfactory sensitivity. Irrespective of their genetic starting point, however, taste and smell receptors die as people get older and are not replaced. This explains the observation that as people get older, they complain that food does not taste as good as it used to taste. The food is the same but older people have a lower ability to smell and taste it.

The effectiveness of taste and smell can also be temporarily reduced by a heavy cold, or by excessive exposure to a particular taste or smell resulting in over-stimulation of the relevant receptors to a point at which they no longer fire.

There are hazards which are detectable by smell and the absence of a sense of smell, or anything which masks the sense of smell, such as a heavy cold or heavy smoking, will prevent the identification of these hazards.

In the case of domestic gas supplies in the UK, an offensive smell is deliberately added to the gas as a warning. Domestic gas manufactured from coal did smell unpleasant and people became used to associating escaping gas and the consequent dangers with this offensive smell. So-called natural gas, from the North Sea oil fields, for example, does not smell, so traces of chemicals (called mercaptans) are added so that escapes can still be detected by smell.

Another problem with detecting hazards by smell is that people can very quickly get used to a particular smell. People who live surrounded by cats, for example, soon stop noticing the less attractive odours created by their feline friends. In industry, people exposed to, for example, hydrogen sulphide for a period will fairly quickly be unable to smell it.

In human reliability terms, this means that if people are constantly exposed to a smell, even if that smell indicates the presence of a hazardous gas or vapour, they will soon cease to perceive that smell, and often this is within a matter of minutes.

The position is further confused because humans find some smells pleasant and others unpleasant. Unfortunately, there is no reliable relationship between the unpleasantness of a smell and the hazards of the chemical producing it. This is illustrated by example in Table 14.1.

Unpleasant smell	Toxic	Hydrogen sulphide, pyridine
Unpleasant smell	Safe	Surgical spirit, methylated spirit
Pleasant smell	Toxic	Benzene, aniline
Pleasant smell	Safe	Ethyl acetate and most lower esters

Table 14.1
Pleasantness of
smell and toxicity

Note that the term 'safe' as used in the table is relative, depending as it does on the degree of exposure, and the pleasantness or unpleasantness of a smell is subjective; what one person finds pleasant, another might find unpleasant.

Determining the pleasantness or otherwise of a smell or taste is a function of the smell and taste areas in the brain. As with the other senses, the perception of taste and smell takes place in brain areas linked to the taste and smell receptors and also to other regions of the brain.

This explains why, as with vision and hearing, the senses of smell and taste can be trained. A wine taster, who should more accurately be called a wine smeller, has been trained in this way. The presented information for the wine taster, the wine in his or her mouth, is the same as for the non-wine taster but what the wine taster has learned is different and, hence, the perceived information for the two is completely different.

Some hazards are detectable by a trained sense of smell. For example, some machine operators can smell when a machine is running hot and take action before any harm is done. The relevant smell is usually generated by the partly volatile oils used for lubrication which become more volatile when their temperature increases.

Compared with vision and hearing, the range of tasks where smell and taste are relevant is limited. However, in some cases, unless there are suitable detection instruments in use, the sense of smell may give the first warning (for example, fire) and in other cases it may give the only warning (for example, gas leaks).

There are tests available which can be used to check the effectiveness of an individual's sense of smell and these should be used to screen people where hazard detection by smell is a critical element of a task.

There are two other limitations of the sense of smell which it is necessary to consider before moving on to the sense of touch.

- **All of our senses can only detect changes that happen faster than a certain rate.** With vision and hearing, their sensitivity is such that the speed of change which they cannot detect is too slow to be of much practical significance, but this is not the case with the sense of smell. Imagine that there is a container of some toxic and volatile chemical in the same room as people. If there is a small leak in this container, or the lid is not properly closed, the chemical will evaporate into the room and could eventually build up to harmful levels. Even if this chemical has a characteristic smell, the rate at which the smell increases, as its concentration in the air increases, may be too slow for the people in the room to detect. However, someone walking into that room from outside would detect the presence of the chemical immediately. It is this insensitivity to slow change which explains the fact that people often notice domestic gas leaks only when they return home. If the leak is a small one, they will not notice the slow build-up of mercaptans in the air. It also explains why visitors can often smell gas when the resident cannot. The phenomenon of being unable to detect something that is changing too slowly is known as the 'just noticeable difference' or JND.

- **It is necessary for there to be a minimum concentration of a chemical in the atmosphere before humans can detect its presence by smell.** The amount necessary varies from chemical to chemical, but the amount necessary for detection may be greater than the amount which causes harm to humans. In these circumstances, suitable measuring instruments and audible or visible alarms are needed to protect people who may be exposed.

Touch

The sensors for the sense of touch are largely confined to the skin of the outer surface of the body, the inside of the nose and the mouth, and the tongue. There are touch receptors inside the body but these are of a rather special kind and they will not be dealt with here.

Figure 14.7 shows a simplified cross-section of a small piece of skin with the various types of receptor labelled. Like the taste buds, which respond only to one type of taste, the receptors in the skin respond to only one type of stimulation. However, there are various sorts of receptor, the most common of which respond to pressure, heat, cold, touch and the movement of hairs. It is various combinations of firings of these different receptors which the brain translates into how a surface 'feels'. For example, a piece of marble or glass would usually cause cold receptors to fire and, as the finger tips are moved over it, all the touch receptors in the skin in contact with the surface fire at the same time because the surface is smooth. A rough surface would cause only some touch receptors to fire, that is those in contact with the raised part of the surface.

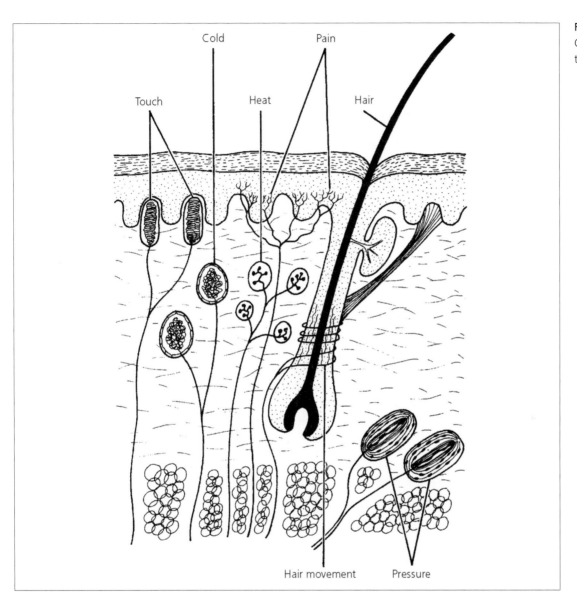

Figure 14.7
Cross-section through skin

The sensitivity of the skin to touch varies markedly from one part of the body to another. The most sensitive areas are the tips of the fingers and the tip of the tongue with very low sensitivity in, for example, the skin on the back.

There is a simple experiment which can be done to verify this. Get a pair of dividers and give them to a friend. Ask him or her to set the dividers with their points at different distances apart and then touch your back with them, either with one point, or two points simultaneously. You then have to say whether you have been touched by one or two points. A typical finding is that the divider points have to be around 75 mm apart before the difference between being touched by one or two points can be reliably detected. Now close your eyes and get your friend to repeat the experiment on the tip of your finger. It is generally the case that it is not possible to get the points of the divider close enough together for a person to confuse being touched by one or two points.

If the friend was heavy-handed at any point during the previous experiment, you will also have encountered the other main type of receptor in the skin, the pain receptors. If these receptors are stimulated, the brain interprets this as pain. The pain receptors respond to a wide variety of different stimuli, including excessive heat, cold, pressure or damage to the skin itself. There are two special points to note about the pain receptors and how they work.

- **The nerves which link the pain receptors to the brain are special ones which operate much faster than the normal touch receptors.** If, for example, a person puts a hand on a hot surface, the usual reaction is to withdraw it again very quickly and this is done before it is 'known' that the surface is hot. This is because the signals for pain have got to the brain and the brain has sent instructions to the hand before the signals for 'hot' have got to the brain. It is only after the hand has been removed from the hot surface that the person thinks about how hot the surface was.
- **The pain receptors can be fired, and continue to be fired, by chemicals in the skin or other parts of the body.** These chemicals are released by damaged tissue so that, for example, the pain from a burn continues for a long period after the burn was sustained. Pain of this sort serves a useful function by reminding individuals that a particular part of the body is damaged and needs care and attention. However, there are many aspects of pain which are not understood and which appear to serve no useful function, for example, phantom pain in limbs which have been amputated.

As with other senses, the sense of touch can be trained and this is important in a number of occupations including assessing the quality of various sorts of cloth and grading vegetable oils. Many photographers when printing photographs in a darkroom discriminate between the two main chemical processes (developing and fixing) by the difference in feel between the two solutions used. This discrimination is not simply a matter of knowing that the solution which feels slightly 'soapy' is the developer; someone printing photographs for the first time cannot feel any difference between the two solutions.*

There are certain jobs where the sense of touch is used to avoid hazards. For example, it is possible to monitor the temperature of engines by touching their casings and experienced engine operators can discriminate small changes in temperature which indicate that something may be going wrong.

Another example, again involving engine operators, is when they rub the engine oil between their finger and thumb. The amount of 'grittiness' detected when doing this is again an indicator of possible problems.

Like the sense of smell, the sense of touch habituates quickly. The most common habituation to touch is to clothing. Once people have put on their clothes, they are rarely aware of them touching their body, although they can make themselves aware of them if they wish. How this conscious change of awareness operates will be dealt with in Part 2.2 (Chapter 28).

Other sorts of touch habituation also occur. It is a frequent observation that people who do a lot of cooking habituate to hot surfaces so that they can, for example, handle hot plates which would be painful for someone not so habituated. This may be, in part, because they have literally developed a thick skin but there is also an element of purely psychological habituation.

In terms of human reliability of the sense of touch, the sorts of issue raised for the other senses also apply to touch. If discrimination of hazards by touch is required for safe working, then appropriate training must be given and steps must be taken to avoid habituation. In addition, it must be ensured that there is no masking of the required sense of touch by, for example, inappropriate gloves.

---

* The photographer will, of course, be wearing appropriate rubber gloves, but the soapiness can still be detected even when the hands are properly protected.

Balance

Even with their eyes shut, people can stand up straight, know whether they are standing up or lying down, and know whether they are stationary or moving. The information for making these assessments is provided by the organs of balance and the areas of the brain responsible for interpreting the information from these organs.

Unlike the other senses, the organs of balance do not take in information from the outside world, but generate their own information. There are two types of balance organs. The first, the vestibular organs, enable the position of the head to be determined – for example, whether it is upright, tilted to one side, or tilted forward or back. The second, the semi-circular canals, which are immediately above the vestibular organs, enable the speed and direction of movement of the head to be determined. There is one of each type of organ located in each ear, near the cochlea (see Figure 14.4). How each type of organ operates will now be described.

Position of the head

Figure 14.8 shows, in a simplified form, the mechanism which signals the position of the head to the brain. In essence, it is a small bone cup (macula) lined with sensitive hairs and filled with jelly, in which there are small granules of chalk (otoliths). When the head is at rest in an upright position, these granules stimulate the hairs in a characteristic pattern and the brain interprets this pattern of stimulation as the head being level. When the head is tilted, the pattern of stimulation changes and the brain interprets each pattern as a particular position of the head. The mechanism is illustrated in Figure 14.9. For simplicity, Figure 14.9 shows cross-sections through the macula but in real life it is three-dimensional so that tilts of the head in any direction can be assessed.

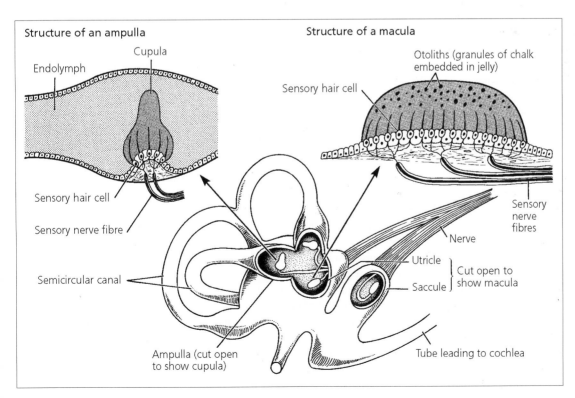

Figure 14.8
The organs of balance

Figure 14.9
How a macula
works

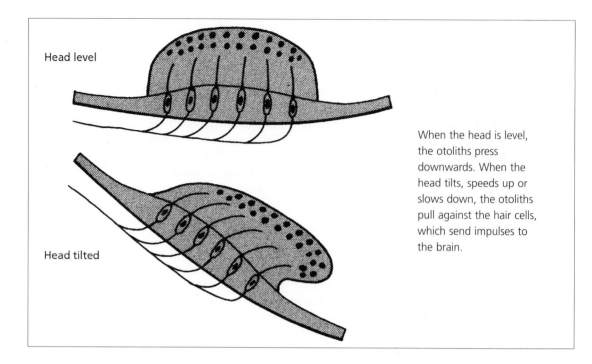

Head level

Head tilted

When the head is level,
the otoliths press
downwards. When the
head tilts, speeds up or
slows down, the otoliths
pull against the hair cells,
which send impulses to
the brain.

Under most circumstances, the information on the position of the head is supplemented by visual information. That is, the position of known horizontals (floors and ceilings) and verticals (walls) is providing information which supports the information from the vestibular organs. People who have lost this particular aspect of their sense of balance can use these visual cues to compensate for the loss. However, even when this loss has occurred and there are no visual cues because, for example, it is dark, people can still maintain some balance by using information from the proprioceptive senses, which will be dealt with later.

Movement of the head

The second aspect of the sense of balance concerns the direction and rate of movement of the head. This information is generated by organs known as the semi-circular canals (see Figure 14.4) and there is one of these organs in each ear. The semi-circular canals provide information on head movement, either if the whole body, including the head, is moving as is the case when people are walking or travelling in a car, or if the head alone is moving, as is the case when people nod or shake their head.

The semi-circular canals operate in a similar manner to the balance organ just described, in that the inside of each semi-circular canal contains sensory hair cells (contained in groups in a cupula) and the canals themselves are filled with fluid (endolymph). When the head is moved, the fluid inside the canals also moves but, due to inertia, the fluid moves more slowly than the walls of the canal. This causes a drag effect on the cupula and causes the hair cells to fire and the rate of firing depends on the strength of the drag effect which, in turn, depends on the rate of acceleration of the head (see Figure 14.10).

If it is not clear how this works, think instead of a glass of water. If you turn a glass of water very slowly, all the water in the glass will also turn. However, if you turn it very quickly, the water will not move as quickly as the glass. It is possible to experiment with this by floating a small piece of paper on the surface of the water and turning the glass through various distances at a variety of accelerations.

Now imagine that there were flexible hairs on the inside of the glass rather than the piece of paper floating on the water. If the glass was turned very slowly, the water would turn too, and the hairs would not bend. If

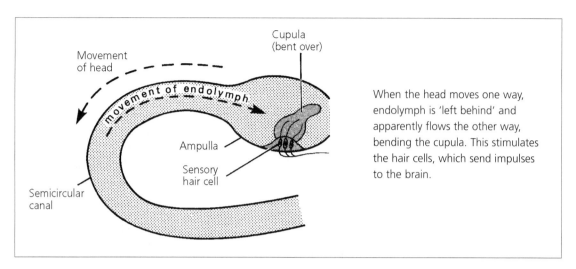

Figure 14.10
How semi-circular canals work

When the head moves one way, endolymph is 'left behind' and apparently flows the other way, bending the cupula. This stimulates the hair cells, which send impulses to the brain.

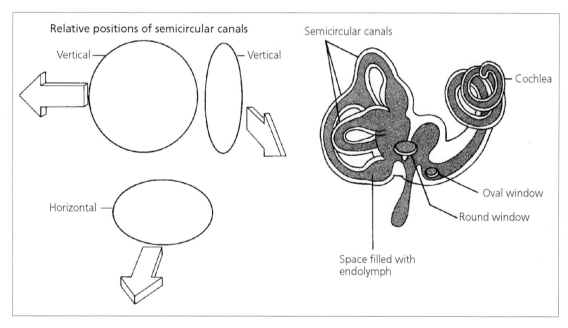

Figure 14.11
Diagrammatic representation of the semi-circular canals

the glass was turned quickly, so that the glass was moving faster than the water it contained, the tips of the hairs would bend because they would be dragged through the slow-moving water.

Thus, the rate of head acceleration is detected by the degree of drag exerted on the hair cells, which increase their rate of firing the more they are forced to bend. However, the direction of head movement is determined by the brain in a different manner. There are three semi-circular canals at right angles to each other and this is shown in diagrammatic form in Figure 14.11. A more realistic representation of how the semi-circular canals look from the outside was given in Figures 14.4 and 14.8.

If the head moves horizontally, the main drag effect will be on the horizontal semi-circular canal. If the head moves vertically as it would, for example, when going up or down in a lift, the main drag effect will be on the vertical semi-circular canal, and so on. Because there are three canals at right angles to each other, a movement of the head in any direction can be thought of as a pattern of drag effects in the three canals taken together. The brain translates patterns of drag effects into direction of head movement and combines this with the strengths of the various drag effects to add information on the rate of acceleration to the information on direction.

The operation of the semi-circular canals explains the common illusion that occurs after spinning round for a time. When you stop spinning, the world appears to spin in the opposite direction. This happens because the fluid in the semi-circular canals continues to move after the head has stopped spinning. Since the brain can only interpret movement of this liquid as head movement, you feel that you are spinning in the opposite direction to the previous real movement.

Apart from exceptional cases like the spinning illusion, the information from vision and from the organs of balance reinforce each other, and the information from the organs of balance only becomes really important when there is no visual information available – for example, when people have to move around in the dark.

There are various medical conditions which result in failures in the organs of balance and a standard medical test is to check whether people can remain standing upright when they have closed their eyes. Certain occupations, like diving or piloting an aircraft, may require a fully functional sense of balance and, where this is the case, appropriate tests should be carried out as a matter of routine.

However, being able to stand upright is only partly dependent on the sense of balance since it is also necessary to know where our limbs are and what they are doing. This information is provided by the proprioceptors.

The proprioceptors

In each muscle and tendon of the body, there are special nerve endings known collectively as proprioceptors, which are best thought of as small pressure pads. As the muscles move, different receptors are squeezed and this causes them to send signals to the brain via the sensory nerve. This is shown in a simplified form in Figure 14.12, which also shows that there are other nerve endings in the muscles (motor nerves). These are used to move the limbs and their function is dealt with later in this section.

As a muscle contracts (for example, in the way the biceps would contract to lift the forearm), the proprioceptors are squeezed and send signals to the brain. Thus, the state of relaxation or contraction of each muscle can be added to the brain's overall picture of what the limbs are doing. Since movement of muscles affects surrounding areas, other pressure receptors such as those under the skin (see Figure 14.7) also fire and add to the information available to the brain.

Figure 14.12
The proprioceptors in a muscle (adapted from Kroemer and Grandjean[42])

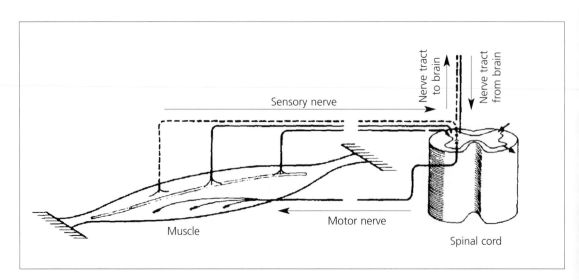

Most of the time, the information from the proprioceptors is dealt with by the brain in an automatic or subconscious manner. It is rare that people are consciously aware of the position of their limbs or the state of tension in their muscles. However, people can make themselves consciously aware. For example, there will probably be an unnecessary amount of tension in your shoulder and neck muscles at this moment. If you think about it, however, you can deliberately relax this tension by 'dropping' your shoulders, although it is likely to return soon after you stop making the effort to relax. A number of taught methods of relaxation involve exercises which aim to make the relaxed state the usual one, rather than the more common state of tension.

So far the emphasis has been on what the brain 'knows' about the state of the muscles and limbs, but the brain is also responsible for controlling the muscles and limbs. Each muscle has embedded in it a number of nerve endings which, when they fire, cause the muscle to contract. The firing of these nerve endings is under the control of the brain via motor nerves (see Figure 14.12). For example, if you want to move your right hand, the brain fires, in the appropriate sequence, the nerve endings in the various muscles required for this movement.

As can be imagined, the number of instructions which have to be issued to individual muscles is very high and the sequence of these instructions is extremely complex. This is why children have to learn to walk and adults have to learn complex tasks such as playing a musical instrument or driving a car. The learning process involves the brain determining which are the appropriate instructions and what sequences are necessary to obtain the desired effect.

The information travelling from the brain to the muscles and from the muscles and joints back to the brain forms a loop. The brain issues an instruction to a muscle to contract and the proprioceptors in the muscle itself, and the joint it influences, send back information to the brain about the effects of this contraction. This loop (for a simple grasping operation) is illustrated in Figure 14.13.

This feeding back of information, known as a feedback loop, is extremely important in terms of human reliability because of the way it enables actions to become automatic.

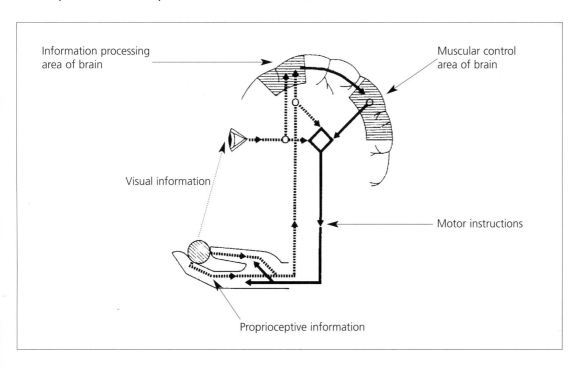

Information processing area of brain

Muscular control area of brain

Visual information

Motor instructions

Proprioceptive information

Figure 14.13
Links between eye, muscles and brain (adapted from Kroemer and Grandjean[42])

Think again about learning to drive a car. The first time learner drivers have to put their foot on the accelerator pedal, they have to look for the pedal and visually guide their foot onto it. In these circumstances, there are two lots of feedback: how the movement of the foot looks and how it feels (as is the case in Figure 14.13). When the movement has been repeated a number of times, the learner drivers know how the required movement feels and can eventually carry it out by 'feel' alone. With further practice, the required movement becomes smoother, faster and more accurate. The brain has learned the most efficient sequence of instructions and issues this as a single burst of commands.

There are, therefore, three identifiable stages in learning a particular movement or sequence of movements:
1. initial attempts under visual control (visual feedback loop)
2. further practice without visual control, or occasional correction under visual control (proprioceptive feedback loop)
3. complete automation of the movement (no feedback loop).

If human beings could not reach this third stage of completely automatic movement, many of our complex practical skills would not be possible. The amount of information processing required for skills like driving, playing musical instruments and typing would make them impossible without a high degree of automation. However, this major benefit is associated with a rich source of hazardous events. This is because when one of these highly automated movements has been initiated, it is extremely difficult, if not impossible, to stop it.

When movements have been practised so many times that they have become automatic, they are referred to as ballistic. This is a recognition of the fact that once they have been initiated, they are impossible to modify or control and the word 'ballistic' is used here in the same sense as it is used in 'ballistic missile'. A bullet from a gun or a simple rocket are examples of ballistic missiles.

Ballistic movements are, therefore, well-learned physical movements where a single instruction is issued by the brain – for example, 'put foot on brake pedal' – and all the complex, detailed instructions to individual muscles are issued automatically without the need for any further thought, or for visual or proprioceptive feedback.

Many of the ballistic movements required for car driving can create hazards. For example, in a car with a manual gearbox, one of the required ballistic movements is to put the left foot on the clutch pedal when a gear change is needed. If a person changes from driving a car with a manual gearbox to driving a car with an automatic gearbox, then this movement is no longer required. However, the need for this movement is usually so ingrained that, in an emergency, the driver used to a manual gearbox will make this movement even when the car being driven has an automatic gearbox and a brake pedal where the clutch pedal is situated in manual gearbox cars. Because of the force normally applied to the clutch pedal, this can have serious side effects resulting from too fierce braking. Some drivers who have to change frequently between automatic and manual gearbox cars make a habit of tucking their left foot under the seat while driving automatic cars.

We used driving in the example above because it is familiar to most people, but there are few occupations involving physical activity which do not depend for their efficient performance on ballistic movements.

In order to maintain human reliability as far as ballistic movements are concerned, it is necessary to monitor changes to machinery and working practices. For example, when a new machine is installed for use by people trained on a similar machine, the people concerned will have to learn new ballistic movements. This is usually done quite successfully for the movements required for routine operation, because these are recognised as being necessary changes. However, attention should also be paid to less frequently required movements, including those required for emergency responses. The old machine might have had the emergency stop button on the left while the new machine has this button on the right. Unless a deliberate effort is made to learn the new emergency response, the operator will revert to an inappropriate ballistic movement in an emergency and there will be a delay in shutting off the machine.

By their very nature, ballistic movements have to be learned very thoroughly before they become ballistic. However, even this very thorough learning cannot ensure that they will be carried out completely accurately every time. Human beings have an innate variability in carrying out actions – in other words, no matter how much they try to make an action accurate, there will be some variation. If this were not the case, there would be little point in games like darts or golf – everyone would hit the bull's eye or sink the putt every time they wanted. What practice does is to reduce this innate variability.

Ballistic movements, because they are highly practised, have probably the least variability of all movements. Nevertheless, they do have some and although 999 times out of 1,000 the ballistic movement to put a car into gear will be successful, on the thousandth time, innate variability will take over and the driver will get it wrong.

When considering human reliability, it is extremely important to take into account the existence of this variability, both in terms of likely direction and likely extent. To take a simple example, if a chef is slicing vegetables, the direction of variability of most importance will be towards and away from the fingers holding the vegetable. The extent of the variability will be unimportant if the direction is away from the hand; all that will result is a thinner slice of carrot than usual. If the direction is towards the hand, then the same amount of variability could result in a cut finger.

It is generally the case that innate variability conforms to a pattern in that the more extreme the variation, the less frequently it occurs. However, human beings, unlike machines, are not 'designed' to repeat movements with complete accuracy. When designing tasks or risk control measures, this fact must be taken into account. It is not enough to assume that a particular action can be repeated large numbers of times with a small amount of variability. The less frequent occurrence of highly inaccurate movements must also be taken into account. This aspect of reliability is dealt with in HSG48.[43]

Summary

This chapter dealt with the human senses, how they work, how they can go wrong and how failures can be detected. It has also considered the role of the brain in perception and pointed out that the sense organs alone are not sufficient for perception to occur. Without the relevant brain activity, there would be no perception.

15: The individual – psychology

Introduction

This chapter deals with the following topics:
- **Fatigue, boredom and stress.** In this section the causes and effects of fatigue are described, together with the relationships between boredom and stress and fatigue. There is also a separate section on the causes and effects of stress.
- **Attitudes, motivation, personality and intelligence.** These four topics cover the major sources of psychological difference between individuals. Each of them is dealt with in a separate section. There is much debate as to whether these individual differences are genetically determined – for example, whether people are born with a particular level of intelligence – or arise from the way in which people are brought up. The chapter begins, therefore, with a discussion of individual differences and their sources.

Individual differences

It is obvious that people differ from each other and, as has been seen, the easiest differences to observe are physical differences such as height, hair colour and the small variations which make each person's face unique. However, there are physiological differences which are just as important but more difficult to see. These include fitness, strength, eyesight, hearing and blood group.

A physiological difference such as strength or fitness can only be inferred from how a person behaves and usually special measuring techniques are required for accurate assessment. It may be adequate in general conversation to say that people 'look fit' but this would be an inadequate assessment if they were being sent off to run a marathon.

Individuals also differ in psychological aspects such as attitudes, motivation, personality and intelligence. As with physical fitness, these mental attributes cannot be seen and they have to be measured in special ways. It is possible to say that someone 'looks intelligent' but, like someone 'looking fit', this is an inadequate assessment.

Mental differences between people have to be inferred from what they say and do, but there are two important points to make about this type of inference.
- **The inference can be made either from what people say and do in real life or what they say and do in some artificial circumstances.** For example, it is possible to assess a person's intelligence or personality from how they behave at work, or by how they complete an intelligence test or personality questionnaire. The real-life assessment is obviously preferable but if, for example, it is necessary to compare the intelligence of a number of people who are doing different jobs, it would be better to give them all the same task to do, even if this was an artificial intelligence test.
- **Saying and doing are not always distinguishable.** The most obvious case is when people shake their head instead of saying no, but there is a whole range of 'doing' which can also 'say' things. It is usually referred to as body language and can be used to convey quite complex messages. The confusion between saying and doing also works in the other direction. If people are asked questions designed to assess their intelligence, they can speak the answers and, depending on how many they get correct, an assessment can be made of their intelligence. In these circumstances, the saying is really doing since it is a demonstration of a person's level of intelligence. Contrast this with what might happen if we asked people how intelligent they thought they were. They could answer 'Not very' or 'I'm a genius' but in real terms we would know no more about these people's intelligence than we knew before. If this distinction is not immediately clear,

remember that some intelligence tests involve written answers, or solving mechanical puzzles, and these types of test are more obviously 'doing'.

This differentiation of saying and doing is important when considering mental differences such as attitudes and motivation, because people can 'say one thing and do another'. When assessing differences between people, it is, therefore, important to distinguish between assessments based on some measurement of 'doing' and those which are based only on what people say about themselves. Some examples are given below.

- A person can claim to be able to drive a particular kind of fork-lift truck. Only by observing the person driving this type of fork-lift truck can this claim be checked.
- A person can claim to like children (or cats or dogs) but this can only be checked by observing his or her behaviour when confronted by children (or cats or dogs). This sort of assessment is further complicated by the fact that people's behaviour is likely to be different if they know they are being observed, so that an accurate assessment requires observation when the person concerned does not know that they are being assessed.
- A person can claim not to be motivated by money, but the sorts of jobs they take, and the way they carry them out, may suggest that this is not, in fact, the case.

The main sources of individual difference, for example, personality and attitudes, will be discussed later in the chapter but we need first to consider how these individual differences arise.

Sources of individual differences

There are two main sources of differences between individuals:

- **Their genetic make-up – that is the features they inherit from their parents.** In terms of physical features, children of short parents will tend to be shorter than the children of tall parents, and if both parents have black hair, it is more likely that their children will also have black hair. Although the evidence is less clear, certain mental attributes are also likely to be passed from parent to offspring.
- **What happens to them throughout their life, including their education and training, and particularly what happens to them in childhood.** A child of Chinese parents will grow up speaking English if he or she is brought up by English-speaking parents, and *vice versa*. However, the language learned is only one type of difference. Every individual, from the time he or she is born, has a unique set of experiences that contribute to that individual's mental make-up and because these experiences are different, the individuals are different.

There is a continuing (and sometimes heated) debate about the relative importance of genetic influence and the influence of experience. This is usually referred to as the 'nature/nurture' or 'heredity/environment' debate, and it is extremely difficult to obtain conclusive evidence about the relative strength of the two factors. Attempts to resolve the issue have usually involved studies of identical twins, who are known to have exactly the same genetic material. This arises because identical twins occur when a fertilised egg divides and forms two separate foetuses, which must, therefore, have the same genetic material. Non-identical twins occur when two separate eggs are fertilised simultaneously, and these twins are genetically no more similar than two children born to the same parents at different times.

In theory, therefore, all differences between identical twins should be due to experience, and there have been a number of studies of identical twins who have been separated at birth. For a variety of reasons, these studies have not been conclusive and, for all practical purposes, inherited features and features arising as a result of experience are inextricably mixed. From now on, therefore, unless otherwise stated, differences between individuals will be assumed to arise from a mixture of genetic and environmental influences.*

So far, the emphasis has been on differences between individuals which are relatively stable over time and allow statements such as 'he is even-tempered', 'she is lively', or 'they are an intelligent lot'.

However, being generally even-tempered does not mean that a person never gets angry; normally lively people can have dull days, and the fact that people are intelligent does not mean that they never do stupid things. Superimposed on every individual's basic character and normal ways of behaving there are fluctuations which can arise for a number of reasons:
- fatigue, boredom and stress influence people's behaviour
- the environment a person is in, including the people in that environment, influence an individual's behaviour
- drugs, including alcohol, nicotine and certain medicines, influence people's behaviour either directly or during the withdrawal phase for addictive substances
- illness, whether mental or physical, influences how people behave from time to time.

When considering human reliability, it is necessary to take into account all of these factors. People who are usually reliable may fail:
- in environments which are very different from their usual ones
- when they are taking drugs, even quite common ones such as antihistamines
- when they are ill.

However, all of these aspects of human reliability are complete studies in themselves and will not be followed up in this chapter. The important thing to note is that these factors might be in operation in any given circumstances or at any particular time.

Fatigue, boredom and stress have not been included in the paragraph above because these three factors taken together are probably the most important sources of failures in human reliability. They are, therefore, dealt with in more detail in the notes which follow.

Fatigue, boredom and stress

These three personal factors are dealt with together because they are linked. For example, people are more likely to be bored if they are tired and being tired makes them less able to cope with stress. Fatigue is dealt with first in the following notes, together with the part played by boredom and stress in creating feelings of fatigue. There is then a more detailed discussion of stress as a separate topic.

Fatigue
It is convenient to identify two different types of fatigue: muscular fatigue, which arises when a particular muscle or group of muscles is overused, and general fatigue, which involves the whole body. General fatigue involves the nervous system as well as muscles and it is, therefore, also known as nervous fatigue. As with most things to do with human beings, however, the two types of fatigue can be related; extended periods of writing can produce writer's cramp (muscular fatigue) plus general fatigue due to the intellectual effort involved.

---

* Non-identical twins, although they have different genetic make-ups, are more similar than siblings born separately because they share more experiences.

Muscular fatigue

With muscular fatigue, resting the muscles involved will usually restore things to normal fairly quickly, but the severity of muscular fatigue depends on three things:

- the period of time over which the activity is carried out
- the intensity with which the activity is carried out
- the state of fatigue of the muscles before the activity started.

Take the simple physical exercise of doing press-ups as an example. Most people can do half a dozen press-ups at a rate of around four per minute without any noticeable muscular fatigue. However, if this activity is carried on for 12 or 20 or more press-ups, the muscles in the arms lose strength and become fatigued. The period over which people can do press-ups can be extended by reducing the rate at which they are done – that is, by reducing the intensity of the activity. Typically, people would be able to do many more press-ups at a rate of one per minute than they could do at a rate of four per minute. If the rate was reduced even further to, say, one every 10 minutes, most people could go on indefinitely or until they got bored.

The final factor, the state of the muscles at the start of an activity, can also be illustrated using the press-ups example. If a person started doing press-ups now at a rate of four per minute, he or she might be able to do around 10 before fatigue set in. If the person rested for five minutes, he or she might then only be able to do a further five before fatigue set in because the muscles were already at a high level of fatigue when the second batch was started. These three factors influencing the extent of fatigue (duration, intensity and starting state) must all be taken into account when designing jobs and systems of work for maximum reliability.

In general, human reliability diminishes with increased fatigue, so anything which minimises fatigue increases reliability. This relationship will be dealt with in more detail after we have described general, or nervous, fatigue.

As we mentioned earlier, muscular fatigue is removed by resting the muscles involved, but this is not a straightforward relationship. The complication is that the amount of work done and the amount of rest required is not a straight line relationship.

Figure 15.1

Fuel consumption at different speeds

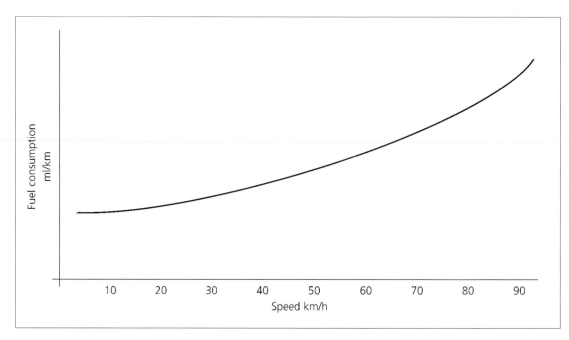

What this means in practice is that if it takes 10 minutes to recover from a set amount of work, it will take more than 20 minutes to recover from twice that amount of work. Readers may be familiar with this sort of relationship from a vehicle's fuel consumption at different speeds as illustrated in Figure 15.1.

Figure 15.1 illustrates what most drivers already know: the faster they go, the greater their car's fuel consumption. However, in most cars with an engine size between one and two litres, the increase in fuel consumption between 30 km/h and around 90 km/h is relatively modest. Above 90 km/h, the fuel consumption increases rapidly, which is why most fuel consumption figures supplied by manufacturers are for a sustained 90 km/h.

Having explained the relationship between speed and fuel consumption, it is now possible to go back to amount of work and fatigue. Figure 15.2 is a general representation of the relationship between amount of work and the time required to recover from this work. Note that there are no numbers given in the figure. There are ways of measuring amounts of work and recovery time, but we need not deal with them here. There are some points to note about the implications of Figure 15.2:

- **For small amounts of work, the recovery time is very short.** If individuals hold their arms outstretched for 10 seconds, there will be little fatigue and recovery will be rapid. If they hold their arms outstretched for 10 minutes, there will be significant fatigue and recovery time will be much more than 60 times that required to recover from the 10 seconds of outstretched arms.

- **For large amounts of work, the recovery time increases very rapidly and, if the amount of work is extended too far, the required recovery time will be off the graph.** In effect, if people do too much work, they may not be able to recover at all. For example, extreme overworking of particular muscles can result in irreversible damage to those muscles and, if the muscles concerned are in the heart, the result can be fatal.

- **The amount of work done depends on two main factors: the intensity of the work and its duration** – that is, how hard a person works and for how long. For most activities, it is possible to calculate, for a specified individual, intensity and duration rates which can be sustained without harm or undue fatigue, and these rates can be used in, for example, task design procedures. However, the effect on recovery time of a specified amount of work will also depend on the state of fatigue at the beginning of the task and the person's general level of fitness.

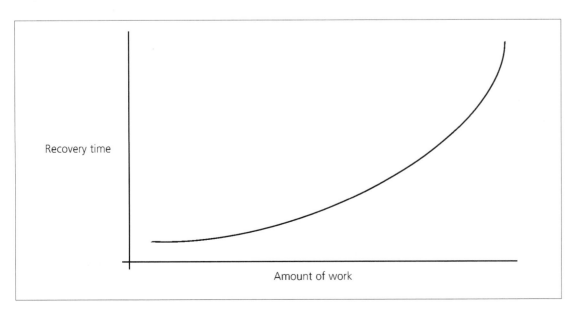

Figure 15.2
The relationship between the amount of work and recovery time

The most important practical implication of Figure 15.2 is that beyond a certain level, further work is counterproductive since the recovery times required are so long. This is why work breaks and the pace of work are so important, and these will be considered again in the context of general fatigue.

Note that the body has natural defences against overworking its muscles. Physical effort results in the build-up of lactic acid in the muscle tissues and this, together with a number of other factors, prevents the continued use of the muscle. Thus, there is a natural defence against using a muscle for too long or at too high an intensity. People also experience pain in overused muscles which acts as a reminder that these muscles should be treated with care.

General or nervous fatigue

Muscular fatigue depends largely on the amount of work done and the starting condition of the muscles concerned, but general fatigue is more complex. For example, it is possible to feel generally fatigued without having done significant amounts of work, and it is also possible to carry on working even when feeling very fatigued.

These apparent anomalies arise because while objective fatigue depends on how much work a person does, how fatigued a person feels – subjective fatigue – depends on various factors. In order to understand how this can be, it is necessary to know something of how certain parts of the brain work and, since the fatigue mechanism is rather complex, the principle will be illustrated by describing the thirst mechanism, which operates in a similar way but is much simpler.

When people feel thirsty, it is because a certain part of the brain is sending out signals which make them feel this way. What starts this brain activity is usually a fall in the amount of water in the body tissue, generally due to sweating or taking in too much salt. Water 'sensors' throughout the body send signals to the thirst centre in the brain which, in turn, sends out signals which prompt the person to drink, and drinking a glass of water will stop the signals from the thirst centre.

These signals from the brain stop long before the water reaches the body tissue; the presence of water in the stomach stops the signals, even though the water 'sensors' are still signalling a lack of water. If this were not the case, people would continue to drink throughout the 10 or 15 minutes required for the water to move from the stomach to the dehydrated body tissue. It is the need for this type of control which makes the thirst centre in the brain essential and humans have similar centres for all sorts of body activities, including hunger.

For most people, most of the time, the thirst centre is just another part of the brain's automatic functioning which is taken for granted. However, if the thirst centre is activated artificially, as it might be during a brain operation, the person reports feelings of thirst, and if the thirst centre is damaged (for example, by disease), the person concerned cannot control their fluid intake in the normal way.

The general principles illustrated by the thirst mechanism is that there are brain centres which, when they send out signals, cause specific feelings and that these brain centres are stimulated to send out their signals by activities elsewhere in the body. These general principles apply to fatigue but our overall feeling of fatigue depends on the relative states of two separate mechanisms: the inhibitory system and the reticular activating system. We will now consider each of these systems separately and then describe the causes and effects of fatigue.

The inhibitory system

There is a fatigue centre in the brain which, when it sends out signals, causes the person to feel fatigued. This centre is known as the inhibitory system (IS) and the more active this centre is, the sleepier and more fatigued the person feels.

Rest and, in particular, sleep will reduce the activity in the IS. By the end of a normal night's sleep, the IS activity is very low and fatigue is no longer felt. As the day goes on, activity in the IS builds up again, thus increasing feelings of fatigue. However, this build-up during the day is not necessarily smooth. Eating food or drinking alcohol can cause a temporary increase in IS activity which diminishes as the food or alcohol is digested.

All this information about brain centres and the IS may seem just an elaborate way of explaining the obvious fact that we generally feel more fatigued as the day goes on. However, its importance cannot be appreciated until the role of another brain centre is explained.

The reticular activating system

The other brain centre is known as the reticular activating system (RAS). The reticular part refers to where in the brain the centre is located. The RAS operates in the opposite way to the IS in that, when the RAS is sending out signals, the person feels active and alert. The effect of rest and sleep on the RAS is also the opposite to their effect on the IS. During rest and sleep, the activity in the RAS builds up so that the person feels more active and alert after sleep or rest.

Thus, there are two opposing brain centres, the IS and the RAS, which influence feelings of fatigue and for most of the time these two systems operate in a 'see-saw' manner. During a typical day, the activity in the IS will increase and the activity in the RAS will decrease. During a typical night's sleep, the activity in the RAS will increase and the activity in the IS will decrease. Our subjective feeling of fatigue at any particular time will, however, depend on the state of the two systems taken together. A typical daily cycle of IS and RAS activity is shown in Figure 15.3.

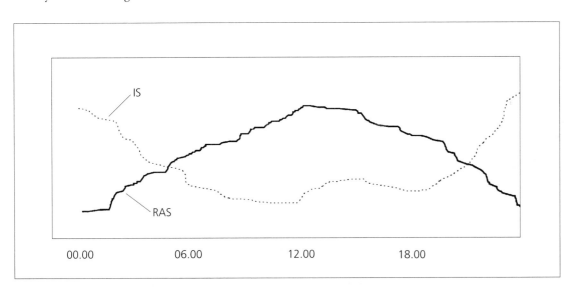

Figure 15.3
Typical daily IS and RAS activity

The IS and RAS, like most centres in the brain, are connected to many other areas of the brain which are capable of influencing their activity. For example, it is harder to get to sleep when hungry or thirsty because the hunger and thirst centres send signals to the RAS which increase its activity. These different linkages in the brain explain various features of fatigue:

- **Jobs with low information content or low sensory input, usually described as being boring or monotonous, typically produce feelings of fatigue.** The reason for this is that if the brain is receiving too little stimulation, activity in the RAS declines so that the normal balance between the RAS and the IS is upset. Since the RAS activity is now lower than the IS activity, the person feels fatigued and sleepy. However, boredom or monotony are very subjective things. What one person finds boring, another can find fascinating, and what one person finds monotonous, another may find relaxing.

- **People tend to daydream when they find a job boring or monotonous.** Daydreaming is a way of generating information within the brain which will maintain activity in the RAS, thus keeping at bay feelings of fatigue and boredom. The problem with daydreaming, from the point of view of human reliability, is that it can shut out information on hazards coming from the outside world.
- **Feelings of fatigue, even quite strong ones, will disappear quickly if something sufficiently interesting happens or a crisis occurs.** In these circumstances, the RAS is overstimulated so that its level of activity is greater than the level of activity in the IS.

It is important to remember that the activities in the RAS and IS are independent. An increase in RAS activity due to some crisis does not reduce the activity in the IS, it merely overwhelms its effects temporarily. Readers can probably identify from their own experience occasions when this temporary overwhelming of the IS activity has occurred. Table 15.1 summarises the subjective feelings associated with extreme levels of activity in the IS and RAS.

Table 15.1
Subjective feelings associated with levels of IS and RAS activity

Activity in RAS	Activity in IS	
	High	Low
High	Feelings of fatigue but a drive to do things despite this. One of the typical stress patterns	Active and alert
Low	Fatigued and sleepy	No feelings of fatigue but no inclination to do anything. Typical boredom or apathy pattern

Causes of fatigue

Having looked at the mechanisms underlying feelings of fatigue, it is now possible to move on to the factors which cause such feelings and the ways in which they can be minimised. This is an important issue in human reliability because, in general, people make more errors when they are feeling fatigued.

If everything else were equal, the amount of general fatigue produced by mental or physical work would follow the relationship described in Figure 15.2. That is, general fatigue would depend on the intensity of the work done, its duration and the state of fatigue at the time the work was started. However, 'everything else' is rarely equal with human beings and the relatively straightforward relationship between amount of work and general fatigue is complicated in a number of ways.

- **The physical environment in which the work is being done can influence the amount of fatigue the work produces.** Certain aspects of the physical environment cause fatigue because they are outside the limits within which the body is 'designed' to operate. For example, working in extreme heat or extreme cold is more fatiguing because the body has to work harder to maintain the stable internal temperature it requires. Working in high noise levels or in very dull light is also more fatiguing because the sense organs and the brain have to work harder in order to obtain the information they require.
- **The psychological state of the person influences the amount of fatigue felt.** In general, people find things they enjoy doing less fatiguing than things they dislike, although in objective terms the amounts of work involved may be the same. Similarly, routine work is usually more fatiguing than varied work, and work that people find stressful is more fatiguing than work that is not stressful. Note, however, that all these things depend on the person involved; what one person likes, or finds interesting or non-stressing, may have the opposite characteristics for another person.

- **There will be differences within an individual from time to time which affect the amount of fatigue experienced from a given amount of work.** For example, things away from work may be worrying the person concerned or putting them under some form of stress. Alternatively, things may be going so well away from work that work is less fatiguing than normal.
- **The physiological state of the person influences the amount of fatigue.** Illness, disease, pain and inadequate nutrition are all factors which can increase the fatigue produced by work. Similarly, drugs that influence the physiological balance, for example antihistamines or alcohol, can increase fatigue. On the other hand, regular exercise and stimulant drugs can reduce the amount of fatigue experienced during work.

The four things listed above combine with the intensity, duration and starting state of fatigue to determine how much fatigue is actually experienced as a result of work done.

It is important to recognise that all of these factors can operate together, because unless the true cause of fatigue is recognised, little can be done to ameliorate it. For example, if a person is experiencing fatigue because they find a job boring, the fatigue will not be cured by more sleep. In the context of human reliability at work, there will be factors which can be influenced at the workplace, such as ensuring a comfortable environment and adequate rest breaks, and factors which cannot.

Some of the factors unaffected by action at the workplace are:
- domestic worry or stress
- ill health not caused by the work being done
- second jobs – for example 'moonlighting' – or voluntary work
- strenuous hobbies
- drug taking (including certain prescribed drugs) and alcohol consumption.

The effects of fatigue

Whether the cause of fatigue is the actual amount of work done, or an indirect cause like boredom or stress, the effects will be similar. In general terms, the brain acts in order to reduce the amount of information being processed. The fatigued person finds it difficult to concentrate, loses interest in what is going on in the environment, and eventually the eyes close and sleep follows.

Normally, a more or less extended period of sleep will restore the balance of the IS and RAS to the alert, wakeful state. However, the methods used by individuals to reduce their information input when fatigued is very variable, both from person to person and within the same person from time to time. General methods include:
- unwillingness to start new tasks, especially if these are perceived as onerous, boring or unimportant
- a narrowing of attention to information only of direct relevance to the person's current concern
- mildly antisocial behaviour and irritability which is used to distance people who are an actual or potential source of information.

Stress

There are many different ways of defining stress and many approaches to assessing its causes. However, one general approach is to assume that stress arises when a person has a goal they want to achieve and something is preventing them reaching that goal. The 'something' is referred to as a stressor, that is something which induces stress.

A simple example would arise if a person was due to attend an important meeting and was held up in a traffic jam. In these circumstances, the goal is arriving at the meeting on time and the stressor is the traffic jam.

Using the simple model of stress just described, it is apparent that there are two ways of reducing stress.
- **Remove the stressor.** In the traffic jam example given above, there would be no stressor if the road had been clear.
- **Change the goal.** If the person can telephone and reschedule the appointment for an hour later, the traffic jam may still be annoying but will be much less stressful (for the next hour anyway) because it is no longer preventing the person from reaching his or her goal.

However, since individuals vary so much in their goals, what individuals find stressful also varies widely; some examples are:
- If a person's goal is to have a quiet life, too much responsibility will be stressful. If a person is ambitious and wants to get on in life, then too little responsibility will be stressful.
- If a person likes his or her job, and has no wish for a job elsewhere, then possible redundancy will be stressful. If a person dislikes his or her job and sees opportunities for doing other things, then redundancy may be seen as an opportunity.
- If a person likes company, then a job where he or she has to work in isolation will be stressful. For the person who likes being on his or her own, having to work with other people will be stressful.
- If a person likes to learn new things and have new challenges, a routine job will be stressful. For the person who wants a routine job, for whatever reason, one that requires learning new things and meeting new challenges will be stressful.

In order to combat stress, therefore, it is necessary to work at the level of an individual's requirements. Before stress can be combated, however, it is necessary to identify that a person is under stress – in other words, to identify the symptoms of stress – and these are described next.

The symptoms of stress are probably familiar to everyone, since everyone will have suffered from at least mild stress at some time. When a person is prevented by a stressor from reaching a goal, one of the things that the body does is to produce, and release into the bloodstream, the hormone adrenaline. The main function of adrenaline is to prepare the body for additional activity; originally it was to prepare the body for fighting or running away when the person was threatened. This threat response still functions in humans and any stress will result in the production of adrenaline, which has two main effects:
- it speeds up the heart rate
- it reduces the blood flow to the digestive system and slows down or stops digestion temporarily.

It is these effects which result in a thumping heart and feeling sick after being frightened. Adrenaline produced in response to a threat remains in the bloodstream and affects the heart rate and digestive system until it is removed by physical activity. In the course of flight or fight, this activity would normally occur soon after the adrenaline was released and the level of adrenaline in the bloodstream would quickly be returned to normal.

In the modern world, however, there is normally no appropriate physical reaction to a stressor so that people who are under long-term stress have a persistently high level of adrenaline in their blood. This means that they have too high a heart rate and a poorly functioning digestive system. In the long term, this can lead to heart disease and a variety of digestive disorders, including stomach ulcers.

Note that the only effective way of removing adrenaline from the bloodstream is physical activity so that, effectively, physical exercise is the only way of avoiding the heart and digestive problems associated with stress.

Identification and treatment of stress is an important aspect of maintaining human reliability, since a person under stress is less likely to be able to deal with the hazards in the environment.

However, beyond a general awareness of the importance of individual differences in susceptibility to stress, and the likely symptoms, there is rarely anything that the non-specialist can do by way of treatment. There are two main reasons for this.

- **Treatment requires either control of the stressors or the ability to change an individual's goals.** If the stressors are work-related, then control may be possible, but there are many stressors outside work (such as domestic difficulties) over which no control is possible as far as employers are concerned. Similarly, it is debatable whether altering an individual's non-work goals is the legitimate concern of the employer.
- **Treatment for the symptoms of stress is the province of the medical services.** Psychiatric or drug-based treatments may be necessary for the more extreme psychological symptoms such as depression or chronic anxiety, and drug- or surgery-based treatments may be necessary for the more extreme physiological symptoms such as heart or digestive disorders. Unfortunately, there is a split in modern society between the sources of stress and the sources of treatment for stress. The medical services rarely have control over the stressors, and those responsible for the stressors may not be aware of the effects the stressors are having.

In general, these two points are a strong argument for an important contribution by the organisation's doctors to stress reduction since these doctors are more likely to be able to identify work-related stressors and to influence their reduction.

Despite the problems associated with the identification and treatment of stress, it is becoming an important topic. This is because there is evidence, at least in the UK, that changes in work patterns – for example, longer hours of work, more intense work and higher levels of responsibility – are creating higher levels of stress than was the case in the past.

As a result of this, there are reports of individuals having to give up work for stress-related reasons and making claims against their employers for the damage sustained from work-induced stress. In effect, stress at work is now becoming another type of work-related injury or ill health. In the UK, the HSE has recognised this and has produced guidance on managing stress.[44] This guidance is based on Management Standards which are designed to simplify risk assessment for stress, encourage working partnerships between employers, employees and employees' representatives, and provide a yardstick for gauging performance in tackling the causes of stress. The Management Standards (www.hse.gov.uk/stress/standards) cover the following sources of stress at work:

- demands, such as workload, work patterns and the work environment
- control, such as how much say the person has in the way they do their work
- support, such as the encouragement, sponsorship and resources provided by the organisation, line management and colleagues
- relationships, such as promoting positive working to avoid conflict and dealing with unacceptable behaviour
- role, such as whether people understand their role within the organisation and whether the organisation ensures that they do not have conflicting roles
- change, such as how organisational changes (large or small) are managed and communicated in the organisation.

Attitudes, motivation, personality and intelligence

These four factors are major sources of individual difference, but they share a number of characteristics and we will deal with these before considering each factor in turn.

- **The attitudes, motivation, personality and intelligence of an individual at a particular time are determined by many things which have happened to that individual in the past.** Similarly, what is happening to an individual now will influence that individual's attitudes, motivation, personality and intelligence in the future. When considering any one of these factors, therefore, it is rarely possible to say how a particular set of attitudes, a particular aspect of motivation, a particular personality trait, or a particular level of intelligence has arisen.

- **All four factors are subject to the problems of measurement discussed above.** That is, it is possible to assess, for example, a person's attitudes either by asking about them, or by observing a person's behaviour and inferring attitudes from behaviour. These two measurement techniques can produce very different results.
- **Except in a few, fairly limited, cases, it is difficult to make any significant change in a person's attitudes, motivation, personality or intelligence.** Even action designed specifically to change, for example, a person's motivational make-up can have the opposite effect to the one desired. Where the action is applied to more than one individual, the same action will almost certainly produce different results from individual to individual.

However, despite these difficulties, there is value in knowing something about attitudes, motivation, personality and intelligence since it enables a more enlightened approach to any problems of human reliability involving these factors.

Attitudes

When it is said that a person has a certain attitude towards something, the implication is that the person has more than just an opinion or a view on a subject. This implication is that the person is willing to behave in ways which are consistent with his or her attitudes.

The problem with attitudes is that it is easy to express an attitude but behave in a contrary way. In many cases the expressed attitude and the attitude inferred from observed behaviour do not match, but this mismatch can only be identified if people are asked about their attitudes and a sample of relevant behaviour is observed.

There are formal methods of assessing people's attitudes by asking. Attitude surveys in industry, for example, are used to assess people's attitudes towards work or their levels of job satisfaction, and market research surveys are used to measure people's attitudes towards existing or proposed products. The value of such surveys is, however, always in question until evidence can be collected which shows that people's behaviour is consistent with their expressed attitudes.

In the field of safety, for example, there are very few people who do not express the attitude that safety is important and that activities to create safe working should be encouraged. However, when these people's behaviour is examined in detail, it is often not consistent with their expressed attitudes. Some examples are given below.

Senior managers:
- do not attend health and safety committee meetings
- do not allocate adequate funds for safety
- do not support work stoppages called for safety reasons.

Middle managers:
- do not enforce safety measures
- do not comply with safety measures they have set for other people
- do not allow time off for safety training
- do not support safety when there is a conflict between safety and productivity.

Operatives:
- do not wear the specified PPE
- do not follow laid-down safety procedures
- do not report hazards, minor injuries or near misses.

Health and safety professionals:
- do not obtain competences in, for example, investigation and analytical techniques
- do not prepare adequate safety plans
- do not maintain an advisory role when there is pressure to police safety
- take on safety responsibilities which should be discharged by managers.

These contradictions between expressed attitudes and attitudes inferred from behaviour illustrate the problems encountered when it is felt necessary to 'change people's attitudes'. It is relatively easy to change people's expressed attitudes since there are very good reasons for them saying what they think the other people want to hear. However, changing expressed attitudes is not, in most cases, what is required. The real objective is to bring a person's behaviour into line with the correct expressed attitude. Table 15.2 illustrates some common relationships between behaviour and attitudes.

		Attitude	
		Undesirable	Desirable
Behaviour	Undesirable	I do not attend health and safety committee meetings because I think they are a waste of time. I do not wear a hard hat because I do not believe they are effective protection	I think health and safety committee meetings are a good idea but I do not have the time to attend them. I think hard hats are effective protection but I do not wear mine
	Desirable	I attend health and safety committee meetings but I think they are a waste of time. I wear my hard hat but I do not think it provides useful protection	I attend health and safety committee meetings because I think they are a good idea. I wear my hard hat because I think it will protect me

Table 15.2
Relationships between behaviour and attitudes

As far as health and safety is concerned, there is an argument that attitudes should be ignored and all efforts should focus on changing behaviour. This is often described as the 'behavioural safety approach' or simply as 'behavioural safety'. There is some sense in this argument since, for example, if people wear their PPE, it does not matter whether or not they think wearing it is a good idea. The problem arises when people are required to wear PPE in circumstances where there is no enforcement. If they do not believe that wearing PPE is a good idea, they are unlikely to wear it in the absence of enforcement.

There is also evidence that enforcing certain forms of behaviour will in itself produce changes in attitudes towards that behaviour. People who give up smoking, for example, often end up with a powerful anti-smoking attitude, and statements like 'I didn't want to do this when it was first introduced, but I see the sense in it now' indicate a change in attitude towards the behaviour concerned.

In an ideal world, everyone will be in the bottom right cell of Table 15.2 and they will be doing the correct things because they believe that they are the correct things to do. In these circumstances, there is said to be a strong, positive 'safety culture'. Safety culture is discussed in more detail in Chapter 21.

Motivation

Most of the time, when people do things, it is for a reason. When this reason is under the control of the people concerned, that is, they have not been asked or ordered to do something, the reason is usually referred to as 'motivation'. For example, if someone is eating, it is usually because they are motivated by hunger, although other psychological states such as depression can cause people to eat. Note, however, that the motivation is inferred from the behaviour and this inference may be incorrect. The alternative approach is to ask people about their motivation but this may also produce incorrect answers since people do not always understand their own motivation – and they can also lie about it. This is, of course, a similar problem to the measurement of attitudes which has just been discussed. Motivations can be relatively straightforward ones like hunger or thirst, or very complex ones like loyalty or ambition. However, the essential idea of any motivation is the same; if a person has a particular motivation, he or she will behave in one way rather than another.

In view of an individual's complexity, this basic idea is mostly inadequate since individuals rarely act on the basis of a single motivation. Hunger may be the motivation to eat but there will be a whole range of other, more complex, motivations which determine where they eat, what they eat, and in whose company they eat.

Similarly, most people work because they need to make money but other motivations will influence the job they choose to do, the organisation for which they work, and the geographical location of their workplace.

Despite these difficulties with motivation, there have been many attempts to produce theories of motivation and these are considered briefly in Chapter 26.

For most people, there will be a few, broad, long-term motivations which, other things being equal, will direct the choices they make. Different people will have different motivations of this sort and they are likely to change slowly as the person gets older. Examples of such broad motivations are ambition, desire for companionship, desire to help others, desire to maintain a quiet life, desire to travel, or the need to create something new.

Superimposed on these broad motivations there are various factors which complicate the issue. Examples of these factors are:

- **Broad motives may be contradictory.** For example, the desire to live well may contradict the desire to save money in order to satisfy a motivation for security.
- **Short-term motives may override long-term motives.** For example, the short-term desire to smoke a cigarette may override the long-term motivation to stay healthy.
- **Motivations change with time.** What motivated people to do things when they were children no longer has the power to motivate them now; and what motivates people now may not motivate them in five years' time.
- **Demands of the environment or other people may run counter to broad motivations.** People may be motivated to learn but because of lack of time or funds, they may not be able to do this to the extent they wish.

In the context of human reliability, motivation is relevant since most people have a broad, general motivation to avoid injury and damage to their health. As pointed out above, however, there are factors which can override this broad motivation. In order to ensure long-term human reliability, tasks must be designed in such a way that they do not allow other motivations to conflict with, or override, the basic motivation to avoid injury and ill health.

Some examples of how such conflicts can arise are:

- piecework, where people are paid according to how much they produce, which puts speed of production into conflict with safety of operation
- PPE that is uncomfortable to wear, which puts the motivation for comfort in conflict with the motivation for safety
- internal accounting systems that make managers pay for health and safety training, PPE and so on but not for accidents, which are paid for by insurance; these systems allow the motivation for good financial performance to override the motivation for safe performance.

The key thing to remember about motivation is that there are very few people who are motivated to injure themselves or to take actions that will make them ill. What has to be done to promote human reliability is to ensure that no motivations are introduced that will have this effect.

Personality

People make statements such as 'he does not have much personality' or 'she is a personality'. However, psychologists and others involved in studying human personality do not use 'personality' in these ways. Rather, psychologists use personality to describe the general characteristics of how an individual thinks and behaves.

It is easy to observe that some people are livelier than others, some are more sociable than others, and some are more thoughtful than others. What psychologists do is to attempt to identify the main ways in which people's personalities differ (in liveliness, sociability, thoughtfulness and so on). They then devise ways of quantifying these differences so that they can say, for example, how much more lively one person is than another, or how much more sociable.

As with other aspects of human beings, personality measurements can be based on what people say ('I think I am very sociable'), or what they do, for example, by measuring how much time they spend voluntarily with other people. As with attitudes and motivation, the idea behind measuring personality is that a person with a particular type of personality is more likely to behave in certain ways. In general this will be true, but sociable people do not spend all of their time with other people, and lively people can have dull days. There have been various theories of personality put forward and some are described in Chapter 26. However, one aspect of personality which is particularly relevant to human reliability is people's tendency to take risks.

Let us start with a simple experiment. Imagine that you go into a casino and the manager tells you that you are the casino's millionth customer, and to celebrate this you are to be given £1,000 with which to gamble. However, there are rules laid down for your gambling.

- You can only make one bet and you must bet the whole of the £1,000 at roulette.
- You have a choice of only three bets:
  - red or black so that you have a 50:50 chance of leaving the casino with £2,000
  - a block of four numbers so that you have a 1 in 9 chance of leaving the casino with £10,000
  - a single number so that you have a 1 in 36 chance of leaving the casino with £37,000.*
- You must make the bet – you cannot leave with your £1,000.

Which bet would you choose and why? Can you think of any circumstances which would make you choose a different bet? Which bets do you think people you know would choose?

What this simple experiment illustrates is that, in these imaginary circumstances, different people have different levels of risk they prefer to accept. They also have a variety of reasons for accepting that level of risk, and a variety of circumstances which will cause them to change their preferred level.

These sorts of difference can also be observed in real life with, for example, people's choice of hobbies or their preferred speed of driving. In theory, this should make ensuring human reliability much easier since all that should be necessary is to identify those people who like taking risks and avoid employing them. In practice, this simple solution does not work, and there are several reasons for this.

1. **It is difficult to identify the risk takers.** People are unlikely to tell potential employers that they are risk takers if they think that this will reduce the chance of their being employed.
2. **Risk-taking behaviour is situation-specific.** For example, people may prefer high risk hobbies but not take risks at work, or *vice versa*. The person who would not dream of taking a risk while driving may be a keen hang-glider or mountain climber.
3. **An individual's risk-taking behaviour varies from time to time and is influenced by a large number of other factors.** For example, the normally cautious driver may take risks when late for an appointment or annoyed by other road users, and the normally cautious worker may take risks to ensure that an important job is completed on time.

---

* The house zeroes have been suppressed for the purposes of this bet. There are 36 numbers on a roulette wheel (1 to 36), half of which are red and half black, hence the odds quoted above.

If the aim is to ensure human reliability, then these differences in risk taking (both between individuals and within the same individual from time to time) must be taken into account. Those factors which increase people's likelihood of taking risks have to be identified and, if possible, removed. If they cannot be removed, then measures must be taken to ensure that no harm comes to individuals in cases of failure.

Intelligence

Intelligence can be defined in a number of ways, but for the purposes of this chapter it is best thought of as an individual's ability to process information and deal with abstract concepts. However, it is usual to divide intelligence into a number of different types depending on the sorts of information and abstract concepts involved. Typical subdivisions of intelligence include:

- **verbal intelligence** – the ability to deal with the written and spoken word
- **numerical intelligence** – essentially the ability to deal with numbers, but it is often extended to more general mathematical concepts
- **spatial intelligence** – the ability to deal with the two- and three-dimensional relationships between objects.

Some people argue that emotional intelligence should be included in this list. Emotional intelligence is the ability to manage your emotions in a healthy and productive manner and to have a positive influence on other people's emotions.

Within each category, intelligence is demonstrated by such things as being able to remember the relevant information and concepts (for example, people's vocabulary is one measure of their verbal intelligence) and being able to reason using the information and concepts.

However, there are major areas of human activity which are not included in most classifications of intelligence and these are often referred to as aptitudes. For example, people may have an aptitude for music or mechanics. It is important to appreciate that the various subdivisions of intelligence and the various aptitudes are largely independent of each other. For example, a person can have an extremely high musical aptitude without demonstrating any other signs of above average intelligence.

As with other individual differences, intelligence is determined by genetic make-up and an individual's experiences but, since there is no clear evidence linking intelligence and reliability, a more detailed discussion of intelligence will be left until Chapter 26.

Summary

This chapter dealt with the main sources of individual differences – heredity and environment – the linked concepts of fatigue, boredom and stress, and four major dimensions of individual difference – personality, attitudes, motivation and intelligence.

In the next chapter, we will consider how individuals interact with their environment, starting with their interactions with machines and work.

16: The human factors environment

Introduction

So far the emphasis has been on the individual in isolation, although we have mentioned at various points that individuals are influenced by a number aspects of their physical and social environment. This chapter considers in more detail certain aspects of an individual's environment and the effects these can have on the individual's physical, physiological and psychological wellbeing. The aspects of the human factors environment to be dealt with are:

- machines and work
- groups
- organisations
- society.

Each topic will be discussed primarily from the point of view of the effect it has on the individual, but remember the caveat on mutuality of effects described in Chapter 13. Although such things as groups and organisations affect individuals, individuals also affect the groups and organisations of which they are part.

Machines and work

Introduction

Human beings are incredibly adaptable. They can do a whole host of things in the short term that are damaging to their physical and mental wellbeing in the long term. People smoke and drink too much alcohol because they give a short-term benefit. Similarly, they work in high levels of noise or environments contaminated with toxic chemicals because these activities bring medium-term benefits, such as a wage packet. In both cases, people trade off short- or medium-term benefits against long-term costs.

Arguably, this may be acceptable if the individuals concerned have accurate information on the costs and benefits. If adults choose to smoke knowing that it causes cancer and that the cancer is lethal, then that is their choice. Similarly for adults who choose to drink too much alcohol and risk cirrhosis of the liver. The problems arise when people are forced to do things that are damaging to their health or mental wellbeing without being provided with full information on what the harms might be. Unfortunately, society in general, and work in particular, has a long history of imposing these sorts of risk on people and, in the notes which follow, machinery and work will be considered in this context. Later in the chapter, society and risk are considered in more detail.

It is possible to consider a person and the work that person is doing, including any machinery or equipment required for the work, as a system or systemic unit. Systems are abstract concepts (which are dealt with in detail in Chapter 21) but for current purposes, a system can be defined as something intended to achieve a specific aim. Thus, in industry, there can be a widget-producing system or systems for cleaning windows or delivering training or whatever else an organisation's products and services may be.

Each system will be made up of a number of elements – for example, raw materials, machines, people, energy supplies and finished products or services. If what is necessary for the creation of a product or service is thought of as a 'systemic unit' or system, rather than individual elements, this provides a number of options for thinking about the risks involved and the ways of reducing these risks.

Consider a hypothetical example: the systemic unit required to produce widgets. This unit will need at least the following:

- raw materials
- energy
- consistency – all the widgets have to be identical, or at least within specification
- self-monitoring – all the widgets have to be checked for match against the specification.

If widget production is thought of in this abstract way, it is possible to consider how best to employ human beings in the system. The arguments might be as follows:
- **raw materials** – with a few minor exceptions, human beings are hopeless as a source of raw materials
- **energy** – humans are good at supplying energy for short periods but they can only supply limited amounts, and if they have to supply more than these limited amounts, they become damaged
- **consistency** – humans are hopelessly inconsistent when compared with machines
- **self-monitoring** – humans are quite good at spotting errors but they cannot do this consistently (see the previous point).

Viewed in this abstract way, the logical conclusion appears to be that humans should not really have a role in widget production at all. Yet, in the vast majority of manufacturing, people are employed in all but the first of the four areas listed above, and it is useful to identify why this is the case. In system terms, there would appear to be three main reasons:
- **Humans are relatively cheap.** This is particularly the case when they are already employed and no training is thought to be necessary for the task they are to undertake.
- **Humans are adaptable.** Unlike machines, which are usually designed for a restricted range of tasks, human beings can carry out a wide variety of tasks.
- **Humans have error recovery potential.** If a machine goes wrong, it stays wrong until someone fixes it. Humans can correct their own errors, and errors by machines and other people.

Unfortunately, the benefits of using people in work systems can result in the humans being used in inappropriate ways. The problem arises because people can do things which are essentially damaging to them. For example, they can work in noise levels which are making them deaf, in polluted atmospheres which are likely to give them cancer, and in extremes of heat and cold which damage their general health. What is required, therefore, are machine and work systems which use the advantages of human beings without causing them harm in the process, and the devising of such systems is one of the main functions of ergonomics.

Ergonomics

The central role of ergonomics has been described in simple terms as 'fitting the job to the man'. This is in contrast to the (more usual) way of doing things, which is to design a machine or job and then expect the people who will operate the machine or do the job to adapt as appropriate.

Perhaps the most famous example of inappropriate machine design is the lathe which led to the design of 'Cranfield Man'. Researchers at Cranfield studied the layout of the displays and controls of a particular type of lathe and designed the ideal man for operating this lathe. The layout of the lathe controls and the shape of Cranfield Man are illustrated in Figure 16.1.

In order to fit the job to the person successfully, two angles of approach are required, and these are covered by two branches of ergonomics.
- **Research ergonomics.** This is the study of people and, in particular, their natural limits. The question the research ergonomist is trying to answer is: 'What can people do, day after day, for a lifetime, without sustaining damage?' Obviously this question is subdivided, with different groups of ergonomists working on, for example, lifting weights, arm movements, leg movements, and various aspects of the environment such as temperature and humidity. An increasingly important branch of research ergonomics is cognitive

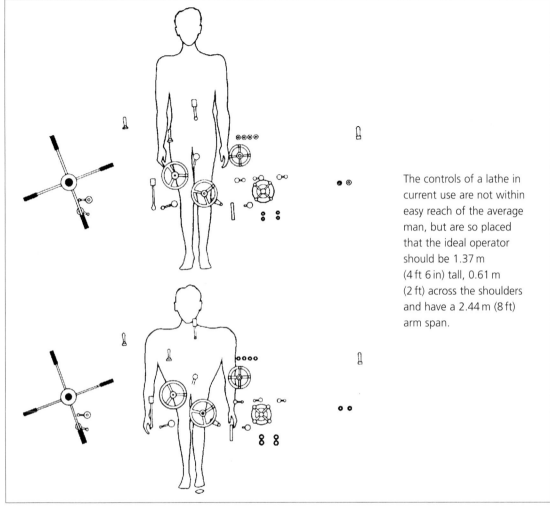

Figure 16.1
Cranfield Man

The controls of a lathe in current use are not within easy reach of the average man, but are so placed that the ideal operator should be 1.37 m (4 ft 6 in) tall, 0.61 m (2 ft) across the shoulders and have a 2.44 m (8 ft) arm span.

ergonomics, which looks at individuals' capacity for processing information, particularly in the context of computer software. Essentially, cognitive ergonomists are looking at the stress effects of having to process information too quickly, or in ways which are good for computers but not for human beings.

- **Applied ergonomics.** This is the design of machines and work systems which achieve the aim of the system (making widgets or getting words onto a hard disk) but do not demand of the human beings in the system physical or mental efforts which would be to their long-term detriment. It is the applied ergonomists who 'fit the job to the person'. However, in order to do this successfully, they have to know what the 'person' is capable of doing comfortably, and they obtain the necessary data from the research ergonomists.

A particular branch of ergonomics is anthropometry which, literally translated, means 'man measurement'. Essentially, anthropometrics is collecting data on how people vary in a wide range of physical dimensions such as height, weight, hand size and physical strength. These data are then used by a range of people for a variety of purposes including the following:

- **Dimensions for furniture.** Most furniture is designed for the 'average' person as determined by anthropometry. Where furniture is adjustable, the range of adjustment required will also be such that it accommodates specified percentages of the population, again as determined by anthropometry.
- **Machine guards.** Various aspects of machine guards rely for their effectiveness on dimensions such as arm reach, finger length and thickness of the wrist being within certain limits. Anthropometry provides the machine guard manufacturers with the data they need to ensure that these dimensions are correct.
- **PPE.** On a commercial level, anthropometry provides PPE manufacturers with the data they need to calculate how many pairs of shoes or gloves of a particular size they should manufacture. (This is true of most clothing manufacturers, not just those manufacturing PPE.) On a safety level, anthropometry will provide manufacturers with data on such things as typical shapes of faces that are needed for the accurate manufacture of breathing apparatus, masks and goggles.

Some typical anthropometric data are given in Tables 16.1 and 16.2 by way of illustration. Note that SD in Table 16.2 is an abbreviation of Standard Deviation. This is a statistical term explained in Chapter 19.

Using the data from research ergonomics, and considering all the system elements required to complete a task, including machinery, equipment, environment and people, it should be possible to identify a systemic unit or system. Working at the level of this systemic unit, the requirements in the human element can be

Table 16.1
Average anthropometric data (in millimetres) estimated for 20 regions of the world – adapted from Jurgens et al.[45]

	1. Stature		8. Sitting height		15. Knee height, sitting	
	Females	Males	Females	Males	Females	Males
North America	1650	1790	880	930	500	550
Latin America Indian population European/Negroid population	1480 1620	1620 1750	800 860	850 930	445 480	495 540
Europe North Central East South east France Iberia	1690 1660 1630 1620 1630 1600	1810 1770 1750 1730 1770 1710	900 880 870 860 860 850	950 940 910 900 930 890	500 500 510 460 490 480	550 550 550 535 540 520
Africa North West South east	1610 1530 1570	1690 1670 1680	840 790 820	870 820 860	500 480 495	535 530 540
Near East	1610	1710	850	890	490	520
India North South	1540 1500	1670 1620	820 800	870 820	490 470	530 510
South China	1520	1660	790	840	460	505
Japan	1590	1720	860	920	395	515
Australia European extraction	1670	1770	880	930	525	570

		Males		Females	
		Mean	SD	Mean	SD
1. Length	US soldiers	194	10	181	10
	US Vietnamese	177	12	165	9
	Japanese	–	–	–	–
	Chinese, Hong Kong	–	–	–	–
	British	189	10	174	9
	Germans	189	9	175	9
2. Breadth at knuckles	US soldiers	90	4	80	4
	US Vietnamese	79	7	71	4
	Japanese	–	–	90	5
	Chinese, Hong Kong	–	–	92	5
	British	87	5	76	4
	Germans	88	5	78	4
3. Maximal breadth	US soldiers	–	–	–	–
	US Vietnamese	100	6	87	6
	Japanese	–	–	–	–
	Chinese, Hong Kong	–	–	–	–
	British	105	5	92	5
	Germans	107	6	94	6
4. Circumference at knuckles	US soldiers	214	10	186	9
	US Vietnamese	–	–	–	–
	Japanese	–	–	–	–
	Chinese, Hong Kong	–	–	–	–
	British	–	–	–	–
	Germans	–	–	–	–
5. Wrist circumference	US soldiers	174	8	151	7
	US Vietnamese	163	15	137	18
	Japanese	–	–	–	–
	Chinese, Hong Kong	–	–	–	–
	British	–	–	–	–
	Germans	–	–	–	–

Table 16.2
Hand and wrist sizes (in millimetres) – adapted from Jurgens et al.[45]

specified, such that they do not cause long-term damage to the human, and the remainder of the system elements modified to suit these human requirements.

This is the ideal approach, but what tends to happen in practice is that machines and work are designed without an understanding of ergonomics and it is assumed that, because people can adapt in the short term and work without apparent harm, there will be no long-term harm. In health and safety terms, this has important consequences. First, the long-term harm being caused will in itself be of concern, such as deafness, back injury, work-related upper limb disorders or stress. Second, having to work in ways which are causing long-term harm causes a general degradation in performance which may make the person more likely to sustain an injury or suffer other forms of ill health. For these reasons, ergonomically sound machine and work systems can make a major contribution to reducing risk in organisations.

In the UK, the HSE has long recognised the safety implications of poor ergonomics and the practical difficulties organisations encounter in their attempts to improve ergonomics. To assist organisations, the HSE has provided guidance[43] which:

- explains how human error and behaviour can impact on health and safety
- shows how human behaviour and other factors in the workplace can affect the physical and mental health of workers
- provides practical ideas on what you can do to identify, assess and control risks arising from the human factor
- includes illustrative case studies to show how other organisations have tackled different human problems at work.

Groups

Introduction

Many tasks are carried out by, or on behalf of, groups; therefore, the influence of groups on an individual's behaviour is an important topic in risk management. For example, in extreme cases, a person predisposed to work safely, on joining a group where it is usual to work in unsafe ways, may choose to work as the rest of the group does in order to gain acceptance by the group.

This section will deal with some of the features of a group which can influence the behaviour of its members. However, first we will describe the main features of groups. Note, however, that industry groups are now often referred to as teams and these two terms are interchangeable for the purposes of this section.

Characteristics of groups

A group is a collection of people who are defined as, or consider themselves as, a group. It is usual to identify two main types of group:

- **Formal groups.** These are usually set up for a reason: that is they are a defined group. Typical formal groups include health and safety committees, boards of directors, product teams and working parties set up to prepare, for example, an organisation's emergency plan.
- **Informal groups.** These are usually formed by the mutual consent of the people concerned, usually to satisfy social needs, and they are groups because the people in them consider themselves as a group. Typical informal or social groups at work include people who share a table in the canteen to discuss mutual interests and people who meet to play cards or sports in the breaks.

Formal groups may be set up to perform a particular task, such as preparing an emergency plan, in which case the group has a natural 'life span'. However, other formal groups, such as boards of directors and health and safety committees, have no such terminal point.

Irrespective of the time span involved, however, newly formed groups typically go through the following phases:

1. **Forming.** The group members come together for the first time and go through the social rituals associated with introductions.
2. **Storming.** There is an initial period of conflict and hostility, which may be so brief as to be hardly noticeable, or it may be so extended that the group never gets anything useful done. Storming may also recur throughout the life of the group but usually in a less marked form. If the group is to be effective, the storming phase should end with the establishment of sufficient trust among the group members for the next phase to begin.
3. **Norming.** As a result of discussions, group members agree to behave in certain ways – that is they agree group norms. Once norms have been established, they then influence the behaviour of the group members during the final phase, and since they have this influence, they are discussed in more detail later in the section.

4. **Performing.** This is the stage during which the group works to meet its objectives (mainly formal groups), or behaves in ways which provide group members with whatever it was they joined the group to receive.

At all of these stages, certain features of the group will determine the extent to which the group influences the individuals within it. Some of the more important of these features are described below, but when reading about them remember that there will be mutual influence: not only will the group influence the individual, the individual will influence the group.

Group size

A group can consist of two or more individuals, but the evidence suggests that a group of between five and seven is both the optimum size for effective performing and also has the most influence on the people in the group. The reasons for this appear to be as follows:

- **Groups smaller than the optimum number do not have sufficient variety.** This lack of variety covers both the knowledge and skills required to carry out tasks (the performing stage) and the personal attributes of the group members which make it possible for each individual to find like-minded group members for affiliation and support. If there are only two people in a group (a dyad) and they are socially incompatible, the group may not get beyond the storming stage.
- **Groups larger than the optimum number tend to fragment into subgroups.** These subgroups may not be obvious to casual observers, or even to the group members themselves. There are methods of measuring the structures within groups and one of these is described in Chapter 29, but for now we can simply state that it is possible to identify subgroups. Individuals who are members of one of these subgroups will tend to be influenced by the subgroup's norms rather than by the norms of the group as a whole.

Group size has obvious practical implications for groups such as health and safety committees, where the size is usually larger than the optimum. It requires great skill on the part of the person chairing such a group to ensure that it does not become two groups in opposition to each other, rather than a single group trying to resolve particular problems.

Group norms

Group norms are the rules and standards by which the group governs its members' behaviour. These norms may be explicit and for some groups they may even be written down. However, the majority of group norms are implicit and not articulated by the group members.

Group norms cover a wide variety of acceptable and unacceptable behaviours, depending on the group concerned, but typical examples include the following:

- **Dress.** This may be what is considered 'fashionable' for social groups, or acceptable dress for groups of managers, or what PPE is worn by members of a work group.
- **Speed of working.** This may be a general norm related to untimed tasks or very specific norms where a pay structure such as piecework is involved. In general, group members appear to agree among themselves on a 'fair' speed of work for the rewards being received and exert pressure on the group members who work slower or faster than the agreed rate.
- **Hours of work.** In a similar way to the way groups agree speed of work, they can also agree hours of work. This is a particular issue with management in the UK, where group norms are now significantly above contracted hours in many organisations. In these organisations, group pressure is such that anyone working only for the contracted working hours is excluded from the 'management group'.
- **Work methods.** Group members can agree among themselves ways of working and impose these work methods on everyone in the group, including people who join the group after the norms have been agreed.

This can be a particular problem in health and safety if the methods of work agreed involve 'taking short cuts' or other practices with high risks associated with them.

Group norms, of whatever type, can be strong or weak depending on the circumstances. In general, where the norms are weak, the group will tolerate limited deviation from the norms, but little deviation, if any, is tolerated when the group norms are strong. A number of factors influence whether or not a group norm will be strong and typical ones include:

- **The importance of the subject of the norm to the members of the group.** Norms which are perceived to be related to pay, promotion and job security are examples which are important to the majority of group members.
- **The perceived status of the group.** Groups which are perceived to be of a high status will exert a strong influence on group members to remain in the group, and there are likely to be people wishing to join the group. Norms in a high status group can, therefore, be very strong, since it is unlikely that anyone will wish to go against them and so lose their group membership or opportunity to join the group.
- **The cohesiveness of the group.** Highly cohesive groups are ones in which the norms are agreed by all of the group members. Because everyone had a say in the formation of the norms, they are more likely to abide by them.
- **The quantity and quality of communication within the group.** Communication within the group influences cohesiveness in that the opportunities to agree norms are limited if there is restricted communication. However, good communication also enables deviation from norms to be identified and corrected so that the norms do not become 'diluted' or fall into disuse.
- **Peripheral benefits from the group.** These benefits include social interaction, support, friendship, opportunities to learn from others (or have one's opinions confirmed) and protection. Different individuals will perceive different aspects of the group's activities as peripheral benefits. For example, some will see the group as an opportunity to express opinions, while others will see it as an opportunity to listen to other members' opinions. However, where such peripheral benefits are perceived, they have a similar effect on group norms as status does. That is, individuals will conform closely to the group norms because they do not want to lose the benefits they receive from being a member of the group.

There is a final point on groups, which concerns the relative status of the group members. In formal work groups, this may be defined in terms of level in the management hierarchy, while in informal groups, it may be defined in terms of age or level of skill. However it is defined, groups consisting of members of similar status are often referred to as 'peer groups'. Since most people spend most of their time in peer groups, they are a particularly important category to target when trying to influence group behaviour. The topic of influencing group behaviour will be left to Chapter 29.

Organisations

Introduction

The vast majority of the type of risk management dealt with in this book takes place in organisations and, if risk management is to be effective, it is necessary to know how organisations are structured and how they function. This section is, therefore, devoted to the structure and function of organisations and the various aspects of these topics which have to be taken into account if risk management in an organisation is to be carried out effectively.

The section begins with a definition of the characteristics of an organisation and then considers certain aspects of the formal structure of organisations. There is then a consideration of organisations as systems and a brief note on organisational culture.

Definition of organisations

In his book *Organizational behavior*,[46] Duncan defined an organisation as:

> ... a collection of interacting and interdependent individuals who work towards common goals and whose relationships are determined according to a certain structure.

This definition is a useful one, although, as might be expected, it is not accepted by everyone, since it highlights two particular characteristics of organisations:

- Organisations consist of people interacting with each other and the 'hardware' associated with organisations, such as buildings and machinery, is peripheral to this interaction of people.
- The people in the organisation must be working towards a common goal or goals. While it is unlikely that everyone in the organisation will have the same perception of these goals, or have the same priorities for achieving them, there must be some degree of commonality to justify labelling the group of people an organisation.

There are, however, two important aspects of organisations which are not made explicit in Duncan's definition:

- Organisations are not confined to industrial or commercial organisations. Schools, colleges, social clubs, hospitals and so on are all organisations.
- Although organisations may have a formal structure, perhaps set out in the form of an organisation chart (see below), there will also be informal groups and structures within organisations and these may be as important as the formal structures in determining how the organisations function.

This second point should be borne in mind when reading the description of formal structures of organisations which follows.

Formal structures of organisations

Early work on the formal structure of organisations was based on the assumption that people who worked in organisations behaved in a completely rational manner. However, this approach, described as 'scientific management' and originally developed by F W Taylor[47] in the early 20th century, took a very limited view of rationality in that it was primarily concerned with economic rewards and sanctions. There was relatively little account taken of people's social or other needs.

On the basis of these assumptions about rational behaviour in organisations, the preferred structure for an organisation was the bureaucratic model, which had the following features:

- division of labour, with different people performing different tasks
- definition of the organisation in terms of roles (what people do), not in terms of named individuals
- a hierarchy of authority with most authority allocated to the roles at the 'top' of the hierarchy
- written rules, regulations and procedures
- rational application of the written rules, regulations and procedures.

It is possible that this sort of bureaucratic model was an attempt to mimic military styles of management or the early bureaucracies which functioned because people had to do what they were told. With hindsight, it appears that these early proponents of management theory were attempting to argue that 'rational man' was as amenable to a disciplined approach as 'military man' or the serfs, slaves and other underclasses which made early bureaucracies possible.

Fortunately, to describe an organisation as bureaucratic is now more of a criticism than a commendation and there are few organisations these days which aspire to being true bureaucracies. However, certain characteristics of bureaucratic organisations are still important and these will be considered by illustrating them in the context of organisation charts.

Organisation charts

Most organisations use some form of organisation chart. In smaller organisations, this is usually a single diagram of the type illustrated in Figures 16.2 and 16.3.

Figure 16.2
Organisation chart for a small service organisation

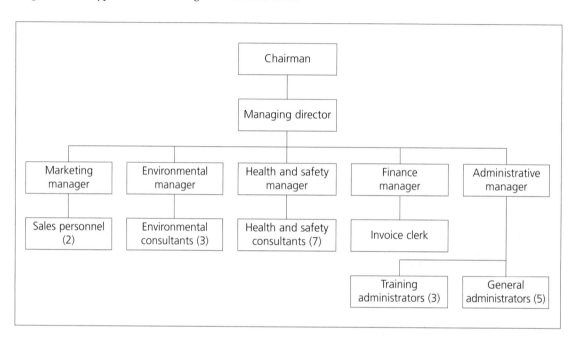

In larger organisations, it is not practical to record all the required information on a single sheet so that some form of hierarchy of charts is required. A typical hierarchy of charts is illustrated in Figures 16.4 to 16.6.

Figures 16.2 to 16.6 illustrate the most commonly used type of organisation chart – that is, describing hierarchical organisations. There have been attempts to avoid the impression of a hierarchy by laying out the charts in other ways, and examples of these are shown in Figures 16.7 and 16.8.

However, the use of these alternative forms of organisation chart is rare and we will illustrate the various aspects of organisational structure using the more conventional hierarchical organisation chart.

Organisational roles

The organisation charts identify roles and, only as a subsidiary function, the people who currently fill these roles. Indeed, an organisation chart which showed only individuals' names would be of limited value. However, organisation charts describe the intention, which may or may not be carried out in practice. You may have come across organisations where particular jobs are done by the person most competent or most willing to do them, rather than by the person identified on the organisation chart as the person who should do them.

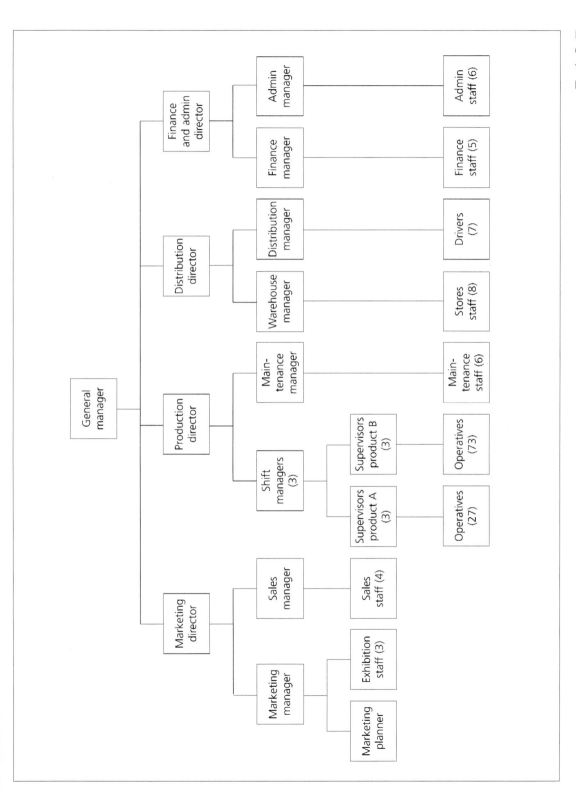

Figure 16.3
Organisation chart
for a medium-sized
production company

Figure 16.4
Top level
organisation chart
for a large
organisation
(Chart 1)

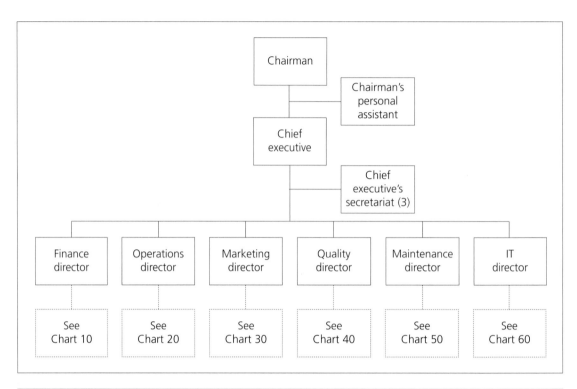

Figure 16.5
Intermediate level
organisation chart
for a large
organisation
(Chart 60)

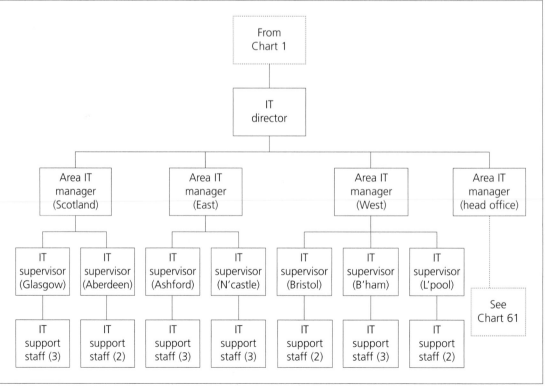

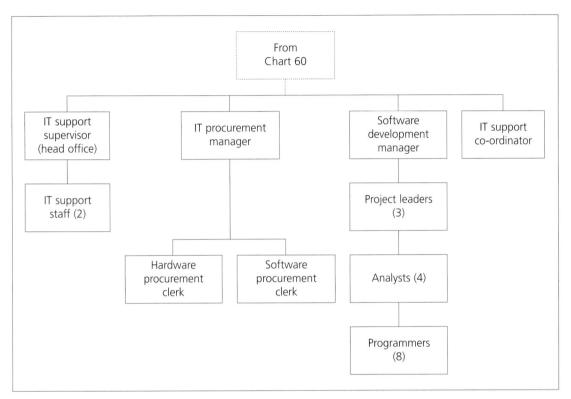

Figure 16.6
Final level
organisation chart
for a large
organisation
(Chart 61)

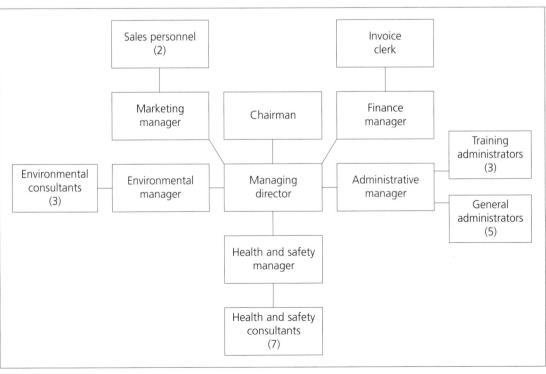

Figure 16.7
Non-hierarchical
organisation chart
for a small service
company

Figure 16.8
Non-hierarchical
organisation chart
for a small
production company

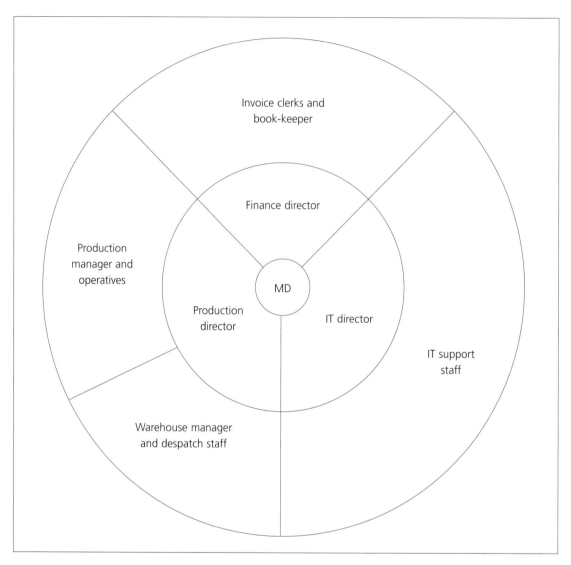

All organisational roles (including managerial roles) will have safety responsibilities. In practice, all the elements in the risk management system are managerial responsibilities. Even in those cases, such as audit, where external resources are used, the responsibility for ensuring that the necessary work is done remains with managers.

However, the health and safety adviser is a special case and HSG65[1] provides a summary of the role and functions of health and safety advisers as follows:

Health and safety advisers need to have the **status** and **competence** to advise management and employees or their representatives with authority and independence. They are well placed to advise on:

• formulating and developing health and safety policies, not just for existing activities but also with respect to new acquisitions or processes;

- how organisations can promote a positive health and safety culture and secure the effective implementation of health and safety policy;
- planning for health and safety including the setting of realistic short- and long-term objectives, deciding priorities and establishing adequate systems and performance standards;
- day-to-day implementation and monitoring of policy and plans including accident and incident investigation, reporting and analysis; and
- review of performance and audit of the whole health and safety management system.

To do this properly, health and safety advisers need to:
- be properly trained and suitably qualified (the Occupational Health and Safety Lead Body* standards offer one route to demonstrating competence);
- maintain adequate information systems on topics including civil and criminal law, health and safety management, and technical advances;
- interpret the law in the context of their own organisation;
- be involved in establishing organisational arrangements, systems and risk control standards relating to hardware and human performance, by advising line management on matters such as legal and technical standards;
- establish and maintain procedures for reporting, investigating, recording and analysing accidents and incidents;
- establish and maintain procedures, including monitoring and other means such as review and auditing, to ensure senior managers get a true picture of how well health and safety is being managed (where a benchmarking role may be especially valuable); and
- present their advice independently and effectively.

Relationships within the organisation
Health and safety advisers:
- support the provision of authoritative and independent advice;
- have a direct reporting line to directors on matters of policy and the authority to stop work if it contravenes agreed standards and puts people at risk of injury; and
- have responsibility for professional standards and systems. On large sites or in a group of companies, they may also have line management responsibility for other health and safety professionals.

Relationships outside the organisation
Health and safety advisers liaise with a wide range of bodies and individuals including: local authority environmental health officers and licensing officials, architects and consultants, the Fire Service, contractors, insurance companies, clients and customers, HSE, the public, equipment suppliers, HM Coroner or the Procurator Fiscal, the media, the police, general practitioners, and occupational health specialists and services.
[Bold text above is bold in HSG65.]

While this is a useful summary of the role and functions of health and safety professionals, it should be noted that there is no reference to any role in risk assessment.

---

* The Occupational Health and Safety Lead Body has been replaced by a new body serving the same functions – the Employment National Training Organisation. See reference 96 in the References section at the end of the book.

Division of labour

The vertical divisions of the organisational chart represent division of the labour required by the organisation into a series of specialisms. This is a very common method of setting up an organisation's structure, but it does have the weakness that gaps and overlaps may be inherent in the structure, or develop as the organisation evolves.

Historically, at least in the UK, this type of structure has caused demarcation disputes where particular jobs were only carried out by one or two people who could, effectively, bring production to a halt by withdrawing their labour. To overcome this problem, and the gaps and overlaps already mentioned, several different approaches to division of labour have been adopted, including:

- **Multi-skilling.** This operates at the individual level with, for example, operatives being provided with the competences required to do their own maintenance work and quality control checks.
- **Matrix management.** This operates across all or part of an organisation, with people reporting to different managers for different parts of their work, and managers having different types of responsibility in different work areas.
- **Product or service teams.** Since in most organisations one of the primary goals is to create products or provide services, some organisations adopt a structure in which a team of people is responsible for all aspects of the product or service and call on the services of specialists within the organisation, in a similar way to that which they might use to call in services from outside the organisation. In effect, a customer–supplier relationship is set up. This can be a very useful structure, since it keeps support services, such as maintenance and finance, focused on the organisation's goals, rather than allowing them to become an end in themselves.

Lines of responsibility

Organisation charts define lines of responsibility or chains of command in the organisation, but it is notable that these lines run only vertically. This means that, in theory, senior personnel are responsible only for people 'below' them in the organisation chart and no-one in the organisation reports to more than one person. This is usually not true in practice and organisation charts are often liberally sprinkled with 'dotted line' responsibilities in an attempt to illustrate the complex lines of responsibility required in real life.

Communication

While the lines of responsibility on an organisation chart imply communication, organisation charts are very poor at describing the communications required for an organisation to operate effectively. It would not, for example, be practical to have all lateral communications in an organisation routed through the single person at the top of a hierarchical organisation chart, although this is what these types of chart imply.

It is, however, possible to draw up communications charts which describe the flow of information in an organisation. This can be done either as a description of how communications should be, or a communications audit can be carried out to establish what actually happens and a chart drawn up on the basis of the data collected. Typically, a communications audit will look at a sample of communications and establish the following types of information:

- the origin and destination of the communication
- the type of communication, for example face-to-face, telephone, fax or email
- the content of the message
- the purpose of the message and whether this purpose was achieved.

In practical terms, a communications audit is usually carried out to identify resources which are being wasted on creating and processing unnecessary data, and opportunities which are being missed because communication is inadequate in some way.

Major dimensions of organisational structure

So far, a number of aspects of organisational structure have been identified, including hierarchy, roles, chains of command and communication. Organisations vary widely with respect to these aspects of structure, and in many other ways. Various people have tried to establish whether there are key dimensions of structure and one major study[48] arrived at the following as the major dimensions of organisational structure:

* **specialisation** – the extent to which particular tasks and roles are allocated to individuals within the organisation
* **standardisation** – the extent to which the organisation has standard procedures
* **formalisation** – the extent to which procedures are written down
* **centralisation** – the extent to which authority and decision making are restricted to those at the top of the hierarchy
* **configuration** – whether the chains of command are long or short.

Organisations as systems

While formal systems theory will not be dealt with until Part 2.1, it is useful at this point to deal with organisations as systems, including organisations as part of wider, social systems. This will be done by looking at two system diagrams:

1. an open system view of an organisation which shows the inputs to the organisation, the outputs from it and, within the organisation itself, how there are separate formal, technology and social subsystems (Figure 16.9)
2. a more detailed system diagram illustrating the dynamics of an organisation and the factors which have to be taken into account if an organisation is to be considered part of the wider society (Figure 16.10).

Systems and systems theory are discussed in more detail in Chapter 18.

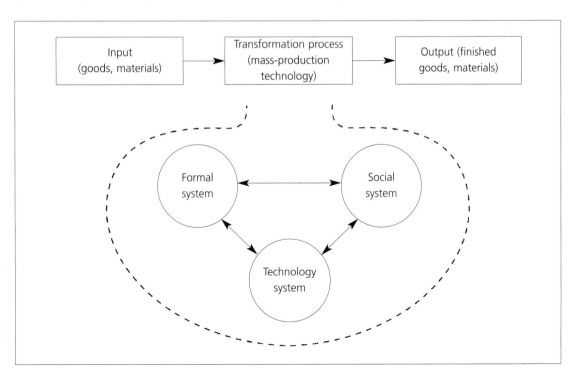

Figure 16.9
An open system view of an organisation (from Arnold et al.[49])

Figure 16.10
Kotter's model of
organisational
dynamics (from
Arnold *et al.*[49])

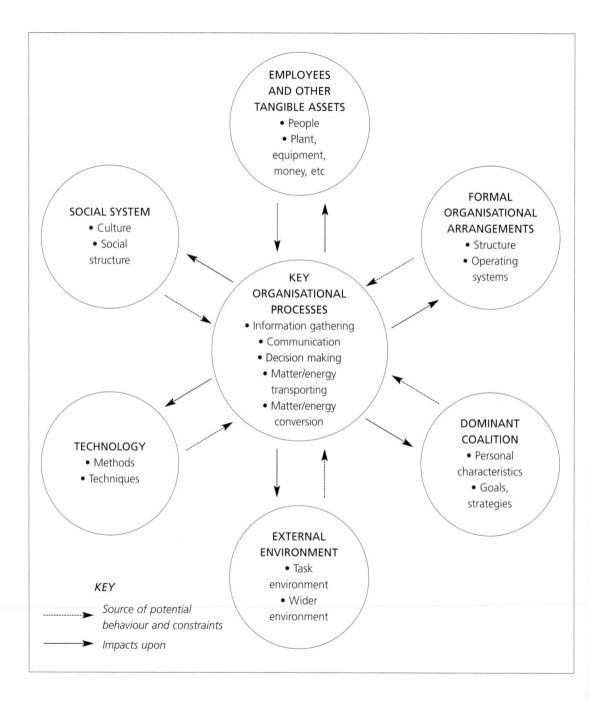

The second model, illustrated in Figure 16.10, is a much more detailed one and can be used as a checklist of the factors which have to be taken into account when considering organisations and how they operate.

The concept of organisations as systems is dealt with in more detail in Chapter 18. We will end the current discussion of organisations with some brief notes on organisational culture.

Organisational culture

The safety culture of an organisation is an important topic which is dealt with in Chapter 21. For now it is enough to know that safety culture has been defined by the HSE[1] as:

> ...the product of individual and group values, attitudes, competencies and patterns of behaviour that determine the commitment to, and the style and proficiency of, an organisation's health and safety programmes. Organisations with a positive safety culture are characterised by communications founded on mutual trust, by shared perceptions of the importance of safety, and by confidence in the efficacy of preventive measures.

However, an organisation's safety culture is only a subset of its overall culture. It may be that desired changes in safety culture can only be brought about by changing the overall culture. For example, a basic requirement of a good safety culture is communications which are based on mutual trust. While it may be possible to establish and maintain mutual trust on safety matters while distrust on other matters remains, this seems unlikely to be sustainable in the long term. Where, therefore, there is mistrust in the general culture of an organisation, this will probably have to be addressed before progress can be made on this aspect of safety culture.

Handy[50] has identified four main types of organisational culture:
* **the power culture**, which depends on a central power source and involves few rules and little bureaucracy
* **the role culture**, which works by logic and rationality and is, essentially, a bureaucracy
* **the task culture**, which is task or project-oriented and is typified by matrix management
* **the person culture**, in which the organisation exists only to serve the individuals making it up.

Handy describes these cultures in detail and their implications for the organisations concerned. An important contributor to organisational culture is the style of management and leadership within the organisation, and one of the main symptoms of a poor organisational culture is conflict. These topics are the subject of the next two subsections.

Management style and leadership

Management and leadership are not necessarily synonymous. In general, managers are people who organise the work of others so that certain tasks are accomplished or objectives met. A leader, on the other hand, is in Handy's definition, 'someone who is able to develop and communicate a vision which gives meaning to the work of others'.[50] This difference can have important implications in organisations where there is a formally appointed manager of a group and an informal leader of that group and the two are in conflict. However, for the purposes of this subsection, it will be assumed that the managerial role and the leadership role are fulfilled by the same person and management style and leadership style will be used interchangeably.

As might be expected, there are numerous theories of management and leadership. However, Handy has grouped them into three basic categories as follows.

Trait theories

These theories are based on the assumption that good leaders will exhibit certain traits. Studies quoted by Handy have identified the following traits as being characteristic of good leaders:
* **intelligence**: above average, but not of genius level, and good at solving abstract and complex problems
* **initiative**: independent and inventive, with the capacity to perceive the need for action and the urge to take action
* **self-assurance**: self-confident, with reasonably high self-ratings on competence and aspiration levels
* **the helicopter factor**: the ability to rise above the particulars of a situation and perceive its relations to the overall environment.

However, it has also been found that there are successful leaders who do not possess these traits.

Style theories

These theories are based on the assumption that good leaders will exhibit certain leadership styles. Various styles have been identified, including the following classifications:

- autocratic *versus* democratic
- leader control, through shared control to group control
- telling, selling, participating and delegating.

As with trait theories, successful leaders have been found in all categories and it appears that different styles are successful in different circumstances, hence the third group of theories.

Contingency theories

These theories try to take into account factors other than the traits or style of the leader. On the basis of his review of these theories, Handy suggests that three other factors are important:

- **The subordinates.** Different people respond differently to different leadership styles. Successful leaders may be ones who can alter their style to suit particular groups of subordinates.
- **The task.** Different leadership styles are appropriate for different types of task. For example, autocratic leadership is more likely to be appropriate for simple tasks which have to be completed in a short timescale.
- **The environment.** This is the organisational environment, not the physical environment. It includes such factors as the seniority of the leader and the organisational norms. For example, if the organisational norm is constant criticism and little praise, this limits the conforming leaders' range of styles.

As with most aspects of life which involve humans and the interactions between them, it is unlikely that there will be a single theory which encompasses all relevant factors. However, a knowledge of the sorts of factors which should be taken into account is useful for anyone who has to deal with organisational issues.

Conflict in organisations

Conflict in organisations can take many forms. It is useful to begin by considering three types of conflict.

- **Conflict within an individual.** This, in the organisational context, is called role conflict, since it arises when an individual has two or more roles to fulfil and their requirements conflict with each other. A common role conflict is safety versus production, where a manager's role includes ensuring that his or her personnel work safely, but there is pressure from senior management to get them to work quickly.
- **Conflict between individuals.** These sorts of conflicts arise for a variety of reasons, some of which may be 'personal' rather than 'organisational'. For example, the conflict may be due to something which is happening away from work but the effect of the conflict influences behaviour at work.
- **Conflict between groups.** This sort of conflict arises either because the groups contain individuals who are in conflict and this conflict is generalised to other group members, or because the goals of the groups are in conflict.

Conflict usually arises because individuals or groups have competing goals, or one group has a goal whose achievement would preclude a second group from achieving its goal. Conflict resolution techniques are, therefore, based on:

- accurate identification of the goals which are producing the conflict
- attempts to find ways of satisfying both sets of goals
- attempts to get one or both groups to change their goals so that there is no longer a conflict.

Effective conflict resolutions require a range of skills, including investigation skills (to identify the conflicting goals), creativity (to identify alternative goals), negotiation skills (to keep the groups talking), and, in extreme cases, counselling skills, since conflict can be extremely stressful for the individuals involved.

Society

Introduction

Organisations do not operate in isolation; they are all part of a society. This society will influence the ways in which the organisation operates through such things as legislation, enforcement agencies and public opinion. Society also influences groups and individuals, and these types of influence will also be considered in this section. First, we will briefly consider what the term 'society' covers.

A society is a group of individuals interacting in a large number of ways. When discussing a particular society, it can help to subdivide it and deal with each subdivision separately, and we will consider one such subdivision in a moment. However, there are usually overlaps between the subgroups, and they interact with each other in various ways, so it is also necessary to consider the implications of these overlaps and interactions.

Some of the major subdivisions of society that are relevant to risk management are:

- **legislators** – the people who decide on the contents of legislation. In the UK, this is done by Parliaments and, in the case of health and safety legislation, there is input from the HSE
- **enforcers** – the people who enforce the legislation. In the UK, this is primarily done by the HSE and enforcing agencies in local authorities
- **arbitrators** – the people who make judgments in the case of disputes. In the UK, they include courts, tribunals and other arbitration agencies. The decisions made by some arbitrators have a legal standing, so these arbitrators' role overlaps with the role of the legislators
- **business owners and managers** – including shareholders, people who own businesses under other legal arrangements, the managers employed by owners to run their businesses, and the self-employed
- **organisations representing business owners and managers** – including, in the UK, the Confederation of British Industry, the Institute of Directors, Chambers of Commerce, and various organisations representing particular industry sectors or sizes of business
- **employees** – essentially, everyone who works in businesses, but it is usual to divide them into categories, such as managerial and non-managerial
- **organisations representing employees** – in the UK, usually trade unions
- **insurance companies** – in the UK, certain types of insurance are legally required but organisations, groups and individuals can also take out other types of insurance. In health and safety, the compulsory nature of some insurances gives insurance companies a role in risk management
- **'pressure groups'** – a generic description of organisations set up specifically to promote a particular point of view. Notable pressure groups at present are those working to save the natural environment and its flora and fauna
- **customers and suppliers** – most organisations and individuals are both customers and suppliers, or contractees and contractors, but they may have different priorities and behave differently when they switch between the two roles
- **the public** – in effect, everyone, hence the need to discuss overlaps.

Societies are complex because of the extensive overlaps between the various subgroups and the variety of links between them. The position is made more complicated by circular links. For example, the legislators decide what the law will be and therefore how the public should behave, but the public elects the legislators.

Similarly, enforcing authorities uphold the law but they also advise the legislators on the laws they will have to enforce.

It is tempting (and probably true) to say that the effect of any change in society will be unpredictable, but it is possible to look back over recent changes and try to tease out some trends which are likely to have implications for risk management. This section ends, therefore, will a brief list of such trends.

- **The increasing standard of living of the general public.** This appears to be having the effect of reducing people's willingness to take risks with their long-term wellbeing. Intuitively, this would seem sound since, if life is good, people do not want it threatened. In risk management terms, this would seem to predict a workforce which will become increasingly motivated to work in ways which minimise the risk of injury and ill health.
- **The increasing pressure for productivity.** This appears to be a necessary accompaniment to the increasing standard of living. Unfortunately, its effect is often in direct opposition to that resulting from an increased standard of living.
- **An increasing concern with environmental matters.** Public opinion, perhaps spurred on by the activities of pressure groups, is influencing organisations, which now have to pay more attention to any effects they may be having on the environment. In risk management terms, this is likely to lead to the inclusion of environmental risks in the general portfolio of risks to be dealt with.
- **In the UK, a move from 'prescriptive' legislation to 'goal-setting' legislation.** Modern legislation is moving away from telling organisations what they have to do, such as 'put a specific type of guard on a particular class of machine', towards telling organisations what they have to achieve, such as 'prevent people being injured by your machines'. This appears to be having a number of effects on organisations and the health and safety profession, since it now requires a much higher level of expertise in risk management techniques in order to comply with legislative requirements.
- **Contractee–contractor relationships.** There is a trend in the UK towards a greater use of contractors, often as a result of 'outsourcing', that is arranging for a complete organisational function such as information technology or catering to be carried out by contractors rather than by direct employees of the organisation. This has been accompanied, perhaps coincidentally, by a greater awareness of the legal obligations organisations have towards their contractors with respect to health and safety. Many large organisations now audit the safety management systems of their contractors and some even provide help and guidance where these are required.

The points raised above are, however, only a sample of recent changes. There are many more which might be explored, including the decline in membership of trade unions, the increasing cost of insurance, the large increase in car use, and the changes in risk management techniques associated with more effective use of risk assessment.

Changes in any aspect of society will always have the potential to affect risk management but, unfortunately, it is often only possible to identify the actual effects with hindsight.

Summary

This chapter considered a number of factors which affect the individual, beginning with machinery, equipment and work, followed by an increasingly wide circle of social environment that is groups, organisations and society.

Part 2.1:
Risk management – advanced

17: Part 2.1 – overview

Introduction

Part 2.1 deals with the more advanced aspects of risk management contained in the NEBOSH Diploma syllabus and the knowledge requirements for certain aspects of the NVQ/SVQ (see Appendices 1 and 2). It also addresses topics covered in some university health and safety syllabuses (see Appendix 3). This chapter gives an outline of the risk management material covered in this part.

- **Chapter 18: Management systems.** The concept of an HSMS was introduced in Part 1.1 and was illustrated using a number of examples of published HSMSs. However, there are HSMSs other than those described in Part 1.1, and there are management systems for other processes, such as environmental and quality management, which are relevant to risk management. These additional systems are dealt with in this chapter. In addition, the chapter discusses how different management systems – safety, health, quality and environment for example – can be integrated, either via total quality management or by using some other form of integration. The chapter ends with a brief note on the implementation of management systems.

- **Chapter 19: Measuring performance.** This chapter deals with a number of issues that are relevant to measuring performance in the context of risk management, but the emphasis is on quantitative measurements. The chapter begins with a discussion of the terms 'reactive' and 'active', and an overview of the types and nature of data which can be collected for measurement purposes. It then moves on to a description of how numerical data can be used incorrectly if the nature of the underlying numerical scales is not accurately identified. However, the majority of the chapter is devoted to methods for presenting and analysing numerical data, beginning with descriptive statistics. There are then sections on summary statistics, probability, the calculation of accident and ill health rates, trend analysis and epidemiology.

- **Chapter 20: Advanced accident investigation and risk assessment.** These two topics are dealt with in a single chapter because, at the advanced level, there are major overlaps between them. This is not surprising when we remember that accident investigation is used to find out what actually caused an accident, while risk assessment is used to find out what could cause an accident. Thus, both sets of techniques are concerned with the identification of accident causes. Chapter 20 begins with a brief look at two accident investigation techniques: Events and Causal Factors Analysis (ECFA) and the Management Oversight and Risk Tree (MORT). However, most of the chapter is devoted to risk assessment, starting with a discussion of various conceptual problems associated with risk assessment, and the practical difficulties these create for the application of risk assessment techniques. The chapter ends with a brief overview of four of the advanced risk assessment techniques: Hazard and Operability studies (HAZOPs), Failure Modes and Effects Analysis (FMEA), Event Tree Analysis (ETA) and Fault Tree Analysis (FTA).

- **Chapter 21: Advanced risk control techniques.** This chapter is divided into three broad sections, each relevant to some aspect of advanced risk control techniques. The first section covers how the effectiveness of risk control measures can be estimated in terms of risk reduction and how an organisation's reliance on risk control measures can be estimated. However, since the practical effectiveness of risk control measures is crucially dependent on their reliability, this section also discusses how the reliability of risk control measures can be estimated. The second section of the chapter deals with methods which can be used to generate ideas for risk control measures. It can be argued that many risks continue to be dealt with inadequately because too little creative thought is put into choosing or designing control measures. This could be remedied by the application of techniques for stimulating creative thinking and two such techniques, brainstorming and systems thinking, are described in this chapter. The final section in the chapter deals with the broad issue of safety culture which can be seen as the most advanced type of risk control measure currently available.

- **Chapter 22: Emergency planning.** This chapter has two main aims: first, to describe what is meant by an emergency and, second, to describe planning techniques. For the purposes of the chapter, we use the planning technique described in Annex D of BS 18004 and describe how it is used in the context of emergency planning.
- **Chapter 23: Advanced audit and review.** The basic techniques of review and audit were dealt with in Part 1.1. Chapter 23 builds on these basic techniques by describing the more advanced techniques available.
- **Chapter 24: Financial issues.** It may be the case that the most effective motivation to do something about managing risk effectively is a demonstration that effective risk management leads to improvements in an organisation's financial performance. However, in order to do this, a range of financial data on the costs of risk management functions will have to be collected and analysed. In addition, the savings, in terms of reductions in losses, will have to be calculated or estimated. With these two types of data available, a cost–benefit analysis can be carried out to determine whether the costs of risk management activities are outweighed by the benefits these activities produce.

18: Management systems

Introduction

In Chapter 4, we introduced five published HSMSs, and two of them were described in detail. However, there are other HSMSs which were not mentioned earlier, and there are other management systems which have had an influence on health and safety management. This chapter will, therefore, begin with an overview of management systems.

It is apparent from the material presented in Chapter 4 that there is a high degree of overlap between the various HSMSs, and, as will be seen in this chapter, there is even more overlap between the environmental management system (ISO 14001[52]) and the OH&SMS used in BS OHSAS 18001. From the overview of management systems to be presented later in this chapter, it will also be seen that there is, in fact, a high degree of overlap between all management systems. Because of this overlap, it makes sense to ask why organisations have different management systems for health and safety, environment and quality. The second part of the chapter is, therefore, a discussion of the arguments for and against integrated management systems.

Any management system relies on people for its implementation, so the chapter ends with a brief discussion of the implementation aspects of management systems.

Management systems

It is useful to deal first with three general aspects of management systems: their sources, the topics they cover and certification. We will then describe particular management systems in more detail.

The sources of management systems

Management systems are devised and promulgated by a number of different types of organisation, including:

- international or national standard setting bodies – these have produced, for example, the ISO 9000 series,[51] BS 18004,[2] and BS OHSAS 18001[3]
- authoritative agencies such as the UK's HSE, which has produced HSG65,[1] and the ILO, which has produced OSH 2001[9]
- commercial organisations, which produce 'branded' management systems in a variety of formats[7]
- academic organisations or individuals.

As we have seen, there is, at least in the UK, an overlap between the first two categories and, in practice, they all overlap since commercial organisations adopt ideas from published standards and authoritative bodies, and academics influence, or attempt to influence, all the others. The main difference between the categories is that commercial organisations and, to an increasing extent these days, academic institutions, provide a range of products and services such as training to assist with the implementation of their management systems.

The topics covered by management systems

Certain management systems deal with specific topics, such as health and safety (OHSAS 18001 and HSG65, for example), quality (ISO 9001[5]) or the environment (ISO 14001[52]), and these specific management systems will be dealt with in more detail later. However, there are management systems which attempt to cover all topics and these are usually known as total quality management systems (TQMSs), total loss control systems (TLCSs) or integrated management systems (IMSs). Effectively, TQMSs, TLCSs and IMSs attempt to integrate health, safety, environment and quality and, in addition, elements such as security, vandalism, and 'off the job' health and safety. That is, they include anything which might degrade product or service quality, or contribute to the losses sustained by an organisation. There are practical problems with this integrated approach, which will be discussed in the section on integration later in the chapter.

Certification of management systems

The fact that an organisation claims to operate a management system is, of itself, not a guarantee that the management system is adequate or effective. This has long been recognised in the case of product and service quality and has led to the development and implementation of a certificatable quality management standard, ISO 9001. However, for there to be a valid certification process, there are several requirements:

- there must be a written standard for the management system with which the organisation seeking certification must comply
- there must be companies employing disinterested and competent auditors who can assess whether an organisation is complying with the written standard; this part of the process is usually referred to as external audit, or simply audit
- there must be an authoritative body which verifies that the auditors remain disinterested and competent, and accredits the organisation that employs them to issue certificates of conformity to the relevant Standard; in the UK, this function is performed by the United Kingdom Accreditation Service (UKAS).

All of these are in place in the UK for product and service quality management (based on ISO 9001), for environmental management (based on ISO 14001), and health and safety management (based on BS OHSAS 18001). Details of organisations that are accredited to issue certificates of conformity are available on the UKAS website, www.ukas.com.

Note that BS 18004 is not a certificatable standard. Its foreword states: 'BS 18004 takes the form of guidance and recommendations. It should not be quoted as if it were a specification.'[2] Since HSG65 is guidance, it too is not a certificatable standard.

Grouping of management systems

For the purposes of this chapter, the management systems can be divided into four groups:
1. HSMSs
2. environmental management systems
3. quality management systems
4. integrated management systems.

HSMSs

The health and safety professional in the UK needs a detailed knowledge of the main HSMSs (HSG65 and the BS OHSAS 18000 series), and they should be studied from the source material (see the References at the end of the book). However, a number of alternative HSMSs have been published and sources for a selection of these are also given at the end of the book. At various points in this section, particular aspects of HSG65, BS 18001 and BS OHSAS 18004 will be discussed and referred to. Readers who are unfamiliar with these texts should read them before proceeding with the remainder of this section.

Environmental management systems

The most influential environmental management system in the UK is published by the BSI and is generally known as ISO 14001.[52] This standard consists of four clauses preceded by an introduction and followed by two annexes and a bibliography.

The introduction to the standard deals with a number of background issues and describes the model for the environmental management system. The model is very similar to the one described in Part 1.1 during the discussion of BS OHSAS 18001. However, the actual model used in ISO 14001 is reproduced in Figure 18.1 and the meaning of the various elements is discussed later in this section, when the clauses making up ISO 14001 are described.

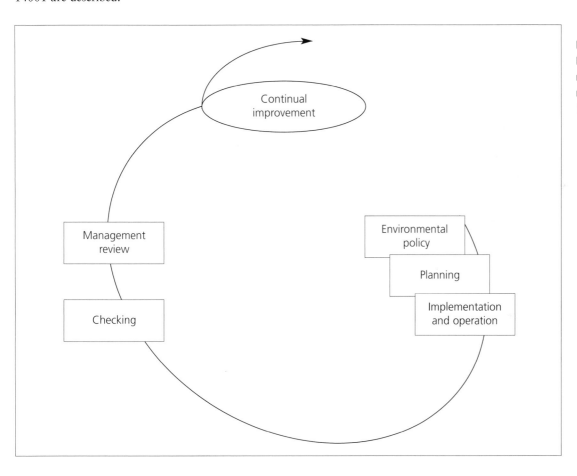

Figure 18.1
Environmental management system model for ISO 14001

The introduction also contains a number of caveats and qualifications, and a selection of these is quoted below.

> The success of the system depends on commitment from all levels and functions of the organization, and especially from top management.
>
> Guidance on supporting environmental management techniques is contained in other International Standards.
>
> This International Standard contains only those requirements that can be objectively audited. Those organizations requiring more general guidance on a broad range of environmental management system issues are referred to ISO 14004.[53]
>
> This International Standard does not establish absolute requirements for environmental performance beyond the commitments, in the environmental policy, to comply with applicable legal requirements and with other requirements to which the organization subscribes, to prevention of

pollution and to continual improvement. Thus, two organizations carrying out similar operations but having different environmental performance can both conform to its requirements.

This International Standard does not include requirements specific to other management systems, such as those for quality, occupational health and safety, financial or risk management, though its elements can be aligned or integrated with those other management systems. It is possible for an organization to adapt its existing management system(s) in order to establish an environmental management system that conforms to the requirements of this International Standard.

Note that, as in quality management, where organisations define the level of quality and conform to it, in ISO 14001 organisations define the level of environmental performance and comply with it, so 'two organizations carrying out similar operations but having different environmental performance can both conform to its requirements'.

The penultimate paragraph deals with the integration of ISO 14001 with other management systems, including occupational safety and health, and the use of these other management systems as a basis for environmental management.

Clause 1 – Scope
The core of the scope clause is as follows:

This International Standard is applicable to any organization that wishes to:
a) establish, implement, maintain and improve an environmental management system
b) assure itself of conformity with its stated environmental policy
c) demonstrate conformity with this International Standard by
    1) making a self-determination and self-declaration, or
    2) seeking confirmation of its conformance by parties having an interest in the organization, such as customers, or
    3) seeking confirmation of its self-declaration by a party external to the organization, or
    4) seeking certification/registration of its environmental management system by an external organization.

Clause 2 is 'Normative references' but none is given.

Clause 3 – Definitions
Twenty terms are defined but some of them, for example 'organization', are not specific to environmental issues. The following definitions are of interest:

Environment – surroundings in which an organization operates, including air, water, land, natural resources, flora, fauna, humans and their interaction
Environmental aspect – element of an organization's activities or products or services that can interact with the environment
Environmental impact – any change to the environment, whether adverse or beneficial, wholly or partly resulting from an organization's environmental aspects.

It follows from the definition of environmental impact that claiming to reduce the environmental impact of an organisation's activities is not necessarily a good thing; the relevant claims should be to reduce the adverse environmental impact or increase the beneficial environmental impact.

Environmental performance – measurable results of an organization's management of its environmental aspects

Prevention of pollution – use of processes, practices, techniques, materials, products, services or energy to avoid, reduce or control (separately or in combination) the creation, emission or discharge of any type of pollutant or waste, in order to reduce adverse environmental impacts

Note: Prevention of pollution can include source reduction or elimination, process, product or service changes, efficient use of resources, material and energy substitution, reuse, recovery, recycling, reclamation and treatment.

Continual improvement – recurring process of enhancing the environmental management system in order to achieve improvements in overall environmental performance consistent with the organization's environmental policy.

Clause 4 – Environmental management system requirements

There are six subclauses:

4.1   General requirements
4.2   Environmental policy
4.3   Planning
4.4   Implementation and operation
4.5   Checking
4.6   Management review.

The first subclause states that: 'The organization shall establish, document, implement, maintain and continually improve an environmental management system in accordance with the requirements of this International Standard and determine how it will fulfil these requirements.' The remaining subclauses are detailed statements of the requirements this system should meet.

4.2 Environmental policy

This subclause consists of a list of fairly standard policy requirements, but the last item specifies that the policy should be 'available to the public'.

As with other management systems, the environmental policy element of ISO 14001 does not match the other elements, since the policy is the output of a process, whereas all the other elements are processes.

4.3 Planning

This subclause mostly deals with standard planning procedures, such as identifying the environmental impact of an organisation's activities, setting objectives and targets, and having environmental management programmes. However, there are also requirements to 'establish, implement and maintain a procedure(s) (a) to identify and have access to the applicable legal requirements and other requirements to which the organization subscribes related to its environmental aspects, and (b) to determine how these requirements apply to its environmental aspects'.

4.4 Implementation and operation

This subclause deals with 'Resources, roles, responsibility and authority', 'Competence, training and awareness', 'Communication', 'Documentation', 'Control of documents', 'Operational control' and

'Emergency preparedness and response', all of which are self-explanatory. In effect, this subclause deals with the same issues that are addressed in the 'Implementation and operation' element of BS OHSAS 18001 and the 'Organising' and 'Implementing' elements of HSG65.

4.5 Checking

This subclause is equivalent to the 'Checking' element of BS OHSAS 18001 and the 'Active monitoring' element of HSG65 but there is more emphasis on recording the results of checks, and the corrective action taken. However, there is a major difference from HSG65 in that 'Internal audit' is included in this subclause, not treated as a separate element as in HSG65.

The subdivisions of the Checking subclause are 'Measuring and monitoring', 'Evaluation of compliance' and 'Nonconformity, corrective action and preventive action'. There is no equivalent to the incident investigation section in the corresponding elements of HSG65 and BS OHSAS 18001.

4.6 Management review

This clause describes the inputs to and outputs from management review and states that: 'Top management shall review the organization's environmental management system, at planned intervals, to ensure its continuing suitability, adequacy and effectiveness.'

The Standard also contains two informative annexes:

A. 'Guidance in the use of this International Standard'. This annex is more extensive than clause 4 and gives useful explanatory notes and lists of examples.

B. 'Links between ISO 14001:2004 and ISO 9001:2000'.* This annex contains tables comparing the two Standards and the content of these tables is dealt with later in this chapter.

There is also a bibliography which gives a list of other International Standards to which users of the Standard may need to refer. The first two of these are quality standards, which are dealt with later in this chapter, and the second two are ISO 14004:2004, *Environmental management systems – General guidelines on principles, systems and support techniques*,[53] and ISO 19011:2002, *Guidelines for quality and/or environmental management system auditing*.[54] ISO 19011 is dealt with in Chapter 23.

Quality management systems

Although these systems are widely referred to as 'quality' management systems, it is important to realise that they use 'quality' in a sense which does not correspond with the usual meaning of the word. A typical definition of quality is 'the degree of excellence of a thing'[54] but in quality management systems, quality is used to mean something akin to 'fit for purpose'. An organisation running a quality management system is not necessarily concerned with the absolute quality of its products or services or, indeed, improving their absolute quality. Instead, it is concerned with identifying the requirements of its customers with respect to quality and consistently meeting these requirements. If a customer's requirement is for a lower quality (usually cheaper) item, then the quality management system is used to specify the new requirements accurately and ensure that the customer's new needs are met consistently. Continual improvement, in quality terms, is not, therefore, a continual improvement in the absolute quality of the products or services. It is rather a continual improvement in the accuracy with which customers' requirements are identified and the consistency with which they are met.

---

* Now ISO 9001:2008 – see ISO 9001 section later in this chapter.

The ISO 9000 series

The best-known and most influential quality management system in the UK is the ISO 9000 series. This series was extensively revised and republished in 2000 and to distinguish the revised series from its predecessor, it is often referred to as ISO 9000:2000. However, since the description and discussion which follows are confined to the new series, it will be referred to as ISO 9000 throughout. The documentation in the series is described in ISO 9000 as follows:

> ISO 9000 describes fundamentals of quality management systems and specifies the terminology for quality management systems.
>
> ISO 9001 specifies requirements for a quality management system where an organization needs to demonstrate its ability to provide products that fulfil customer and applicable regulatory requirements and aims to enhance customer satisfaction.
>
> ISO 9004 provides guidelines that consider both the effectiveness and efficiency of the quality management system. The aim of this standard is improvement of the performance of the organization and satisfaction of customers and other interested parties.
>
> ISO 19011 provides guidance on auditing quality and environmental management systems.*

The notes which follow summarise the contents of ISO 9000, ISO 9001 and ISO 9004. ISO 19011 is dealt with in Chapter 23.

ISO 9000

This Standard has three clauses and an annex:

1. Scope
2. Fundamentals of quality management systems
3. Terms and definitions
A. Methodology used in the development of the vocabulary

The scope of the ISO 9000 family is very wide-ranging, as the Standard itself says:

> This international Standard is applicable to the following:
> a) organizations seeking advantage through the implementation of a quality management system;
> b) organizations seeking confidence from their suppliers that their product requirements will be satisfied;
> c) users of the products;
> d) those concerned with a mutual understanding of the terminology used in quality management (e.g. suppliers, customers, regulators);
> e) those internal or external to the organization who assess the quality management system or audit it for conformity with the requirements of ISO 9001 (e.g. auditors, regulators, certification/registration bodies);
> f) those internal or external to the organization who give advice or training on the quality management system appropriate to that organization;
> g) developers of related standards.

Clause 2, 'Fundamentals of quality management systems', has 12 subclauses. Key points from each can be summarised as follows:

---

* Now ISO 19001:2011 *Guidelines for auditing management systems* – see Chapter 23 for details.

2.1 – 'Rationale for quality management systems'. The primary reason quoted for implementing a quality management system is 'enhancing customer satisfaction'.

2.2 – 'Requirements for quality management systems and requirements for products'. This subclause points out that the 'ISO 9000 family distinguishes between requirements for quality management systems and requirements for products'. This is an important distinction since there are references in ISO 9000 to 'continual improvement' but these refer to improvements in the quality management system, not improvements in the quality of the product.

2.3 – 'Quality management system approach'. This subclause describes the steps required to develop, implement, maintain and improve a quality management system. The steps listed are:

a) determining the needs and expectations of customers and other interested parties;
b) establishing the quality policy and quality objectives of the organization;
c) determining the processes and responsibilities necessary to attain the quality objectives;
d) determining and providing the resources necessary to attain the quality objectives;
e) establishing methods to measure the effectiveness and efficiency of each process;
f) applying these measures to determine the effectiveness and efficiency of each process;
g) determining means of preventing nonconformities and eliminating their causes;
h) establishing and applying a process for continual improvement of the quality management system.

Figure 18.2
Model of a process-based quality management system

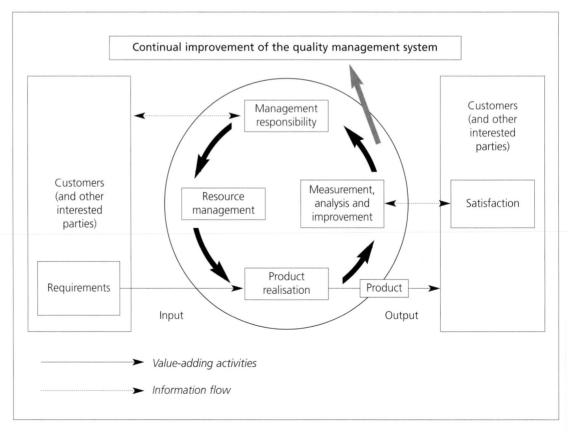

2.4 – 'The process approach'. A process is defined as 'Any activity, or set of activities, that uses resources to transform inputs to outputs'. The subclause is illustrated with a figure which is reproduced here as Figure 18.2.

2.5 – 'Quality policy and quality objectives'. This subclause states that the quality policy and quality objectives 'are established to provide a focus to direct the organization'.

2.6 – 'Role of top management within the quality management system'. Top management is given detailed and comprehensive roles in the quality management system and, since these might beneficially be included in an HSMS, they are quoted in full:

a) to establish and maintain the quality policy and quality objectives of the organization;

b) to promote the quality policy and policy objectives throughout the organization to increase awareness, motivation and involvement;

c) to ensure focus on customer requirements throughout the organization;

d) to ensure that appropriate processes are implemented to enable requirements of customers and other interested parties to be fulfilled and quality objectives to be achieved;

e) to ensure that an effective and efficient quality management system is established, implemented and maintained to achieve these quality objectives;

f) to ensure the availability of necessary resources;

g) to review the quality management system periodically;

h) to decide on actions regarding the quality policy and quality objectives;

i) to decide on actions for improvement of the quality management system.

2.7 – 'Documentation'. This subclause is divided into the value of documentation and the type of documentation to be used. It is pointed out that documentation 'enables communication of intent and consistency of action' but that generation of documentation 'should not be an end in itself but should be a value-adding activity'. The types of documentation to be used in a quality management system are defined as 'manuals', 'plans', 'specifications', 'guidelines', 'work instructions and drawings', and 'records'. Note that the last of these, 'records', has, in the past, been described as a different category, as in 'documents and records', rather than as a subset of documents.

2.8 – 'Evaluating quality management systems'. This subclause has four sections: evaluating processes, auditing, reviewing and self-assessment. However, the evaluation of processes can be carried out using data collected during auditing, reviewing or self-assessment.

In ISO 9000, 'audit is used to determine the extent to which the quality management system requirements are fulfilled' and three types of audit are identified:

• First-party audits are conducted by, or on behalf of, the organization itself for internal purposes and can form the basis for an organization's self-declaration of conformity.

• Second-party audits are conducted by customers of the organization or by other persons on behalf of the customer.

• Third-party audits are conducted by external independent organizations. Such organizations, usually accredited, provide certification or registration of conformity with requirements such as those of ISO 9001.

Review is a top management procedure for 'evaluations of the suitability, adequacy, effectiveness and efficiency of the quality management system with respect to the quality policy and quality objectives'. Among the inputs to review are audit reports and the output is 'determination of the need for action'.

Self-assessment is a 'comprehensive and systematic review of the organization's activities and results referenced against the quality management system or a model of excellence'. (Models of excellence are dealt with later in this chapter.)

2.9 – 'Continual improvement'. The 'aim of continual improvement of a quality management system is to increase the probability of enhancing the satisfaction of customers and other interested parties'. It should be noted that the aim is not to improve product quality and that the success of the improvement is probabilistic. These points are discussed later in this chapter in the context of HSMSs.

2.10 – 'Role of statistical techniques'. This is a rather abstractly worded section which contains an important point for HSMSs. The key point is that the use of statistical techniques 'can help in understanding variability' and 'facilitate better use of data to assist in decision making'. It can be argued that a high proportion of accidents occur as a result of variability in actions, so any techniques designed to assist with the analysis of this variability are to be welcomed.

2.11 – 'Quality management systems and other management system focuses'. The gist of this section is that other management systems, including environmental and occupational safety and health management systems, could be integrated with the quality management system.

2.12 – 'Relationship between quality management systems and excellence models'. This section points out that the difference between the ISO 9000 family and excellence models (see later in this chapter) is that quality management systems are primarily intended for use within an organisation while the excellence models are primarily concerned with comparisons between organisations – that is, benchmarking.

Clause 3 of ISO 9000 deals with 'Terms and definitions'. The clause is subdivided into 10 subclauses and, to give an idea of the range of terms and definitions included, they are listed here:

3.1	terms relating to quality
3.2	terms relating to management
3.3	terms relating to organization
3.4	terms relating to process and product
3.5	terms relating to characteristics
3.6	terms relating to conformity
3.7	terms relating to documentation
3.8	terms relating to examination
3.9	terms relating to audit
3.10	terms relating to quality assurance for measurement processes.

The majority of these terms and definitions are primarily relevant in the context of the ISO 9000 family, but some are relevant in the context of HSMSs, and these are described below.

- 'Continual improvement' is 'recurring activity to increase the ability to fulfil requirements' which are defined as 'need or expectation that is stated, generally implied, or obligatory'. The important point to note is that continual improvement does not require improvement in the quality of the product or service. In the context of HSMSs, therefore, continual improvement does not necessarily imply improvements such as reduction in accident rates or days lost through ill health.
- 'Effectiveness' is the 'extent to which planned activities are realized and planned results achieved'.
- 'Efficiency' is the 'relationship between the result achieved and the resources used'.

Note that the last two concepts are used in this book in, for example, the discussions of financial issues (Chapter 24).

- 'Process' is defined as a 'set of interrelated or interacting activities which transform inputs into outputs'.

- 'Preventive action' is defined as 'action to eliminate the causes of a potential nonconformity or other undesirable potential situation'.
- 'Corrective action' is defined as 'action to eliminate the cause of detected nonconformity or other undesirable situation'.

Note that these definitions of preventive and corrective appear to mean that the same action can be either preventive or corrective. In safety terms, a machine guard is a corrective action if it is fitted after someone has lost an arm in the machine, but it is a preventive action if it is fitted before an accident. While it is useful to discriminate between the preventive and corrective processes (risk assessment and accident investigation, in safety terms), it would seem more appropriate, since the outcomes of the processes are identical (risk control measures), not to make a distinction between these outcomes on the basis of their origin.

- 'Document' is defined as 'information and its supporting medium'.
- 'Record' is defined as 'document stating results achieved or providing evidence of activities performed'.
- 'Inspection' is defined as 'conformity evaluation by observation and judgement accompanied as appropriate by measurement, testing and gauging'.

The terms relating to auditing are also relevant to HSMSs, but these will be dealt with in the section on auditing in Chapter 23.

Most of the remainder of ISO 9000 consists of an informative annex, 'Methodology used in the development of the vocabulary'. The bulk of this annex contains diagrams of 'concept relationships' which illustrate how the various concepts defined in clause 3 relate to each other. Since it is particularly relevant to HSMSs, 'Figure A12 – Concepts relating to audit' is reproduced overleaf as Figure 18.3 to illustrate the form of these diagrams. ISO 9000 ends with a bibliography and an index.

ISO 9001
This standard consists of eight clauses and two informative annexes:
1. Scope
2. Normative references
3. Terms and definitions
4. Quality management system
5. Management commitment
6. Resource management
7. Product realization
8. Measurement, analysis and improvement
A. Correspondence between ISO 9001:2008 and ISO 14001:2004
B. Changes between ISO 9001:2000 and ISO 9001:2008

These clauses and annexes are preceded by an introduction (clause 0). The notes which follow describe the main contents of each section.

0 Introduction
The introduction to ISO 9001 makes certain general points, which have already been dealt with in the description of ISO 9000, and then moves on to deal with the process approach, relationships with ISO 9004 and compatibility with other management systems.

Figure 18.3
Concepts relating to audit (Figure A12 from ISO 9000)[51]

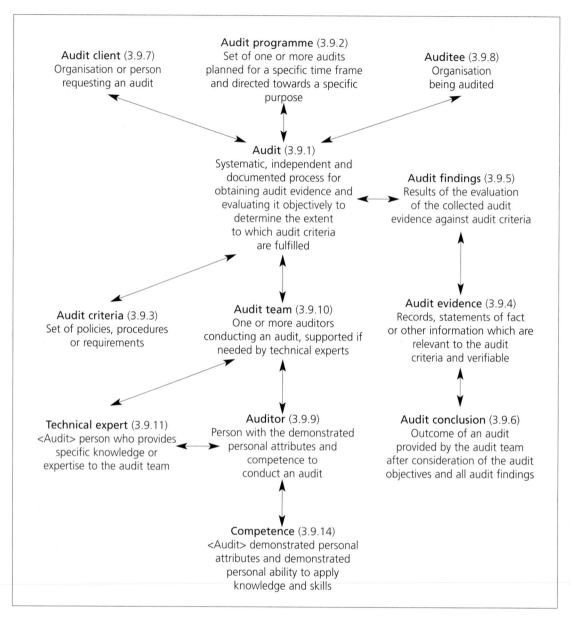

0.2 – 'Process approach'. This subclause is an extension of subclause 2.5 in ISO 9000 and is illustrated with the same diagram (see Figure 18.2). However, it introduces 'the methodology known as Plan-Do-Check-Act (PDCA)' as applied to processes. This methodology is described as follows:

Plan: establish the objectives and processes necessary to deliver results in accordance with customer requirements and the organization's policies.

Do: implement the process.

Check: monitor and measure processes and product against policies, objectives and requirements for the product and report the results.

Act: take actions to continually improve process performance.

0.3 – 'Relationship with ISO 9004'. This clause explains that in 2008 ISO 9004 was being revised and that the new edition would provide 'a wider focus on quality management than ISO 9001'. The revised edition of ISO 9004 was published in 2009 and it is described later in this chapter.

0.4 'Compatibility with other management systems'. This clause states that: 'During the development of this International Standard, due consideration was given to the provisions of ISO 14001:2004 to enhance the compatibility of the two standards for the benefit of the user community.' This is important for HSMSs since, as has been pointed out earlier, ISO 14001 aligns closely with BS OHSAS 18001. Thus, there is a trend for increasing compatibility between quality management systems and HSMSs. The extent of this compatibility is illustrated later in this chapter in Table 18.5.

1 Scope

This clause is divided into 1.1 – 'General' and 1.2 – 'Application'. The essence of subclause 1.1 is that ISO 9001 is for an organisation that '(a) needs to demonstrate its ability to consistently provide product that meets customer and applicable regulatory requirements, and (b) aims to enhance customer satisfaction...'.

Subclause 1.2 points out that the requirements of ISO 9001 are 'intended to be applicable to all organizations regardless of type, size and product provided'. However, organisations can request exclusions of the requirements contained within clause 7 – 'Product realization' (see below).

2 Normative references

The only normative reference is ISO 9000, described earlier in this section.

3 Terms and definitions

These are as set out in ISO 9000, but ISO 9001 states that 'wherever the term "product" occurs, it can also mean "services" '.

4 Quality management system

This clause has two subclauses:

4.1 – 'General requirements'. This is mainly a summary of the requirements which appear in more detail in later clauses. However, there is an important stipulation on outsourcing processes that affect 'product conformity to requirements' in that 'the organization shall ensure control over such processes'.

4.2 – 'Documentation requirements'. This subclause sets out requirements for a quality policy, quality objectives, a quality manual, documented procedures and quality documents, including records. It begins with a list of general requirements (4.2.1) and then sets the requirements for a quality manual (4.2.2). The remaining two subclauses deal with control of documents (4.2.3) and control of records (4.2.4).

5 Management commitment

This clause deals with 'Management commitment' (5.1), 'Customer focus' (5.2), 'Quality policy' (5.3), 'Planning' (5.4), 'Responsibility, authority and communication' (5.5) and 'Management review' (5.6).

Subclause 5.1 requires that top management demonstrates its commitment by communicating to the organisation 'the importance of meeting customer as well as statutory and regulatory requirements', the importance of 'establishing the quality policy', 'ensuring that quality objectives are established', 'conducting management reviews' and 'ensuring the availability of resources'.

Subclause 5.2 consists of one sentence: 'Top management shall ensure that customer requirements are determined and fulfilled with the aim of enhancing customer satisfaction.'

Subclause 5.3 sets out the requirements for the quality policy, which should be 'appropriate to the purposes of the organization', include 'a commitment to comply with requirements and continually improve the effectiveness of the quality management system', provide 'a framework for establishing and reviewing quality objectives', be 'communicated and understood within the organization' and be 'reviewed for continuing suitability'.

Subclause 5.4 deals with a number of issues. Quality objectives have to be set for relevant functions and levels and these objectives have to be measurable (5.4.1). There has to be planning for the quality management system, the meeting of quality objectives and ensuring the integrity of the quality management system during change (5.4.2).

Subclause 5.5 requires that 'responsibilities and authorities are defined and communicated within the organization' (5.5.1). In addition, a 'member of management' must be appointed as, effectively, a quality manager (5.5.2) and 'appropriate communication processes' must be established and communication must take place 'regarding the effectiveness of the quality management system' (5.5.3).

Subclause 5.6 deals with management review. It specifies that 'top management' at 'planned intervals' shall review the quality management system 'to ensure its continuing suitability, adequacy and effectiveness'. The review must include 'assessing opportunities for improvement and the need for changes to the quality management system, including the quality policy and quality objectives'. Records from management reviews must be maintained (5.6.1). In 5.6.2 and 5.6.3, the inputs to, and outputs from, review are dealt with. These are conventional inputs and outputs and match those described in the review section in Chapter 23.

6 Resources management

This clause requires organisations to 'determine and provide' the resources needed to 'implement and maintain the quality management system and continually improve its effectiveness' and 'enhance customer satisfaction by meeting customer requirements' (6.1). Subclause 6.2 deals with human resources and it requires that personnel 'performing work affecting conformity to requirements' be 'competent' (6.2.1). There are also requirements to analyse competence needs, provide the required competences, evaluate the effectiveness of competence provision, ensure that 'personnel are aware of the relevance and importance of their activities and how they contribute to the achievement of the quality objectives' and keep records that are relevant to competence (6.2.2). Subclause 6.3 requires organisations to 'determine, provide and maintain the infrastructure needed to achieve conformity to product requirements'. Infrastructure includes buildings, workplaces, utilities, process equipment (hardware and software), and supporting services such as transport and communication. Subclause 6.4 requires organisations to 'determine and manage the work environment needed to achieve conformity to product requirements'.

7 Product realization

This is a detailed clause with six main subclauses:

7.1   Planning of product realization
7.2   Customer-related processes
7.3   Design and development
7.4   Purchasing
7.5   Production and service provision
7.6   Control of monitoring and measuring equipment.

Subclause 7.1 requires that organisations 'plan and develop the processes needed for product realization' and gives details of what such a plan should contain. The planning should include the specification of the records which will be required.

Subclause 7.2 has three main sets of requirements. First, the organisation must determine product requirements, including 'requirements specified by the customer' and 'statutory and regulatory requirements' (7.2.1). Second, the product requirements must be reviewed at various stages including 'prior to the organization's commitment to supply a product to a customer' and when 'product requirements are changed'.

Records of the results of reviews and 'actions arising from the review' must be maintained (7.2.2). Third, organisations must 'determine and implement effective arrangements for communicating with customers in relation to (a) product information, (b) enquiries, contracts or order handling, including amendments, and (c) customer feedback, including customer complaints' (7.2.3).

Subclause 7.3 deals with design and development and has seven subdivisions. Organisations are required to control the design and development of product and, where different groups are involved in the design and development, 'manage the interfaces between different groups' (7.3.1). The next two subdivisions deal, respectively, with design and development inputs and outputs. Inputs must be determined and records of them maintained; typical inputs include 'functional and performance requirements' and 'statutory and regulatory requirements' (7.3.2). The outputs must be in a form which enables verification (see below) and they must meet a number of other requirements, including specifying 'the characteristics of the product that are essential for its safe and proper use' (7.3.3). Review of design and development is required at 'suitable' stages and the results of these reviews must be recorded (7.3.4).

The next two subdivisions deal with carrying out and recording verification and validation. There is a difference between the two.

Verification is ensuring that 'the design and development outputs have met the design and development input requirements' (7.3.5). Validation is ensuring that the product 'is capable of meeting the requirements for the specified application or intended use, where known' (7.3.6).

Subsection 7.3.7 deals with purchasing and it has three subsections: the purchasing process, purchasing information, and verification of purchased product.

Organisations have to 'ensure that purchased product conforms to specified purchase requirements' and 'evaluate and select suppliers based on their ability to supply product in accordance with the organization's requirements' (7.4.1). The subsection on purchasing information sets out the information which should be available before a product is purchased and requires that the adequacy of this information be ensured before its communication to a supplier (7.4.2). Subsection 7.4.3 requires that purchased product be verified (see above).

Subclause 7.5 deals with 'Production and service provision' and it has five subsections.

Organisations are required to 'plan and carry out production and service provision under controlled conditions' (7.5.1) and 'validate any processes for production and service provision where the resulting output cannot be verified by subsequent monitoring or measurement' (7.5.2). Where appropriate, products must be identifiable and traceable (7.5.3). Subsection 7.5.4 requires that organisations 'exercise care with customer property' and subsection 7.5.5 requires that organisations 'preserve the product during internal processing and delivery to the intended destination'.

Subclause 7.6 deals with the control of monitoring and measuring equipment. The requirements are to 'determine the monitoring and measurement to be undertaken', 'establish processes to ensure that the monitoring and measurement can be carried out', and ensure that monitoring equipment is properly calibrated. An important point is that 'the ability of computer software to satisfy the intended application shall be confirmed'.

8 Measurement, analysis and improvement

The last of the clauses has five subclauses:

8.1   General
8.2   Monitoring and measurement
8.3   Control of nonconforming product
8.4   Analysis of data
8.5   Improvement.

Subclause 8.1 lists general requirements to 'plan and implement the monitoring, measurement, analysis and improvement processes needed (a) to demonstrate conformity to product requirements, (b) to ensure conformity of the quality management system, and (c) to continually improve the effectiveness of the quality management system'.

Subclause 8.2 deals with various requirements for monitoring and, where applicable, measurement. As appropriate, corrective action has to be taken where monitoring or measurement identifies nonconformity. There are four subclauses dealing with customer satisfaction (8.2.1), internal audit (8.2.2), monitoring and measurement of processes (8.2.3), and monitoring and measurement of product (8.2.4).

Subclause 8.3 deals with the control of nonconforming product. This must be 'identified and controlled to prevent its unintended use or delivery' and four alternative ways of dealing with nonconforming product are given. That is, 'eliminate the detected nonconformity', 'authorizing its use, release or acceptance under concession', 'taking action to preclude its original intended use or application' or 'taking appropriate action... when nonconforming product is detected after delivery or use has started'. Records of nonconformities and action to deal with them must be maintained and nonconforming product must be subject to reverification.

Subclause 8.4 requires organisations to 'determine, collect and analyse appropriate data to demonstrate the suitability and effectiveness of the quality management system and to evaluate where continual improvement of the effectiveness of the quality management system can be made'.

Subclause 8.5 requires that organisations 'continually improve the effectiveness of the quality management system through the use of the quality policy, quality objectives, audit results, analysis of data, corrective and preventive actions and management review' (8.5.1). Subsection 8.5.2 deals with corrective action, which is 'action to eliminate the cause of nonconformities in order to prevent recurrence' and subsection 8.5.3 deals with preventive action, which is 'action to eliminate the causes of potential nonconformities in order to prevent their occurrence'. As has been noted, these two clauses specify action which, in safety terms, would be relevant to incident investigation (corrective action) and risk assessment (preventive action).

Annex A (Informative)

This annex consists of tables describing the correspondence between ISO 9001:2008 and ISO 14001:2004. The correspondence between ISO 9001 and ISO 14001:2004 is summarised in Table 18.5 later in this chapter, together with the correspondence between these two management systems and BS OHSAS 18001.

Annex B (Informative)

This annex consists of tables describing the changes between ISO 9001:2000 and ISO 9001:2008. These tables were of use to organisations moving from the old to the new version of ISO 9001.

Bibliography

The bibliography consists of 22 items, and details of relevant ones are given in the References section at the end of the book. There are also four references for websites.

ISO 9004

ISO 9004 'provides guidance to support the achievement of sustained success for any organization in a complex, demanding, and ever-changing environment, by a quality management approach... This International Standard provides a wider focus on quality management than ISO 9001; it addresses the needs and expectations of all relevant interested parties and provides guidance for the systematic and continual improvement of the organization's overall performance.' The contents of ISO 9004 are:

1 Scope
2 Normative references
3 Terms and definitions

The Scope section includes the statement: 'This International Standard is not intended for certification, regulatory or contractual use.' The only reference included in the 'Normative references' section is ISO 9001.

Section 3 contains two definitions:

- 'sustained success' is defined as the '(organization) result of the ability of an organization to achieve and maintain its objectives in the long term'
- 'organization's environment' is defined as the 'combination of internal and external factors and conditions that can affect the achievement of an organization's objectives and its behaviour towards interested parties'.

Sections 4 to 9 set out requirements for the aspects of a quality management system identified in the title of the section with a strong emphasis on analysis of the organization's environment. At various points there are references to risk management but the primary focus is on business risks.

Annex A describes a self-assessment tool for use by 'operational managers and process owners to obtain an in-depth overview of the organization's behaviour and current performance'. Self-assessment is described as 'a comprehensive and systematic review of an organization's activities and results, referenced against a chosen standard'.

The self-assessment tool 'uses five maturity levels, which can be extended to include additional levels or otherwise customized as needed'. The use of the tool involves comparing the organisation's current performance with the criteria set out in tables A.1 to A.7, each of which deals with one of the sections of the Standard, beginning with section 4.

Tables 18.1 and 18.2 contain samples of these criteria taken from, respectively, Table A.3 – Strategy and policy and Table A.8 – Monitoring, measurement, analysis and review.

Annex B describes eight quality management principles and 'gives the standardized descriptions of the principles'. In addition, it 'provides examples of the benefits derived from their use and of actions that managers typically take in applying the principles to improve their organization's performance'.

The eight principles are:

- customer focus
- leadership
- involvement of people
- process approach
- system approach to management
- continual improvement
- factual approach to decision making
- mutually beneficial supplier relationships.

The text of the 'factual approach to decision making' is reproduced below by way of illustrating the content of this Annex:

Table 18.1

First part of the 'Strategy and policy self assessment' (Table A.3 from ISO 9004)

Subclause	Maturity level				
	Level 1	Level 2	Level 3	Level 4	Level 5
5.1 (Strategy and policy) General 5.2 Strategy and policy formulation	The planning process is organized in an *ad hoc* manner. Strategy, policies and objectives are only partly defined. Inputs into policy and strategy formulation are *ad hoc*, and only product and financially related aspects are formulated.	A structured process for the formulation of strategy and policies is in place. The process of strategy and policy formulation includes an analysis of the needs and expectations of customers, along with an analysis of statutory and regulatory requirements.	The process of strategy and policy formulation has evolved to include an analysis of the needs and expectations of a broader range of interested parties. Plans are developed after addressing the needs and expectations of relevant interested parties. The planning process includes consideration of changing external trends and the needs of interested parties; it makes necessary re-alignments when needed. Beneficial outcomes can be linked to past strategic approaches.	Strategy, policies and objectives are formulated in a structured manner. Strategy and policies cover aspects relating to relevant interested parties. The outcome of the organization's processes for strategy and policy formulation are consistent with the needs of its interested parties. Threats, opportunities and availability of resources are evaluated and considered before plans are confirmed. Structured and periodic reviews of planning processes are in place.	It can be demonstrated that strategies have resulted in the achievement of the organization's objectives and optimization of the needs of interested parties. Interested parties are engaged in and contributing to the organization's success; there is confidence that the level of their contribution will be maintained. There is confidence that successes will be sustained. Effective monitoring and reporting mechanisms are in place, including feedback from interested parties for the planning process.

Effective decisions are based on the analysis of data and information.

a Key benefits
- informed decisions,
- an increased ability to demonstrate the effectiveness of past decisions through reference to factual records,
- increased ability to review, challenge and change opinions and decisions.

b Applying the principle of factual approach to decision making typically leads to
- ensuring that data and information are sufficiently accurate and reliable,
- making data accessible to those who need it [*sic*],
- analysing data and information using valid methods,
- making decisions and taking action based on factual analysis, balanced with experience and intuition.

Subclause	Maturity level				
	Level 1	Level 2	Level 3	Level 4	Level 5
8.1 (Monitoring, measurement, analysis and review) General 8.2 Monitoring	Monitoring is performed on a sporadic basis, with no processes in place. The focus of monitoring is on products. Action is triggered by product problems or management problems (ie crisis situations). While information about applicable statutory and regulatory requirements is collected, changes in the requirements are only determined in an *ad hoc* way.	A monitoring process is performed periodically. The focus of monitoring is on customers. Customer needs and expectations are monitored systematically. Changes in statutory and regulatory requirements are tracked systematically through formal designed mechanisms.	The monitoring process is evaluated regularly to improve its effectiveness. The focus of monitoring is on suppliers, with a limited focus on people and other interested parties. Feedback from key suppliers and partners is gathered in a planned manner. Feedback from people is gathered by default only. Current process capabilities are monitored. The processes for tracking statutory and regulatory requirements are effective and efficient.	The monitoring process is performed in a systematic and planned way, and includes cross-checks with external data sources. Resource requirements are assessed in a systematic and planned way, over time. Feedback from employees and customers is gathered through professionally conducted surveys and other mechanisms, such as focus groups.	The monitoring process delivers reliable data and trends. The focus of monitoring is on trends within the organization's activity sector, technologies and the labour situation, with optimization of the use and development of resources. Changes that are taking place, or are expected, in economic policies, product demands, technologies, environmental protection, or in social and cultural issues, which could have an impact on the organization's performance, are monitored in a planned way.

Table 18.2
First part of the 'Monitoring, measurement, analysis and review self-assessment' (Table A.6 from ISO 9004)

Other HSMSs

Five HSMSs were described in Part 1.1, but there are numerous others, including the following:
- **Commercial HSMSs** – these are systems designed and marketed by commercial organisations, either as HSMSs *per se* (for example, the International Safety Management Organisation's 'Safety Management Systems') or in support of an incident investigation tool (for example, the Management Oversight and Risk Tree (MORT) system – see below), or in support of an audit system (for example, the International Safety Rating System (ISRS)[56] and the Complete Health and Safety Evaluation (CHASE) system[57]).
- **'Academic' systems** – produced by academics and published either as books or as papers in learned journals.

However, one published HSMSs is especially relevant because of its use in advanced accident investigation techniques. This system is the Management Oversight and Risk Tree and its basic elements are described below. Its use in advanced accident investigation is dealt with in Chapter 20.

Management Oversight and Risk Tree (MORT)

Although the best known use for MORT is probably its role as an accident investigation tool, it is, effectively, an HSMS set out in the form of a logic diagram. This logic diagram is based on fault tree analysis principles, which are dealt with in Chapter 20. However, at this point it is only necessary to know that the MORT system is based on the structure illustrated in Figure 18.4.

Figure 18.4

Main elements in the MORT HSMS

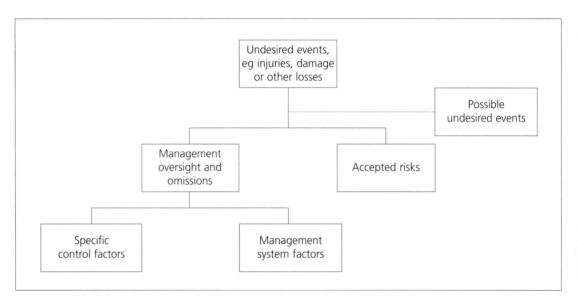

The terminology used in MORT is rather confusing to UK readers: for example, 'LTA' in MORT is not 'lost-time accident' but 'less than adequate'. For this reason, the terminology used in Figure 18.4 is not the MORT terminology but a 'translation' which will make more sense to UK readers.

The 'undesired events' include events which lead to all of the sorts of loss described earlier in this book (see Chapter 8), including quality and environmental losses. The MORT logic diagram can be used for events which have happened, or for events which could happen, that is 'possible undesired events'. Thus, at least in theory, MORT can be used for both accident investigation (events which have happened) and risk assessment (events which could happen).

The 'management oversight and omissions' element is the core of the MORT HSMS. It consists of around 100 generic problems arising from about 1,500 basic causes. By implication, an HSMS is effective if there are procedures in place to solve all the generic problems and deal with all the basic causes. To assist in identifying whether or not procedures are effective, there are thousands of criteria which can be used to help in making a judgment of effectiveness.

The 'accepted risks' element includes all those risks which, with knowledge, an organisation chooses to do no more about. In effect, this element is a formal recognition of the idea that there is no such thing as a risk-free organisation and that some risks have to be accepted. The major difference in MORT is that these risks are made explicit, together with their acceptance by the organisation.

The 'specific control factors' element covers a number of issues, but they mostly arise from the way in which MORT defines an accident as an unwanted flow of energy or an environmental condition that results in adverse consequences.

This definition is fundamentally sound in that there can be no loss without some energy transfer. Yet, this is the same sort of observation as the one which equates all economic activity to energy usage. It is true, but it is so abstract, and so general, that it is difficult to use in practice. However, for the disciplined thinker, the MORT procedure of identifying energy transfers, and the barriers to these transfers that can be used to prevent losses, provides a rigorous approach to thinking about HSMS requirements. In addition, the final element, 'management system factors', lists large numbers of aspects of an HSMS which might be relevant, so a lot of the required analysis has already been done.

The MORT approach to health and safety management arrives at more or less the same point as other HSMSs but provides a different route. The other HSMSs that we have described set out what has to be done and what is to be achieved. The MORT HSMS starts with what has gone, or is likely to go, wrong and describes what has to be done in order to protect against this.

Integrated management systems

Integrated management systems appear to have evolved along three main lines:
1. Identify all the losses which an organisation might sustain, and apply loss management principles to all these losses, including losses arising from failures in quality. This is known as 'total loss control' (TLC) and, as it is no longer widely used, will not be considered further here.
2. Identify all the threats to product or service quality and apply quality management techniques to all these threats, including threats to health and safety. This is known as 'total quality management' (TQM), and is described below.
3. Set up a separate management system with its own framework and terminology that can accommodate the requirements of all management systems, including quality, environment and health and safety. Such an approach is described in the Publicly Available Specification (PAS) 99 *Specification of common management system requirements as a framework for integration*.[27] This type of management system will be referred to as an Integrated Management System (IMS).

In theory, any of the approaches can be successful but in practice there have been major difficulties in implementing all of them. These difficulties arise for a number of reasons and, rather than list the reasons here, they will be identified and discussed later in the chapter in the context of integrating management systems since, in effect, TLC, TQM and IMS are attempts to create integrated management systems. However, the principles underlying TQM systems are dealt with first, with more information on the IMS afterwards.

Total quality management systems

TQM is defined in a number of ways and three examples are:

> The way the organisation is managed to achieve business excellence based upon fundamental principles, which include: customer focus, involvement and empowerment of people and teams, business process management and prevention based systems, continuous improvement. (European Foundation for Quality Management, *Self-assessment – guidelines for companies*[58])

> ...a way of managing an organisation so that every job, every process, is carried out right first time and every time. It affects everyone. (Department of Trade and Industry, *Total quality management and effective leadership – a strategic overview*[60])

A process of management that enables a business to continuously improve its ability to meet and surpass the needs and expectations of its stakeholders by focusing its business processes onto clearly defined goals, policies and strategies and controlling them by systematic measurement and review within an organisation and culture that provides positive leadership towards fulfilling the business mission by involving, supporting, empowering and recognising the achievements of its people in their endeavours. (HSE, *Total quality management and the management of health and safety*[61])

However, in the UK, the most commonly used model for TQM is the European Model for Total Quality Management (EFQM), also known as the European Model for Business Excellence (EMBE). The main elements of this model are illustrated in Figure 18.5.

Figure 18.5
The European Model for Business Excellence

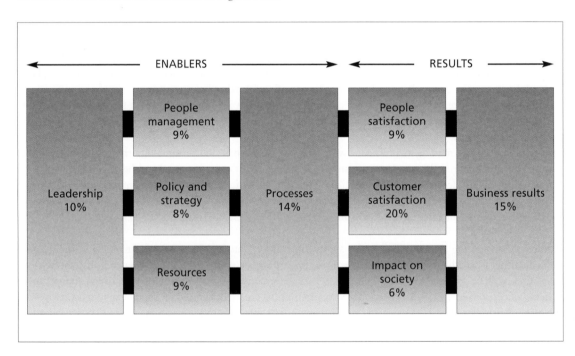

The EMBE is used for assessing an organisation's progress towards business excellence and the various elements in the EMBE model are allocated percentages which indicate their relative importance (see Figure 18.5). Organisations have to collect evidence to demonstrate their level of performance on each of the elements. More details on the elements, and a selection of the categories of evidence which may be required, are described in the HSE report, *Total quality management and the management of health and safety*.[61] The relevant information from this report is reproduced in Table 18.3.

This study, from which the information in Table 18.3 was abstracted, involved the collection of data from 24 companies that used TQM principles in their core business and, among other things, identified the extent to which these principles were applied to the management of health and safety. The general finding was that there was little application of TQM principles to health and safety, except in those circumstances where health and safety were essential to the core business. The authors of the report identified a number of barriers to the use of TQM principles in the context of health and safety and, since these barriers will have to be removed if TQM is to be successfully applied, it is valuable for the health and safety professional to be aware of them.

Description of element	Element characteristics
Leadership	
The behaviour of managers in leading the organisation towards Total Quality. How the executive team and all other managers inspire, drive and reflect Total Quality as the organisation's fundamental process for continuous improvement	Visible involvement in leading Total Quality. A consistent culture for quality. Timely recognition and reward of effort and achievement. Providing appropriate resources and assistance. Involvement with customers and suppliers. Active promotion of Total Quality outside the organisation
People management	
The management of the organisation's people. How the organisation releases the full potential of its people to improve its business continuously	People resources are planned and improved. Skills and capabilities are preserved and developed. People and teams agree targets and review performance. Involvement of everyone in continuous improvement. Empowerment. Effective communication, vertically and laterally
Policy and strategy	
The organisation's mission, values, vision and strategic direction and the manner in which it achieves them. How the organisation's policy and strategy reflect the concept of Total Quality and how the principles of Total Quality are used in the deployment, review and improvement of policy and strategy	Formulated on the concepts of Total Quality, eg continuous improvement, preventive action, doing things right, zero defects. Communicated internally and externally. Regularly reviewed and updated
Resources	
The management, utilisation and preservation of resources. How the organisation's resources are effectively deployed in support of policy and strategy	Financial risk is managed. Access to appropriate and relevant information to do the job. Supplier relationships align with policy and strategy. Optimum use of assets. Minimising waste. Minimising environmental impact [sic]. Technology is exploited and harnessed in support of process improvement
Processes	
The management of all value adding activities within the organisation. How processes are identified, reviewed and if necessary revised to ensure continuous improvement of the organisation's activities	Identification and management of critical processes. Standards of operation established by reference to stakeholders. Challenging targets for improvement set by benchmarking and review of past experience. Performance is monitored. Creativity in the process design and improvement. Process changes are controlled
People satisfaction	
What the organisation is achieving in relation to the satisfaction of its people	Measurement of employee perceptions through the use of a variety of techniques that include focus groups, employee surveys, suggestion schemes, etc

Table 18.3
EFQM description of elements and element characteristics

Table 18.3
continued

Description of element	Element characteristics
Customer satisfaction	
What the organisation is achieving in relation to its external customers	Measurement of the customer's perception of service. Use of secondary indicators of service, such as warranty payments, delivery performance, corrective action arising from complaints, etc.
Impact on society	
What the organisation is achieving in satisfying the needs and expectations of the community at large. This includes perceptions of the organisation's approach to the quality of life, the environment and to the preservation of global resources and the organisation's own internal measures	Measurement of the community's perceptions of the activity of the business. Assessing the degree of involvement in the local community. Monitoring waste. Monitoring sources of community risk (eg pollution, hazards, noise, smell). Monitoring enforcement notices applied by regulators
Business results	
What the organisation is achieving in relation to its planned business objectives and in satisfying the needs and expectations of everyone with a financial interest or stake in the organisation	Monitoring of profit and loss account items (eg sales, margins and net profit). Monitoring of balance sheet account items (eg total assets, working capital, shareholder funds). Focus on credit ratings as a potential measure of success. Measuring improvements in long-term shareholder value

The main barriers identified are, therefore, summarised next:

- TQM is driven predominantly by a positive motive to delight the customer and a drive to improve performance continuously in that respect. Health and safety is driven more by a negative motive to comply with legislation and to avoid being penalised, by 'keeping the wolf from the door'.
- Executive values, knowledge and leadership are focused more on the positive aspects of total quality than the perceived negative aspects of health and safety, to the extent that there could be a leadership vacuum at the executive level in respect of health and safety.
- The business focuses on a narrow range of stakeholders' needs and expectations, predominantly associated with product and finance.
- In TQM the measurement of performance tends to be focused on measuring and improving customer satisfaction. In health and safety performance measurement tends to be focused on quantifying and reducing the frequency of failure or value of loss. This can be less effective than using proactive measures in identifying and stimulating improvement activity.
- There is a lack of knowledge and skills in the discipline of process management.

The authors of this HSE report also identified a number of root causes of these barriers, and they summarised these as follows.

People factors

- A leadership style for the management of health and safety that is inconsistent with that used in a total quality environment.
- A leadership vacuum at executive level in respect of health and safety.
- Executive values do not include health and safety or executives do not understand the significance of health and safety to their business.

- A limited identification at executive level with the wider needs and expectations of stakeholders.
- The view that investment in health and safety is necessary primarily to comply with legislation and that it does not generate income or profit.
- A limited meaning applied to the term 'health and safety management', which in turn encourages a blinkered and functional approach to the subject.
- Personal fears and insecurity, often generated by organisational change programmes.
- Assumptions that health and safety is being managed by someone else, because it always has been.
- A negative attitude to the principle of taking personal responsibility for safety behaviour and management.
- A low calibre of safety personnel and others responsible for generating improvements in safety performance.
- Deficiencies in process management, safety management and change management skills.

Process factors
- Inadequate processes for developing, deploying and reviewing health and safety policy.
- Difficulties in measuring health and safety performance.
- Use of inappropriate measures that do not support the potential of TQM to release the opportunities for improvement in health and safety performance.
- Ineffective and inadequate communication about health and safety matters.
- Different house styles for health and safety and total quality documentation.

Organisational factors
- An overbearing culture of compliance.
- Deficiencies in the organisation's approach to TQM.
- Difficulties in controlling contractors and expanding the core business's approach to TQM into its contractors' organisations.
- A functional approach to the organisation and management of health and safety that encourages different aspects of safety to be managed independently, and a singular and limited role for the health and safety specialist.
- Changing business priorities diverting leadership from the underlying change programmes associated with TQM and health and safety management.

External factors
- The size of the business is perceived to be too small to warrant the ideal approach to health and safety management.
- The nature of the business is such that health and safety is not seen to be critical to the overall success of the business.
- The approach the health and safety profession has adopted to their subject.
- Legislation is functional, in that separate legislation has in the past been drawn up for product safety, employee workplace safety, environment safety and so on.

It is obvious from this summary that the HSE study addresses a number of fundamental issues, and it is therefore included in the Further Reading at the end of the book. Notice, too, that many of the issues raised are ones which have been addressed elsewhere in this book.

The sorts of barrier identified in the HSE study are also relevant to the integration of management systems, since they will have to be removed or overcome for many aspects of integration. In a later section, integrating management systems is considered in more detail.

IMS

In PAS 99 the approach to integrated management systems is summarised in a diagram and this is reproduced below as Figure 18.6. It can be seen from Figure 18.6 that the PAS 99 approach has many elements in common with the management systems already described in this chapter. However, from the point of view of health and safety it is worth noting the following:

- The aspects and impacts terminology from ISO 14001 is used rather than the hazards and risks terminology used in health and safety.
- The definition of risk used is the 'likelihood of an event occurring that will have an impact on objectives'. However, Note 1 expands on this with: 'Risk is normally determined in terms of combination of the likelihood of an event and its consequences.' The adequacy of this definition is discussed in Chapter 20.
- Contingency planning is included. This goes beyond the emergency planning in BS OHSAS 18001 since it includes a requirement to 'consider the continuity of the business operations'.
- There is more detail on management of resources, particularly the management of human resources.
- There is no element dealing explicitly with incident investigation. Such an element would, of course, be essential in an IMS that included health and safety but PAS 99 deals with this in the Introduction, where it states that 'organizations using this PAS should include as input, the specific requirements of management system standards or specifications to which they subscribe'.

Annex A of PAS 99 provides guidance on some of the elements shown in Figure 18.6. However, more detailed guidance is provided in two booklets, *IMS: The framework*[63] and *IMS: Implementing and operating*,[59] both of which predate PAS 99 but are consistent with its principles.

Since the publication of these two booklets, supplementary booklets have been published dealing with the integration of various aspects of management, including good governance and the excellence model. Details of these are given in the Further Reading section at the end of the book.

Figure 18.6
The PAS 99 approach to integrated management systems

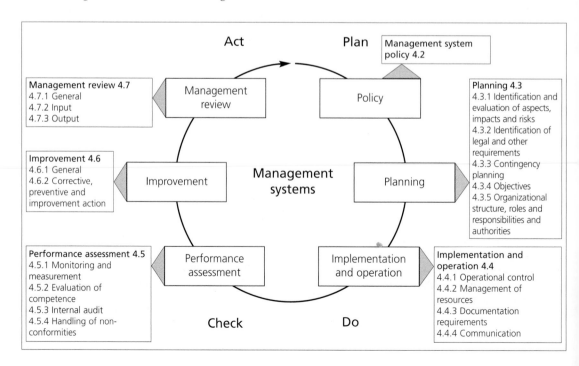

Integrating management systems

The majority of organisations have a number of management systems which run in parallel. Typically, there is a quality management system, an HSMS and an environmental management system, and in this section the discussion will be limited to these three. However, the principles discussed apply to any other management systems – for example, for security or personnel – that may be in place.

An IMS is, at first sight, a desirable aim and there have been various attempts to explain why it proves so difficult in practice. One such attempt was that undertaken by the Institution of Occupational Safety and Health (IOSH) in the UK,[62] and the notes which follow are a summary of its results. While integration of management systems seems, on superficial consideration, to be a desirable objective, there are arguments for and against it. Table 18.4 summarises the cases for and against integration.

For integration
• the objectives and processes of all management systems are essentially the same
• a well-planned IMS is likely to operate more cost-effectively than separate systems, and allow decision-making that best addresses the overall needs of your organisation
• integration should reduce duplication, for example in personnel, meetings, record-keeping software, audits and paperwork
• integration should reduce the risk that resolving problems in one discipline will create new ones in another
• an IMS offers the prospect of more rewarding career opportunities for specialists in each discipline. However, additional training will be required for those given new responsibilities
• it may be easier to develop existing quality procedures within an IMS to incorporate the needs of health and safety and environmental management
• IMS reviews can help ensure each element develops at the same rate. In contrast, independent systems could develop at different rates, leading to incompatibility
• it is easier to bring together expertise in each discipline to address specific issues. This would promote the exchange of fruitful initiatives (eg employee and supply chain surveys) and techniques (eg risk assessment and problem-solving methodologies) between the disciplines. Moreover, all the specialists, working together, are likely to arrive at optimum solutions that take fully into account the needs of each discipline
• it may be easier to link the IMS with management arrangements for other purposes, eg product safety, information systems and security – in cases where these are not already part of an IMS
• an IMS should minimise distortions in resource allocations in separate systems associated with: 　○ a determination to retain current priorities, despite contrary evidence 　○ personnel responsible for one management system being more effective champions of their discipline 　○ variations in the immediacy and precision of feedback – for example, quality assurance feedback is usually rapid and statistically reliable, whereas there may be a time delay of several years before an organisation has statistically significant evidence of the effectiveness of a health and safety initiative
• a positive culture and strengths in one function may usefully be carried over to the others.

Against integration
• the existing systems may simply work well. A process of integration could threaten the coherence and consistency of current arrangements that have the support of everyone involved
• relevant specialists can continue to concentrate solely on their core area of expertise and so further specialist training may not be needed
• an IMS can become over-centralised and over-complex, and lack the capacity to consider local needs and constraints enough. Employers and employees who are sceptical of the excessive bureaucracy in their existing management systems may fear this could worsen under integration
• the time during which you are planning and implementing an integrated system is a period of organisational vulnerability. Existing procedures may lapse, or be found wanting, at the moment when key personnel are focusing attention on the development of new systems

Table 18.4
Summary of the case for and against integration

Table 18.4
continued

- system requirements may vary across the topics covered. For instance, you may need a simple quality system, but a more complex health and safety or environmental management system. In this case, the IMS could introduce unreasonable bureaucracy into quality management (for example, in an organisation that manufactures a simple product to a customer specification, but uses dangerous machinery and creates toxic waste). By way of contrast, a computer software company would need a highly sophisticated quality management system, but comparatively simple health and safety and environmental management systems. Once again, it may not be appropriate to integrate in these circumstances
- there may be distortions in IMS coherence because BS EN ISO environment and quality management standards are internationally recognised and certificatable, but the OHSAS 18001 *Occupational health and safety management systems – specification*, though certificatable, is not internationally recognised
- health and safety and environmental management are underpinned by UK statute, while quality management system requirements are largely determined by customer specification
- you may not wish to alter existing health and safety, environment and quality reporting lines and/or board-level accountabilities, which an IMS might require. Also, if a specialist is given seniority over areas outside their competence, more competent peers and subordinates may feel resentment
- it is possible that rivalries in terms of discipline pre-eminence and resource allocation may impair the collective operation of an integrated system
- regulators and single-topic auditors may have difficulty evaluating their part of the IMS when it is (quite properly) interwoven with other parts of no concern to the evaluator. In contrast, auditing all elements of an IMS at the same time requires an audit team competent in all aspects of the system and may be time-consuming and demanding for the auditee
- a negative culture or flaws in one system area may unwittingly be carried over to the others.

Because of the range of issues which may affect a decision on whether integration is appropriate for a particular organisation, IOSH's guidance recommends that no final decision should be made until the organisation has:

- reviewed the overall business case for an IMS
- reviewed the adequacy of existing arrangements and future needs of each management system which would form part of the IMS
- identified the key competences required for each management system element within the IMS for those who will design and continually improve the IMS structure and contents, and for those who will implement and operate it
- decided on the phasing and extent of integration. It is possible, for example, to start to integrate at the policy and strategic planning levels, and also within 'sharp end' operational procedures and systems. However, you may wish to maintain separate procedures in the short term for specific tasks, such as energy conservation, quality control techniques, and statistical analysis of health and safety performance data. You will need to determine how best to use existing health and safety, environment and quality management departments within an integrated system
- consulted widely throughout the organisation. Many employees will have extra work to do to implement an IMS and their participation and support is essential
- obtained the enthusiastic support of top management for the IMS, and their commitment that appropriate resources will be made available
- studied the recommendations of any relevant industry-specific IMS standards, and considered whether you need to take external advice
- decided on the measurable criteria that you will use to monitor and review the effectiveness of the IMS, and completed a baseline survey so that you can readily assess future changes. Such criteria should be linked to the business case for an IMS.

It is also the view of those who prepared the guidance that, as part of the process of integration, organisations should decide:

- on the choice of an overall IMS model. If you adopt the BS EN ISO 9000 series approach, take care, because it is the least generic of the standards, and does not include explicit consideration of risk assessment. Some organisations have developed quality systems that follow too slavishly the sequences of topics given in earlier versions of that standard
- how to retain the integrity and effective functioning of existing systems while you develop and implement the new system
- whether you need to pilot parts of the IMS to confirm that they are effective before you introduce them
- how to introduce an IMS using appropriate organisational change management processes and skills
- how you are going to analyse training needs and delivery to ensure adequate competence
- how you will introduce a continuing programme designed to retain the commitment of all employees involved.

It can be seen from the issues raised in the IOSH guidance that the process of integration in a given organisation, even if it is seen as desirable, is unlikely to be achieved easily.

However, it may become easier as the various standards become more congruent. This increasing congruence has been noticeable over successive drafts of the three Standards (ISO 9001, ISO 14001 and OHSAS 18001). The current level of congruence is shown in Table 18.5, which serves two functions:

- it provides a useful summary of the contents of the three documents
- it shows that ISO 14001 and OHSAS 18001 are very closely aligned, with ISO 9001 differing in content from these two Standards.

Clause	BS OHSAS 18001:2007	Clause	ISO 14001:2004	Clause	ISO 9001:2008
				0	Introduction
				0.1	General
				0.2	Process approach
				0.3	Relationship with ISO 9004
				0.4	Compatibility with other management systems
1	Scope	1	Scope	1	Scope
				1.1	General
				1.2	Application
2	Reference publications	2	Normative references	2	Normative references
3	Terms and definitions	3	Terms and definitions	3	Terms and definitions
4	OH&S management system requirements	4	Environmental management system requirements	4	Quality management system

Table 18.5 Correspondence between BS OHSAS 18001:2007, ISO 14001:2004 and ISO 9001:2008

Table 18.5
continued

Clause	BS OHSAS 18001:2007	Clause	ISO 14001:2004	Clause	ISO 9001:2008
4.1	General requirements	4.1	General requirements	4.1 5.5.1	General requirements Responsibility and authority
4.2	OH&S policy	4.2	Environmental policy	5.1 5.3 8.5.1	Management commitment Quality policy Continual improvement
4.3	Planning	4.3	Planning	5.4	Planning
4.3.1	Hazard identification, risk assessment and determining controls	4.3.1	Environmental aspects	5.2 7.2.1 7.2.2	Customer focus Determination of requirements related to the product Review of requirements related to the product
4.3.2	Legal and other requirements	4.3.2	Legal and other requirements	5.2 7.2.1	Customer focus Determination of requirements related to the product
4.3.3	Objectives and programme(s)	4.3.3	Objectives, targets and programme(s)	5.4.1 5.4.2 8.5.1	Quality objectives Quality management system planning Continual improvement
4.4	Implementation and operation	4.4	Implementation and operation	7 7.1	Product realization Planning of product realization
4.4.1	Resources, roles, responsibility, accountability and authority	4.4.1	Resources, roles, responsibility and authority	5.1 5.5.1 5.5.2 6.3	Management commitment Responsibility and authority Management representative Infrastructure
4.4.2	Competence, training and awareness	4.4.2	Competence, training and awareness	6.2.1 6.2.2	General Competence, awareness and training
4.4.3	Communication, participation and consultation	4.4.3	Communication	5.5.3 7.2.3	Internal communication Customer communication
4.4.4	Documentation	4.4.4	Documentation	4.2 4.2.1	Documentation requirements General
4.4.5	Control of documents	4.4.5	Control of documents	4.2.3	Control of documents

Clause	BS OHSAS 18001:2007	Clause	ISO 14001:2004	Clause	ISO 9001:2000
4.4.6	Operational control	4.4.6	Operational control	7.1	Planning of product realization
				7.2.1	Determination of requirements related to the product
				7.2.2	Review of requirements related to the product
				7.3.1	Design and development planning
				7.3.2	Design and development inputs
				7.3.3	Design and development outputs
				7.3.4	Design and development review
				7.3.5	Design and development verification
				7.3.6	Design and development validation
				7.3.7	Control of development and design changes
				7.4.1	Purchasing process
				7.4.2	Purchasing information
				7.4.3	Verification of purchased product
				7.5.1	Control of production and service provision
				7.5.2	Validation of processes for production and service provision
				7.5.3	Identification and traceability
				7.5.4	Customer property
				7.5.5	Preservation of product
4.4.7	Emergency preparedness and response	4.4.7	Emergency preparedness and response	8.3	Control of nonconforming product
4.5	Checking	4.5	Checking	8	Measurement, analysis and improvement
4.5.1	Performance measurement and monitoring	4.5.1	Monitoring and measurement	7.6	Control of monitoring and measuring equipment
				8	Measurement, analysis and improvement
				8.1	General
				8.2	Monitoring and measurement
				8.2.3	Monitoring and measurement of processes

Table 18.5
continued

Table 18.5
continued

Clause	BS OHSAS 18001:2007	Clause	ISO 14001:2004	Clause	ISO 9001:2000
				8.2.4	Monitoring and measurement of product
				8.4	Analysis of data
4.5.2	Evaluation of compliance	4.5.2	Evaluation of compliance	8.2.3	Monitoring and measurement of processes
				8.2.4	Monitoring and measurement of product
4.5.3	Incident investigation, nonconformity, corrective action and preventive action	4.5.3	Nonconformity, corrective action and preventive action	8.3	Control of nonconforming product
				8.5.2	Corrective action
				8.5.3	Preventive action
4.5.4	Control of records	4.5.4	Control of records	4.2.4	Control of records
4.5.5	Internal audit	4.5.5	Internal audit	8.2.2	Internal audit
4.6	Management review	4.6	Management review	4.6	Customer focus
				5.6	Management review
				5.6.1	General
				5.6.2	Review input
				5.6.3	Review output
				8.5.1	Continual improvement
Annex A	Correspondence between OHSAS 18001:2007, ISO 14001:2004 and ISO 9001:2000	Annex A	Guidance on the use of this International Standard	Annex A	Correspondence between ISO 9000:2008 and ISO 14001:2004
Annex B	Correspondence between OHSAS 18001:2007, OHSAS 18002* and ILO-OSH:2001	Annex B	Correspondence between ISO 14001:2004 and ISO 9000:2000	Annex B	Correspondence between ISO 9000:2000 and ISO 9001:2008
	Bibliography		Bibliography		Bibliography

Implementing management systems

It can be argued that a poor management system, effectively implemented, is likely to have more impact than an excellent management system poorly implemented.

It is probably the case that there are organisations that are managing health and safety effectively without any written management system. There are probably also organisations that have impeccable written management systems but where the actual management of health and safety is very poor. It is not necessarily the case, therefore, that there is a perfect positive correlation between the quality of the written HSMS and the quality of health and safety management.

It would appear that the major advantage of introducing an HSMS arises from the information and training which accompanies this introduction. That is, managers in the organisation are informed, perhaps for the first

---

* OHSAS 18002 was published as guidance to OHSAS 18001:1999. It was replaced by OHSAS 18002:2008.[86]

time, that health and safety can be managed effectively, and they are given the competences required for this effective management. The major advantage of an ongoing HSMS is the in-built feedback loops in such a system that should, in theory, make it self-perpetuating. However, the effectiveness of these feedback loops will depend on the efforts of the management and, in particular, the senior management. If this effort is not made, the HSMS will gradually degrade, until things are as they were prior to the implementation of the system.

As far as choosing between different management systems is concerned, it seems likely that practically any management system will do, so long as it is implemented and maintained effectively. If there is no effective maintenance of the system, even an ideal HSMS, if there is such a thing, will fail. It may be the case that organisations which keep changing their HSMSs, or keep introducing 'flavour of the month' safety initiatives, do so because it is easier, and more interesting, to do this than to commit themselves to the day-to-day monitoring and corrective action required to maintain the existing HSMS.

Summary

This chapter, and the associated essential study material, has introduced a number of influential management systems. A detailed knowledge of these systems, and their requirements, will enable health and safety professionals to review their own HSMS and suggest ways in which it can be improved. Health and safety professionals will also be able to consider the feasibility and advisability of integrating their organisation's HSMS with other management systems in operation in their organisation.

19: Measuring performance

Introduction

Drucker, the famous management consultant, said: 'If you cannot measure it, you cannot manage it.'

While this may be an extreme view, there is no doubt that the accurate measurement of relevant aspects of performance increases managers' and health and safety professionals' scope for effective identification of problems. This chapter will begin, therefore, with a consideration of what would constitute relevant aspects of performance as far as health and safety are concerned.

However, the range of performance measures which can be identified varies widely in the type of data used and how and when these data are collected. For this reason, we need to look at the nature of data types and some of the problems which may be encountered when using particular types of data.

The chapter ends with a description of the basic statistical techniques which can be used for the analysis and presentation of numerical data.

However, before dealing with performance measures, we need to look at the relationship between monitoring and measuring.

Monitoring and measuring

As is the case with many terms in health and safety, monitoring and measuring are not used consistently in the literature and the terms are not defined in ISO 9000 or ISO 14001 or BS OHSAS 18001. For the purposes of this chapter, we will make the following distinction:
- Monitoring is used to mean checks on conformity with standards so that corrective action can be taken where necessary.
- Measuring is used to mean the collection of data intended to be used for subsequent analysis.*

While monitoring and measuring activities may be carried out simultaneously by, for example, the use of monitoring checklists (see Chapter 10), they are identifiably different functions. How monitoring and measuring are dealt with in the main published HSMSs is summarised below.

HSG65[1]
In Appendix 1 of HSG65, measuring is defined as:

> ... the collection of information about the implementation and effectiveness of plans and standards. This involves various checking or 'monitoring' activities.

It is not clear why 'monitoring' is in quotation marks. In the chapter on measuring performance, HSG65 identifies two types of monitoring:

- **Active** systems which monitor the design, development, installation and operation of management arrangements, RCSs and workplace precautions;

---

* At the time of the latest revision (April 2012), an ISO draft guide was available that contained definitions of monitoring and measuring – ISO/DGuide 83 *High level structure and identical text for management system standards and common core management system terms and definitions*. We will include these definitions when the document is published.

- **Reactive** systems which monitor accidents, ill health, incidents and other evidence of deficient health and safety performance. [Bold in the original.]

However, seven of the 11 pages in the 'Measuring performance' chapter are devoted to investigation, which has nothing to do with monitoring or measuring! Further information on the active monitoring system includes the statement that it:

> ... gives an organisation feedback on its performance **before** an accident, incident or ill health. It includes monitoring the achievement of specific plans and objectives, the operation of the health and safety management system, and compliance with performance standards. [Bold in the original.]

The further information on reactive monitoring systems includes:

> Reactive systems, by definition, are triggered after an event and include identifying and reporting: injuries and cases of ill health (including monitoring of sickness absence records); other losses, such as damage to property; incidents, including those with the **potential** to cause injury, ill health or loss; hazards; weaknesses or omissions in performance standards. [Bold in the original.]

Notice that the last three of the categories mentioned above are all before an accident, ill health or other loss. That is, they meet the primary criterion for active monitoring described in HSG65.

BS 18004[2]

BS 18004 uses the terms proactive and reactive monitoring, which are applied to leading and lagging performance indicators. The relevant definitions from Annex J, which deals with measuring and monitoring, are:

> **Proactive monitoring** – timely routine and periodic checks:
> i)   that OH&S plans have been implemented;
> ii) to determine the level of conformity with OH&S management systems; and
> iii) that seek evidence of harm that has not otherwise come to the attention of the organization via reactive monitoring.
> **Reactive monitoring** – structured responses to OH&S management system failures, including hazardous events and cases of ill health.
> **Leading performance indicator** – data on conformity or nonconformity with the performance requirements of OH&S plans and the organization's OH&S management system generally. Note: the data result mainly from proactive monitoring.
> **Lagging performance indicator** – exclusive data on the prevalence of hazardous events, ie incidents and accidents, and of occupational ill health. Note: the data result mainly from reactive monitoring.

Annex J also contains lists of leading and lagging performance indicators (see Table 19.1). The content of this table will be discussed later in this section.

BS OHSAS 18001[3]

BS OHSAS 18001 requires organisations to have monitoring and measuring procedures that provide for:

Leading performance indicator data	Table 19.1
	Performance indicator data from BS 18004

Leading performance indicator data

- an appropriate safety policy has been written
- a safety policy has been communicated
- a director with health and safety responsibilities has been appointed
- OH&S specialist workers have been appointed
- the extent of influence of OH&S specialists
- the extent to which plans have been implemented
- worker perceptions of management commitment to OH&S
- number of top managers' OH&S inspection tours
- frequency and effectiveness of OH&S committee meetings
- frequency and effectiveness of worker OH&S briefings
- number of worker suggestions for OH&S improvements
- time to implement action on suggestions
- number of personnel trained in OH&S
- workers' understanding of risks and risk controls
- number of risk assessments completed as a proportion of those required
- extent of compliance with risk controls
- extent of compliance with statutory requirements
- workers' attitudes to risks and risk controls
- house-keeping standards
- personal exposure sampling reports
- workplace exposure levels (eg noise and dust, fumes)
- use of PPE
- worker safety representatives and representatives of worker safety have been appointed and are able to exercise their powers

Lagging performance indicator data

- health surveillance reports
- worker absences due to illness (occupationally-related or non-occupationally related)
- cases of occupational diseases or conditions, such as dermatitis, deafness, work-related upper limb disorders; stress, asbestosis, occupationally-induced cancers
- near-misses
- damage only accidents
- reportable dangerous occurrences
- lost-time accidents, when at least one work shift (or other time period) is lost by a person as a result of an accident injury
- reportable accidents involving absence from work for more than three days
- reportable major injuries
- fatal accidents

Data that can be either leading or lagging

- complaints made by the workers
- indicators demonstrating that the organization's OH&S objectives have been achieved
- criticisms made by regulatory agency staff
- regulatory agency enforcement action
- complaints made by workers who are not direct employees of the organization or by members of the public

a) both qualitative and quantitative measures, appropriate to the needs of the organisation;
b) monitoring of the extent to which the organization's OH&S objectives are met;
c) monitoring the effectiveness of controls (for health as well as for safety);
d) proactive measures of performance that monitor conformance with the OH&S programme(s), controls and operational criteria;
e) reactive measures of performance to monitor accidents, ill health, incidents (including accidents, near misses etc), and other historical evidence of deficient OH&S performance.

These summaries from HSG65, BS 18004 and BS OHSAS 18001 illustrate that there are two quite distinct uses of the terms 'reactive', 'active' and 'proactive'.

In HSG65 and BS 18004 the terms are used to describe types of monitoring. Reactive monitoring is the collection of accident, ill health and related data – broadly what were described as incident or loss data in Part 1.1. In HSG65, active monitoring is used to describe the monitoring of what were referred to as conformity data in Part 1.1. In BS 18004, proactive monitoring is used to describe this activity.

In BS OHSAS 18001, the terms are used to describe types of measure, with reactive measures equating to incident or loss data and proactive measures equating to conformity data.

The use of the terms 'reactive', 'active' and 'proactive' monitoring was avoided in Part 1.1 because of this confusion and the notes which follow are an explanation of why the terms were not used.

Logically, reactive and active (or proactive) describe general methods of monitoring. Reactive monitoring involves the people who are doing the monitoring waiting until something is brought to their attention; this was referred to as 'passive monitoring' in Part 1.1. Accident and hazard reporting systems fall into this category. Active monitoring requires that the people who are doing the monitoring take action to collect the required data. Interviews to collect data on minor injuries and hazard identification tours fall into this category. This was referred to as 'active monitoring' in Part 1.1.

However, BS 18004 and HSG65 confuse the way things are monitored and what is being monitored, so that, for example, reactive monitoring is defined by the type of data collected rather than by the method used.

It is likely that the confusion arises because, historically, it was assumed that accidents and other losses would be reported so that there was no need for active monitoring of these data. While this may have been true for fatalities, major injuries and other major losses, it has never been true for accidents with less serious outcomes.* The other side of the coin was the historical assumption that the failure to comply with health and safety procedures would never be reported and would have to be identified during inspections. This is an assumption which was probably justified in the majority of cases!

Irrespective of the reasons for these uses of reactive and active, the uses themselves create difficulties for the adoption of a logical approach to the measurement of health and safety performance.

We need to separate the methods of data collection (passive and active) from the types of data to be collected. Table 19.2 shows, for a range of data types, how this might be done. It can be seen from the examples in Table 19.2 that all the forms of passive monitoring involve waiting until someone else does something, while all the forms of active monitoring are under the control of the people doing the monitoring.

Any of the data types listed in Table 19.1 can be monitored actively or passively; it is a useful exercise to work through the data types listed and consider how each type of monitoring could be applied.

If you are not yet convinced, you may wish to consider the following examples:

---

* This is recognised in BS 18004 where, as we have seen, the definition of proactive monitoring includes checks 'that seek evidence of harm that has not otherwise come to the attention of the organization via reactive monitoring'.

Data type	Passive monitoring	Active monitoring
Minor injuries	Minor injury report forms or accident book	Questioning people about minor injuries they have sustained or witnessed
Damage incidents	Damage report forms	Asset inspections to identify damage
Customer complaints	Publish telephone number for complaints	Telephone customers and ask whether they have any complaints
Employee stress	Wait for employees to complain of stress or report stress-related illnesses	Conduct stress surveys at appropriate intervals
Various medical conditions, eg lead poisoning	Wait for employees to report symptoms	Health surveillance
Damage to hearing	Wait for employees to report symptoms	Audiometry at appropriate intervals
Hazards	Hazard report forms	Safety tours or other inspections to identify hazards
Non-compliance with legal requirements	Wait for action by the relevant inspectorate	Carry out or procure a legal compliance audit

Table 19.2
Data types and reactive and active monitoring

- In the UK, the HSE is aware that major injuries are significantly under-reported (passive monitoring) so they undertake additional (active) monitoring in an attempt to identify the extent of this under-reporting.
- In the UK, the police authorities are aware that various crimes are under-reported (passive monitoring) so they undertake crime surveys (active monitoring) to obtain a more accurate picture of the extent of crime.

The use of 'reactive' and 'proactive' (or 'active') to describe data types (or measures as they are referred to in the HSMSs) is also inappropriate. As we saw earlier, HSG65 has data types in both categories and BS 18004 has data types which can be both leading and lagging performance indicators. In BS OHSAS 18001 it is not clear why such things as unsafe conditions (proactive measure) are not 'historical evidence of deficient OH&S performance' (reactive measure).

It is unfortunate that active and reactive have totally different meanings in quality management and health and safety management, and it is even more unfortunate that the meanings assigned in health and safety management are inappropriate in all the major HSMSs. As with other terminology in the health and safety world, it is likely to take some time for more sensible definitions to be adopted.

The next section of the chapter considers data types more systematically.

The nature of data types

We can appreciate from the list given in the previous section that a wide range of data types is available for measuring health and safety performance. However, it is not necessary to discuss each type of measure individually, since they can all be classified on certain main dimensions:

- whether the data are on a causation continuum
- whether they are objective or subjective

- whether they are reliable or not
- whether they are valid or not
- whether they are qualitative or quantitative.

We will discuss each of these aspects of data types before considering the particular problems associated with the use of numeric data.

Causation continuum

Various approaches to accident causation imply that there is a causation continuum. Examples of these range from the simple, as in the various versions of the Domino Theory described in Chapter 9, to the complex, as described by Hale.[64]

Figure 19.1 shows an illustrative causation continuum which can be used to make the following points:
- Losses are the end point in a long sequence which has its ultimate origin in individual differences.
- As an organisation's health and safety management evolves, it will move from focusing on losses to focusing on the precursors of losses.

Figure 19.1
An illustrative causation continuum

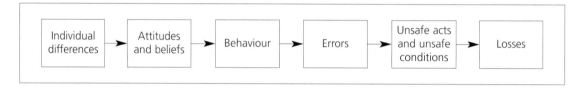

It could be the case, therefore, that an organisation committed to continual improvement which has reduced losses to an irreducible minimum has to adopt a management model in the form shown in Figure 19.2.

Figure 19.2
A management model based on safety culture

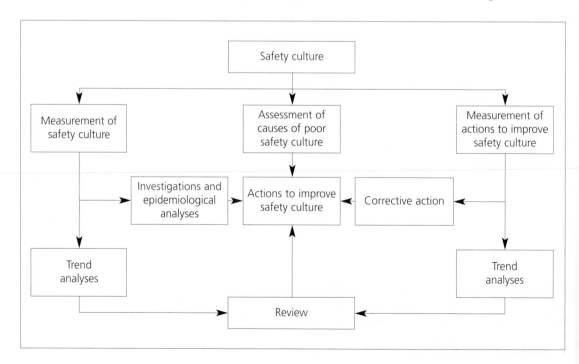

Objective or subjective

Most people have an intuitive idea of what is meant by objective and subjective; they might illustrate the difference with an example such as:

- the temperature of a room in degrees Celsius is an objective measure
- whether people think the room is warm or cold is a subjective measure.

However, even a simple example like this is fraught with problems. The objective measure of temperature is only valid for a particular point in the room and there may be several degrees' difference between the temperature at the floor and the temperature at the ceiling. In addition, it is only valid for a particular period of time. This means that if half a dozen people are sent into a room one after the other, each could obtain a different measurement for the temperature of the room. In contrast, everyone might agree on the subjective measurement of warmth if the room temperature was 30° C.

These problems illustrate that, while it may be possible to differentiate between objective and subjective measures, it is not always the case that the objective measure is the more useful.

Reliability and validity

These two aspects of measurement can be summarised as follows.

1. **Reliability.** This is the extent to which the same type of measurement will produce the same result when used repeatedly. The concept is probably best explained with a simple example. Suppose that it is necessary to measure the dimensions of a rectangular field. If this was done several times using a surveyor's tape measure, the variation between the different measurements would probably be very small. However, if it was done using a 50 cm ruler, the variation between measurements is likely to be much larger. In technical terms, it is said that the former way of measuring is more reliable since it is more likely to produce the same result with repeated use.

2. **Validity.** This is the extent to which a measurement identifies what it purports to identify. Again, this can be illustrated with a simple example. If 'thermal comfort' is defined in terms of the temperature in a room, it might be said that a room temperature between 18° C and 20° C is 'comfortable'. However, people who were having to carry out strenuous physical exercise in that room would find it too warm, and people who were waiting around for a medical examination dressed only in a towel would find it too cold. Thus, room temperature, while it may be an indicator of thermal comfort, is not a particularly valid measure, since it does not take into account other relevant issues such as humidity, the activity being carried out and the clothes being worn.

Reliability, validity and the subjective–objective dimension are all interrelated, but not in any simple manner. An apparently objective measure such as room temperature may not be valid in the context in which it is used, and an objective measure, like length in metres, need not be reliable. In contrast, certain apparently subjective measures, such as measures of attitudes, can be both valid and reliable.

All this means that, when trying to set up ways of measuring health and safety performance, there should ideally be measures which are objective, reliable and valid, but that reliability and validity are more important in the majority of circumstances.

Qualitative and quantitative

The straightforward definition of quantitative is that it is a measure expressed in numbers, in contrast to a qualitative measure which is expressed in words. However, both quantitative and qualitative measures must be based on some underlying scale, and usually the nature of this scale is more important than the numbers or words used to express it. Two examples are given next.

1. **Weight.** The weight of an object can be expressed quantitatively in kilograms or qualitatively with words

such as heavy or light. Thus, a qualitatively expressed weight scale, such as that used to classify boxers into lightweight, heavyweight and so on, has a quantitative underpinning.

2. **The taste of foods.** The taste of foods can be qualitatively rated from poor to excellent, with various points in between. However, it is also possible to allocate numbers to the points on this scale, for example 1 = 'poor' and 5 = 'excellent', but the fact that numbers are being used does not make the underlying scale any more objective.

It is important, therefore, when discussing the difference between quantitative and qualitative, to make a distinction between the nature of the underlying scale and the description used for this scale. If this is not done, fundamental errors can be made in the interpretation of the numerical data. In the next section, we consider the question of different types of numerical scale in more detail.

Numerical scales

The sorts of problems which can arise without an adequate understanding of numerical scales are illustrated below through a number of simple examples.

One need not be less than two

There are three departments in an imaginary factory: production, administration and sales. Two employees write critical reports on the health and safety record of the factory and since they do not wish to identify the three departments, they label them instead. One employee labels them A, B and C and the other employee labels them 1, 2 and 3. Now it is obvious in this case that:

1 is not less than 2 and 2 is not less than 3.

In this case, one employee uses the numbers only to 'name' the departments in the same way as the other employee used letters. Numbers are also used to 'name', for example, houses, employees who use clocking-in machines, and cars (numbers are part of the registration). The use of numbers to name things is known as **nominal scaling** and no arithmetic can be legitimately carried out using this type of number – they are just labels.

Numbers do not always add up

In the imaginary factory, the managing director wants to know which of the three departments came out best in the report prepared by the two employees. For his own use, therefore, he numbers the departments so that the one with the best health and safety record is number 1, number 2 has the next best record and number 3 has the worst. In this case the numbers mean slightly more than they did with nominal scaling because they represent a meaningful ordering of the departments, so that it can be said that:

1 is 'better than' 2 and 2 is 'better than' 3

But it is obvious that:

1 plus 2 still does not equal 3 ($1 + 2 \neq 3$)
1 plus 3 still does not equal 4 ($1 + 3 \neq 4$)
2 plus 3 still does not equal 5 ($2 + 3 \neq 5$)

However, notice two important things about this use of numbers:

1. The direction of the scale is arbitrary. If the managing director passed the report over to the organisation's health and safety professional for action, the health and safety professional might find it more convenient to reverse the ordering on the grounds that the department with the worst health and safety record is in most need of attention and should, therefore, be number 1.
2. The numbers say nothing about how much better or worse one department is than another. For example, the health and safety record in the production department may be very poor while those of the administration and sales departments are quite good and nearly equal.

In cases like this, the numbers are being used to represent the order of the departments on some scale of use to the person concerned and, in everyday life, numbers are used to order a variety of things including football leagues, and first- and second-class railway carriages.*

The use of numbers to order things is known as **ordinal scaling**. Again, no arithmetic can legitimately be carried out on the numbers.

Twenty is not always twice ten

On studying the accident records in the imaginary factory, the health and safety professional finds that most of the accidents in the production department occur in the winter, and that in the winter, the average temperature in the production department is 10°C while in the other two departments it is 20°C. The health and safety professional concludes that the lower temperature in the production department is a contributory factor to the higher accident rate. However, in the request to the managing director for additional heating, the health and safety professional slips up by writing 'in the winter, the average temperature in the sales department is twice as high as the average temperature in the production department'.

It may not be immediately obvious why this is incorrect but if the two temperatures are converted to degrees Fahrenheit, the reason will be obvious:

$$10°C = 50°F$$
$$20°C = 68°F$$

If the measurements had been taken with a Fahrenheit thermometer, no-one would have been tempted to say that the higher temperature was twice the lower. The mistake arises because the two scales have different starting points (with reference to the freezing point of water, the starting points are 0°C and 32°F) and different 'distances' between the points on the scale, that is, a rise of one degree Celsius is a greater rise in temperature than a rise of one degree Fahrenheit. This is because the interval between the freezing point and boiling point of water is divided into 100 units on the Celsius scale and into 180 units on the Fahrenheit scale.

However, the scales do have one important thing in common: a rise in temperature of one degree Celsius represents the same rise in temperature whether it is from 0°C to 1°C or from 99°C to 100°C. Thus, the intervals on the scales are equal within the scales, which is not the case with ordinal scales.

Where a scale has an arbitrary zero and equal intervals on the scale are equivalent, it is known as **interval scaling**. Many arithmetic operations can be carried out on interval scales but, as was seen in the temperature example, we must take care.

Numbers can 'sound funny'

There are several other types of scaling, most of which most people will never need to use. However, one scale, the logarithmic scale, is used in the bel scale for sound. As most managers and health and safety professionals probably

---

* Changes in railway terminology in the UK have renamed 'passengers' as 'customers' and 'second class' as 'standard class'. However, the general principle applies despite the railway's euphemisms.

already know, logarithmic scales have their own arithmetical rules and using 'normal' addition and subtraction will not give the correct answers. The logarithmic scale is probably the best-known illustration of how important it is to know the type of scale being used.

Numbers are numbers are numbers

Most of the time the scale in use is not one of the scales already described, but a **ratio scale**. Ratio scales have an absolute zero, and equal intervals on the scale. A common ratio scale is length. We can use this to illustrate that all normal arithmetical operations can be carried out when using ratio scales:

3 metres is longer than 2 metres and 2 metres is longer than 1 metre
1 metre + 2 metres = 3 metres; 1 metre + 3 metres = 4 metres; and 2 metres + 3 metres = 5 metres
6 metres is twice 3 metres

The difference between the ratio scale and the interval scale can be illustrated by converting metres to feet in the same way as Celsius was converted to Fahrenheit in the earlier example.

1.828 m = 6 ft
0.914 m = 3 ft

It is important to know what sorts of scales are being dealt with for the following reasons:
1. The type of scaling being used has implications for the sorts of analyses which can be carried out.
2. The type of scaling sometimes has important practical implications. Work out what would have happened if the managing director of the imaginary factory had told the boiler man to double the temperature in the production department during the winter – and the old boiler man did not believe in this new-fangled Celsius.

This means that whenever we use a new set of numbers, or unfamiliar measurements, it is essential to determine which sort of scale we are dealing with before beginning to draw conclusions on the basis of the numbers.

Having dealt with the possible problems with numerical data, we can now move on to the next section of the chapter, which deals with the presentation and analysis of numerical data.

Presentation and analysis of numerical data

There are numerous textbooks on statistics which deal with the presentation and analysis of numerical data, and a selection of these is given in the further reading section at the end of the book. However, there are three fundamental aspects of statistics which are of general relevance to the health and safety professional: descriptive statistics, summary statistics and probability. In addition, the health and safety professional needs to know how to apply these fundamentals to trend and epidemiological analysis. The remainder of this chapter is, therefore, devoted to a brief overview of these topics.

Descriptive statistics

When information is collected in the form of numbers, it is usual to want to do one, or both, of two things with the information:
1. pass the information on to others
2. analyse the information in ways which provide additional insights into the nature of the information.

This section is concerned with the first of these aims, that is, presenting numbers accurately, concisely and in ways which make it easy to understand what the numbers are 'saying'. A subsidiary aim is to try to make the presentation look neat and attractive.

Suppose you need to show others the numbers of accidents incurred by the staff in a factory department. Table 19.3 shows the numbers of accidents abstracted from the department's accident records. This obviously gives all the information required, but the person reading the table has to do a lot of work to interpret the information in the table. For example, it will be necessary to count the names to find out how many people are in the department and count the numbers of zeros, ones, twos, and so on, to find out how many people had a specified number of accidents. And does the person reading the table really need to know everyone's name and initial?

Name	Number of accidents	Name	Number of accidents
Smith, J	0	Bowler, M	1
Johnson, T	1	Cheshire, A	0
Lyons, W	2	Kelley, C	0
Arkwright, L	0	Ransom, R	1
Davies, N	0	Telford, E	0
Smith, N	1	Smith, W	0
Girer, A	0	Owen, N	0
Jones, E	0	Young, F	4
Jones, P	3	Edwards, K	0
Thompson, B	0	Fowler, T	0
Evans, J	1	Sawyer, N	0
Manson, W	0	Miles, L	1
Reid, M	0	Wright, K	0
Peters, N	0	Neil, B	1
Hutton, K	0	Thurlow, A	0
Wright, T	2	Mackay, I	0
Murphy, L	3	Vincent, H	2
Stewart, I	0	Murphy, P	0
Cartwright, C	0	Parsons, M	0
Broad, T	1	Kelley, W	0

Table 19.3 Numbers of accidents incurred by each person in a factory department

It is possible to recast Table 19.3 so that the important information stands out more clearly. This is done in Table 19.4.

Number of accidents	Number of people having specified number of accidents
0	26
1	8
2	3
3	2
4	1

Table 19.4 Numbers of accidents incurred by the 40 people in a factory department

This is obviously more concise than Table 19.3 and, by sacrificing some information (names and initials), it makes the point more clearly. However, the point could be made more strongly and readers saved even more work if percentage figures were included, as has been done in Table 19.5, which shows readers what they need to know in the most concise and relevant way.

Number of accidents	Number of people having specified number of accidents	% of people having specified number of accidents
0	26	65
1	8	20
2	3	7.5
3	2	5
4	1	2.5

It may be preferable to present the information not as a table but as a diagram. In general, diagrams should be used as much as possible because most people find it much easier to understand a diagram than a series of columns of figures. However, there are two big disadvantages to the use of diagrams.

1. If there is no access to computer facilities, diagrams take a good deal of time to prepare (compared with the time required to produce a table).
2. Diagrams can be used to present data in ways which are not strictly truthful. This, of course, will apply only to other people's diagrams but some examples of dishonest diagrams are included below as a warning of what you may encounter.

The next section of this chapter describes some of the commoner forms of presenting numbers diagrammatically.

Pie charts

If you need to present the data given in Table 19.5 as a diagram, it is possible to use a pie chart. With this technique, the area inside a circle (the pie) is considered to be 100 per cent and the pie is divided up into slices corresponding in size to the percentages of the things it is intended to show. So for the data from Table 19.5, the pie chart would be as shown in Figure 19.3.

Figure 19.3

How many people have how many accidents?

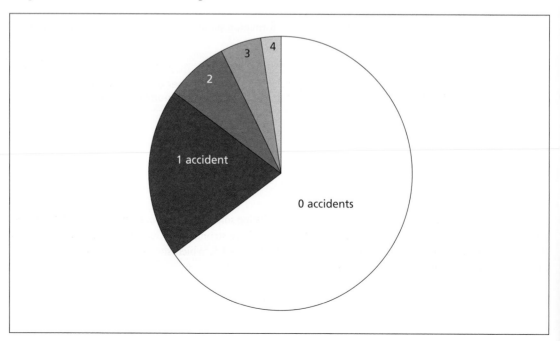

Because it has been noted that the data used to make up Figure 19.3 come from Table 19.5, Figure 19.3 is an honest diagram. But if it were standing on its own, it would be dishonest in that there would be no way of knowing the number of people involved. The diagram could be a nationwide study involving thousands of people or the modest mythical study it really is.

This introduces two important points about the headings or titles to be used for tables and diagrams:
1. the heading should describe briefly what is in a table or figure
2. the heading should give any additional information necessary to interpret the table or figure.

This can result in rather long-winded headings but it is much better for readers to have a full description in the heading than to have to wade through pages of text to find out what the tables or figures are all about. A good rule of thumb when devising headings is that the heading alone should give a clear idea of what the table or figure contains.

Using percentages when preparing pie charts is sufficiently accurate for most purposes and has the advantage of not requiring special equipment. However, a more accurate method is to use a protractor and divide the circle into degrees instead of percentages.

The circumference of a circle is divided into 360 degrees and the relative sizes of the 'slices' of a pie chart can be expressed in numbers of degrees with the whole pie equal to 360 degrees. This is done by expressing the size of a 'slice' as a fraction of the whole pie in the original units and multiplying by 360 to convert to degrees (in the same way that the fractions were multiplied by 100 to convert to percentages). Using a protractor, the circumference of the circle is divided up to give the 'slices' and the pie chart completed in the same way as when percentages are used. However, it is rarely necessary to prepare pie charts manually, as spreadsheet programs usually provide the relevant facilities, for example the chart wizard in Excel.

Ideographs*

Ideographs are a form of diagrammatic representation of numbers which depend for their effect on choosing an appropriate pictorial substitute for the numbers it is intended to portray. For example, if we needed to present as an ideograph the information already used in Table 19.4, we could use the ideograph shown in Figure 19.4.

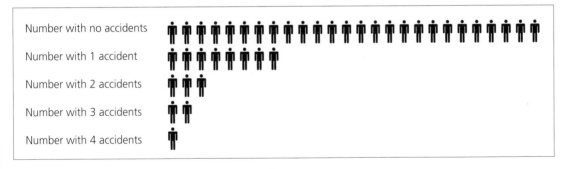

Figure 19.4 is, of course, a very simple ideograph and not a particularly useful way of presenting the information (except when, as here, it is used as an illustration). Ideographs come into their own, however, when it is necessary to superimpose two sets of numbers on each other. For example, if both men and women worked in the factory department used in Figure 19.4, it would be possible to identify the two groups by using different symbols.

---

* Ideographs are also known as ideograms.

Even more complex ideographs can use symbols to illustrate two or more sets of numbers with actual numbers superimposed on the symbols to illustrate additional points. This can often be a useful technique when a complex argument has to be put over (see Figure 19.5 below) but it should be remembered that the main aim of using any descriptive statistics is to give readers an overall view, and this will be lost if the diagrams are too complex.

The use of different symbols and superimposed additional information on an ideograph can be illustrated by extending Figure 19.4. If we needed to show, for the people making up Figure 19.4, first, the proportions of men and women, and second, whether they were employed on packing or on assembly, this might be done in the way illustrated in Figure 19.5.

Figure 19.5
Numbers of men and women involved in assembly and packing having a specified number of accidents (n = 40, whole department)

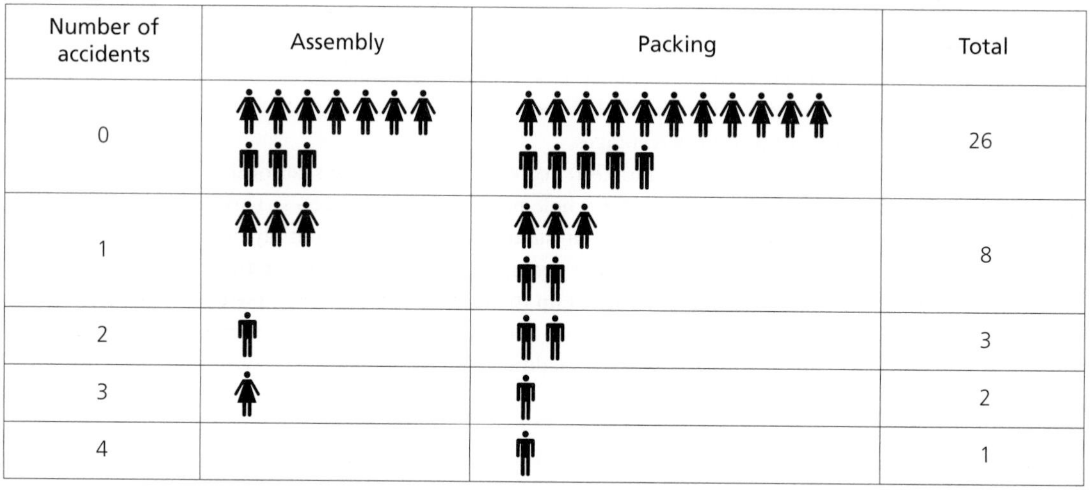

Number of accidents	Assembly	Packing	Total
0	🚶🚶🚶🚶🚶🚶🚶🚶🚶	🚶🚶🚶🚶🚶🚶🚶🚶🚶🚶🚶🚶🚶	26
1	🚶🚶🚶	🚶🚶🚶🚶🚶	8
2	🚶	🚶🚶	3
3	🚶	🚶	2
4		🚶	1

Figure 19.5, like Figure 19.4, is unnecessarily elaborate for the amount of information it is intended to convey. However, it illustrates how several sorts of information can be contained in the same ideograph.

When preparing ideographs it is important to remember who will be making use of the information. Fellows of the Royal Statistical Society may rather resent having legions of stick men and women marching over reports submitted to them and, on a more mundane level, it is doubtful whether Figure 19.5 above would make much impression on a board of directors. They would probably want to know why health and safety professionals or managers did not have something better to do with their time.

This leads nicely to a final point about producing ideographs. They can be very time-consuming to produce so, before starting, it is advisable to work out how long it will take. If it is necessary to embark on drawing 20,000 stick people, then it might be better to sit back and devise ways of reducing the work. In this case, one obvious way would be to let each stick person represent 1,000 people rather than one.

As with other descriptive statistics, ideographs can be used dishonestly as well as honestly. An enemy of the health and safety professional finds out that expenditure on health and safety posters has doubled each year for the last three years and he represents this as shown in Figure 19.6.

Now Figure 19.6 is honest, if it is examined closely. It shows that expenditure was £200 in 2008, £400 in 2009 and £800 in 2010. That is a ratio of 1:2:4. This ratio is shown in the ideograph by the height of the bars. However, what impresses readers immediately when they look at the diagram are the areas of the bars and these are in the ratio 1:4:16. This means that anyone skipping quickly through a report containing this ideograph gets a mistaken view of the increase in expenditure on health and safety posters.

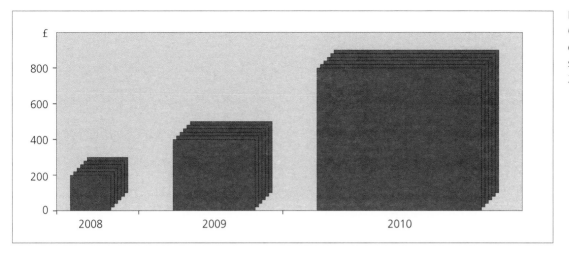

Figure 19.6
Organisation's expenditure on safety posters, 2008–2010

Histograms*

Figure 19.6 is redrawn more honestly as a histogram in Figure 19.7.

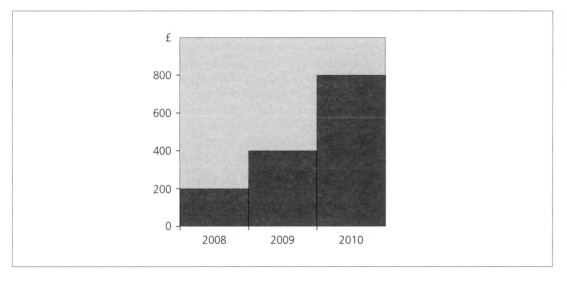

Figure 19.7
Organisation's expenditure on safety posters, 2008–2010

In Figure 19.7, because the three columns are of the same width, there is no dishonesty due to misleading area ratios. However, histograms can be made dishonest in other ways and these will be described later. As one other example of histograms, the ideograph shown in Figure 19.4 is redrawn as a histogram in Figure 19.8.

Figure 19.8 is obviously a more appropriate way of presenting the information than Figure 19.5. (In practice, if Figure 19.5 is turned on its side the height of the columns of stick figures are in the same ratio as the height of the columns in the histogram.) In general, it is true that when we want to show the relationship between two sets of numbers (in this case, numbers of people and number of accidents), the histogram is the simplest descriptive form to use and it has a general acceptance for a wide range of readers.

---

* Histograms are also known as bar charts.

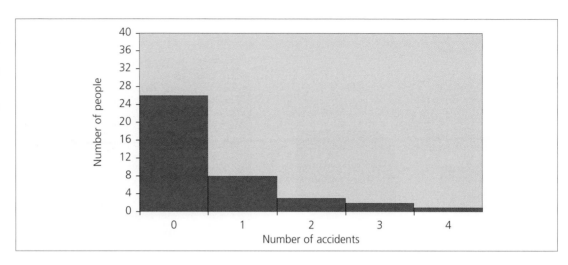

Turning now to the ways in which histograms can be used to misrepresent information, the main techniques are illustrated in Figure 19.9 using the information presented in Figure 19.7.

By combining A and B in Figure 19.9, it is possible to make the histograms even more extreme. For example, by combining A1 and B2 it is possible to make the increase in expenditure look very small indeed.

Suppressing the zero is perhaps the most commonly used trick to make histograms look as if they are supporting the argument put forward. The technique is particularly effective when the numbers to be compared are large in comparison to the differences between them. For example, suppose that the expenditure on safety posters in 2008, 2009 and 2010 was £500, £600 and £700 respectively. In Figure 19.10, the 'honest' histogram is shown at A, the histogram with the suppressed zero at B.

The main purpose of this section was not to teach you how to draw pie charts, ideographs and histograms. Anyone capable of reading this book is capable of working out for himself or herself how to do this. The main purpose of the section was to show the ways in which information can be falsified to suit the argument being put forward. A more important exercise is one which you can do for yourself, and this is to spend some time looking at descriptive statistics in newspapers, reports and so on, and trying to spot any distortions which may have been used.

Summary statistics

The previous section dealt with ways of describing numerical information in tables and in diagrammatic forms. This section deals with numerical ways of summarising numerical information. This may sound nonsensical but it works on the principle adopted for summarising a report or a novel – a smaller number of words is used to give the main features of a much larger set of words. With numerical information, a few numbers are used to give the main features of a much larger set of numbers.

Here is an example to illustrate the main points.

Suppose that we want to assess the general level of health and safety knowledge of a group of six senior managers and their 30 deputies. This is done by drawing up a list of nine questions covering a variety of health and safety topics and getting the 36 managers to answer the questions under test conditions. The result of this is 36 test scores, one from each manager, and these results might be presented as shown in Table 19.6. This is the sort of table which would be produced if the scores were simply recorded in the order in which the tests were marked – the typical starting point for an analysis of this type.

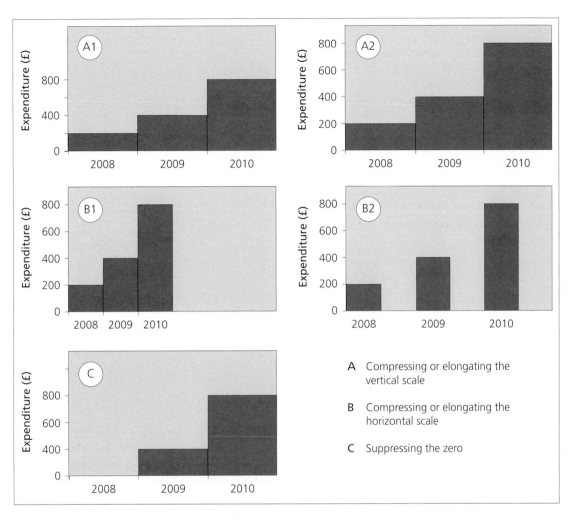

Figure 19.9
The main ways of distorting the information presented in a histogram

A   Compressing or elongating the vertical scale

B   Compressing or elongating the horizontal scale

C   Suppressing the zero

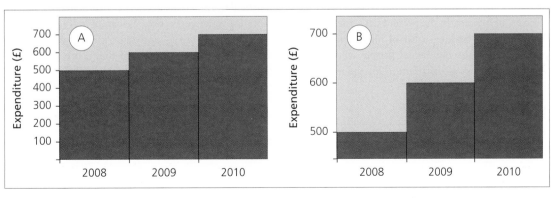

Figure 19.10
Histograms showing the effect of suppressing the zero

Table 19.6
Results from 36 managers who answered a nine-question safety quiz

8	5	4	6	7	1
4	3	2	6	7	5
5	1	4	9	5	6
6	7	3	5	3	8
4	5	5	9	2	4
3	5	6	4	7	6

Table 19.6 obviously gives all the test scores, but it could hardly be called a summary. It would obviously be preferable to put the scores in some sort of order, as has been done in Table 19.7.

Table 19.7
Results from 36 managers who answered a nine-question safety quiz

1	1	2	2	3	3
3	3	4	4	4	4
4	4	5	5	5	5
5	5	5	5	6	6
6	6	6	6	7	7
7	7	8	8	9	9

Table 19.7 is an improvement on Table 19.6, in that the scores are now in increasing order. The scores have been ordered from the lowest to the highest, putting together all the scores which are the same. This procedure is known in statistical jargon as **ranking** and it is a procedure which is used extensively in the more advanced statistical techniques. However, although Table 19.7 is an improvement on Table 19.6, it is not much better as far as summaries go, since 36 numbers are still being used to summarise 36 numbers.

A much better summary can be produced by drawing up a **frequency distribution**. In general a frequency distribution identifies the number of occurrences of a particular value (the frequency) on some dimension which is of interest. In the present case, the dimension of interest is the quiz score. We can prepare a frequency distribution by making a list of all the possible scores (in this case 1 to 9) and then writing, against each possible score, the number of people who got that score – that is, the frequency of that score. Table 19.8 shows the frequency distribution for the information presented in Table 19.7.

Table 19.8
Numbers of managers getting particular scores in a safety quiz ($n = 36$)

Score (x)	1	2	3	4	5	6	7	8	9
Frequency (f) (ie number of people with each score x)	2	2	4	6	8	6	4	2	2

Considered as a summary of the information, Table 19.8 is an obvious improvement on the earlier tables, since only 18 numbers have been used to summarise the original 36 without losing any information at all (in other words, Table 19.7 could be reconstituted from Table 19.8 if required). A 50 per cent saving in numbers used is worth having, but the main advantage of the frequency distribution is that the quantity of numbers required remains the same however many people take the test. For example, if 1,000 people were to take the test, all the information collected could be put into Table 19.8; all that would change would be the size of the frequencies. But imagine how long it would take to make up a table like Table 19.7 for 1,000 people.

Although frequency tables are very useful and have to be prepared as a first step in many statistical procedures, they are only useful as a summary when the number of scores is relatively small. If there had been 100 items on the health and safety quiz, apart from the problem of getting the managers to do it, there would have been a major problem in drawing up a frequency table since there would have been 100 possible scores along the top line. When this is the case, mathematical methods of summarising are more appropriate and these mathematical methods are described next, beginning with the average.

Most people think that they know already what an average is, but statisticians do not use the term average. What is commonly known as the average statisticians refer to as the **mean** (or, more accurately, the **arithmetic mean**). The mean is one measure of central tendency. It takes into account all the available scores and gives one number indicating the centre of the range of scores. This may be difficult to imagine but it is very easy to imagine scores typical of the low and high ends of the range of scores (in this case, 1 and 9 respectively) so why not a score typical of the middle? It is also easy to imagine how the typical scores for the ends of the range might change by, for example, all the managers knowing a lot about health and safety when the lowest score might perhaps be 5 or, more likely, all the managers being ignorant of health and safety, when the highest score might be 6 or 7. So, if typical scores for the ends of the range can change under this sort of influence, why not the typical score for the middle?

Moving on now to the calculation of the mean, this is a relatively simple procedure; the only complicated thing is learning the mathematical formulae involved. In order to calculate the mean for a set of scores, first add up all the scores in the set and then divide this total by the number of scores making up the set:

$$\text{Mean} = \frac{\text{Sum of all the scores in the set}}{\text{Number of scores making up the set}}$$

This is written algebraically as:

$$\bar{x} \text{ (pronounced 'x bar')} = \frac{\sum x}{n}$$

As an example, the mean of the scores given in Table 19.7 is calculated below. First sum (add up) all the scores:

$$\sum x = 1+1+2+2+3+3+3+3+4+4+4+4+4+4+5+5+5+5+5+5+5+5+6+6+6+6+6+6+7+7+$$
$$7+7+8+8+9+9 = 180$$

Then divide the sum of all the scores ($\sum x$) by the number of scores in the set (n), so that:

$$\bar{x} = \frac{180}{36}$$

$$\therefore \bar{x} = 5$$

This is a perfectly acceptable way of calculating the mean but it is rather long-winded. If a frequency distribution of the sort shown in Table 19.8 has been prepared, the calculation can be shortened appreciably by using the following equation:

$$\bar{x} = \frac{\sum fx}{n}$$

where $\sum fx$ means first multiply each score (x) by its frequency (f) and sum the resulting products.* To work through the example given in Table 19.8:

$$\bar{x} = \frac{(2 \times 1) + (2 \times 2) + (4 \times 3) + (6 \times 4) + (8 \times 5) + (6 \times 6) + (4 \times 7) + (2 \times 8) + (2 \times 9)}{36}$$

$$\bar{x} = \frac{2 + 4 + 12 + 24 + 40 + 36 + 28 + 16 + 18}{36}$$

$$\bar{x} = \frac{180}{36}$$

$$\therefore \bar{x} = 5$$

Note that for the same set of scores $\bar{x}$ will be the same, irrespective of which method of calculation is used.

The mean provides one measure of central tendency for the set of scores for which it has been calculated, and it has the advantage of doing it in only one number. However, the mean has certain disadvantages, one of which can be seen by looking separately at the scores of the six senior managers and those of their 30 deputies. These scores are shown in Table 19.9.

Table 19.9
Scores of six senior managers and their 30 deputies in a safety quiz

Score (x)	1	2	3	4	5	6	7	8	9	n
Senior managers (f_S)				1	4	1				6
Deputy managers (f_D)	2	2	4	5	4	5	4	2	2	30

From Table 19.9, we can calculate that the mean score of both the senior managers and their deputies is 5, but it is obvious that the mean score does not tell the whole story. As well as a measure of central tendency, we need some identification of the range of scores. This could be done simply by quoting the mean, followed by the highest and lowest scores:

Senior managers: $\bar{x} = 5$ (range: low 4 to high 6)

Deputy managers: $\bar{x} = 5$ (range: low 1 to high 9)

For some purposes, this would be sufficient. However, just quoting the range and mean can give a misleading picture. Table 19.10 shows how, with the same mean and range of scores, the frequency distribution can vary.

So, although the range of scores can be a useful summary statistic, we also need some measure which indicates the nature of the distribution of frequencies within the range of scores. The measure most commonly used for this purpose is the standard deviation. If the mean of a set of scores has been calculated, each score can be subtracted from the mean to give a measure of 'how far' each score is from the mean. If all these distances

---

* Remember that when numbers are added, the result is their sum; when they are multiplied, the result is their product.

Score (x)	1	2	3	4	5	6	7	8	9	n	$\bar{x}$
Distribution 1 (f)	1	1	1	1	1	1	1	1	1	9	5
Distribution 2 (f)	1	-	-	-	7	-	-	-	1	9	5
Distribution 3 (f)	4	-	-	-	1	-	-	-	4	9	5
Distribution 4 (f)	2	2	-	-	1	-	-	2	2	9	5
Distribution 5 (f)	1	-	-	2	3	2	-	-	1	9	5
Distribution 6 (f)	1	1	2	-	1	-	2	1	1	9	5
	and so on										

Table 19.10 Variations in frequency distributions having the same mean and range

from the mean are added up and divided by the number of scores, it gives the mean distance of the scores from the mean. This is the basic idea behind the standard deviation, but for various reasons the actual calculations involved are slightly more complex. The steps involved in calculating the standard deviation for the information shown in Table 19.9 are shown in Table 19.11, which is followed by an explanation of these steps.

Score (x)	Frequency (f)	fx	$d\,(x-\bar{x})$	d^2	$f\,d^2$
1	2	2	−4	16	32
2	2	4	−3	9	18
3	4	12	−2	4	16
4	6	24	−1	1	6
5	8	40	0	0	0
6	6	36	+1	1	6
7	4	28	+2	4	16
8	2	16	+3	9	18
9	2	18	+4	16	32

$$\bar{x} = \frac{\sum fx}{n} \qquad\qquad sd = \sqrt{\frac{\sum fd^2}{n}}$$

$$\bar{x} = \frac{180}{36} \qquad\qquad sd = \sqrt{\frac{144}{36}}$$

$$\bar{x} = 5 \qquad\qquad sd = \sqrt{4}$$

$$sd = 2$$

Table 19.11 Steps in calculating the standard deviation

The steps involved in calculating standard deviations are:

1. Calculate the mean: $\bar{x} = \dfrac{\sum fx}{n}$

2. Subtract the mean from each score to give a difference (d): $d = (x - \bar{x})$
3. Square each difference: $d^2 = d \times d$
4. Multiply each d^2 by its associated frequency (f): $f d^2$
5. Sum all the values of $f d^2$ and divide by n; this gives the variance (v): $v = \dfrac{\sum f d^2}{n}$

6. take the square root of the variance to give the standard deviation (sd): $sd = \sqrt{\dfrac{\sum f d^2}{n}}$

From Table 19.11, it can be seen why it is necessary to square the differences and then correct for this at the end by taking a square root in order to get the standard deviation. If the differences were simply added up, the answer would be zero because of the + and – signs. By squaring the differences, all the signs are made positive (since $- \times - = +$).

In order to show that the standard deviation tells us more about a distribution than the mean and the range, the standard deviations for the six distributions originally shown in Table 19.10 have been calculated and are given in Table 19.12.

Table 19.12
Variations in standard deviations of distributions with the same mean and range

Score (x)	1	2	3	4	5	6	7	8	9	$\bar{x}$	Range of scores	sd
Distribution 1 (f)	1	1	1	1	1	1	1	1	1	5	1–9	2.58
Distribution 2 (f)	1	-	-	-	7	-	-	-	1	5	1–9	1.89
Distribution 3 (f)	4	-	-	-	1	-	-	-	4	5	1–9	3.77
Distribution 4 (f)	2	2	-	-	1	-	-	2	2	5	1–9	3.33
Distribution 5 (f)	1	-	-	2	3	2	-	-	1	5	1–9	2.00
Distribution 6 (f)	1	1	2	-	1	-	2	1	1	5	1–9	2.71

Calculating the standard deviation for a frequency distribution therefore provides more information about the distribution than just the mean and the range of scores. Its disadvantages are that it takes a little while to calculate and requires a little effort to interpret it properly.

To go back now to the mean, we mentioned above that there were certain disadvantages associated with it. One of these, the fact that it says nothing about the nature of the frequency distribution, has just been dealt with by using the standard deviation. A second problem associated with the mean is that it might not always be the most appropriate measure of central tendency.

When a histogram is drawn of the frequency distribution shown in Table 19.8, the result would be Figure 19.11. When the mean is drawn in on Figure 19.11, it becomes obvious that it is an appropriate measure of central tendency. However, if a different type of measurement was being used, the mean might not be the most appropriate measure of central tendency. This can be illustrated by accident costings. Table 19.13 shows details of 10 accidents which were chosen at random from a total of 254 which had occurred in an organisation over a period of one year. Notice that the accidents have been arranged in ascending order by cost – that is, they have been ranked by cost.

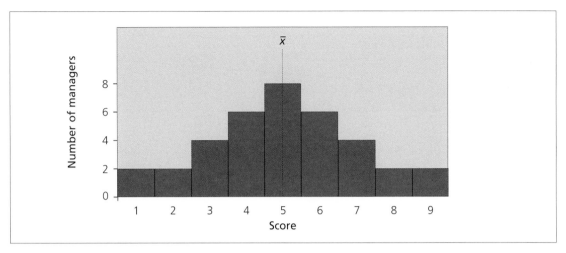

Figure 19.11
Numbers of managers getting particular scores on a safety quiz (*n* = 36)

Accident number	Cost	Accident number	Cost
1	£66.28	6	£206.94
2	£71.74	7	£234.00
3	£111.78	8	£308.64
4	£130.70	9	£567.28
5	£195.46	10	£2,097.04

Table 19.13
Costs of 10 accidents selected at random from 254 accidents

The mean cost of these accidents is £398.99, but close inspection of this mean in relation to all of the 10 accidents shows that only two accidents cost more than the mean while the other eight cost less. This is in contrast to the previous case shown in Figure 19.11, where, of those scores that were not on the mean, half were below it and half above. It seems, therefore, that in the case of the accident costs used in Table 19.13, the mean is not an appropriate measure of central tendency. This creates two questions: what is the appropriate measure and why is the mean not appropriate?

The appropriate measure in this case is the **median**. The median is simply the middle score or measure (after ranking) – so in this case it is the measure which occurs halfway between accident number 5 and accident number 6, that is (£195.46 + £206.94) ÷ 2 = £201.20. If there had been an odd number of accidents, say nine, then the median would have been the cost of the fifth accident, assuming of course that they had been ranked by cost.

The second question ('Why is the mean not appropriate?') can best be answered by plotting the accident costs shown in Table 19.13 in the form of a graph – this has been done in Figure 19.12.

When Figure 19.11 is compared with Figure 19.12, the most obvious difference is that the histogram in Figure 19.11 is symmetrical, while the line joining the costs of the 10 accidents is patently not symmetrical. It is this lack of symmetry which produces the difference between the mean and the median.

Earlier in this section we pointed out that statisticians avoid using the term 'average'. This is because both the mean and the median, since they are both measures of central tendency, can be called an average. Unfortunately, too few people know this and it allows for one of the commonest pieces of statistical dishonesty. For example, the information on the costs of the 10 accidents listed above could be used to impress on a board of directors how costly accidents can be. If this were the aim, the argument would be:

Figure 19.12
Costs of 10 accidents selected at random from 254 accidents

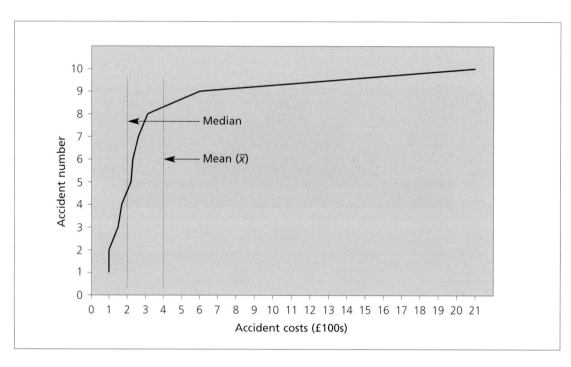

1. the average cost of these 10 accidents was £398.99
2. there was a total of 254 accidents during the year
3. therefore, accidents for the year cost the organisation 254 × £398.99...
4. ... which gives a total cost of £101,343.46.

However, if the intention was to underplay the cost of accidents, the argument would be:
1. the average cost of these 10 accidents was £201.20
2. there was a total of 254 accidents during the year
3. therefore, accidents for the year cost the organisation 254 × £201.20...
4. ... which gives a total cost of £51,104.80.

Both of these arguments can claim to be honest because average is a word which can be used for a number of different things. This is why statisticians stick to mean and median. As an exercise, it is possible to look through any newspaper, periodical or report and find how many times average is used without specifying which average. As can be seen from the example above, it can make quite a difference.

There is also a third meaning which average commonly takes. This is the **mode**. The mode is simply that measure or score which occurs most frequently in a set of measures or scores. To illustrate the use of the mode, we can go back to the histogram showing the numbers of people having different numbers of accidents which was given in Figure 19.8. It is reproduced as Figure 19.13. Here the measure or score which most frequently occurs is zero accidents – so the mode is zero.

At first reading the three measures of central tendency (the mean, median and mode) may be rather confusing. Why are there three measures? And when should each one be used?

The answer to the first question ('Why have three measures?') is relatively simple and in two parts. First, the different measures of central tendency will give a more or less 'true' summary of the information available depending on the nature of that information. Second, it may be necessary to use a measure of central tendency

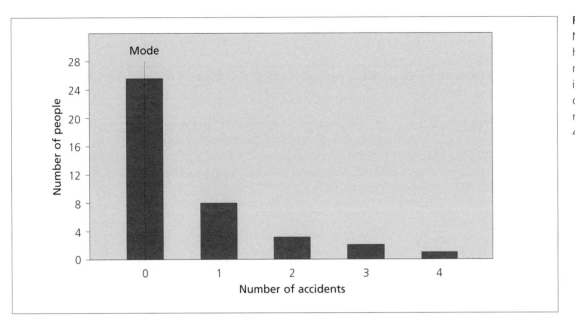

Figure 19.13
Numbers of people having a specified number of accidents in a factory department (total number of people = 40)

for different purposes. An example of honest/dishonest use was given above for the mean and median of the accident costs.

In general, the choice of a particular measure of central tendency will depend on the use to which it will be put. Using Figure 19.13 as an example, it is possible to use this information to predict two things.

- how many accidents there will be in the same department in an equivalent period of time
- the number of accidents a person chosen at random from that department is most likely to have during the same equivalent period of time.

In order to deal with the first prediction, it is possible to add up the accidents which have occurred in the department and assume that the number will be roughly the same in the equivalent period of time. So:

$$\text{Expected number of accidents} = (8 \times 1) + (3 \times 2) + (2 \times 3) + (1 \times 4)$$
$$= 8 + 6 + 6 + 4$$
$$= 24$$

But it should be noted that the same result can be obtained by using the mean. So:

$$\bar{x} \text{ accidents} = 0.6$$
$$\therefore \text{ expected number of accidents} = 40 \times 0.6 = 24$$

In this example, either method of calculation is convenient, but if it is assumed that the number of personnel in the department will go up to, say, 42 in the second period of time, then using the mean is essential.

If, however, it is required to estimate the number of accidents a person chosen at random will have in an equivalent period of time, then the mean is obviously not appropriate. One of the few things it is possible to be certain about is that no-one is going to have 0.6 of an accident! If one person was chosen at random from the department, the best estimate that can be made is that he or she will have zero accidents, which is the mode.

Remember two key things about measures of central tendency:
- where the distribution of the information is symmetrical (as in Figure 19.11), the mean, median and mode are identical
- if in doubt on which one to quote in a particular context, then quote all three. No-one can then make an accusation of a cover-up.

The more advanced calculators, and spreadsheet programs such as Excel, have an inbuilt function for calculating measures of central tendency and the standard deviation. This means that there is no need to use the manual calculations just described. However, it is essential to know how these calculations are done if you are to have an understanding of their use. This use of calculator and spreadsheet functions also applies to the calculations described in the rest of this chapter.

Probability

Describing what we mean by probability is difficult, but most people have an intuitive understanding of what it means, and in this section this intuitive idea will be put on a more formal footing. In addition, there will be an outline of some of the elementary techniques associated with working out probabilities.

The two main types of probability

For the purposes of this book, you need to know about two main types of probability: *a priori* probability and empirical probability.

A priori probability

Suppose that two marbles, one black and one white but otherwise identical, are put into a bag. The bag is shaken up and someone, without looking into the bag, takes out one marble. Intuitively, it would be expected that the probability of taking out the white marble is equal to the probability of taking out the black marble – that is, there is a 50 per cent chance of getting the black marble and a 50 per cent chance of getting the white one.

The probability can be worked out in these circumstances because the circumstances themselves have been 'engineered'. If we needed to have, say, a 1 in 10 chance of picking a black marble, then one black marble and nine white marbles would be put in the bag and, in theory, any probability can be set up in this way. The important feature here is that it is possible to predict the probability before the event and, hence, this type of probability is referred to as *a priori* probability.

Many gambling games depend on pre-determined probabilities of this type. For example, in roulette the probabilities of all the various possible bets (red *versus* black, odd *versus* even, one of the numbers from 1 to 36, and so on) are all worked out before the game starts and on these probabilities are based the odds which are offered on any particular bet. Probabilities can also be worked out for card games and those involving dice but these are more complex and often the difference between a good gambler and a bad gambler is the ability to work out these complex odds quickly.

Empirical probability

Unfortunately, in real life it is not possible to set up probabilities in advance. The chances of a person having an accident on a newly installed machine, or the chances of someone or something falling from a particular set of scaffolding, cannot be set up as though they were balls in a bag. The reason these probabilities cannot be worked out before the event is that the causes of the event are so numerous, and interrelated in so many different ways, that it is impossible to cope with them all. However, what can be done is to record what happens over a period of time in similar circumstances, work out probabilities from these records and assume that these probabilities will hold in the future. This is referred to as empirical probability.

In general, empirical probabilities are calculated using the formula:

$$\text{probability} = \frac{\text{the number of times a thing happens}}{\text{the number of times the thing could possibly happen}}$$

An example of how the formula can be used will help to make it clearer.

If the concentration of a toxic gas at a particular location in a factory was measured 200 times and it was found that the level was over the recommended limit 16 times in 200, it is possible to work out the empirical probability of the next measurement being over the safe limit as follows:

$$\text{probability} = \frac{\text{Number of times value was over safe limit}}{\text{Number of times measurement was made}}$$

$$\text{probability} = \frac{16}{200}$$

$$\text{probability} = 0.08$$

This procedure sounds pretty hit and miss, and in some circumstances it is. But if records are made over long periods of time and large numbers of people or observations are used, it is possible to arrive at a probability which will be reasonably accurate in the future. This procedure is the one used by insurance companies in assessing rates of life insurance, car insurance and so on, and the proof of its success lies in the fact that most of these companies remain solvent.

A major problem with empirical probabilities arises when there is no method of working out how many times a thing could happen. For example, the following equations do not make sense:

$$\text{probability of an accident} = \frac{\text{Number of times an accident happened}}{\text{Number of times an accident could have happened}}$$

$$\text{probability of a car breakdown} = \frac{\text{Number of times car broke down}}{\text{Number of times car could have broken down}}$$

However, these sorts of probabilities are important in a number of health and safety areas and we will consider how they can be estimated later.

Numerical description of probability values

Table 19.14 shows various ways in which probability values can be described numerically. The first two use the convention that the probability of something which is absolutely certain is equal to 1 and the probability of something which cannot happen is equal to 0.

Anything between an absolute certainty and something which cannot happen is described as a fraction (either a common fraction or a decimal fraction) with the size of the fraction decreasing as the thing becomes less likely. The second way of describing probability numerically is to convert the decimal fractions into percentages.

Examples of probability values could, therefore, be written as follows:

Table 19.14
Diagrammatic representation of the main methods of expressing probability

Verbal expression	Example	Probability expressed as a...			How to set up that chance of drawing a black marble from a bag
		common fraction between 0 and 1	decimal fraction between 0 and 1	percentage	
Absolute certainty	That the sun will rise	1	1.0	100%	10 black marbles 0 white marbles
		$\frac{9}{10}$	0.9	90%	9 black marbles 1 white marble
		$\frac{8}{10}$	0.8	80%	8 black marbles 2 white marbles
		$\frac{7}{10}$	0.7	70%	7 black marbles 3 white marbles
		$\frac{6}{10}$	0.6	60%	6 black marbles 4 white marbles
Even chance	That an unbiased coin will fall tails	$\frac{5}{10}$	0.5	50%	5 black marbles 5 white marbles
		$\frac{4}{10}$	0.4	40%	4 black marbles 6 white marbles
		$\frac{3}{10}$	0.3	30%	3 black marbles 7 white marbles
		$\frac{2}{10}$	0.2	20%	2 black marbles 8 white marbles
		$\frac{1}{10}$	0.1	10%	1 black marble 9 white marbles
Absolute impossibility	That someone will jump over St Paul's Cathedral	0	0.0	0%	0 black marbles 10 white marbles

$$p = 0.5 = \tfrac{1}{2} = 50\%$$

$$p = 0.75 = \tfrac{3}{4} = 75\%$$

In the majority of examples in this section, the decimal fraction notation will be used, but do not be surprised if one of the other two crops up from time to time. The final column of Table 19.14 shows how the *a priori* probability described in that row can be set up using the marbles in a bag technique mentioned at the beginning of the section.

Note that when probabilities are very small, for example, 0.00005, it is more convenient to use a different notation:

$$0.5 \text{ can be written as } 5 \times \tfrac{1}{10}$$
$$\tfrac{1}{10} \text{ can be written as } 10^{-1}$$
$$\text{Therefore, } 0.5 \text{ can be written as } 5 \times 10^{-1}$$

Similarly,

0.032 can be written as 32 x $\frac{1}{1000}$

$\frac{1}{1000}$ can be written as 10^{-3}

Therefore, 0.032 can be written as 32×10^{-3} or, more commonly, 3.2×10^{-2}

With this notation, very small values can be written without the need for zeros following the decimal point; for example, 0.00005 would be written as 5×10^{-5}.

An equivalent notation can be used for very large values, for example:

5,000 can be written as 5×10^{3}

2,340,000 can be written as 2.34×10^{6}

and so on.

Probability calculations

There is a wide range of calculations which can be carried out using probabilities but, for the present purposes, only two need be considered.

1. The calculation of probability in circumstances where the probability increases. For example, the increasing probability of getting a head if a coin is tossed repeatedly.
2. The calculation of probability in circumstances where the probability decreases. For example, the decreasing probability of getting two heads on two successive throws of a coin, three heads on three successive throws, and so on.

One of the first, and most difficult, things to learn in working with probability is how to establish whether a probability is increasing or decreasing but the examples given later in the section will help with this.

Increasing probability

If a thing can happen in several ways, then the probability of it happening is the sum of the probabilities of the different ways it can happen. That is:

$$P = p_1 + p_2 + p_3 + p_4 \ldots$$

If a coin is tossed twice, what is the chance of getting at least one head? There are four possible ways the coins can fall:

First throw	Second throw
Head	Head
Head	Tail
Tail	Head
Tail	Tail

These possible ways must add up to a probability of 1.0 (since nothing else can happen) and since each one is as likely as the other, the probability of any one combination = ¼ = 0.25.

Since three of the four ways give at least one head, the probability, on tossing two coins, of getting at least one head is:

$$p = p_{HH} + p_{HT} + p_{TH}$$
$$p = \tfrac{1}{4} + \tfrac{1}{4} + \tfrac{1}{4} = \tfrac{3}{4} = 0.75$$

However, it is often the case that the probabilities of all of the possible events is not known, or it would require too much effort to work them all out. In these cases, an alternative method of calculation is used. We can illustrate this method using the same example of throwing a coin twice.

First, calculate the probability of **not** getting what is wanted on the first throw. This is $1 - 0.5 = 0.5$ and this value will be referred to as q. Second, calculate the number of throws. In the present case, this is 2 and this value will be referred to as n. The probability of getting a head in two throws is then:

$$p = 1 - q^n$$
$$p = 1 - 0.5^2$$
$$p = 1 - 0.25$$
$$p = 0.75$$

This, fortunately, is the same answer as was obtained using the addition method.

Another example will make it clear why this second method of calculation is to be preferred. If it is required to calculate the probability of getting at least one six on two successive throws of a die,* the addition method could be used, but it would be rather tedious. It is very quick using the second method.

Probability of not throwing a six, $q = \tfrac{5}{6}$
Number of throws, $n = 2$
$$p = 1 - (\tfrac{5}{6})^2$$
$$p = 1 - \tfrac{25}{36}$$
$$p = \tfrac{11}{36}$$

Decreasing probability

The probability of two or more things happening together is the product of the probabilities of those things happening separately. That is:

$$P = p_1 \times p_2 \times p_3 \times p_4 \ldots$$

If two dice are rolled, what is the probability that both the first die and the second die will show six? The probability of getting a six on rolling one die is $\tfrac{1}{6}$ ($= 0.17$), since each of the six faces of the die is equally likely to show. The probability of getting two sixes when two dice are rolled is therefore:

$$0.17 \times 0.17 = 0.029$$

Calculating empirical probabilities

As we have seen, in order to calculate an empirical probability, we need to know how many times a 'thing' could happen, and in health and safety, this may not be a sensible question to ask. Nevertheless, for subjects such as component reliability, human reliability, and quantified risk assessment, it is necessary to make the best estimate possible of empirical probabilities.

The remainder of this section will, therefore, consider some of the techniques used to deal with these estimates, starting with making estimates of component reliability.

---

* Die is the singular of dice, as in 'the die is cast'.

Estimating component reliability

Suppose that it is necessary to know the probability of a light bulb failing. It is not possible to calculate this from the behaviour of a single light bulb, but it can be calculated from the behaviour of a number of light bulbs as follows:

1. switch on a large sample (say 10,000) of new light bulbs and start a timer
2. as each light bulb burns out, record the time at which it failed
3. keep doing this until all of the light bulbs have failed.

The data collected in this study can be arranged in a number of different ways but, for present purposes, it is most useful to group the data into the numbers of bulbs which failed in each hour from the time the bulbs were switched on. However, if the bulbs have a design life of 10,000 hours, then some of them are likely to burn for 14,000 hours or more. For this reason, it is more convenient to use time periods longer than one hour. For example, it is possible to group the data into the number of bulbs which failed in the first 100 hours, the second 100 hours, and so on. When the data are arranged in this way, it is possible to plot a graph where the y axis is the number of bulbs which failed in each 100-hour period and the x axis is the time over which the trial was run. This has been done in Figure 19.14, where, for convenience of display, the time on the x axis is in units of 1,000 hours.*

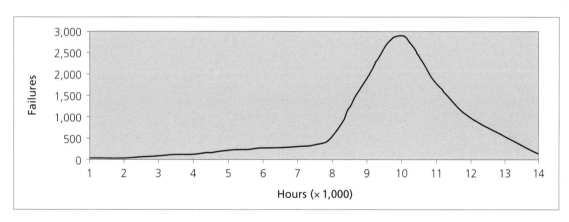

Figure 19.14
Time of failure of 10,000 light bulbs

Figure 19.14 shows that the majority of failures occur around the design life, but a few fail early, and some go on for much longer. From these data it would be possible to calculate the probability of failure in any given hour by dividing the number of light bulbs which failed in that hour by 10,000 (the number of failures which could have occurred). However, since the majority of these probabilities will be zero, and all of them will be small, it is more useful to prepare a cumulative plot of probability against time, as shown in Figure 19.15.

The curve in Figure 19.15 is calculated from the data used to prepare Figure 19.14 as follows:

1. The probability of failure in each time period is calculated by dividing the number of light bulbs which failed in that time period by 10,000 (the number which could have failed).
2. The probabilities for each time period are added together (cumulated) such that the probability in any given time period is the probability for that time period plus the probabilities for all of the earlier time periods. For example, the cumulative probability for the fourth time period is the probabilities for the first four time periods added together.

---

* The 'graph' shown in Figure 19.14 is based on data which would first have been plotted as a histogram. The rationale for the curve shown in the figure is that the failures have a mathematical distribution as yet unknown. The concept of a mathematical distribution is described later in the chapter.

Figure 19.15
Cumulative
probability of light
bulb failure

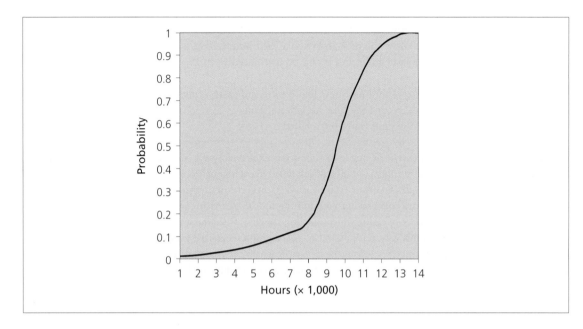

Figure 19.15 has a practical value in that it can be used to read off the probability of failure of the light bulb for any period of time for which it is intended to use it. For example, if we decided to run a bulb for 8,000 hours, we would know that there is a probability of about 0.15 that it would fail in service. If we decided to run a bulb for 13,000 hours, then there is a probability of about 0.99 that it would fail in service.

Probabilities and rates

The simple example above illustrates one technique which can be used to calculate empirical probabilities but, as will be seen, there are others. However, it is important to make a distinction between calculated empirical probabilities and 'rates' which are used as an approximation for calculated empirical probabilities when the relevant data are not available or cannot be collected.

Readers may have come across calculations of probability which take the following form:

> A machine breaks down, on average, three times per year, therefore
> the probability of a breakdown in any one month is 0.25 (3 ÷ 12)

If no better estimate of probability is available, it may be acceptable to use this value, but great care should be taken since it is a rate, and not a 'true' probability. The reason for this is that the calculation relies on the truth of the assumption that the machine cannot break down more than 12 times in a year. That is, the calculation is:

$$\frac{\text{The number of times a breakdown occurred (3)}}{\text{The number of times a breakdown could have occurred (12)}}$$

Since, in theory, the machine could break down more than 12 times in one year, one of the fundamental laws of probability, that probability cannot be greater than 1, is being contravened.

Despite this limitation, there are various circumstances in which rates are used as though they were probabilities and examples will be given in Chapter 20 in the context of advanced risk assessment techniques.

Their use in these circumstances is justified since, as we saw in Chapter 5, risk assessment relies heavily on the subjective judgments of people using the best information they have available. It can, therefore, be argued that using available data on rates to assist in arriving at a probability is a justifiable approach.

Estimating human error

Estimating human error is similar to estimating hardware reliability, in that there are two basic approaches: one estimating 'true' probabilities and one estimating rates. The two approaches are described next.

The usual method of estimating probabilities of human error is to observe a sample of individuals carrying out the same task a number of times. The empirical probability is then calculated as follows:

$$\frac{\text{Number of times an error was made during the task}}{\text{Number of times the task was carried out}}$$

The accuracy of these probabilities depends critically on specifying the errors and the tasks so that, for example, it is never possible to make two errors during the same task. There are published tables of typical error probabilities of this type (see Table 21.3 for an example).

Error rates are usually calculated on a time base with, for example, estimates of errors per hour or per day. However, in some circumstances, other bases are possible, such as error rates per mile driven for driving tasks.

Accident and ill health rates

In Part 1.1, the commonly used frequency, incidence and severity rates were described together with a number of caveats about their use. We also noted that these common rates were used mainly because the data required for their calculation are relatively easy to obtain, rather than because they are particularly useful. What would be more useful would be rates which meet the following criteria:
- they can be used for any loss, not just injury accidents
- they are based on better estimates of risk.

These criteria can be met if we remember that rates are essentially dimensionless quantities which can be calculated for any purpose, and in any way, that suits the person using them. As long as the author of the rate is clear about the data being used, and the method of calculation, anything can be included in a rate.

For example, the first criterion mentioned above (any loss) can result in any of the following being used in a rate calculation:
- number of days lost through sickness, or number of incidences of sickness absence, either in general or as a result of one or more specific medical conditions
- number of hours' downtime, or number of incidents which resulted in downtime
- number of damage-only accidents, or the costs of repair and replacement as a result of damage-only accidents.

The most general item in this category would, of course, be financial cost with, for example, a currency measure being used in the equation. This would enable qualitatively different types of incident to be used in a single rate.

Alternative measures of risk are usually more difficult to find, but the problem is simplified if separate rates are calculated for limited ranges of activities. It is probably impossible, with the current state of knowledge on risk assessment, to find a good measure of risk for a complete factory, for example. However, better measures of risk than numbers employed or hours worked can be found for specific tasks or operations within an organisation or associated with the organisation's activities.

Finding suitable measures of risk usually requires a little ingenuity on the part of the person wishing to create an appropriate rate but, in circumstances where the risk is proportional to the amount of work done, a variety of options is possible. Some examples are given next:

- where the amount of work done can be measured in units produced, for example number of car body panels pressed or number of flower pots packed, then the rate can be based on the number of units
- where the amount of work done is proportional to the amount of energy used in the process (for example in smelting or welding), then the losses can be divided by the relevant electricity (or gas, or coal) bills for the production in different periods
- where the amount of work done is proportional to the volume of material produced, then volume can be used as the basis for the rate.

Other rates can be calculated on such things as the number of miles driven, the number of keystrokes required for data input and, in general, any measure which is proportional to the amount of work done and, hence, the risk.

These ideas for rates may appear odd but if they provide a better estimate of risk, they will give a better measure of how well risk is being managed and, statistically, they are no more outrageous than dividing the number of accidents by the number of people employed.

Any rates of the types suggested above can be used for trend analysis and comparisons as described in Part 1.1 but the same caveats apply. The next section, however, deals with trend analysis is more detail.

Trend analysis

As we pointed out in the section on trend analysis in Chapter 8, it is important to distinguish between fluctuations in accident numbers which are due to random variation and those which represent a real trend, either upwards or downwards. Determining whether a trend is real (or, in statistical jargon, significant) is not a straightforward process; hence, it has this section to itself.

In order to understand the principles involved, it is necessary to explain two more statistical ideas – sampling from a population and variability of measurements – and this will be done by going back to the example used earlier of drawing marbles from a bag.

Suppose there are 50 black marbles and 50 white marbles in a bag. If 10 marbles are drawn out, without looking into the bag, the expectation would be 5 black marbles and 5 white marbles. However, it would not be particularly surprising if the draw resulted in 6 black marbles and 4 white, or *vice versa* – and even if the draw resulted in 9 black marbles and 1 white marble, no 'law of chance' has been violated. However, if the draw did result in a 9:1 ratio, it might be sensible to check that there really were 50 black and 50 white marbles in the bag. This could, obviously, be done by taking all of the marbles out of the bag and counting them. However, another option is to put back the 10 already taken out, shake the bag, and draw out another 10. If this second 10 had nearer to a 5:5 ratio, it might be assumed all was well with the bag, or a third 10 might be drawn to provide more evidence.

This example of the marbles in a bag illustrates the two statistical ideas mentioned above. The first of these is sampling. The 10 marbles drawn are a sample of the 100 marbles in the bag which make up the population and if the first 10 are replaced (so that the population remains the same) and another 10 drawn, this is a second sample. The other statistical idea is that of variability of measurements in that, although a 5:5 ratio is expected in a sample, a certain amount of variability around this ratio is also expected.

Having introduced the ideas of sampling and variability of measurement in terms of marbles, it is now possible to return to accidents.

The accidents for a particular month (or week) can be thought of as being a sample of accidents drawn from the population of accidents for the whole year (or 10 years, or forever). Given that one month's accidents are a sample, therefore, a certain amount of variability in measurement from one sample to the next would be expected. That is, there would be no surprise if the January total were more or less than the February total. The problem now becomes one of deciding how much difference there should be between the two figures before it is possible to say that the difference is not due to variability of measurement. Or, to go back to the marbles in the bag, when is it necessary to count the marbles in the bag?

Fortunately this question can be answered relatively easily, both for the marbles in the bag and for accidents, and the principles are the same in both cases. However, the marbles are easier to explain so we will use them as an introduction before returning to accidents.

The binomial distribution

When there are 50 black marbles and 50 white marbles in a bag, it is known that the *a priori* probability of drawing a black marble equals the probability of drawing a white marble and both probabilities are $p = 0.5$. If a large number of samples of 10 marbles are drawn, replacing the 10 marbles of one sample before drawing the next sample, a symmetrical distribution of the form shown in Figure 19.16 would be obtained.

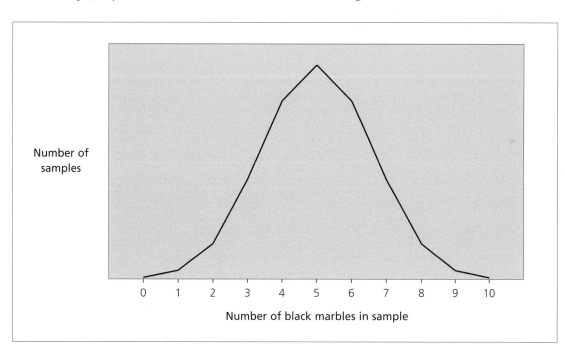

Number of samples

Number of black marbles in sample

Figure 19.16
Distribution of sampling variability for samples of 10 marbles drawn from a population of 50 white and 50 black marbles

The reason it is possible to work out this distribution of sampling error is because the *a priori* probabilities of 0, 1, 2, 3 and so on black marbles in a sample can be calculated using an appropriate equation.

The idea of a distribution was introduced earlier in the chapter in the context of light bulb failures, where we noted that the graph had been drawn assuming that there was an as yet unknown mathematical distribution.

It is often possible to calculate the equation underlying a distribution, although it may not be a trivial task. However, in many circumstances, such as the present marbles example, the work has already been done and it is known that the relevant distribution is the binomial distribution. Other circumstances where the equation may already have been worked out include those distributions which follow a 'normal' curve* and those which follow a Poisson curve. The latter distribution is particularly important for accident data and will be discussed in more detail later.

Returning to the marbles example, the equation for the binomial distribution is:

$$(p + q)^n$$

where p = the probability of an event happening, $q = 1 - p$, and n is the number of trials or samples.

The relevant values are calculated by expanding the binomial and the first three values of the expansion are given below:

$$(p + q)^1 = p + q$$

$$(p + q)^2 = p^2 + 2pq + q^2$$

$$(p + q)^3 = p^3 + 3p^2q + 3pq^2 + q^3$$

In the present case, p (the probability of getting a black marble) = q (the probability of not getting a black marble) = 0.5, and n, the sample size, is 10. When the binomial of value $(0.5 + 0.5)^{10}$ is expanded, the results shown in Table 19.15 are obtained.

Table 19.15

Expected numbers of samples (from a total of 1,024) having a specified number of black marbles (binomial distribution, $p = q =$ 0.5, $n = 10$)

Number of black marbles in a sample	Expected number of samples	Probability of sample occurring (%)
0	1	0.098
1	10	0.977
2	45	4.395
3	120	11.719
4	210	20.508
5	252	24.609
6	210	20.508
7	120	11.719
8	45	4.395
9	10	0.977
10	1	0.098
	1024	100.002

---

* The normal curve is so called because it matches so many distributions found in nature. It is also known as the 'bell' distribution because of its characteristic bell-like shape, and as the Gaussian distribution because it was devised by the mathematician Carl Friedrich Gauss.

It is not necessary to understand the mathematics of this calculation, so long as it is accepted that we can make such a calculation. The equivalent calculation for accident samples will be explained later in this section.

From Table 19.15, it can be seen that there is only a 1 in 1,000 chance of getting no black marbles in a sample of 10 and, similarly, only a 1 in 1,000 chance of getting 10 black marbles. The most likely outcome, as would be expected, is that there will be five black and five white marbles in the sample, the chances here being nearly 1 in 4. It should also be noted that the chance of there being 3, 4, 5, 6 or 7 black marbles is, by the addition of probabilities, $11.7 + 20.5 + 24.6 + 20.5 + 11.7 = 89$ per cent.

Having calculated these probabilities, it is now possible to go back to the question of when it is appropriate to start counting the marbles in the bag. Since the probability of any given sample occurring is now known, the data are available on which to base a decision. The steps in the decision making could be:

- if no-one has tampered with the bag of marbles, there are 50 black marbles and 50 white marbles in the population
- if a sample of 10 marbles is drawn and there are no black marbles in the sample, the odds against this are 1,000 to 1.

With those sorts of odds, it would seem a better bet that someone has been surreptitiously removing black marbles and that the population has been changed.

However, if the person drawing the marbles was naturally suspicious, or had good reason to suspect that someone was tampering with the bag of marbles, he or she might decide that odds of 100 to 1 justified an investigation, in which case they would count all the marbles if a particular sample had 0, 1, 9 or 10 black marbles. Taking this argument to extremes, the real paranoiac might want to count the marbles if there were not 4, 5 or 6 black marbles in a sample.

When we decide, as we did above, that there is a level of probability beyond which a sample will not be accepted as representative of the supposed population from which it is drawn, in statistical jargon a level of significance is being set. It is being said, in effect, that, if the probability of occurrence of a particular sample were less than 5 per cent (or 1 per cent, or 0.1 per cent), the matter should be investigated further.

Note that the level of significance is set depending on factors which have nothing to do with the calculation of the probabilities. For example, if the drawing of marbles was being used as a gambling game, so that people risked losing their money, the significance levels might be set as follows:

- 5 per cent if the person was betting small sums of money or could afford to lose money.
- 1 per cent if the person was betting quite large sums of money or could not afford to lose much money
- 0.1 per cent if the person betting was a police officer who was going to arrest the person drawing the marbles for fraudulent gambling.

It is important to remember that when a statistician says that something is 'significant', this can only be interpreted if the level of significance is known. Significant at, for example, the 1 per cent level still means that the observed result could have occurred by chance 1 time in 100.

The Poisson distribution

As we mentioned above, the accidents for a week, or a month, can be considered as a sample. Therefore, when the monthly accident figures have been collected for one year, there are 12 samples available. Similarly, if weekly accident figures were collected for six months, there would be 26 samples available. It is then possible to calculate a mean value for monthly or weekly accident figures and this would give an expected number of accidents for the following month or week. In addition, this expected number can be used in the same way as the *a priori* probabilities in the example of the marbles to calculate the expected variability of measurement around this mean. Perhaps the best way to explain this is by working through an example. Table 19.16 shows the number of accidents recorded each month in a factory in 2010.

Table 19.16
Number of accidents
in each calendar
month for 2010

Month	Number of accidents	Month	Number of accidents
January	5	July	4
February	3	August	2
March	4	September	2
April	3	October	3
May	3	November	5
June	4	December	4
Total	**22**	**Total**	**20**

Total number of accidents in 12 months = 42

Mean $(\bar{x}) = \dfrac{42}{12} = 3.5$

The mean number of accidents per month over 12 months is therefore 3.5 and, if the accidents for 2010 and 2011 are considered as a single population, in January 2011, the reasonable expectation would be to have around 3.5 accidents, although, even if there were between 2 and 5 accidents, this would be comparable with the 2010 figures. However, what conclusions could be reached if there were only 1 accident, or 6 accidents?

What is required is a way of working out the probabilities of given numbers of accidents in a month's sample, in the same way that the probabilities of the number of black marbles in a sample were worked out earlier.

The binomial distribution cannot be used for this purpose because accidents do not have a symmetrical distribution. However, the Poisson distribution can be used to give a reasonable approximation of accident distributions over relatively small numbers of accidents. The equation for the Poisson distribution is given next, followed by explanatory notes.

$$e^{-\bar{x}}\left(1 + \bar{x} + \frac{\bar{x}^2}{2!} + \frac{\bar{x}^3}{3!} + \frac{\bar{x}^4}{4!} \dots \text{ and so on}\right)$$

e is a mathematical constant equal to 2.7183 and $\bar{x}$ is the mean of the distribution (see Table 19.16); $e^{-\bar{x}}$ can be found from mathematical tables or by using an appropriate calculator or computer program. 2!, 3! and 4! are mathematical shorthand for 'two factorial', 'three factorial', 'four factorial' and so on. They mean:

$$2! = 2 \times 1 = 2$$
$$3! = 3 \times 2 \times 1 = 6$$
$$4! = 4 \times 3 \times 2 \times 1 = 24$$
$$5! = 5 \times 4 \times 3 \times 2 \times 1 = 120$$
and so on.

The first term inside the brackets in the equation above – that is, 1 when multiplied by $e^{-\bar{x}}$ – gives the probability of having no accidents. The second term $(\bar{x})$ when multiplied by $e^{-\bar{x}}$ gives the probability of having one accident. The third term $(\frac{\bar{x}^2}{2!})$ gives the probability of having two accidents, and so on.

The use of the Poisson distribution is illustrated by the example in Table 19.17. In this case $e^{-\bar{x}} = e^{-3.5} = 0.0302$, and the various terms within the brackets work out to the values shown in Table 19.17.

Number of accidents	Term	Term equals	$\times e^{-\overline{x}}$	Probability
0	1	1	$\times 0.0302$	0.0302
1	$\overline{x}$	3.5	$\times 0.0302$	0.1057
2	$\dfrac{\overline{x}^2}{2!}$	$\dfrac{12.25}{2} = 6.13$	$\times 0.0302$	0.1851
3	$\dfrac{\overline{x}^3}{3!}$	$\dfrac{42.88}{6} = 7.15$	$\times 0.0302$	0.2158
4	$\dfrac{\overline{x}^4}{4!}$	$\dfrac{150.06}{24} = 6.25$	$\times 0.0302$	0.1888
5	$\dfrac{\overline{x}^5}{5!}$	$\dfrac{525.22}{120} = 4.377$	$\times 0.0302$	0.1322
6	$\dfrac{\overline{x}^6}{6!}$	$\dfrac{1838.27}{720} = 2.5532$	$\times 0.0302$	0.0771
7	$\dfrac{\overline{x}^7}{7!}$	$\dfrac{6433.92}{5040} = 1.2766$	$\times 0.0302$	0.0386
8	$\dfrac{\overline{x}^8}{8!}$	$\dfrac{22518.75}{40320} = 0.5585$	$\times 0.0302$	0.0169
9	$\dfrac{\overline{x}^9}{9!}$	$\dfrac{78815.63}{362880} = 0.2172$	$\times 0.0302$	0.0066

Table 19.17
Probabilities of 0, 1, 2... accidents (Poisson distribution mean = 3.5)

Having worked out the probabilities associated with the variability of the samples, it is possible to draw a number of conclusions from these results. The most important ones for our purposes are as follows:

- if there are no accidents in a particular month, the chance that this was due to variability of measurement is approximately 3 in 100
- the chance of a given sample being between 1 and 6 accidents is 90 in 100 ($0.1057 + 0.1851 + 0.2158 + 0.1888 + 0.1322 + 0.0771$)
- if there are 7 accidents in a month, the chance is approximately 4 in 100 that this is due to variability of measurement; for 8 accidents, the chance is approximately 2 in 100; and for 9 accidents, the chance is approximately 1 in 100.

It is now possible to set a level of significance – in other words, to decide what chance will be acceptable before going back to count the marbles in the bag. In this sort of work, it is common to use 1 in 20 (that is a probability of 5 per cent or $p = 0.05$) as a level beyond which action should be taken. In the present case, therefore, action would be taken if there were 7, 8 or 9 accidents. Similarly, action would be taken if there were no accidents. In either case, it would be necessary to investigate what had caused the significant change in the monthly accident rate.

The remainder of this section suggests some refinements to the basic system just described but before that, we need to make two general points:

1. The procedure may seem to be lot of work just to find out what is obvious, that if there are more accidents, this should be investigated. However, with a mean accident rate of 3.5 per month, fluctuations between 1 and 6 are likely to occur by chance and there is unlikely to be any good reason why these fluctuations take place. Time spent looking for a common cause for these fluctuations is being wasted.

2. The second general point concerns statistics as a whole. Statistics can rarely, if ever, provide the reason for something. They can provide information about when something is wrong, where it is wrong and even how wrong it is, but the only way to find out why it is wrong is to go and investigate.

The section continues with two additions to the basic scheme just described.

Limit lines

When the monthly mean number of accidents for 2010 has been worked out, then, we can calculate the probabilities of 0, 1, 2 ... accidents occurring in a given month. If conditions are reasonably stable in the organisation, then these expectations will hold for 2011, and a chart in the form shown in Figure 19.17 can be drawn up at the start of 2011.

Figure 19.17

Chart for recording the number of accidents in each month for 2011

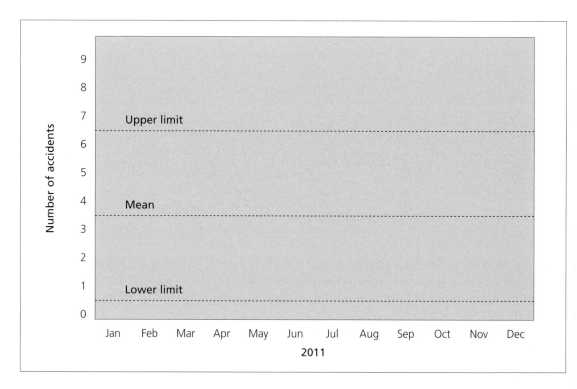

Figure 19.17 is a month by month presentation, of the type commonly used in organisations, of numbers of accidents, with the addition of the three lines shown. These lines are the mean number of accidents per month for 2010 and the upper and lower limits (at around the 5 per cent level of significance) of the variability of measurement expected around this mean on the basis of the Poisson distribution (see the earlier calculation). The lines have been drawn between 6 and 7 and between 0 and 1 in order to make it clear that action need only be taken when the lines are crossed, that is, if the number of accidents is higher than the upper limit, there should be an investigation to find out why things have become worse, and if the number of accidents falls below the lower limit, there should be an investigation to find out why things have become 'safer'.*

---

* There are theoretical problems with using the Poisson distribution in the ways described in this section, and it is debatable whether it is the right distribution to use. However, for practical purposes, it is preferable to calculate random fluctuation limit lines in some way, even if the chosen method is not ideal.

The sort of system illustrated by Figure 19.17 is a simple extension of accident plots already in common use. However, the extra work to calculate the two limit lines makes the technique much more valuable to the practising health and safety professional, and the basic technique can be modified in two main ways to make it even more valuable.

The first way to increase the usefulness of this technique is to base the calculations not on the number of accidents per month, but on some form of frequency rate. This is preferable because if there is a high number of accidents per month, $e^{-\bar{x}}$ becomes very small. For example, e^{-20} is 0.0000000025. However, if a rate based on hours worked is used, it is possible, by choosing an appropriate multiplier, to ensure that $e^{-\bar{x}}$ is kept within reasonable limits.

One frequency rate which can be used in this way is:

$$\frac{\text{Number of three-day loss accidents in year}}{\text{Total number of staff-hours worked in year}} \times 100,000$$

This gives a frequency rate for the year which can then be used as the expected frequency for the following year in the same way that the mean was used with straightforward numbers of accidents. The frequency rate is also used in place of $\bar{x}$ in calculating the expected probabilities of 0, 1, 2 ... accidents in order to determine the limit lines. Once this has been done, a chart similar to that shown in Figure 19.17 can be drawn, substituting 'frequency rate' for 'number of accidents' and omitting the 'mean' label on the middle line.

Each month, instead of plotting numbers of accidents, a monthly frequency rate is plotted. This monthly frequency rate is:

$$\frac{\text{Number of three-day loss accidents in month}}{\text{Number of staff-hours worked in month}} \times 100,000$$

Action will have to be taken when the monthly frequency rate falls outside the limit lines in the same way that action was taken when the number of accidents in a month fell outside the limit lines for accidents. This technique has the additional advantage of taking into account any fluctuation in the number of hours worked from month to month.

The second way in which the usefulness of the system can be increased stems from the use of frequency rates instead of numbers of accidents, which gives a more sensitive measure. There is no need to work in steps from 0 to 1, 1 to 2, 2 to 3 and so on, since it is possible to have frequency rates of 3.4, 1.7, 5.6, and so on. This increased sensitivity enables the use of additional limit lines on a chart as shown in Figure 19.18.

The meanings of these limit lines are:
- **Line 1**: These lines correspond to the limit lines already used in Figure 19.17. If the accident frequency rate is outside these limit lines in any one month, then investigation is required.
- **Line 2**: These lines are used with two consecutive months' accident frequency rates. If the accident frequency rate falls outside these limit lines in two consecutive months (for example May and June or August and September), then investigation is required.
- **Line 3**: These lines are used with three consecutive months' accident frequency rates. If the accident frequency rate falls outside these limits on three consecutive months (for example May, June and July), then investigation is required.

These various limit lines have an obvious use in that the analysts are no longer confined to an 'action or no action' decision on the basis of each month's figures. They are making more use of the information available.

The reason why the limit lines can be drawn for two and three consecutive months has its basis in the theory of sampling which was introduced at the beginning of this section. In simple terms, the larger the

Figure 19.18
Chart for recording monthly accident frequency rates for 2011 (expected frequency calculated on 2010 figures)

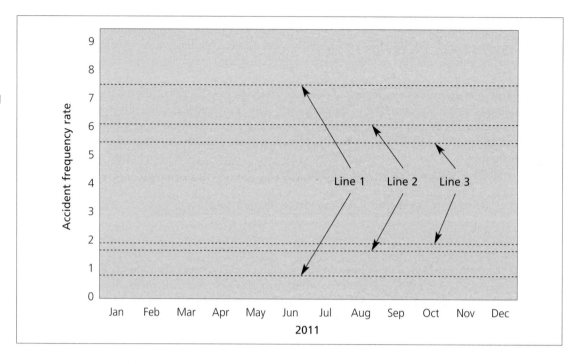

sample drawn, the surer it is possible to be about whether or not it is representative of the population from which it was drawn. The limit lines can, therefore, be drawn closer to the mean value. When two or three months' accidents are used as samples, larger samples are being taken and consequently tighter limits can be put on the variability of measurement.

If it is not clear why taking a larger sample allows limit lines to be drawn more closely, go back to the marbles example and consider the following:

- If a very small sample is drawn – a single marble – all that can be concluded is that there is at least one white (or one black) marble in the sample.
- If a very large sample is drawn – 99 marbles – all that is not known is the colour of the 100th marble.

This idea of larger samples can be taken even further to use all of the records available for a particular year as a series of samples. This would mean that at the end of successive months, the following samples would be available:

End of	One month sample	Running total sample
January	January	January
February	February	January + February
March	March	January + February + March
April	April	January + February + March + April
(and so on)		

This running sample can be used to work out a running total or cumulative frequency rate. For example, the running total frequency rate for April would be:

$$\frac{\text{Total accidents from January to April inclusive}}{\text{Total staff-hours worked from January to April inclusive}} \times 100,000$$

Each month, as accident records become available, the running total frequency rate for that month is calculated and plotted on a chart of the form shown in Figure 19.19. On this chart, the upper and lower limit lines have the same meaning as did Line 1 on Figure 19.18 – in other words, if the running total frequency rate is outside these limits for one month, then action should be taken.

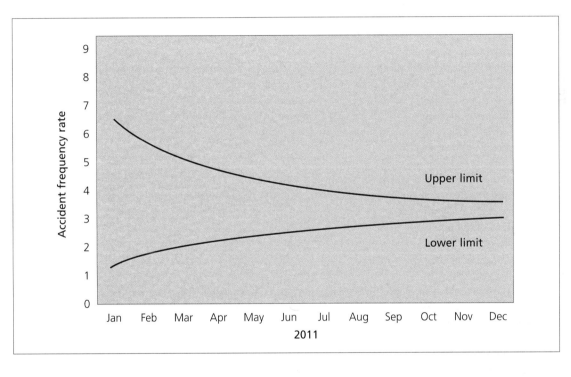

Figure 19.19
Chart for recording monthly running total frequency rates for 2011 (expected frequency calculated on 2010 figures)

It will be noted that the distance between the limit lines decreases from January to December. This is because each successive month's limit lines are based on a larger sample and, as was mentioned above, the bigger the sample taken, the more stringent it is possible to be with the limit lines. It will also be noticed that the limits for January correspond to the two Line 1 limits on Figure 19.18; the limits for February correspond to the two Line 2 limits on Figure 19.18; and the March limits correspond to the two Line 3 limits on Figure 19.18. The reason for this is, of course, that the limit lines in Figure 19.19 are for one-, two- and three-month samples and the January, February and March limit lines on Figure 19.19 are also for one-, two- and three-month samples.

Plotting successively larger samples on charts of the type shown in Figure 19.19 will enable significant trends to be identified over the year and allow such trends to be differentiated from random fluctuations, since the latter will remain within the upper and lower limit lines.

This section has explained the background to sampling theory and given sufficient information for the calculation of limit lines, either in terms of number of accidents or in terms of frequency rates. However, using these techniques requires a certain amount of calculation, which is extremely tedious if done manually. Anyone wishing to make extensive use these techniques should get hold of suitable statistical software. However, a well-specified spreadsheet will have the functions required, for example the POISSON function in Excel, although in a less user-friendly form.

The use of limit lines with frequency rates is discussed in detail by P J Shipp in a British Iron and Steel Research Association open report, *Presentation and use of injury data*.[65] Unfortunately, this report is out of print and difficult to obtain.

Epidemiology

The basic principles underlying epidemiological analysis were introduced in Chapter 9 in the context of the risk management model. We pointed out then that, in essence, epidemiological analysis involves attempting to identify patterns in sets of data. In this section, some of the basic epidemiological techniques are described and illustrated with examples.

Remember that epidemiological analysis requires data on a number of incidents of the same or similar type. For example, epidemiological analysis can be carried out on injuries, occurrences of ill health or, as we saw in the previous chapter, nonconformities in product quality. Irrespective of the type of incident, a range of information will be required on all, or at least a majority, of the incidents. In the case of injury accidents, typical information that may be available for each accident includes:
- the time the accident occurred
- the day of the week on which the accident occurred
- the month in which the accident occurred
- where the accident occurred
- the part of the body injured
- the nature of the injury
- what caused the injury
- the age and sex of the person injured.

However, the injury accident record form used in most organisations has spaces for more items of information than those listed above and, in general, any of the recorded items of information can be included in the epidemiological analysis. Since epidemiological analysis is, in essence, an exploratory technique which attempts to identify patterns in occurrences, there are rarely *a priori* grounds for excluding a particular type of information from the analyses.

Nevertheless, epidemiological analyses can be very time-consuming and, unless appropriate computer software is available, it may be advisable to begin the analyses with the sorts of information listed above.

Even using this restricted list with a relatively small number of incidents can produce valuable results, and some examples are given below from a previously published study.[66,67]

Figure 19.20 shows the numbers of accidents in two of the departments involved in the study plotted by the day of the week on which they occurred.

From Figure 19.20, it can be seen that, in both departments, there were fewer accidents on Mondays. It may be that absenteeism is greater after the weekend so that fewer people work on Mondays, hence fewer accidents; or it may be that less work is done on Mondays (the after-effects of the weekend); or it may be that different, lower-risk work is done on Mondays.

However, it will be remembered that the results of epidemiological analyses rarely identify causes, so additional analyses or investigations are required when a pattern is found. In the case of the pattern shown in Figure 19.20, the following additional work was required:

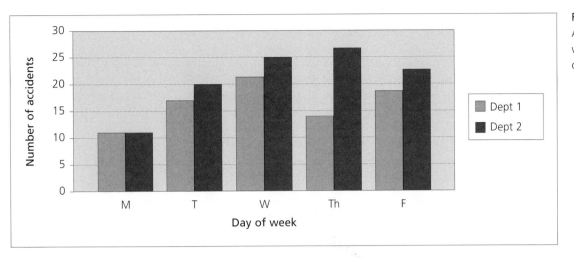

Figure 19.20
Accidents by day of week for two departments

- the calculation of an incidence rate for each day of the week to determine whether the pattern could have been caused by a higher absenteeism rate on Mondays
- the calculation of a rate based on production for each day of the week to determine whether the pattern could have been caused by the amount of work done
- an investigation of the types of work done on each day of the week to determine whether the pattern was caused by variations in risk.

In practice, the reason for there being fewer accidents on Mondays is likely to be a mixture of all of the above, but notice that finding a statistical pattern, in this case fewer accidents on Mondays, is only the start of the sequence. Epidemiological analyses are like maps; they help people find their way, but they do not actually take people anywhere.

It should also be noticed that there are other possible explanations for fewer accidents. For example, it may be that accidents are not being reported properly, or it may be that an effective risk control measure has been identified and put in place.

Turning now to the time of the day, in one of the two departments, accidents occurred evenly over the course of the working day. However, in the other department the pattern was as shown in Figure 19.21.

A priori, the expectation is that, with the exception of the lunch break when no work is done, the accidents would be distributed evenly over time. However, it can be seen from Figure 19.21 that this is not the case.

Subsequent investigation might have identified the following issues:
- although labelled as time of accident occurrence on the accident record form, the time actually recorded was the time of the report for treatment at the work's surgery
- all of the injuries were minor ones which would not have interfered with normal working.
- the works surgery was some distance from the department concerned.

Taking these issues into account, the most likely explanation of the rather complicated distribution shown in Figure 19.21 is that it reflects the arrangements for work breaks. The peaks in accidents may correspond with breaks for drinks and lunch with employees not reporting for treatment until just before, or just after, their normal break time. This is a distribution which has been observed in other studies, and this pattern of reporting has the advantages of not interrupting normal work and extending the usual break period for the injured worker. Whether or not it is an acceptable practice will depend on circumstances. However, the most

Figure 19.21
Time of accident occurrence in one department

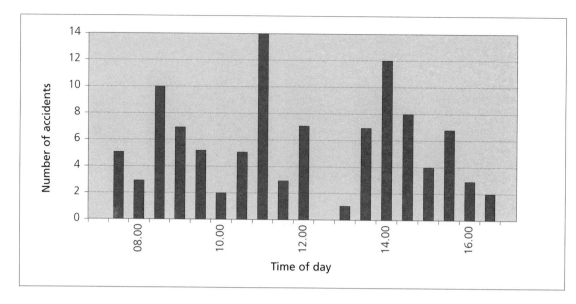

important issue in this case was the fact that, for the department concerned, there were no accurate data on the times at which accidents occurred.

The final illustration concerns the month in which the accidents occurred. Again, in one of the departments, accidents were distributed evenly over the year. However, in the second department the pattern was as shown in Figure 19.22.

Figure 19.22
Month of accident occurrence in one department

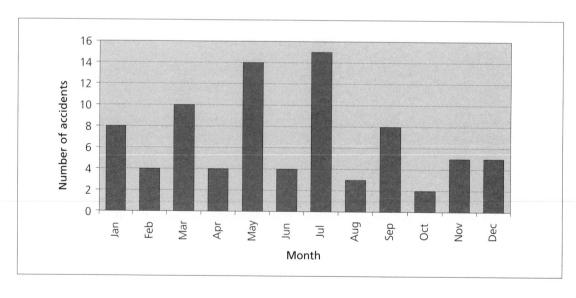

The pattern shown in Figure 19.22 would not be expected, *a priori*, and it was not related to numbers employed, or the amounts of work done, the two factors which are so fundamental to numbers of injury accidents that they should be checked as a matter of routine.

Subsequent investigation suggested that the most likely explanation arose from the department's recruitment policy. Most recruiting was done on a two-month cycle so that at the start of every other month

(January, March, May and so on), there was an influx of untrained workers. Since it is known that untrained workers are more likely to have injury accidents, it was likely that the pattern shown in Figure 19.22 was due to accidents among the new recruits. This hypothesis could quite readily be tested by an appropriate analysis using numbers of days in employment as one of the data dimensions.

In the illustrations above, the patterns identified were from even distribution expected *a priori*. For example, there was no reason to expect that accident numbers would vary widely from month to month in one department. However, with other data dimensions, an even distribution may not be expected and, in these cases, the epidemiological analysis looks for deviations from whatever distribution is to be expected. Two examples are given here:

- **Part of body injured.** The majority of minor injury accidents typically involve injuries to the hands. This is characteristic of a wide variety of different kinds of work and is, of course, because the hands are the parts of the body which are used most. The implication of this is that if an analysis by part of body injured was carried out, it would not be appropriate to go into a more detailed investigation if it was found that the majority of injuries occurred to the hands since this is what is to be expected.
- **Location of the accident.** Different parts of an organisation will be recognised as having different levels of risk associated with working there. For example, shop floors usually have higher levels of risk than offices. For any given site, therefore, there will be a characteristic pattern of accident occurrence which reflects this unequal distribution of risk. Deviations from this characteristic pattern should be investigated, such as increases in office accidents or decreases in shop-floor accidents.

For some items of information, it may be difficult to predict patterns and, where this is the case, it is desirable to establish patterns of injuries over two or three years before comparing the occurrence patterns of recent accidents. The aim of this type of epidemiological analysis is to find deviations from the organisation's norm rather than deviations from some hypothetical or assumed distribution.

As has already been mentioned, unless appropriate computer software is available, epidemiological analysis is very time-consuming. While the data-sorting facilities in spreadsheet programs can be used for epidemiological analyses, specially designed packages are preferable. Computer software development results in new and upgraded packages being released every few months and these are advertised in the health and safety press. Epidemiological analysis or pattern analysis is usually a feature of the better accident recording packages but you should confirm this by obtaining a demonstration copy of the software.

Appropriate computer software speeds up the analysis enormously since all of the preparation of the necessary tables and histograms can be automated. Nevertheless, the last stage of epidemiological analyses – identifying patterns – usually has to be left to the humans. Computer software is wonderful at routine data manipulation but, except in very specialised application, it is very poor at pattern recognition.

Finally, the deviations from what is to be expected which are found during epidemiological analysis vary from quite large (see Figure 19.22) to insignificant, and there are statistical techniques for enabling decisions to be made on whether or not a particular deviation is large enough to be significant in statistical terms. These techniques are outside the scope of this book, but interested readers are referred to the chi-squared test, details of which will be found in most statistics textbooks. However, the patterns which are likely to be worth following up are, like the illustrations given earlier, generally fairly obvious ones.

Summary

This chapter dealt with the important topic of measuring safety performance. It began with a discussion of how measuring is dealt with in the major HSMSs, and then described the various measures which can be used to assess health and safety performance and the nature of the data types which might be available. However, since the best measures of health and safety performance are those which can be quantified, much of the chapter was devoted to numbers and how they can be used.

20: Advanced accident investigation and risk assessment

Introduction

This chapter is divided into two main sections – advanced accident investigation techniques and advanced risk assessment techniques. In turn, each of these sections is subdivided so that the overall layout of the chapter is as follows:

1. Advanced accident investigation
   - Introduction
   - Events and Causal Factors Analysis (ECFA)
   - Management Oversight and Risk Tree (MORT)
   - Accident severity
2. Advanced risk assessment
   - Introduction
   - Hazard and Operability studies (HAZOPs)
   - Failure Modes and Effects Analysis (FMEA)
   - Event Tree Analysis (ETA)
   - Fault Tree Analysis (FTA)

Both accident investigation and risk assessment are skills and, like all skills, they are only as good as the people applying them. It can be argued that many of the features of advanced accident investigation and risk assessment techniques are attempts to force the incompetent to do an adequate job. Basically, the advanced techniques spell out, sometimes in tedious detail, the steps that a competent accident investigator, or risk assessor, would take as a matter of course. However, this is not a criticism of the techniques; anything which forces people to do a better job is to be commended. Rather, it is a criticism of the health and safety professionals who do not learn the basic skills required for accident investigation and risk assessment – this point will be dealt with in more detail in the introductions to the 'Advanced accident investigation' and 'Advanced risk assessment' sections.

However, before moving on to the details of the techniques, we must discuss the reasons for using accident investigation and risk assessment.

Reasons for using accident investigation and risk assessment

In Part 1.1, the reasons for using accident investigation and risk assessment were dealt with in a simplified manner as follows:
- **Accident investigation** is used to find out why an accident occurred and to suggest measures which could be implemented to prevent recurrence.
- **Risk assessment** is used to find out how an accident may occur and to suggest measures which could be implemented to prevent occurrence.

These simple descriptions suited the purposes of Part 1.1, but they gloss over a number of important issues, and these are dealt with in this section.

The first point is that accident investigation and risk assessment have the same outputs: measures to prevent accidents in the future. Since this is the case, it is useful to consider whether accident investigation and risk assessment have common features, as this could help in reducing the confusion of terminology in this area.

A number of common features can be identified:

- The last stages of both processes, suggesting action to prevent accidents, are identical.
- Good accident investigations are rarely restricted to the one set of (possibly unique) circumstances which led to the outcome under investigation. A good investigation will consider similar circumstances which could occur elsewhere in the organisation, and include these in the investigation remit. In other words, a good investigation will include an element of risk assessment.
- Risk assessment can be carried out by identifying possible outcomes, such as injuries and damage, and then working out how these outcomes could arise. That is, risk assessment takes the form of investigating hypothetical accidents.

It can be argued, therefore, that accident investigation and risk assessment are closely related processes; this is a topic which requires further research. As an introduction to this, the following notes are a consideration of the purposes of accident investigation and how it should link with risk assessment.

The 'received wisdom' is that the purpose of accident investigation is to determine the measures which should be taken to prevent an accident happening again, and this was the purpose quoted in Part 1.1. However, in a mature HSMS, the purpose of accident investigation should be to review risk assessments and, in such a management system, accident investigation should be replaced by risk assessment review. This may be considered a contentious statement, so we will explore the reasons for it in some detail.

We argued in Chapter 5 that the rational approach to risk management is for organisations to define what they consider to be acceptable risk and manage their risks in ways which ensure that their criteria for acceptability are met. This implies that organisations are willing to tolerate a certain number of accidents. For example, using the simple risk rating procedure described in Chapter 5 (Tables 5.1 and 5.2), organisations may adopt as their definition of acceptable any hazard with a risk rating of 5 or less. The implication of this is that such organisations are going to tolerate the risks listed in Table 20.1.

Table 20.1
Risk ratings of 5 or less

Category	Likelihood	Severity	Risk	Implications
A	1	5	5	Extremely unlikely fatality
B	5	1	5	Almost certain minor injury
C	1	4	4	Extremely unlikely major injury
D	2	2	4	Unlikely injury resulting in up to three lost days
E	4	1	4	Extremely likely minor injury
F	1	3	3	Extremely unlikely 'three-day' injury
G	3	1	3	Likely minor injury
H	1	2	2	Extremely unlikely injury resulting in up to three lost days
I	2	1	2	Unlikely minor injury
J	1	1	1	Extremely unlikely minor injury

For organisations of any significant size, there will be many thousands, or tens of thousands, of hazards with a risk rating of 5 or less. Therefore, in statistical terms, the following points can be expected:

- **Fatalities** (severity 5) – Extremely unlikely for each hazard but the probability of occurrence increases as the number of hazards in category A in Table 20.1 increases.
- **Major injuries** (severity 4) – Extremely unlikely for each hazard but the probability of occurrence increases as the number of hazards in category C in Table 20.1 increases.
- **'Three-day' injuries** (severity 3) – Extremely unlikely for each hazard but the probability of occurrence increases as the number of hazards in category F in Table 20.1 increases.
- **Up to three days lost** (severity 2) – Unlikely for each hazard but the probability of occurrence increases as the number of hazards in categories D and H in Table 20.1 increases.
- **Minor injuries** (severity 1) – Almost certain (the sum of categories B, E, G, I and J).

When an accident occurs, the question asked should not, therefore, be 'Could this have been prevented?' but rather 'Does this fall within the predicted range?'

This is not a concept with which health and safety professionals in general are comfortable, since it runs counter to the ideas that all accidents are preventable and should be prevented. However, risk is based on probabilities and the concept of acceptable risk assumes that, unless an organisation adopts a very low criterion of acceptability, or is very lucky, accidents will happen. For the fully developed HSMS, the logical approach when an accident happens is as follows:

- **Answer the question: 'Was this accident predicted?'** That is, determine whether the accident which has occurred was included in the list of 'hypothetical accidents' identified during the risk assessment. If the answer is 'no', then the risk assessment must be revised, and an investigation undertaken into the reasons for the failure to include this accident in the list of 'hypothetical accidents'.
- **Answer the question: 'Were the likelihood and severity for the accident estimated accurately?'** Again, if the answer is 'no', the reasons for the inaccurate estimates should be investigated.

If, however, the risk assessment had identified the accident and the estimates of likelihood and severity were correct, then there is no rational reason for doing anything to prevent the accident happening again. If the organisation has decided to accept a specified level of risk, then it has to tolerate the number of accidents which will arise from this risk. If the accidents which happen fall within the level of acceptability, then it is irrational to suggest measures for prevention of recurrence. Thus, when an accident happens, the requirement is for a review of the risk assessment to decide whether the risk which gave rise to the accident was in the acceptable range, rather than a blanket attempt to prevent recurrence.

When considered in this way, accident investigation and risk assessment become even more closely linked, but there are few organisations with sufficiently mature HSMSs to make this connection. For this reason, the remainder of this chapter treats accident investigation and risk assessment as separate topics.

Advanced accident investigation

Introduction
Adequate accident investigation requires the following:
- a clear idea of what is to be achieved by the investigation
- a high level of competence in interviewing and observation
- good analytical skills, particularly those required to analyse often conflicting views of how and why an accident happened
- a high level of creativity in generating possible remedial measures
- a detailed knowledge of human factors and, in particular, individual differences and human reliability.

Unfortunately, in the author's experience, health and safety professionals in general do not meet these criteria. In the worst cases, accident investigations are conducted with a confused mixture of motives, or with the single motive of collecting information which will either protect the health and safety professional's organisation from prosecution, or minimise the amount the organisation will have to pay in compensation. The health and safety professionals involved in this type of investigation use what skills they have to collect evidence in support of the required cause, conveniently ignoring anything which runs counter to their preconceived requirement, or interferes with the interests of their employer. The one saving grace in these circumstances is that the health and safety professionals concerned usually have a very low level of competence so that their scope for doing serious damage to the truth is often limited by the actions of more competent individuals.

Over a period of 30 years, the author has observed health and safety professionals conduct accident investigations, both in the classroom during training and in real life. This experience has identified low levels of skill, particularly at the basic levels of interviewing skills. There appears to be an assumption that because people can talk and listen they are competent interviewers. This is not the case and it is rare that there is anyone who can interview competently who has not received formal training in interview techniques.

The ideal solution to this problem would be to ensure that health and safety professionals have the necessary competences, but there have been attempts to solve the problem in other ways. In particular, organisations have tried to produce tools which will reduce the level of competence required for effective accident investigation. These tools range from more or less detailed accident investigation checklists to computer-based expert systems, and they all have a major advantage in that they make it more likely that an adequate range of basic data on the accident will be collected. However, their effective use is still reliant on the competence of the person using them. Two of these tools, ECFA and MORT, will now be considered.

Events and Causal Factors Analysis

ECFA* is a method of collating data from incident investigations. The output of the method is a chart illustrating the events and causal factors involved in the incident and how these interrelate. A simple ECFA chart is given in Figure 20.1 to illustrate the technique.

The accident described in Figure 20.1 involves a child being hit by a chisel which has been kicked from a scaffold by workers repairing brickwork at a bus station. The sequence of events leading up to the injury is:

1. A mother with her three children is waiting for a bus which has been delayed. She decides to buy the children chocolate and puts money into a chocolate vending machine which is out of order and does not return her money. She sets off to find someone to refund her money, leaving the children unsupervised. One of the children wanders off, sees a pile of sand, ducks under a barrier and begins to play on the sand.
2. Meanwhile, a worker who is repairing brickwork from a scaffold has left a chisel on the scaffold boards. This is accidentally kicked by Steve who is walking past and, since there are no toe boards, the chisel falls from the scaffold. There is a net intended to stop falling objects and the chisel lands on this, but then falls through and hits the child.
3. Since the absence of toe boards and an inadequate net are important factors in this accident, the relevant information is summarised on the bottom lines of the ECFA chart.

The key points from Figure 20.1 are:
- The events leading directly to the outcome (injury to the child) are shown as a linear horizontal sequence towards the middle of the diagram. These are known as the 'primary' events and are recorded in rectangles joined by solid arrows.

---

* ECFA has numerous variants, including Events and Conditional Factors Analysis.

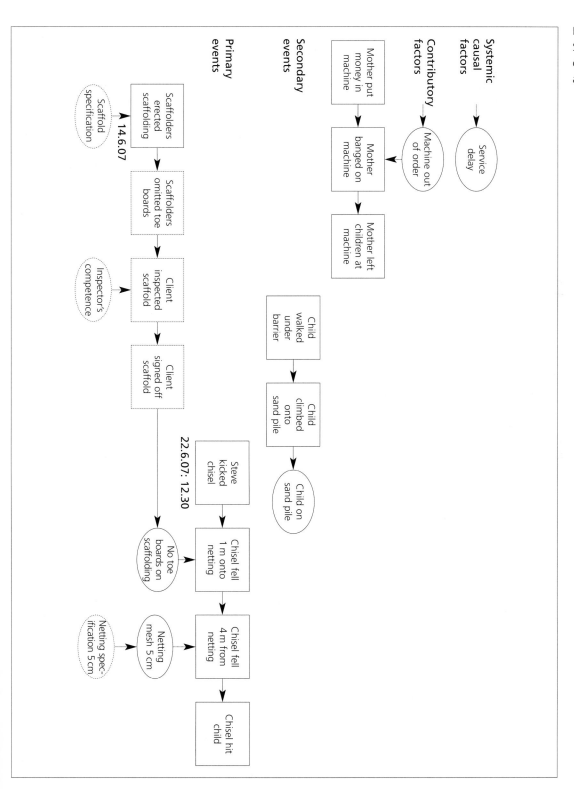

Figure 20.1
Simple events and causal factors analysis chart

- Other relevant events, known as 'secondary' events, are recorded above and below the line of primary events using as many levels as are necessary. These secondary events are also recorded in rectangles and joined by solid arrows.
- Causal factors, also referred to as 'conditions' in the ECFA terminology, are recorded above or below the relevant event sequences. They are, by convention, contained in ellipses (often referred to as ovals) and joined to each other, and to the relevant events, by dotted arrows. It is usual to discriminate between 'contributory factors' and 'systemic factors' although the difference between the two may not be clear cut. In general, systemic factors are conditions that apply because they are inherent in the organisation's structure or arrangements, such as safety rules, while contributory factors are ones which have arisen for other reasons.

The two categories of events (primary and secondary) and conditions (contributory factors and systemic factors) are identified during an investigation and set out on the chart in the general format illustrated in Figure 20.2.

Figure 20.2
General format for ECFA chart

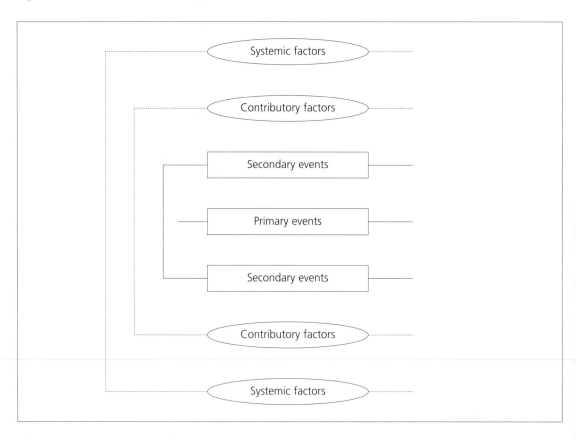

Within this general format, a great deal of flexibility is possible, so the ECFA charting technique can be used to record details of even the most complex incidents. The basic format can also be elaborated in a number of ways, some of which are illustrated in Figure 20.1.

- **Dates and times can be added to the chart to identify when particular events happened or when particular conditions came into, or were in, force.** If appropriate, a 'time line' can be created which shows not only the order of events but also the times between them. Where necessary, the ECFA chart can be structured so that all events and conditions which occurred at the same time are set out, one above the other, vertically on the chart.

- **Events and conditions are recorded in rectangles and ovals with solid outlines when the relevant facts have been established.** When assumptions have to be made, for example, because evidence has been destroyed or witnesses are unable to recall particular items of information, then these events or conditions are recorded in rectangles or ellipses with dotted outlines. Dotted outlines are also used during the investigation process as a means of recording events or conditions which require further investigation, with the dotted outline being changed to a solid one when the appropriate facts have been established.

For the ECFA technique to be of maximum benefit, it is essential that the individual events described on the chart meet certain criteria and these are described next.

- The event must be a true event – something which occurred or happened. If, for example, conditions or states are recorded as events, there will be a break in the logic of the event sequence.
- Each event should be a single occurrence not, for example, 'Steve kicked the chisel and it fell through the netting'. The simplest way of ensuring this is to restrict the description of an event to one noun and one active verb, such as 'Steve kicked the chisel', 'the chisel fell', 'the chisel hit the child'.
- Where possible, events should be quantified, for example 'the chisel fell 4 m', not 'the chisel fell'.

However, even if the individual events meet the criteria listed above, the chart will be of limited value unless the individual elements follow on from each other in a logical sequence. This should be checked carefully for all sequences of events, since illogical sequencing may mean that an event has been omitted, or that a spurious event has been inserted.

When using ECFA during an investigation, it is useful to record each event and causal factor on a separate Post-it note, with different colours for events and causal factors. The Post-its can then be displayed on a suitable surface (for example a table or wall) and moved around as required. This removes the need to redraw the chart as the investigation proceeds. When the chart is complete, the most convenient way of recording it is to use an appropriate software package, such as Visio. The ECFA and related methodologies are not copyright and more information can be found on the internet by searching for the full name of the technique you are interested in.

Competence in the use of ECFA techniques can only be obtained with practice, but using it effectively offers a more structured approach to identifying and recording investigation data. This not only has the potential for improving investigations, it also provides an opportunity for more detailed and wider ranging suggestions for prevention of recurrence. In addition, a well-structured ECFA chart provides a good framework for the preparation of a written report of the investigation.

Management Oversight and Risk Tree (MORT)

The HSMS which forms the basis of MORT was described in Chapter 18, where we pointed out that the MORT system is set out in the form of a fault tree. The underlying rationale of the MORT fault tree is that if none of the basic causes is present, then no accident should happen. It follows from this that if an accident has happened, then one or more of the basic causes must have been present.

Using MORT for advanced accident investigation involves considering the circumstances of the accident in the context of the MORT fault tree. The investigator works systematically through the fault tree identifying, for the accident being investigated, which basic causes were present.

Although this can become a mechanistic process, proper use of the MORT fault tree can have a number of advantages:

- it encourages consideration of a wider range of causes than might otherwise be the case
- it encourages consideration of the management failures as well as the failures of people more closely involved in the accident

- it can identify failures in management systems which, although not relevant to the accident under consideration, have the potential to cause problems in the future.

The practical problems with using MORT for accident investigation are the initial training requirements and the potential for generating large amounts of paperwork. However, as we mentioned before, any effective accident investigation technique requires initial training and any investigation technique using fault tree principles suffers from the paperwork problem.

While it is not possible within the scope of this book to provide a detailed manual for using the MORT approach to accident investigation, the notes which follow give an overview of its main features.

As we noted in Chapter 18, the core of MORT is a detailed fault tree which sets out what the MORT designers believe to be all the potential causal factors for an accident.

As far as MORT is concerned, an accident is an unwanted flow of energy or exposure to an environmental condition that results in adverse consequences. From this definition, the following circumstances are required for an accident to occur:

- a potentially harmful energy flow or environmental condition, the latter being defined as energies which produce injury and damage by interfering with normal energy exchanges
- vulnerable people or objects to which a value is attached
- the absence or failure of a barrier between the energy flow or environmental condition and the vulnerable people or objects.

In the majority of circumstances, energy flows, environmental conditions and vulnerable people and objects are part of the work system and there are relatively few options for their removal. For this reason, much of the MORT analyst's time is devoted to what is referred to as barrier analysis – that is, where in the accident sequence the barriers to energy flows or environmental conditions were less than adequate. Since it is also possible to have unwanted energy flows without people being hurt or objects damaged, MORT uses the terms incident or mishap to describe these circumstances.

The types of barrier referred to in Chapter 8 during the discussion on risk control are included in the MORT concept of barrier, but the range of barriers in the MORT definition is much wider. Haddon[68] described the 10 broad categories of barrier as:

1. prevent the marshalling: do not produce or manufacture the energy
2. reduce the amount: voltages, fuel storage
3. prevent the release: strength of energy containment
4. modify the rate of release: slow down burning rate, speed
5. separate in space or time: electric lines out of reach
6. interpose material barriers: insulation, guards, safety glasses
7. modify shock concentration surfaces: round off and make soft
8. strengthen the target: earthquake-proof structures
9. limit the damage: prompt signals and action, sprinklers
10. rehabilitate persons and objects.

Within this broad framework, MORT reclassifies barriers in a variety of ways and three examples are given here.

1. Barriers for control and barriers for safety. The primary function of some barriers is to control wanted energy flows; examples of these include conductors, approved work methods, job training, disconnecting switches and pressure vessels. Other barriers are, however, primarily intended to control unwanted energy flows and examples of these include PPE, guardrails, safety training, work permits and emergency plans.

2. Classification by location. This classification concerns whether the barrier is in proximity to the energy source (for example, insulation on a cable conducting electricity), in proximity to the person (that is, PPE), or somewhere between the two.

3. Classification by type. Various types of barrier are recognised, including physical barriers, equipment design, warning devices, procedures, work processes, knowledge, skill and supervision.

During MORT analysis every barrier relevant to every energy flow and environmental condition involved in the accident is examined in turn and the reasons for its failure determined. In order to improve the thoroughness of the examination of reasons for failure, MORT provides a generic fault tree for barrier failure analysis. A simplified version of this fault tree is reproduced as Figure 20.3.

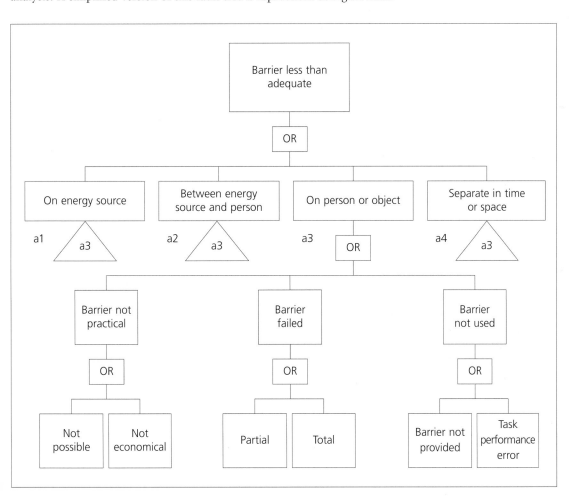

Figure 20.3
Simplified MORT fault tree for barrier failure analysis

Notes on Figure 20.3:
•   In MORT (and FTA) diagrams a special symbol is used for the OR gate (not the word OR). There is also a special symbol for an AND gate. These, and other FTA symbols, are explained in Figure 20.14. However, words are used in Figure 20.3 to make interpretation easier for those unfamiliar with these special symbols.

- The triangles in Figure 20.3 indicate that this section of the fault tree should be developed further, and the code within the triangle indicates which part of the generic fault tree should be used for this purpose. In Figure 20.3, the cross-references are all within the part of the fault tree shown, but the references may be to a different part of the wider MORT fault tree.

The MORT analysts work through this generic fault tree for every energy flow/barrier combination and every environmental condition/barrier combination, identifying which failure (or failures) is relevant for each combination. We can return to the example used to illustrate ECFA earlier in this chapter to illustrate the process:

1. Consider first the fact that a chisel was left on the scaffold board. The barrier to this would have been the work method used, but we do not know whether this barrier failed because the work method was not specified or because a specified work method was not used.
2. The next possible barrier should have been separation in time or space – in other words, people should have been prevented from walking on the scaffold. Again we have insufficient information to judge whether this was a possible barrier and, if it was, what caused this barrier failure.
3. The absence of toe boards (barrier between energy source and person) would also require further analysis, since the data available do not enable a discrimination between 'barrier not practical' and 'barrier not provided'.
4, The netting designed to catch falling objects would have been a 'barrier failed' and it would have been 'partial', since the netting slowed the chisel down but did not prevent it falling further.
5. The barrier which was intended to prevent access to the sand pile was also inadequate, since the child was able to get under it.

In a full analysis, many more barriers would be considered and analysis would cover the sorts of systemic and contributory factors identified in the ECFA example. Although the example being used described only one energy flow (the kinetic energy associated with the chisel), there are, in practice, a wide range of possibly relevant environmental conditions and the sorts of energy flows which may have to be considered in other accidents include electrical, chemical, biological, ionising radiations and non-ionising radiations.

Within MORT, energy sources and environmental conditions are divided into functional or wanted sources or conditions and non-functional or unwanted sources or conditions. The rationale for this classification is that the former category is necessary and has, therefore, to be adequately controlled, while the latter category is unnecessary and can be eliminated.

As with barrier analysis, MORT provides a generic fault tree for the analysis of energy flows and environmental conditions. A simplified version of this fault tree is reproduced as Figure 20.4, to which the following notes apply:

- In the MORT convention, the rectangle with rounded corners is used to indicate an event which is satisfactory.
- Each of the termination points in Figure 20.4 is developed further in the full MORT fault tree and would therefore have a symbol or code beneath it giving the appropriate reference point.

In the example used earlier to illustrate ECFA, the kinetic energy associated with the falling chisel was non-functional and the control was less than adequate.

The other main element of the MORT analysis is the target, referred to as vulnerable people or objects which have a value. The generic fault tree for targets is identical in structure to the generic energy flow fault tree but has two conditions. A simplified form of the target fault tree is reproduced in Figure 20.5, to which the following notes apply:

- The ellipse is used in MORT to indicate that a condition is attached and that the element of the fault tree with the attached condition is only relevant if the condition is satisfied. Thus, vulnerable people or objects

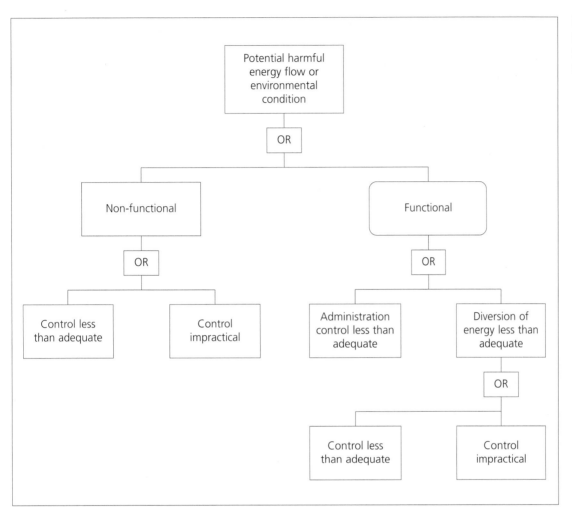

Figure 20.4
Simplified MORT
fault tree for energy
flow analysis

are only relevant if they have a value and the presence of functional personnel is only relevant if the barriers intended to protect them are less than adequate.
- Each of the termination points in Figure 20.5 is developed further in the full MORT fault tree and would therefore have a symbol or code beneath it giving the appropriate reference point.

In the example used to illustrate ECFA, the target (the child) was non-functional and the control was less than adequate. However, had it been a construction worker who had been injured, the analysis would have required the use of the functional branch of the fault tree.

The generic fault trees described so far illustrate the main aspects of the first level of MORT analysis. However, as we saw in the ECFA example, this level of analysis leaves a number of questions unanswered. In these circumstances, MORT analysts use the more detailed fault trees provided to structure further analysis. It is outside the scope of this book to provide the detail required for the full use of MORT, but the example below shows how fault trees are provided for successively more detailed analysis.

Where it is found that the controls on a barrier are less than adequate, the fault tree splits into six branches dealing respectively with less than adequate:

Figure 20.5
Simplified MORT
fault tree for target
analysis

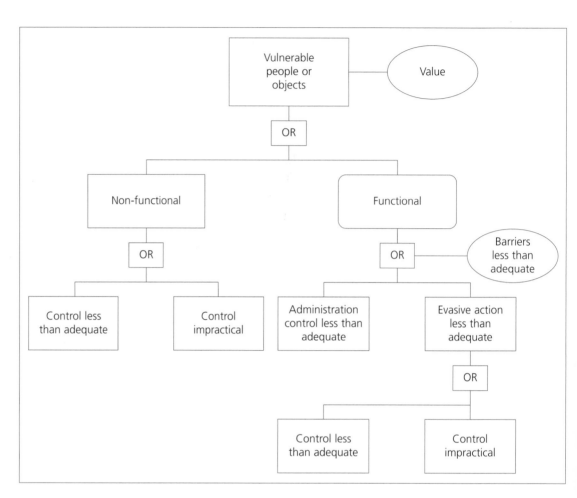

1. technical information systems
2. facility functional operability
3. maintenance
4. inspection
5. supervision
6. higher supervision services.

Where, for example, supervision is found to be less than adequate, the fault tree splits into five branches dealing respectively with:

1. less than adequate help and training
2. less than adequate time
3. less than adequate supervisor transfer plan
4. did not detect or correct hazards
5. performance errors.

The MORT manual and supporting documentation are free from copyright – if you want to study the technique in more detail, you can make use of a wide variety of material available on the internet. However, a search for 'Management Oversight and Risk Tree' in full will provide a more useful list of addresses than a search for 'MORT'.

Accident severity

In Chapter 9 we argued that as much can be learned from investigating individual minor accidents as can be learned from investigating individual major accidents. It was also pointed out that there are many more minor accidents than major accidents, so investigation of minor accidents gives many more opportunities to learn from what has gone wrong. This has been known for some time and various authors have published accident triangles, or accident ratios, of the form shown in Figure 20.6 and Table 20.3.

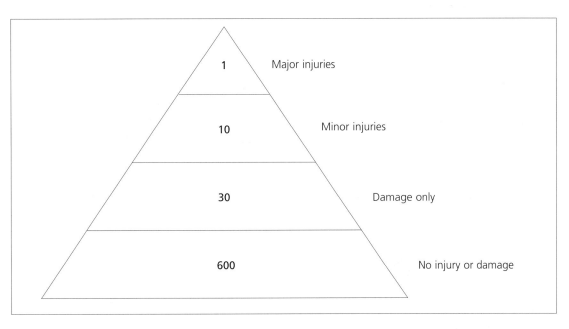

Figure 20.6
Bird's accident triangle[69]

1 — Major injuries

10 — Minor injuries

30 — Damage only

600 — No injury or damage

	Construction	Creamery	Oil platform	Hospital
Over-three-day injury accidents	1	1	1	1
Minor injuries	56+	5	4	10
Non-injury accidents	3750+	148	126	195

Table 20.3
Accident ratios (HSE)[4]

However, there is evidence from research by Hale[64] that major and minor accidents do not have the same range of causes. But this should have limited practical impact on accident investigation, and Hale concludes:

> If minor injury scenarios are tackled it should be because minor injuries are painful and costly enough to prevent in their own right, not because it is believed the actions might control major hazards.

Advanced risk assessment

Introduction

In Part 1.1, we dealt with risk in simple terms and described a basic form of risk assessment. While this was appropriate as introductory material, it oversimplified a number of issues and glossed over a number of fundamental problems with the terminology and techniques of risk assessment.

In this section, therefore, we consider risk and risk assessment in more detail and identify a number of problems, some of which have yet to be solved. Finally, we will consider the implications of this more detailed view of risk for risk management in an organisation. The section has six subdivisions:

1. risk assessment terminology
2. hazard identification
3. risk rating scales from HSG65 and BS 18004
4. severity distributions
5. deciding on the acceptability of risks
6. risk assessment management.

Risk assessment terminology

As we saw in Chapter 5, the terms 'hazard' and 'risk' have no agreed definitions, and there is a lack of clarity about what is meant by 'likelihood' and 'severity' in the risk equation (risk = likelihood × severity). Table 20.4 summarises the definitions of hazard and risk used in a number of authoritative publications.

Table 20.4
Definitions of hazard and risk

Source	Hazard	Risk
OHSAS 18001:1999[71]	Source or situation with the potential to cause harm in terms of injury or ill health, damage to property, damage to the workplace environment, or a combination of these	Combination of the likelihood and consequence(s) of a specified hazardous event occurring. (Note that 'hazardous event' is not defined in OHSAS 18001:1999)
BS OHSAS 18001:2007[3]	Source, situation or act with a potential for harm in terms of human injury or ill health, or a combination of these	Combination of the likelihood of an occurrence of a hazardous event or exposure(s) and the severity of injury or ill health that can be caused by the event or exposure(s)
BS 18004	As BS OHSAS 18001	Gives the BS OHSAS 18001 definition in the text but a different definition is used in Annex E when describing how to estimate risk – see later in this chapter
PAS 99	No definition given	Likelihood of an event occurring that will have an impact on objectives
Management of Health and Safety at Work Regulations 1999[13]	Something with the potential to cause harm (this can include articles, substances, plant or machines, methods of work, the working environment and other aspects of work organisation)	The likelihood of potential harm from [a] hazard being realised. The extent of the risk will depend on: i) the likelihood of that harm occurring ii) the potential severity of that harm, ie of any resultant injury or adverse health effect; and iii) the population which might be affected by the hazard, ie the number of people who might be exposed

Table 20.4
continued

Source	Hazard	Risk
Five steps to risk assessment (first edition)	'Hazard' means anything that can cause harm (eg chemicals, electricity, working from ladders, etc)	The chance, high or low, that somebody will be harmed by the hazard
Five steps to risk assessment (second edition)[72]	Anything that may cause harm such as chemicals, electricity, working from ladders, an open drawer etc	The chance, high or low, that somebody could be harmed by these and other hazards, together with an indication of how serious the harm could be
Fire Safety Risk Assessment Series[73]	Anything that has the potential to cause harm	The chance of that harm occurring
Essentials of health and safety at work first edition	A hazard is anything that can cause harm (eg chemicals, electricity, working from ladders etc)	Risk is the chance (big or small) of harm being done
Essentials of health and safety at work 2006[74]	No definition given	No definition given
HS(G)65[8]	Hazard means the potential to cause: harm including ill health and injury; damage to property, plant, products or the environment; production losses or increased liabilities	The likelihood that a specified undesired event will occur due to the realisation of a hazard by, or during, work activities or by the products and services created by work activities. (Appendix 1) Risk = hazard severity × likelihood of occurrence (of the hazard). (Chapter 4)
HSG65[1]	Hazard means the potential to cause: harm including ill health and injury; damage to property, plant, products or the environment; production losses or increased liabilities	Risk means the likelihood that a specified undesired event will occur due to the realisation of a hazard by, or during, work activities or by the products and services created by work activities. (Appendix 1) Risk = severity of harm × likelihood of occurrence. (Chapter 4)
INDG275[75]*	A hazard is something with the potential to cause harm. The harm will vary in severity – some hazards may cause death, some serious illness or disability, others only cuts and bruises	Risk is the combination of the severity of harm with the likelihood of it happening
Railway safety principles and guidance, part 3A[40]	Hazard means a thing, condition or situation with the potential to cause ill health and/or physical injury to people, damage to property, plant, products or harm the environment	Some combination of the frequency of occurrence, probability of failure and severity of consequence. An additional variable that addresses the probability of failure to recover may be needed for assessing human factor risks

* INDG275 is a summary of HSG65 with the title *Managing health and safety: five steps to success*.

The following points can be drawn from the definitions given in Table 20.4:

- In the majority of the definitions, a hazard is an entity, variously described as 'a source or a situation', 'something' or 'anything'. However, HSG65 does not define **a hazard**; it defines 'hazard', which is a legitimate, if archaic, use which, it could be argued, only serves to confuse matters further.*
- All of the definitions refer to harm, but what is included in harm varies, with some definitions including only harm to humans, while others include damage to property or the environment.
- Some of the definitions of risk have two elements (likelihood and severity or consequence) and some have only one, likelihood or chance. Note that in HS(G)65 and HSG65 risk is defined in both ways!
- The likelihood part of risk is variously defined as the likelihood of 'hazard' occurring, the likelihood of a 'hazardous event', the likelihood of 'harm', and the likelihood of a 'specified undesired event'. These are fundamentally different concepts with, for example, the likelihood of a fatality as a result of a trip (likelihood of a specified undesired event) being very different from the likelihood of tripping (likelihood of a hazardous event), which is also different from the likelihood of all possible harms.
- There is a fundamental change between HS(G)65 and HSG65 in the risk equations used. Hazard severity (as used in HS(G)65) could be a useful concept, but it would have to be measured in different ways. For example, noise above 95dB(A) is a 'more severe' hazard than noise below this level, and 240 volts is a 'more severe' hazard than 110 volts. However, the HS(G)65 definitions are included here to illustrate the fact that even authoritative sources such as the HSE change their collective mind on risk.

It could be argued that these differences in definitions are due to inadequate drafting on the part of the various authors. However, it seems more likely that this lack of agreement arises because one or more of the concepts being described is inadequate in some way.

I have argued[76] that part of the problem arises because there is a tendency to oversimplify what is required for hazard identification. This is reflected in phrases such as 'hazard spotting', which imply that hazard identification is simply a matter of going out and looking at what is there.

However, hazard identification is a perceptual and decision making process that requires a number of mental activities. The mental activities involved in perception and decision making are dealt with in detail in Chapter 28, but in order to understand hazard identification for our current purposes, a brief overview is needed, and this is provided in the next subsection.

Hazard identification

In order to arrive at the conclusion that there is a hazard, four mental activities are required – 'seeing', knowing, reasoning and deciding. These are described briefly below.

'Seeing'

It must be possible to 'see' the thing which constitutes the hazard, or some manifestation of it. More accurately, it must be possible to perceive it, since hearing, touch, smell and taste can serve equally well for hazard identification. Note, however, that some things which are hazardous are not perceptible – for example, most radiations and certain gases.

Knowing

People must know that something has a potential to cause harm. This means that they must know the nature of the harm which could arise and the causal links involved. It is useful to divide this knowledge into two categories:

---

* People used to 'hazard their lives' or put them 'at hazard', but this use of the word has fallen into disuse, except in the expression 'hazard a guess'. It is not clear why the HSE has resurrected it.

1. **World knowledge.** This is knowledge which it is reasonable to expect is possessed by the majority of adults. For example, working at height is dangerous and fire burns.
2. **Domain knowledge.** This is knowledge which is restricted to certain individuals because of the domain in which they work. For example, chemists know about chemical hazards such as mutagenicity, and nuclear engineers know about nuclear hazards.

Reasoning

People must make inferences based on their observation and knowledge. In many cases, this will involve a series of 'What would happen if...?' questions and the reasoning will involve both the generation of the questions and their answers. It could be argued that the main skill in hazard identification is the ability to ask, and answer, the right questions.

Deciding

At the end of the process people must decide whether or not there is a 'hazard'.

These four steps are a gross oversimplification of the mental steps involved in hazard identification, but they are sufficient to make the point that hazard identification, from a psychological point of view, is not a simple process.

If these four steps, or something like them, are a reasonable representation of the hazard identification process, then it could be argued that hazard identification and risk rating are the same process. Essentially, **hazard identification is the process of identifying risks which are greater than zero, or greater than a predefined level** – that is, those risks which are not 'trivial' in the terminology described in Chapter 5. It is possible that much of the confusion over terminology arises because we are trying to find two different sets of words to define hazard identification and risk rating when they are, in practice, the same process. If the definitions of hazard and risk are set out one beneath the other, this similarity becomes obvious.

| Hazard | POTENTIAL | to cause | HARM |
| Risk | LIKELIHOOD | combined with | SEVERITY (of harm) |

It is questionable whether it is useful to make this distinction between 'potential' and 'likelihood' and between 'harm' and 'severity of harm'. It would appear to make more sense to consider the first step in risk assessment – traditionally, hazard identification – as a screening process which identifies risks above a certain level. This is a useful process since there are usually large numbers of 'trivial risks' with decreasing numbers of more severe risks. An initial overview which classifies risks into those which can be ignored and those which require further investigation can be a useful exercise.

It would be preferable to use the term 'risk screening', rather than hazard identification, for this initial step in risk assessment. However, since it is such a commonly used term, 'hazard identification' is still used in this book.

Risk rating scales from HSG65 and BS 18004

HSG65 uses three points on both the likelihood and severity scales and details are given in Table 20.5. BS 18004 contains tables showing a four point scale for likelihood and a three point scale for severity and these tables are reproduced as Table 20.6 and 20.7.

Notice that both HSG65 and BS 18004 use 'likelihood of harm' in the scales, so neither is calculating risk. This is despite BS 18004 defining risk as a combination of the likelihood of a hazardous event and its consequence(s).

Table 20.5
Likelihood and
severity scales from
HSG65

Likelihood	Severity
High Where it is certain, or near certain, that harm will occur	Major Death or major injury (as defined in RIDDOR) or illness causing long-term disability
Medium Where harm will occur often	Serious Injuries or illness causing short-term disability
Low Where harm will seldom occur	Slight All other injuries or illness

Table 20.6
Examples of
categories for
likelihood of harm
(BS 18004, Table E3)

Categories for likelihood of harm	Very likely	Likely	Unlikely	Very unlikely
Typical occurrence	Typically experienced at least once every six months by an individual	Typically experienced once every five years by an individual	Typically experienced once during the working lifetime of an individual	Less than a 1% chance of being experienced by an individual during their working lifetime

Table 20.7
Examples of harm
categories
(BS 18004, Table E2)

Harm category[A] (examples)	Slight harm	Moderate harm	Extreme harm
Health	Nuisance and irritation (e.g. headaches); temporary ill health leading to discomfort (e.g. diarrhoea).	Partial hearing loss; dermatitis; asthma; work-related upper limb disorders; ill health leading to permanent minor disability.	Acute fatal diseases; severe life shortening diseases; permanent substantial disability.
Safety	Superficial injuries; minor cuts and bruises; eye irritation from dust.	Lacerations; burns; concussion; serious sprains; minor fractures.	Fatal injuries; amputations; multiple injuries; major fractures.
[A] The health and safety harm categories are effectively defined by quoting examples and these lists are not exhaustive.			

In the case of HSG65 it does not really matter that risk is not being calculated because of the statement made about the use of the scales:

> **These differ from the detailed risk assessments needed to establish workplace precautions to satisfy legal standards.** (Bold in HSG65)

BS 18004 also contains a table which combines the likelihood and severity scales to give a qualitative description of risk. This table is reproduced as Table 20.8. This table would be a simple risk estimator if 'likelihood of the hazardous event' was substituted for 'likelihood of harm'.

Severity distributions
In Chapter 5, we emphasised that the 'severity' in the likelihood × severity equation is the severity of the most likely outcome in the circumstances. In this section we discuss the reasons for this.

Likelihood of harm (see Table E3)	Severity of harm		
	Slight harm	Moderate harm	Extreme harm
Very unlikely	Very low risk	Very low risk	High risk
Unlikely	Very low risk	Medium risk	Very high risk
Likely	Low risk	High risk	Very high risk
Very likely	Low risk	Very high risk	Very high risk

Table 20.8
A simple risk estimator
(BS 18004, Table E4)

There are three things known, *a priori*, about the severity of harm arising from a hazardous event:
1. there will be a range of possible harms from no measurable harm to catastrophic harm
2. the probability associated with any chosen level of harm can differ from the probability associated with other levels of harm
3. the sum of the probabilities of all possible levels of harm, including no harm, must be 1, since, once a hazardous event occurs, one of them must happen.

Based on these principles, it is possible to draw severity distributions, and some illustrative examples based on an open manhole are given below. If the manhole shaft is deep, with no opportunities for interrupting the fall and with a concrete base, the severity distribution would be as shown in Figure 20.7.

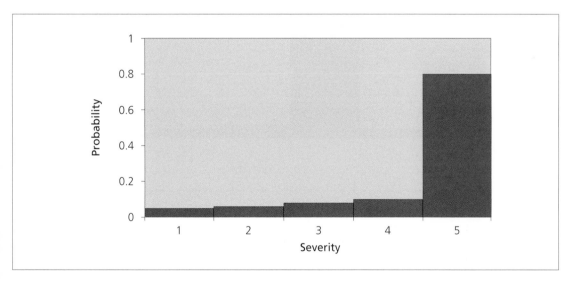

Figure 20.7
Severity distribution for a deep manhole shaft

However, if the manhole shaft was only a metre deep with a soft sand base, the severity distribution would be as shown in Figure 20.8. Manhole shafts with characteristics between these two extremes would have severity distributions between those in Figures 20.7 and 20.8, and examples are shown in Figures 20.9 and 20.10.

There is one more point about the severity distribution. It is possible in theory to use cost on the severity axis, although there will be practical difficulties. Given a cost axis, it will then be possible to amalgamate what are, at present, qualitatively different risks. For example, a hazardous event which could result in injury,

Figure 20.8
Severity distribution
for a shallow
manhole shaft

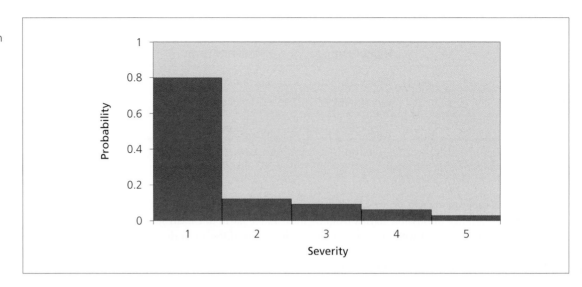

Figure 20.9
Severity distribution
for minor harm

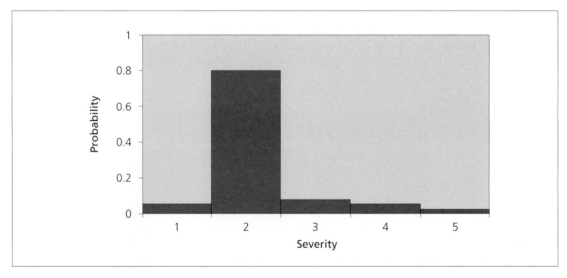

damage to assets and environmental damage could have a single severity distribution on the basis of costs. These sorts of distribution are described in more detail in Chapter 24, which deals with the financial issues associated with risk management.

Deciding on the acceptability of risks

This section deals with two general problems associated with the acceptability of risk, and then focuses on the more specific problems associated with BS 18004.

The first problem is deciding which risk is to be judged acceptable or unacceptable. There are three main possibilities:

1. the risk if there were no risk control measures in place
2. the risk if specified risk control measures were in place and operating as intended
3. the risk if specified control measures were in place, but operating at a realistic level.

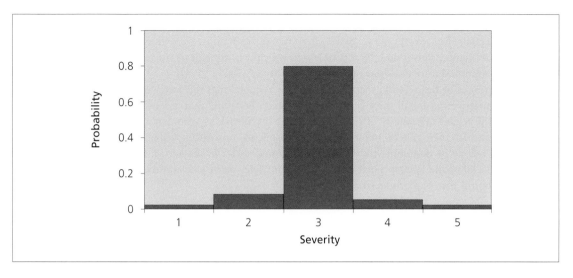

Figure 20.10
Severity distribution
for 'over-three-day'
injury

The distinction between the second and third of these is important, since we know that risk control measures which rely on human behaviour for their effectiveness cannot, in the real world, be effective all the time.

For the purposes of this section, therefore, we will call these three types of risk 'absolute risk', 'theoretical residual risk' and 'practical residual risk'. Where 'residual risk' is used without qualification, it will refer to either or both types of residual risk as appropriate.

The second problem is to establish to whom the risk is acceptable. There are three basic groups of candidates:
1. the people exposed to the risk – this group can be subdivided by the nature of the losses likely to be sustained, for example injury, ill health and financial loss
2. the people who bear the burden of any risk control measures – again, this group can be subdivided by the nature of the burden, for example inconvenience, discomfort, expenditure
3. outside agencies, such as the HSE and courts of law, which may be supposed to be disinterested with respect to losses and burdens.

There is a large degree of overlap between the first and second categories, but it is possible to identify two broad groups:
1. **Those primarily exposed to injury and ill health, and the financial losses arising as a result of these.** The burden of risk control measures for these people will primarily be inconvenience and discomfort, with financial loss where, for example, payment is related to output and the use of risk control measures slows down output. For convenience, we will refer to these people as 'workers' in the discussion which follows.
2. **Those primarily exposed to financial losses such as lost time, compensation payments and damage to assets.** The burden of risk control measures for these people will primarily be the costs associated with implementing and maintaining them. For convenience, we will refer to these people as 'managers' in the discussion which follows.

While in more enlightened organisations the 'workers' have a say in deciding whether or not the risks to which they will be exposed are acceptable, they do not usually have the final say and, in the worst organisations, they have no say at all. It can be argued that this does not matter, since it is in the interest of the 'managers' to spend money on risk control measures to avoid financial losses, but, as will be seen in Chapter 24 which deals with financial issues, this is not necessarily the case. However, even if these financial issues are resolved, there is still the discrepancy between the outcomes for the 'workers' (injury and ill health) and the outcomes for the 'managers' (financial loss). However, this financial loss is usually incurred by the organisation rather than being a personal loss, except in rare cases where managers' bonuses are reduced as a result of poor health and safety performance. This is an ethical problem to which it is difficult to see any solution other than high levels of co-operation during decisions on the acceptability of risks.

For the purposes of this section, therefore, it will be assumed that the acceptability or unacceptability of a risk has been arrived at by all three of the groups listed above and that a consensus has been reached. In addition, we will regard tolerability of risk and acceptability of risk as synonymous.

Having dealt with these general points, it is now possible to consider the difficulties they create in practice and this will be done by describing the problems associated with the simple risk level estimator included in Annex E of BS 18004 and its associated risk control plan. The simple risk estimator was given in Table 20.8 and the risk control plan is reproduced as Table 20.9.

Table 20.9
A simple risk-based control plan
(BS 18004, Table E6)

Risk level	Tolerability: guidance on necessary action and timescale
Very low	These risks are considered acceptable. No further action is necessary other than to ensure that the controls are maintained.
Low	No additional controls are required unless they can be implemented at very low cost (in terms of time, money and effort). Actions to further reduce these risks are assigned low priority. Arrangements should be made to ensure that the controls are maintained.
Medium	Consideration should be given as to whether the risks can be lowered, but the cost of additional risk reduction measures should be taken into account. The risk reduction measures should be implemented within a defined time period. Arrangements should be made to ensure that the controls are maintained, particularly if the risk levels are associated with harmful consequences.
High	Substantial efforts should be made to reduce the risk. Risk reduction measures should be implemented urgently within a defined time period and it might be necessary to consider suspending or restricting the activity, or to apply interim controls, until this has been completed. Considerable resources might have to be allocated to additional controls. Arrangements should be made to ensure that the controls are maintained, particularly if the risk levels are associated with extremely harmful consequences and very harmful consequences.
Very high	These risks are unacceptable. Substantial improvements in risk controls are necessary, so that the risk is reduced to an acceptable level. The work activity should be halted until risk controls are implemented that reduce the risk so that it is no longer very high. If it is not possible to reduce the risk the work should remain prohibited.

The main problem with Table 20.8 is that the estimates in the table ('Very low risk' to 'Very high risk') are not estimates of risk.

Since the estimates from Table 20.8 are not estimates of risk, it follows that the control plan in Table 20.9 is not, as BS 18004 claims, 'risk-based'. However, even if it were, the following problems remain:

- Absolute and residual risk are inextricably mixed, as illustrated by the repeated references to ensuring 'that the controls are maintained' (unless the risk is 'very high', in which case there is no need to maintain the controls, which must surely be an oversight on the part of the authors).
- Levels of harm are used in an extremely confusing manner. For example, the entry for 'medium risk' ends with 'particularly if the risk levels are associated with harmful consequences'. How can there be any risk if there is no association with harmful consequences?

- The high risk level category includes a reference to 'very harmful', which is not a category of severity used in Table E3.

The discussion which follows is an attempt to disentangle these various concepts and suggest how they should be used.

In general, the concept of reasonable practicability should be applied to all risks as a matter of good practice since, as we saw from the severity distributions presented earlier in this chapter, hazardous events have a distribution of outcomes. That is, even a low risk includes a very small likelihood of a very severe outcome. In the UK, this general application of reasonable practicability is included in legislation, such as the Health and Safety at Work etc Act 1974. It is not acceptable to argue, as BS 18004 apparently does, that because a risk is low in absolute terms it is unnecessary to reduce it. If this argument were accepted, there would be no reduction in risk levels in, for example, offices, where the absolute levels of risk are generally low.

As we discussed earlier in this section, the question of whether a risk is acceptable or unacceptable depends on who is making the judgment, and on whose behalf the judgment is being made. There will be cases where extremely high risks are judged acceptable, such as certain risks to which the emergency services and armed forces are exposed. Admittedly, these are exceptional circumstances, but they serve to illustrate the point that a risk rating scale should not use value judgments; the risk ratings themselves are quite subjective enough as it is.

More generally, there are various 'substantial' risks to which people are exposed because of the benefits they bring (for example, driving) or because there is no reasonably practicable way of reducing them (for example, the arrangements for boarding and leaving trains). It is naïve of the BS 18004 authors to say of these risks that 'the work activity should be halted until risk controls are implemented that reduce the risk so that it is no longer very high'.

Acceptability, or tolerability, of risk, because of the various problems outlined above would not, in practice, appear to be a useful concept. What appears to be more useful is a combination of the following:

- **The absolute level of the risk.** This will inevitably be subjective, but at least all the interested parties can be involved in the judgments and the subjectivity will be transparent.
- **The theoretical and practical residual risks.** These, particularly the latter, are what are important as far as exposure to the risk is concerned. They are no less subjective than other risk measures but, as with absolute level of risk, all interested parties can be involved in the judgments.
- **The reasonable practicability of reducing the risk.** Again, this is likely to be subjective and it is complicated by financial issues, but decisions should be reached which are acceptable to all relevant stakeholders.

Because of these various complexities, it may be beneficial to adopt the simple approach described at the end of Chapter 6.

Risk assessment management

As we have seen, risk assessment is, at present, a topic for further research and discussion. However, in the meantime, organisations still have to carry out risk assessments and, in order to do this, they usually adopt one of the following three strategies:

1. **Legislation based.** Risk assessments are carried out to meet the requirements of specific sets of regulations.
2. **Sources of hazard based.** Risk assessments are carried out on the basis of the organisation's assets and activities.
3. **Nature of harm based.** Risk assessments are carried out separately for specific types of harm such as injuries or asset damage.

All of these overlap, but each will be dealt with separately below.

Legislation based

The extreme form of this strategy is having separate risk assessors for each relevant set of regulations. In the UK this would involve, for example, manual handling assessors, COSHH assessors, DSE assessors, and so on. General risk assessors would also be needed for the Management of Health and Safety at Work Regulations, but it is not always clear what their role should be. The advantages of this strategy are that:

- it ensures, at least in theory, legal compliance
- it limits the competences required by particular people. It is easier to train someone to assess just DSE risks than to assess all possible risks.

The disadvantages of this strategy are that:

- it can lead to gaps and overlaps. In theory, any gaps should be dealt with by the general risk assessors but the general risk assessors may not know what has already been done. It also results in overlaps with, for example, the same location being visited repeatedly
- the assessors are trained using different techniques and different terminologies, so there is little 'common ground' or transfer of skills
- there is no method of prioritising risks across categories – for example, a high priority DSE risk may, in absolute terms, be a lower risk than a low priority manual handling risk
- legislation does not deal effectively with certain sources of harm, such as stressors, or with damage to an organisation's assets.

Sources of hazard based

With this strategy, 'owners' of the sources of risk do the risk assessment. Sources of risk can be locations, people, plant, activities, or any other appropriate classification. The advantages of this strategy are that:

- it makes responsibility for risk assessment clear and links it with general management responsibility
- competences can be focused, in that there is no need to train people to assess particular types of risk if their ownership does not cover the relevant sources for that type of risk
- it can be linked to legislation where necessary, so people can 'own' activities such as manual handling or DSE.

The disadvantages of this strategy are that:

- it relies on clear allocation of 'ownership' and this can be a problem (see the discussion on inventory preparation in Chapter 5)
- it is job specific, so retraining in risk assessment may be required if the job changes or the person moves to a different job.

Nature of harm based

With this strategy, risk assessments are carried out separately for possible injuries, ill health, damage to assets, damage to the environment, and so on. The advantages of this strategy are that:

- a person trained in, for example, injury risk assessment should be able to apply this to all sources, so that the competences are transferable from one job to another
- it limits the range of competences required and can provide specialists in health risks, a topic which, it can be argued, receives too little attention at present
- the competences required can be generalised so that, for example, someone trained in injury risk assessment can be more easily trained to assess health risks.

The disadvantages of this strategy are that:

- a high level of competence is required, which may involve extensive training and everyone may not be able to reach the level of competence required

- it is difficult to link to any set of regulations, other than the Management of Health and Safety at Work Regulations, so that it is more difficult to demonstrate compliance with particular sets of regulations.

Future strategy

Few organisations have a formal strategy for risk assessment and risk control. In most organisations, risk assessment has 'grown like Topsy', usually driven by successive introductions of legislative requirements. However, organisations should have a strategy for the future.

A future strategy could be based on one of those already described or a combination of these. However, a unified strategy based on the principles set out in Chapter 5 may be preferred.

Advanced risk assessment techniques*

Introduction

This section deals with a number of techniques which can be used, jointly or separately, to assess the risks associated with a defined system. We will describe four main techniques:

1. **Hazard and Operability studies (HAZOPs).** This technique is a team brainstorming exercise with the main purposes of identifying hazards and, where necessary, suggesting risk control measures to deal with these hazards. It is typically a qualitative procedure, although quantification of the risks associated with identified hazards can be added.
2. **Failure Modes and Effects Analysis (FMEA).** This technique is also a team brainstorming exercise, with the main purposes of identifying the ways in which system hardware can fail and, where necessary, identifying ways of detecting possible failures and the risk control measures necessary to deal with these failures. Unlike HAZOP, it is usual to introduce some form of quantitative assessment of the risks associated with particular failures.
3. **Event Tree Analysis (ETA).** This technique is primarily used to obtain estimates of the probability of an undesired event – for example, an uncontrolled fire – as a result of failures in the relevant detection and control systems, such as smoke detectors and sprinkler systems. The procedure is quantitative, being based on binary logic and the probabilities of the binary options.
4. **Fault Tree Analysis (FTA).** This technique is used to describe, in the form of a diagram, the necessary conditions for a particular event – for example, a fire – to occur. The primary purpose of the technique is to describe the conditions necessary for an event to occur in a way which allows for accurate identification of the most effective risk control measures. FTA can be qualitative or quantitative with, in the latter case, one output being an estimate of the probability of the event.

We can see from these brief descriptions that the techniques have different purposes and the appropriateness of their use will depend on the system we are assessing and what stage we are at in the assessment process. The choice and application of these techniques will depend on a number of factors, and these are discussed in the next subsection. In order to illustrate how these factors apply in the real world, we will use a risk assessment for a domestic gas boiler. This example will also be used to illustrate the four techniques listed above.

Preliminary work

The usual first step in an advanced risk assessment project is to set up a team of people who will carry out the work. While, in theory, any of the techniques could be carried out by an individual working alone, in

---

* The techniques described in this section cover both risk assessment and risk control issues. However, they are conventionally referred to as risk assessment techniques and we will use this terminology in this section.

practice the techniques are more effectively carried out by a team. The composition of a typical team will be:

- a team leader – this person is responsible for the management of the team, ensuring that it meets its aims, makes adequate records, keeps to schedule, and so on
- at least one person competent in the advanced risk assessment technique, or techniques, the team will use; often this person is also the team leader
- team members who, between them, have a detailed knowledge of the system that is the subject of the risk assessment.

When the team has been formed, their usual first tasks are to define accurately the system they will assess and agree their aims. We will deal with these topics next.

Defining the system

It is essential that the team define, record and agree on the system which will be the subject of the advanced risk assessment. This is usually done using appropriate diagrams, flow charts and sketches, annotated where necessary. An example of a simplified system definition for a domestic gas boiler is given in Figure 20.11.

Figure 20.11
Simplified system description for a domestic gas boiler

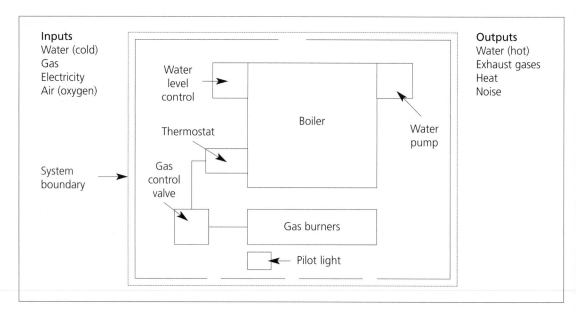

For our current purposes, Figure 20.11 shows only the functional elements of a domestic gas boiler – that is, those elements which are necessary to produce hot water on demand. The notes below describe the main elements in Figure 20.11.

- The boiler holds the water which is to be heated by the gas burners. The temperature of the water in the boiler is monitored by the thermostat which, when the water reaches the required temperature, switches the gas control valve to 'closed'. When the water temperature falls below the preset level, the thermostat opens the gas control valve, the gas flows to the burners and is ignited by the pilot light.
- When a hot water tap (outside the system) is turned on, the water pump takes water from the boiler and pumps it out of the system. The fall in water level in the boiler activates the water level control which lets

cold water into the boiler. This results in a fall in the temperature of the water in the boiler, which activates the thermostat as described above to maintain a supply of hot water in the boiler.

- These various elements are surrounded by a casing which has a vent at the top for exhaust gases and vapours, and smaller vents at the bottom for the inflow of the air which provides the oxygen required for combustion.
- It should be noted that modern gas boilers are more complex than this system and that not all of them operate on the principle of storing a tank of hot water. However, the simple system in Figure 20.11 will serve the main purpose of illustrating the advanced risk assessment techniques.

The operational details of this boiler will be dealt with later in the context of these advanced risk assessment techniques and, for now, it is only necessary to note the following points:

- The system description would normally be more accurate and more detailed than Figure 20.11 and, in practice, if a gas boiler were the subject of an advanced risk assessment exercise, then detailed diagrams would be used. If such diagrams were not available, the risk assessment team would have to have them prepared.
- The inputs to, and outputs from, the system are included in the diagram.
- Only the system elements required to produce hot water are included in Figure 20.11. The elements required for safe operation are dealt with later in the section.
- The system boundary is defined. This is good systems practice but it is also very important in advanced risk assessment, since it defines the scope of the work because, by convention, advanced risk assessment considers only the hazards within the defined system boundary.

This last point is particularly important because advanced risk assessment can be time-consuming and there has to be some mechanism for keeping it in check. If, for example, the system boundary was extended to include any central heating system for which the boiler supplied hot water, then there would be a large increase in the amount of assessment work required. Where the team has clearly defined, and agreed, the system boundary, then there is a rationale for including or excluding the assessment of particular hazards and risks.

Agreeing aims

As with agreeing the system boundaries, agreeing the aims of the advanced risk assessment is done primarily to limit the scope of the exercise. Possible aims for an advanced risk assessment of a domestic gas boiler would include:

1. assessing the risk of an explosion
2. assessing the risk of someone being burned or scalded
3. assessing the risk to the environment
4. assessing the risk of someone being overcome by carbon monoxide
5. assessing all of the risks arising from using the boiler.

An advanced risk assessment team can make its aim as specific or as general as it wishes, but it should consider the likely workload associated with achieving a given aim. There is always, however, the option of changing the group's aim and this is dealt with next.

Iteration

At any point in the advanced risk assessment procedure, it is possible to go back and revise what has already been agreed. For example, findings from a HAZOP may suggest that the system should be redefined or that the aims should be amended. Similarly, the work done during FTA might suggest that further work at the

HAZOP stage is required. It is usually the role of the team leader to manage these issues. During the routine risk assessment work, the team leader should ensure that team members stay within the agreed system and aims. However, where it becomes apparent that one or other of these is no longer appropriate, the team leader should manage any alterations and ensure that they are agreed and documented as rigorously as the initial descriptions were documented.

Choice of techniques

The final stage in the preliminary work is to decide which of the advanced risk assessment techniques is to be used, and in what order. Apart from the fact that the hazard identification techniques (HAZOP and FMEA) should be used first, there are no fixed rules about what procedures should be used, or in what order. However, the following points should be noted:

- FMEA, as its name suggests, is looking at risks associated with failures, and these failures are confined to failures of hardware. However, not all risks arise from hardware failures, hence the need to supplement FMEA with HAZOP, which can be used to identify failures associated with, for example, human error.
- Neither FMEA nor HAZOP is designed to identify hazards which are associated with a system which is operating as intended (known as continuing hazards); they are confined to the identification of hazards associated with system failures or deviations from normal operation. For the identification of continuing hazards, the identification techniques described in Chapter 5 should be used.
- ETA is a specialised technique and may not be appropriate or required for some systems, for example systems without detection and control subsystems.
- FTA is primarily used to determine the most effective risk control measures and, if these measures have been identified earlier in the advanced risk assessment procedure, FTA is likely to be redundant.

However, the health and safety professional needs to know in outline how all four techniques operate. These outline descriptions are given in the next four subsections.

Hazard and Operability studies

Hazard and Operability studies (HAZOPs) are primarily used in the design stage to identify hazards which could occur if the process or operation did not go as planned – in other words, the hazards arising from failures or malfunctions in the system. However, HAZOPs can be used for existing systems.

HAZOP is a qualitative procedure which systematically examines a process by asking questions about what could go wrong. It is generally carried out by a small team of people with knowledge of the system, directed by a group leader experienced in HAZOP. Essentially, HAZOP is a brainstorming exercise and can be very time-consuming if the focus is not kept on significant risks.

The questions asked by the HAZOP team are generated by two sets of key words: property words and guide words.

- **property words** are words chosen to focus attention on how the process operates, for example temperature, pressure and level
- **guide words** are words chosen to focus attention on possible deviations from the design intention.

Property words have to be chosen to suit a particular system, as do guide words. However, the guide words listed below are generally relevant, although others may have to be added for particular risk assessment exercises.

No or **not**	The complete negation of the design intention.
More	Quantitative increase, for example higher temperature.

Less	Quantitative decrease, for example lower temperature.
As well as	Qualitative increase, for example an impurity.
Part of	Qualitative decrease, for example component in mixture missing.
Reverse	The logical opposite of the intention, for example liquid flowing in opposite direction to that intended.
Other than	Complete substitution, for example wrong material.

Once the property words and guide words appropriate to the system have been established, the team works through each combination of guide and property words, brainstorming to decide whether a deviation from the design intention could arise. They then consider possible causes of this particular failure and the consequences if the failure occurred. The results of the exercise are usually recorded in a pre-printed table with the headings listed below.

HAZOP table headings
- Guide word
- Deviation
- Possible causes
- Consequences
- Action required

The 'Action required' column is used to record either risk control measures which will prevent the failure, or requirements for further information. Because HAZOP is a qualitative brainstorming exercise, it requires experience to determine the significant hazards, which should be recorded, and the trivial hazards, which can be left unrecorded. To illustrate how HAZOPs are carried out in practice, the notes which follow describe a partial HAZOP for the domestic gas boiler system.

The first step in the work would be to identify relevant property words and, for the gas boiler, these would include:
- pressure (of gas)
- level (of water in boiler)
- temperature (of water in boiler)
- flow (of oxygen)
  and so on.

Each property word is then considered with each of the guide words to determine whether there is a likely deviation. If a deviation is possible, its consequences are identified and, if these are sufficiently serious, given the aims of the HAZOP, they are recorded, together with their possible causes and any action required. Early in the HAZOP, these actions are likely to include the need to collect further information.

Table 20.10 contains a partial HAZOP for the domestic gas boiler and illustrates how HAZOPs are recorded. Because it is easier to change designs than to change existing equipment and plant, it is particularly valuable to carry out HAZOPs at the design stage and it can be seen from Table 20.10 that the HAZOP has identified the need for additional elements in the gas boiler system. The redesigned system is shown in Figure 20.12.

Note that in 'real life', the redesigned system would be subject to a full HAZOP to check that the new elements in the system were not, in themselves, or in combination with other elements, creating new risks.

Table 20.10
Partial HAZOP for a domestic gas boiler

Domestic gas boiler – to provide hot water on demand

Guide word	Deviation	Possible causes	Consequences	Action required
No	Gas pressure	Supply failure Pipe rupture Pipe blocked Valve failure	Pilot light goes out	Mechanism to prevent gas flowing while pilot light is out (thermocouple) (Possible causes may be outside system)
More		Error by supplier	Pilot light extinguished Too rapid combustion	As above Check effects
No	Level (of water in boiler)	Leaks Supply failure	Overheating	Mechanism to shut down burners if there is no water in boiler
More	Temperature (of water in boiler)	Burner runs for too long	Water boils Pressure build-up Rupture of boiler	Mechanism to shut off burners when water temperature reaches required level (thermostat) Pressure release valve on boiler?
No	Flow (of oxygen into system)	Blocked vents	Pilot light goes out	Mechanism to stop vents being blocked Thermocouple
Less		As above	Partial combustion, carbon monoxide emission	Vent all exhaust gases to atmosphere
As well as		Dusts, other gases in air flow	Depend on dusts, gases	Investigate area for possible dusts, gases
More	Pressure (in boiler)	Boiling of water	Rupture of boiler	Pressure release valve
Reverse		Condensation in steam-filled boiler	Implosion of boiler	Determine required specification for boiler casing
No	Ignition (of gas by pilot light)	Pilot light extinguished Gas pressure too high	Unignited gas in casing	Thermocouple, adequate venting Regulator valve
As well as		Human intervention (match or taper)	Explosion	No entry point in casing for external ignition sources Internal ignition source required (igniter)

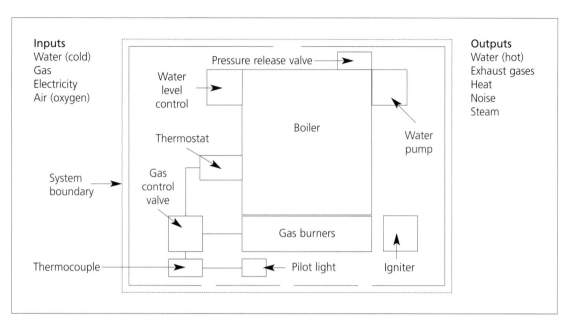

Figure 20.12
System description for domestic gas boiler following HAZOP

Failure Modes and Effects Analysis

Failure Modes and Effects Analysis (FMEA) is used to identify the hazards resulting from failures in hardware. It starts by listing the hardware items and analysing their possible failure modes; it is, therefore, a 'bottom-up' approach. However, FMEA can also be extended to include numerical methods and can, thus, cover preliminary risk rating as well as hazard identification. FMEA starts at the component level and seeks to answer the following questions about each component:

- How can this component fail, that is what are the failure modes?
- What could the effects be if the failure occurred?
- How could the failure mode be detected?
- What would be the risk associated with the effects?

The notes below deal with each of these questions in turn, followed by a description of how FMEA results are recorded. There is then a partial FMEA for the domestic gas boiler to illustrate how the technique is used.

Defining components (and their functions)

The first step in answering this question is to be clear about what constitutes the component. There are two main options:

1. The component can be designated as a 'black box' with a given function, such as a pump (water) or a valve (gas flow control). In these circumstances there is usually a limited number of failure modes, such as 'valve fails open' or 'valve fails closed'.
2. The component may be a more or less complex unit making up part of a larger assembly. For example, the water pump may be subdivided into its control mechanism (on–off switch) and its pumping mechanism. The on–off switch could, in turn, be subdivided into its components (coil, contacts, springs, wiring and so on). At its most detailed, this option treats every discrete physical item as a component.

The level of detail required for a particular FMEA is determined by the need to establish the probability of failure in a particular mode. If, for example, manufacturer's data on the probability of the failure of a pump are available, then this pump can be treated as a 'black box'. Note, however, that the manufacturer will have had to carry out FMEA at the more detailed component level to establish the probability of failure.

Failure modes and effects

Having accurately identified the component and its function, the next steps are to identify the failure modes and their effects. These steps, like HAZOP, are brainstorming exercises.

In FMEA, it is important to identify all conceivable failure modes, not just the likely ones, and the time needed to do this should also be taken into account. For example, if the component is a pump, the failure modes should include failure to operate at the correct time, failure to stop operating at the correct time, and premature operation. When identifying effects, the important point is to ensure that the effects on the system as a whole have been identified. For example, if the component is part of a pump, the failure of this component will have the obvious effect of making the pump inoperable. However, the effect of an inoperable pump on the whole system should also be identified.

Failure detection

During FMEA, the aim is to identify the detection methods which are currently available. However, when the risks are high (see below) recommendations are made as necessary on improving detection methods.

There are two broad categories of detection options:

1. **Detecting the failure mode.** For example, if the failure mode is some mechanical failure (crack, rupture, loose connection and so on), the failure mode itself can be detected.
2. **Detecting one or more of the effects of the failure.** For example, if the failure results in a gas release, the gas in the atmosphere can be detected, or if the failure results in a drop in temperature or pressure, these effects can be detected.

Ideally, the detection method used should detect some precursor of the failure mode and this sort of detection method is particularly important in safety-critical systems. Most modern motor cars are fitted with a number of such detection systems which monitor, for example, the condition of the brakes, aspects of the seat belt operation and the state of readiness of the air bags, and provide drivers with a visual or audible warning when something is wrong with these subsystems. Where such precursor detection methods are not possible on safety-critical subsystems, then measures should be taken to ensure redundancy (effectively a backup subsystem which comes into operation when the first subsystem fails) or that the system 'fails to safety'. This fail to safety option can be illustrated using the car as an example. It would be possible for the precursor detection methods described earlier as being used in modern cars to prevent the car engine running. For example, they could be linked to the fuel pump in such a way that activation of the detection method disconnected the electrical supply to the fuel pump, thus preventing the car being driven, that is, the overall system (the car) would fail to safety.

Risk assessment

This part of the FMEA procedure uses the equation:

$$\text{Risk} = \text{probability of failure mode} \times \text{severity category}$$

In theory, any appropriate range of values could be used for probability and severity, but values based on an American military standard are used most frequently. These are reproduced in Tables 20.11 and 20.12. The combination of severity category and probability level is used to determine a risk priority code, which takes a value of 1, 2 or 3. This is done using a risk assessment map, which is illustrated in Figure 20.13.

Category	Degree	Description
I	Minor	Functional failure of part of a machine or process – no potential for injury
II	Critical	Failure will probably occur without major damage to the system and without serious injury*
III	Major	Major damage to the system and/or potential serious injury to personnel
IV	Catastrophic	Failure causes complete system loss and/or potential for fatal injury

Table 20.11
Severity categories

Level	Probability value	Description	Individual failure mode
A	10^{-1}	Frequent	Likely to occur frequently
B	10^{-2}	Probable	Will occur several times in the life of an item
C	10^{-3}	Occasional	Likely to occur sometime in the life of an item
D	10^{-4}	Remote	Unlikely but possible to occur in the life of an item
E	10^{-5}	Improbable	So unlikely that occurrence may not be experienced

Table 20.12
Probability levels

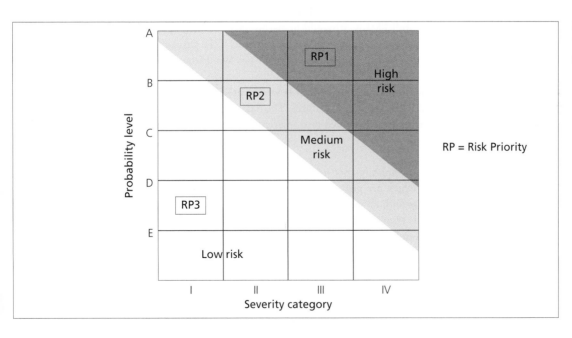

Figure 20.13
Risk assessment map

---

* Note the use of 'probably occur': this probability is not the same as the probability of failure. This is another example of why two probabilities are required in the definition of risk.

As with HAZOP, the results of an FMEA are recorded in a worksheet table which, in the case of FMEA, has columns for details of the component being analysed and answers to the questions listed above. These column headings are shown below, together with notes on the entries required in each column.

FMEA table headings

Component (function)   A brief description of the component (followed by a description of the function of the component).

Failure modes   A list of the way(s) in which the component could fail. Note that each of the remaining columns has to be completed for each of the failure modes identified.

Failure effects   The results if the component were to fail in this mode.

Failure detection method   The way(s) in which this failure, or the effects of this failure, could be detected.

Risk   Estimates of the probability, severity and risk rating for this failure mode.

A sample page of an FMEA worksheet for the domestic gas boiler is given in Figure 20.14. Notice that there are no actions recorded on the FMEA worksheet. Any required actions are recorded on a second type of FMEA table, the FMEA summary. This summary is used to reorganise the failure modes recorded on the FMEA worksheets in priority order according to their Risk Priority Code (RPC) with all the category 1 RPCs first. Once organised in this way, any actions or remarks appropriate to the individual failure modes are recorded. A sample page from a FMEA summary for the domestic gas boiler is given in Figure 20.15.

For ease of description, Figures 20.14 and 20.15 deal only with single-point failure modes. In other words, the RPC estimates are based on the assumption that the system, other than the component being analysed, is functioning as intended. In a full FMEA, it would be necessary to consider concurrent failure modes where two or more system components fail simultaneously. For example, if the gas control valve failed with the valve at open and the pilot light failed simultaneously, unburnt gas would be released into the casing and the atmosphere.

Although FMEA is a time-consuming process, it is widely used because of its effectiveness in identifying hazards and risks associated with component failures. The data it provides can also be used as a basis for the advanced risk assessment techniques described next – Event Tree Analysis and Fault Tree Analysis.

Event Tree Analysis

Event Tree Analysis (ETA) is primarily used to analyse the possible effects and consequences of a failure, unlike HAZOP and FMEA, which are primarily used to identify hazards.

The Event Tree itself starts with an initiating event which might be a fire, release of toxic gas or any other potentially serious event. This type of event will normally have associated with it a number of detection and control mechanisms. For example, in the case of a fire, there would be fire detection systems and sprinkler or other extinguishing systems.

ETA uses binary logic to assess the probabilities of the serious event damaging people or property due to failures in the various detection and control systems. A typical, but simplified, ETA diagram is shown in Figure 20.16, and explanatory notes are given below.

An uncontrolled fire will occur if there is a fire (probability $= P_F$) and the detection system fails (probability $= P_D$). This is outcome 3 in Figure 20.16. Mathematically, this probability will be $P_F \times P_D$.

An uncontrolled fire will also occur if there is a fire, the detection system works, but the sprinkler system fails (probability $= P_S$). This is outcome 2 in Figure 20.16 and has a probability of $P_F \times (1 - P_D) \times P_S$.

By systematically working through the various detection and control systems, and calculating their probabilities of failure, the probability of the undesired consequence can be calculated.

In the simple diagram above, the probability of an uncontrolled fire can be estimated as:

Item	Component (function)	Failure modes	Failure effects	Failure detection method	Risk assessment Severity category	Risk assessment Probability level	Risk assessment RPC*
1.0	Thermostat (switches gas valve to closed when water temperature reaches preset level)	Failure which closes gas valve	No gas supply to burners	Observation (of burners)	I	B	3
		Failure which opens gas valve	Water boils, boiler explodes	Hearing (sound of water boiling)	III	B	1
				Touch (temperature of casing)			
2.0	Gas valve (controls flow of gas to burners – on or off)	Crack in valve casing	Gas release	Observation (smell of gas)	IV	D	2
		Fails with valve closed	No gas supply to boiler	Observation (of burners)	I	D	3
		Fails with valve open	Water boils, boiler explodes	Observation (of burners)	III	D	3
				Hearing (sound of water boiling)			
				Touch (temperature of casing)			
3.0	Thermo-couple (closes gas valve when pilot light is extinguished)	Failure which closes gas valve	No gas supply to burners	Observation (of burners)	I	A	3
		Failure which opens gas valve	Water boils, boiler explodes	Hearing (sound of water boiling)	III	A	1
				Touch (temperature of casing)			
4.0	Water pump	Cracks in seal	Water in casing	Observation	I	B	3
		Fails open	Water runs continuously	Observation	II	D	3
		Fails closed	No hot water supply	Observation	I	D	3

FMEA no.		Page of
Project no.	**Failure Modes and Effects Analysis**	Date
System	Boiler	Prepared by
Subsystem	* RPC = Risk Priority Code 1 = High, 2 = Medium, 3 = Low	Evaluated by

Figure 20.14
Sample page of worksheet for FMEA of a domestic gas boiler

Figure 20.15
Sample FMEA
summary sheet for a
domestic gas boiler

Item	Component	Failure mode	RPC*	Action required/remarks
1.0	Thermostat	Failure which opens gas valve	1	Design change. Any failure in the thermostat should result in the gas valve being closed
2.0	Thermocouple	Failure which opens gas valve	1	Design change. Any failure in the thermocouple should result in the gas valve being closed
3.0	Gas valve	Crack in valve casing	2	Design change. Review probability of valve case cracking to probability level E

FMEA no.				Page of
Project no.		**FMEA Summary**		Date
System	Boiler			Computers
Subsystem		* RPC = Risk Priority Code 1 = High, 2 = Medium, 3 = Low		

Figure 20.16
Simplified Event Tree
Analysis diagram

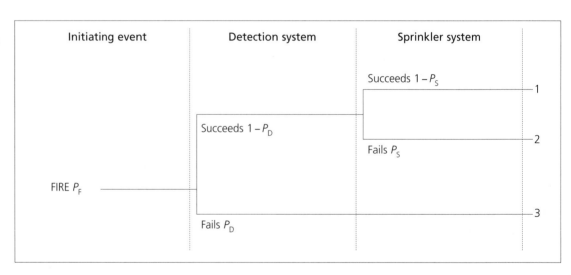

$$(P_F \times P_D) + (P_F \times (1 - P_D) \times P_S)$$

ETA is an extremely effective method of calculating the probabilities of undesired outcomes and it is not restricted to analysing protective systems. Although it has been used successfully in a range of industries, the main practical problem, as with most quantified risk assessment techniques, is establishing the probabilities on which the calculations are based.

To illustrate how ETA might be applied in the case of the domestic gas boiler, Figure 20.17 shows a partial ETA using a release of gas as the initiating event. For the purposes of this exercise, it is assumed that the gas release is of a nature and extent that makes it possible for an explosion to occur if the gas is ignited. The probability of a release of this type is estimated as P_R. The various ways in which the gas and air mixture in the casing could be ignited are then considered using the ETA binary logic.

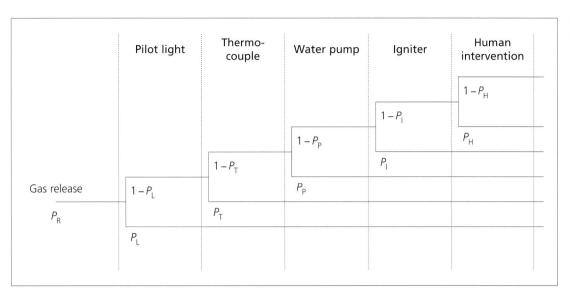

Figure 20.17
Partial ETA for a
domestic gas boiler

When numerical values are assigned to the probabilities in Figure 20.17, it is possible to calculate the probability of an explosion due to the released gas being ignited.

Note that, since it is based on binary logic, ETA cannot take account of partial degradation of system elements and this is one of its limitations. Another limitation is that, like all quantified advanced risk assessment methods, it relies for its effectiveness on accurate estimates of the various probabilities included in the event tree.

Fault Tree Analysis

Fault Tree Analysis (FTA) starts with a possible outcome and systematically identifies how the failures of individual parts of the system and human errors contribute to this type of outcome.

FMEA should normally be carried out as a preliminary to FTA so that the effects of hardware failures are known. Additional data will then have to be collected on possible human errors; these sorts of data will be described in Chapter 21.

The outcomes used in FTA are known as the 'top events', and they are usually the most serious consequence identified during HAZOP or FMEA. Typical top events are explosions, fires, fatalities and serious injuries.

Once a top event has been identified, the fault tree is constructed by identifying all possible combinations and sequences of events which could result in the top event. Logic gates are used to connect the combinations, either with an AND gate or with an OR gate. Sequences of events are identified by setting the fault tree out in graphic form as levels. A simplified fault tree with an explosion as the top event and three levels is shown in Figure 20.18. The dotted lines in Figure 20.18 indicate that the fault tree will have to be developed further along these branches. Where a fault tree cannot be, or has not yet been, developed beyond a certain point because of lack of information, the branch is terminated with a diamond shape, suitably annotated.

Only three levels are shown in this diagram, but in a full fault tree as many levels would be used as were needed. Successive levels are added until a basic cause is reached. These basic causes are recorded in circles and indicate the end point for any given branch of the fault tree.

Fault trees can be analysed qualitatively and quantitatively, depending on the data available. Qualitative analysis involves identifying the most important points at which the fault tree can be interrupted, thus preventing the top event. In general, the best points for interruption are:

Figure 20.18
Simplified fault tree

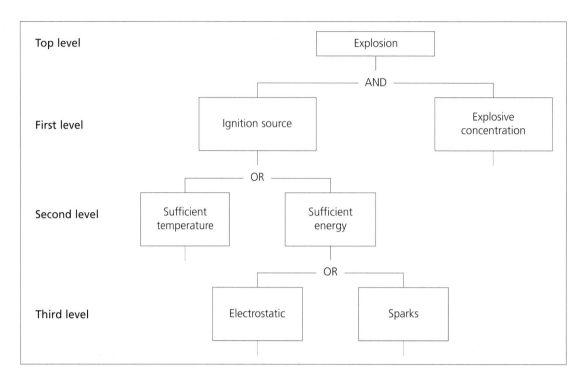

- basic causes at high levels in the fault tree, since these are the least likely to be filtered out at logic gates above them
- AND gates, since removal of any one of the entries to the gate will block this branch of the fault tree.

Where information is available on the probabilities of basic causes, it is possible to carry out a quantified analysis of the fault tree and arrive at an estimate of the probability of the top event. This involves straightforward calculations of probability at each logic gate using the following rules:

AND gates   Multiply the probabilities of the events feeding into the gate
OR gates:
1. For each probability (p), calculate $q = 1 - p$
2. Multiply the values of q
3. Subtract the product of the values of q from 1 to give the overall probability.*

To illustrate how FTA might be applied in the case of the domestic gas boiler, Figure 20.19 shows a partial FTA for an explosion of a gas and air mixture in the boiler casing. Note that in Figure 20.19 the conventional FTA symbols are used; these are explained in Figure 20.20.

---

* See Chapter 19 for a more detailed explanation of calculating increasing probabilities.

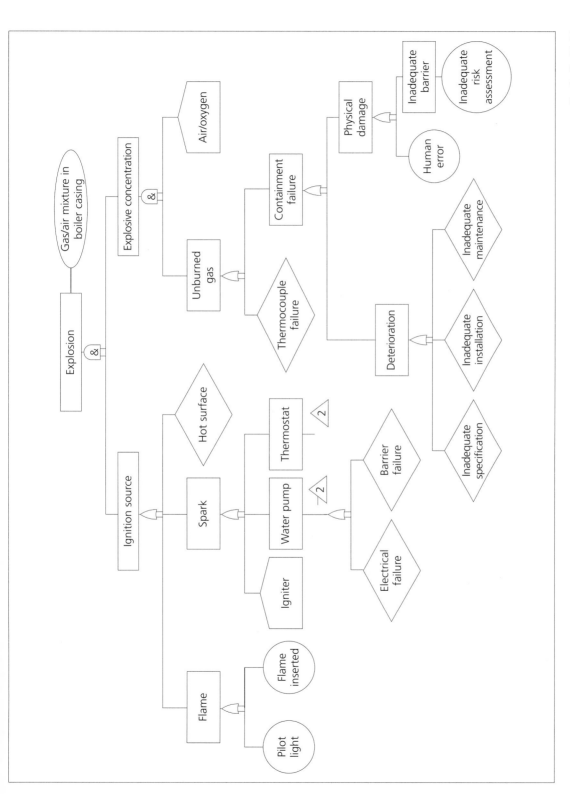

Figure 20.19
Partial FTA for an explosion in a domestic gas boiler casing

Figure 20.20
Sample of FTA symbols

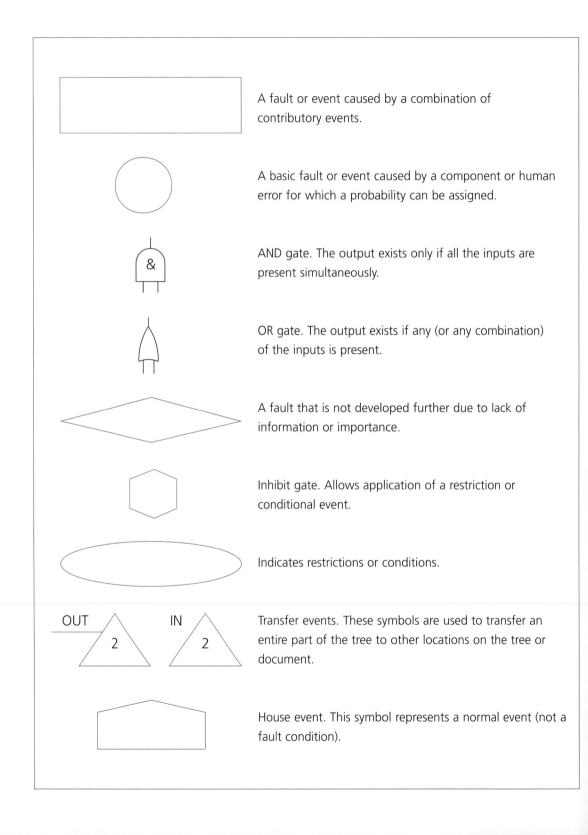

A fault or event caused by a combination of contributory events.

A basic fault or event caused by a component or human error for which a probability can be assigned.

AND gate. The output exists only if all the inputs are present simultaneously.

OR gate. The output exists if any (or any combination) of the inputs is present.

A fault that is not developed further due to lack of information or importance.

Inhibit gate. Allows application of a restriction or conditional event.

Indicates restrictions or conditions.

Transfer events. These symbols are used to transfer an entire part of the tree to other locations on the tree or document.

House event. This symbol represents a normal event (not a fault condition).

Concluding note on risk assessment

This chapter has given a brief introduction to some of the main advanced risk assessment techniques. However, the use of these techniques is a skill and, like all skills, it can only be learned by practice, followed by feedback on performance. No textbook can provide this practice, or the feedback, and, although further information on the advanced risk assessment techniques can be found in the references listed at the end of the book, you will need supervised practice if you are to master the skills involved.

Summary

This chapter dealt with two broad topics: accident investigation and risk assessment. The discussion of each topic began with a review of the relevant conceptual and practical difficulties associated with the use of the techniques and then there was a brief introduction to some of the more advanced techniques relevant to each of the two topics.

21: Advanced risk control techniques

Introduction

The continuance of low levels of risk in an organisation will, in the long term, depend on the effectiveness of the risk control measures chosen, implemented and maintained. However, these three issues are interrelated in that, for example, certain risk control measures which may be chosen are more difficult to maintain than others. In an ideal world, the chosen risk control measures will meet the following criteria:
- they should reduce the risk to near zero
- they should be easy and cheap to implement and maintain
- they should be capable of being maintained with complete reliability.

In order to determine whether the first criterion is being satisfied, it is necessary to have some method of calculating the effect of risk control measures on the level of risk – this will be the subject of the first section of this chapter. However, since the reliability of the risk control measures is also relevant to risk reduction in the long term, the first section of the chapter will also deal with this topic.

One of the main problems in advanced risk control is generating ideas for control measures in those circumstances where it is not immediately obvious that there is a measure which meets the criteria listed above. When this is the case, it is necessary to use techniques which will generate ideas for alternative approaches to risk control.

Two such techniques, brainstorming and systems thinking, are described in the second section of this chapter. The cost of risk control measures will be dealt with in the chapter on financial issues (Chapter 24).

The examples of risk control measures discussed in the first two sections of this chapter will be drawn mainly from the sorts of risk control measure described in earlier chapters (see Chapters 5 and 6). However, in the final section of this chapter, we will argue that many of these types of risk control measure are of limited value because they do not take into account individual differences and attitudinal factors. The chapter ends with a discussion of these issues under the general heading of safety culture.

Measuring risk reduction

If, as we have argued in previous chapters, risk is likelihood multiplied by severity, then it follows that there are only three ways of reducing risk:
1. reduce the likelihood
2. reduce the severity
3. reduce both likelihood and severity.

For any risk control measure being considered, therefore, it should be possible to estimate the effect on both of these scales. If this cannot be done, then it is unlikely that that risk control measure is appropriate. Three simple examples using the numerical risk rating method described in Chapter 5 will illustrate the sorts of calculations which can be made to work out the effect on risk of a risk control measure.

Reducing likelihood
Putting barriers round a manhole.

Without barriers, the risk is, say, $3 \times 5 = 15$. With barriers, the risk is, say, $1 \times 5 = 5$.

Reducing severity

Wearing hard hats on building sites.

Without hard hats, the risk of head injury is, say, $2 \times 5 = 10$. With hard hats, the risk is, say, $2 \times 1 = 2$.

Reducing likelihood and severity

Using low voltage cordless drills rather than 240 V mains drills.

With 240 V mains, the risk of shock is, say, $2 \times 5 = 10$.
With a low voltage cordless drill, the risk of shock is, say, $1 \times 1 = 1$.

Reliance on risk control measures

Organisations that have effectively implemented a risk control programme over a period of time will have reduced a large number of their risks to a relatively low level. They will have done this partly by eliminating hazards, but it is likely that most of the risk reduction has been achieved through risk control measures. Where organisations have reached the point at which it is difficult to reduce risk further with additional control measures, they need alternative strategies to improve the management of risk. Two such strategies are possible: reducing the reliance on risk control measures, which will be dealt with in this section, and increasing the reliability of risk control measures, which we will consider in the next section.

Sometimes, a number of different activities with their associated risk control measures can be rated as having the same risk – some examples of these are given in Table 21.1. The column headings L, S and R in the table stand for likelihood, severity and risk rating.

Table 21.1
Activities and associated risk control measures with the same risk rating

Activity	Risk control	No controls			With controls			Reliance
		L	S	R	L	S	R	
Extended periods of typing (risk of work-related upper limb disorders)	Appropriate workstation, frequent breaks	4	4	16	1	4	4	12
People moving around an office (risk of being hurt in a fall)	Good housekeeping	3	2	6	2	2	4	2
Working in a contaminated atmosphere (health risk from inhalation)	Masks	5	4	20	1	4	4	16

From Table 21.1, we can see that some activities have a higher reliance on risk control measures than others – in other words, the risk rating without risk control measures minus the risk rating with control measures gives a high value. Since it is unlikely that any risk control measures will be completely effective all the time, activities which have a high reliance on risk control measures are prime candidates for the application of the sorts of systems thinking discussed later in this chapter.

The work required to identify activities which have a high reliance on risk control measures can easily be incorporated into the sorts of risk assessment techniques described in Chapter 5. For example, the risk assessment form can be extended to include the sort of data shown in the last seven columns of Table 21.1. An alternative layout used in some organisations is shown in Table 21.2, where, in the likelihood, severity and risk rating columns, the level without control measures is recorded to the left of the stroke and the level with control measures to the right. The reliance value is simply the difference between the two risk ratings.

Activity	Risk control measures	L	S	R	Reliance
		/	/	/	
		/	/	/	
		/	/	/	

Table 21.2
Form for recording reliance on risk control measures

Reliability of risk control measures

As was mentioned above, it is unlikely that risk control measures will be completely effective all the time. In Chapter 6, we described hierarchies of risk control measures which were based, in part, on their reliability.

The calculations made in the previous section assumed that the risk control measures are 100 per cent effective but it is known that, if they rely on human behaviour, this is an unrealistic expectation. Some means of taking into account the influence of human reliability on risk control measures is therefore required.

There are various ways of doing this, but the following simple method will meet our current needs:
1. Estimate the effect of human reliability on the effectiveness of the risk control measure. For example, we may know from observation that hard hats are only being worn 75 per cent of the time.
2. Adjust the relevant scale, or scales, in the risk calculation to take this into account. For example, instead of a severity reduction of 4 (5 – 1) the reduction would be 3 (4 × 75 per cent).
3. Recalculate the risk with the revised value(s), that is 5 – 3 = 2. For example, 2 × 2 = 4 (instead of 2 × 1 = 2), for hard hats.

A similar procedure can be used for hardware reliability where data are available on, for example, failure rates of mechanical components. An alternative method of estimating the effect of human reliability is to adjust the risk score directly; we will consider this method later.

In the example just worked through, we estimated, on the basis of observation, that hard hats would be worn 75 per cent of the time. However, it is not always possible to obtain data by observation, so other sources of reliability data are needed.

There are various publications which give tables of error rates for different activities, and these can be used as a starting point for estimates. Typical error rates are shown in Table 21.3. Not all of the tasks listed in Table 21.3 will be relevant to risk control measures, but they provide an indication of the sorts of data which are available. The data are taken from Swain and Guttmann.[77]

Where relevant error rate data are available, they can be used in a similar way to that just described for wearing hard hats. Consider the case of a safe working procedure which requires, for safety reasons, the following actions:
- connecting a flexible hose
- tightening a nut
- inspecting for a leakage
- reading a pressure gauge.

Table 21.3
Illustrative error rates for a variety of tasks

Task (under no stress or distraction)	Average failure rate per 10,000 occurrences	Task (under no stress or distraction)	Average failure rate per 10,000 occurrences
Read technical instructions	82	Lubricate bolt or plug	21
Read electrical or flow meter	55	Position hand-valves	21
Inspect for loose bolts and clamps	45	Install nuts, bolts and plugs	21
Position multiple-position electrical switch	43	Fill sump with oil	19
Mark position of component	42	Disconnect flexible hose	18
Inspect for bellows distortion	39	Install guard (cover)	17
Install gasket	38	Read time (watch)	17
Inspect for rust and corrosion	37	Verify switch position	17
Install 'O' ring	35	Close hand-valves	17
Record a reading	34	Open hand-valves	15
Inspect for cracks, dents, scratches	33	Position two-position electrical switch	15
Read pressure gauge	31	Verify component removed or installed	12
Tighten nuts, bolts, plugs	30	Remove nuts, bolts and plugs	12
Connect electrical cable (threaded)	28	Remove guard	10
Inspect for leakage	26	Loosen nuts, bolts and plugs	8
Connect flexible hose	25	Verify light illuminated or extinguished	4

If any of these actions is not completed successfully, the safe work procedure fails as a risk control measure.

From the data given in Table 21.3, we can see that the failure rates per 10,000 for the four tasks are, respectively, 25, 30, 26 and 31. If these failure rates are assumed to be independent* in the safe work procedure under consideration, then the likely failure rate is 112 per 10,000 for the complete procedure. This would equate to around a 1 per cent reduction in the reliability of the risk control measure. Where the severity of the outcome is extremely high, as could be the case in nuclear power plants, it is necessary to take reductions in reliability of this magnitude into account.

Remember also that the error rates given in Table 21.3 are 'under no stress or distraction'. Where such factors are, or are likely to be, in operation, then the error rates may be much higher. Error rates are increased by a whole range of 'performance shaping factors' (PSFs), including heat, noise, humidity and speed of required action, as well as stress and distraction.

Where failure rate data are used in conjunction with a limited scale of the type used in the hard hat example, we must judge whether the failure rate obtained from tables is sufficient to warrant an increase in the likelihood scale. However, for most practical purposes, this level of detail is not usually necessary. Since subjective estimates of likelihood and severity are being made, albeit informed estimates, there is no reason why subjective estimates of reliability cannot be made.

What is necessary is to make calculations like the ones just shown, or equivalent estimates, in order to enable systematic comparisons between risk control measures. If the effect a particular risk control measure will have on risk and the reliability of that risk reduction cannot be estimated, then we cannot make informed choices between alternative methods of controlling risk.

However, this implies that alternatives are available – but sometimes alternative forms of risk control measure cannot be identified. Identifying risk control measures is a creative process and, as such, it is best carried out using techniques such as brainstorming and systems thinking. We will therefore consider these techniques next.

---

* That is, the occurrence of one error does not affect the likelihood that any of the others will occur. See the discussion of probability in Chapter 19.

Generating ideas for risk control measures

It can be argued that many risks continue to be dealt with inadequately because too little thought is put into choosing or designing risk control measures. Typical recommendations following risk assessments or accident investigations are in the form: 'more PPE', 'tell them to be more careful', or 'more supervision'.

From what we have learned about human error and human reliability, it is clear that options of this type are unlikely to be successful in the long term (see the discussion in Chapter 29). What we need is a range of options for risk control measures which take into account human error and human reliability, and this requires creative thought.

There is a wide range of techniques for stimulating creative thought; we will consider two here – brainstorming and systems thinking.

Brainstorming

The essence of brainstorming is that people's critical faculties are put in abeyance during the process of creating ideas for solving the problem – in the present case, finding ways of reducing risks. This is not a natural way for most people to work, and the more usual problem-solving strategy is shown in Figure 21.1.

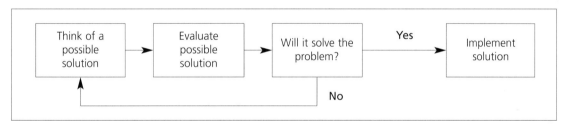

Figure 21.1
Usual problem-solving strategy

The weakness of the strategy shown in Figure 21.1 is that, although it results in a solution, there is no way of demonstrating that it is the best solution. The brainstorming sequence, in contrast, is to:
1. list as many potential solutions as possible
2. evaluate each of the possible solutions
3. select and implement the 'best' solution.

You may be familiar with brainstorming as a group technique, where one person in the group is responsible for recording solutions suggested by anyone in the group, including the recorder. This person is usually also responsible for stopping anyone who goes into 'evaluation mode' during the idea generation stage.

This group method can be very effective, but in some circumstances the group method can restrict creativity. For example, group members may be unwilling to voice suggestions which, at first sight, appear silly or outlandish for fear of being ridiculed by other participants.

Individual brainstorming is also possible, but it requires some self-discipline to stay in the creative mode for long enough to generate a good list of potential solutions. The natural tendency is to revert to the procedure illustrated in Figure 21.1. However, the best defence against this is knowing that it is likely to happen.

Brainstorming, and other creativity enhancing techniques, can be a valuable first step in advanced risk control techniques, since they can result in the generation of innovative solutions to risk reduction problems. However, if they are applied over too restricted an area, they are of limited value. To obtain the maximum benefit from these techniques, they should be used in conjunction with systems thinking. We will deal with systems and systems thinking next.

Systems and systems thinking

Systems and systems thinking cover a number of approaches and together constitute a very broad subject. However, for the purposes of this chapter, we need deal only with the basic features and purposes of a system, since this will provide the understanding necessary to begin applying systems thinking. Figure 21.2 illustrates the main features of a system.

Figure 21.2
The main features of a system

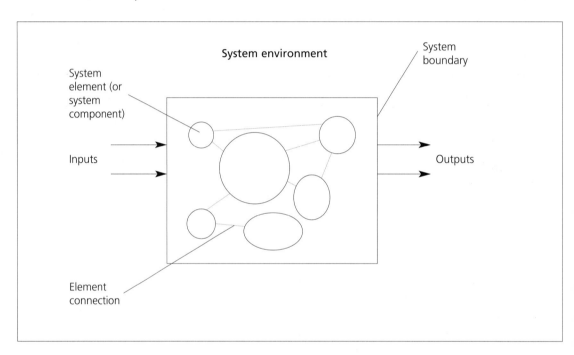

In order to be a true system, the system has to have a system boundary. This need not be a physical boundary, although it may be. For example, a health and safety committee may be considered as a system which has no natural physical boundary. Outside the system boundary is the system environment and, again, this is not just the physical environment. The health and safety committee's environment will include the 'political climate' in the organisation, for example. Within the system boundary there will be a number of system elements – for example, the members of the health and safety committee – which are connected or communicate in various ways. System elements are also called system components.

Irrespective of the type of system, it will have outputs which may be physical (printed minutes of the committee meetings) or abstract (ideas generated during the meeting) and these outputs result from processes which occur within the system boundaries. In the majority of cases, these processes require inputs which, again, may be physical (paper for the minutes) or abstract (problems for the committee to solve).

These main features are common to all systems, and there can be no system if one or more of the features is missing. However, in order to be a true system in the sense in which the word is used by systems theorists, a system must also have the following characteristics:

- **A system has to be 'owned' by someone.** Ownership in systems theory terms is not a matter of rights of possession; rather, it is used in the same sense as is implied when it is said that someone 'owns a problem' or 'owns a responsibility'. System ownership implies that the owner has sufficient interest in, or responsibility for, a system to ensure that the system continues to function or ceases, in an orderly manner, to be a system. The concept of system ownership is an important one in risk management and it will be discussed in more detail later in the chapter.
- **Systems have 'emergent properties'.** This is the systems theory way of saying that systems 'behave' in ways which may not be predictable from an examination of the elements of the system and what is known about their interconnections and communications. It is the system theorist's recognition that the whole is more than the sum of the parts. Emergent properties are important in risk management because some of the emergent properties may be undesirable increases in risk. Some examples of these emergent properties will be discussed later.
- **Adding or removing one or more of the elements in a system changes the system.** This apparently obvious statement has important implications since, if a system is changed, it is effectively a new system and may, therefore, have new emergent properties. The implication of this is that it will not always be possible to predict the effects of adding or removing system elements.
- **Elements are changed by being included in a system.** These change effects range from simple 'wear and tear' of the mechanical elements in a system to complex behavioural changes in people who become elements in an organisational system such as a health and safety committee. Because the inclusion in a system changes the system elements, the system itself may be constantly changing, which is one of the sources of emergent properties.

Systems workers make extensive use of system diagrams and Figure 21.3 illustrates a typical system diagram for a health and safety committee.

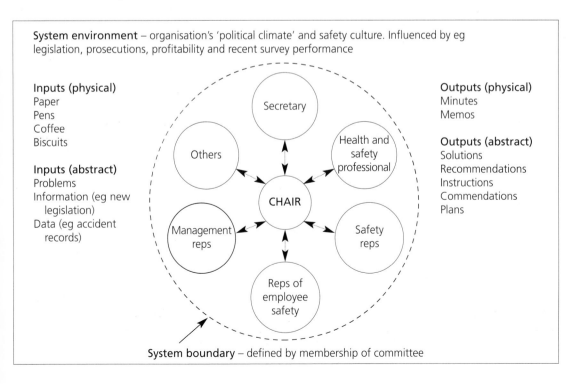

Figure 21.3
System diagram for a health and safety committee

These characteristics give systems some of the features of a living organism and, for those unfamiliar with systems theory, the living organism analogy is a good one. Systems are created, change, do unpredictable things, and 'die' in the fullness of time. As with most analogies, this one breaks down if taken too far, but bearing it in mind when considering systems enables us to think about them at an appropriate level of complexity and abstraction. In other words, it enables systems thinking, which is dealt with next.

Systems thinking

Most people who are trained to think are trained to think systematically – that is, in a style which is often referred to as reductionist. People are taught to break problems down into individual elements, each of which can be dealt with separately – the 'eating the elephant' approach.

For a large number of problems, this is an appropriate approach and in some circumstances it may be the only viable option. However, it is not systems thinking and there is another approach. This is to consider the problem as an emergent property or an unintended output of a system and solve the problem by altering the system. This approach is sometimes referred to as systemic thinking or the 'holistic' approach (as opposed to the reductionist approach described earlier).

Although the term 'systemic' is widely used in systems work, its use is rare in other areas. However, you may be familiar with systemic insecticides. These are chemicals which are taken up by a plant through its leaves and roots and remain for a time in the plant's circulatory system. When insects such as aphids suck liquid from the plant they also ingest the systemic insecticide and are poisoned. In this context, the word systemic is used to identify the fact that the insecticide is in a system (the plant's circulatory system) rather than outside the system as would be the case with insecticides which simply stayed on the outer surface of the plant's leaves and stems.

For readers new to systems theory, it may be difficult to appreciate the distinctions between the various terms used in this section. The notes below summarise the key points.

- **Systematic thinking or the reductionist approach.** The problem is broken down into individual elements and each element is dealt with in isolation. The rationale is that the problem is caused by the failure of a particular element and, by changing the relevant element, the problem will be solved.
- **Systemic thinking or the holistic approach.** The problem is seen as an emergent property of the system and the system is dealt with as a whole. The rationale is that the problem may be caused by the operation of individual system elements and/or the interactions between these elements, and/or the interaction between the system and its environment. In order to solve the problem, therefore, it is insufficient to look only at individual elements and their operation.

Systems thinking in this way has a number of advantages, but for our current purposes the main one is the way in which it can stimulate creative thought.

Probably the single most important step in systems thinking is defining the system that has the undesired emergent property or output. This is critically important because it influences the range of options which are available for solutions. Two problems and their associated systems are described below as examples.

Not wearing hard hats

Possible systems include the individual who is not wearing the hard hat, the work group including this individual, this work group plus its supervisor, the construction site on which they are all working, the construction company which employs them, and everyone already mentioned plus the hard hat manufacturers and the British Standards Institution. Where the emergent property (not wearing a hard hat) is considered as arising from the individual as a system, the opportunities for changing the system to eliminate the undesired emergent property are very limited. However, as the system is successively redefined to become more inclusive, further options become available, including peer group pressure (the work group system), disciplinary action (the supervisor plus work group system), redesign of the hard hats (hard hat manufacturers included in the system) and redefinition of the standards to be met by hard hats (BSI included in the system).

Figures 21.4 and 21.5 show the simplest 'not wearing a hard hat' system and a more inclusive system.

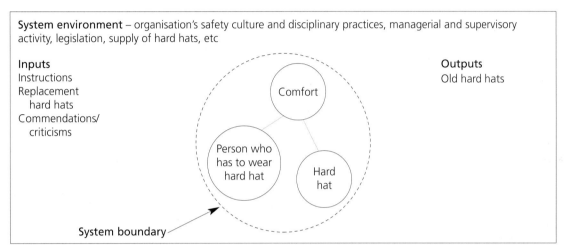

Figure 21.4
Simple system diagram for hard hat wearing

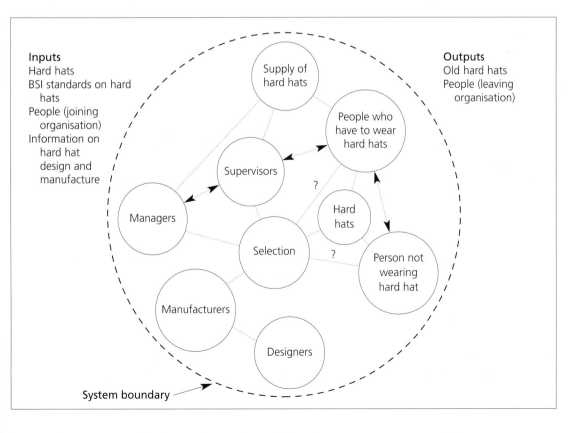

Figure 21.5
Richer system diagram for hard hat wearing

There is a major difference in systems thinking terms between inputs to a system and a system element. For example, disciplinary action can either be considered as an input to a system (in this case the individual as a system) or as an element in a system (the system which includes the supervisor). In general, there can be more control of system elements, which makes it desirable to include important aspects of risk control within a system.

Road traffic accidents not involving pedestrians

The possible systems in this case include, in increasing order of inclusion, the driver of the vehicle, all drivers of vehicles, instructors, road planners and builders, enforcement agencies, local authorities, and the government. Figures 21.6 and 21.7 illustrate a simple and a more inclusive system respectively.

A similar increase in range of options for dealing with undesired emergent properties is available with this increasingly wide definition of the system. Possible actions at the level of the individual are fairly restricted but, as enforcement agencies, local authorities and government are included in the system, many more options become available. These options include redesigning road layouts to remove 'accident blackspots' (road planners and builders included in the system), enforcing drink driving laws (enforcement agencies included in the system) and providing money for options such as speed cameras (the government included in the system).

In Chapter 6, we dealt with choosing risk control measures primarily at the level of fairly restricted systems such as the individual. However, at the more advanced level, risk control measures should be selected on the basis of more inclusive systems. Not only does this provide a greater number of options from which to choose, it also provides the potential for risk control measures that can have a much wider influence. These more complex systems involving people, the equipment they use, and their environment can be referred to as socio-technical systems. As we have seen from the simple examples given earlier, defining systems in this way allows a wider range of human factors to be taken into account.

Systems thinking also provides opportunities for solutions to problems through clear specification of systems. There are several ways of doing this, including:

- **Specifying system aims.** Large numbers of systems evolve in a piecemeal fashion without consideration being given to the intended aims of the system. A clear statement of a system's aims can often result in alternative, or better, ways of achieving these aims, perhaps with a completely different system. For example, it may not be clear whether safety rules are to protect the employees or the employers. Where this is the case, it would be preferable to have two systems with two clearly stated sets of aims.
- **Identifying system outputs.** Some system outputs are intended, such as products and services, and some are not, such as near misses and damage. Typically, intended outputs are well specified and measured but unintended outputs are not. Unintended outputs can be particularly important when they are things which reduce human reliability by, for example, producing fatigue, stress or distraction. Conflicting system aims, for example production speed and safety, may also produce unwanted outputs. These unintended outputs were referred to as spin-offs and side effects in Part 1.1.
- **Identifying system inputs.** Inputs and outputs are obviously linked, but in human reliability terms the important links are between inputs and unintended outputs. In some cases, it is a particular input which produces an unintended output and these are relatively easy to identify. More complex cases occur when it is a combination of inputs, or the absence of an input, which produces the unintended output. Some examples of these links are given in Table 21.4.

There is a variety of other uses for systems and one of these, analysis of systems failures, will be dealt with in the next section. The textbooks listed in the further reading section for this chapter at the end of the book explore the other uses of systems thinking.

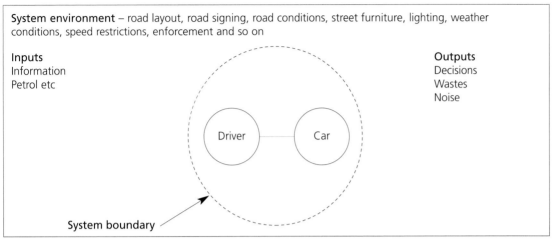

Figure 21.6
Simple system diagram for a road traffic accident

System environment – road layout, road signing, road conditions, street furniture, lighting, weather conditions, speed restrictions, enforcement and so on

Inputs
Information
Petrol etc

Outputs
Decisions
Wastes
Noise

Driver — Car

System boundary

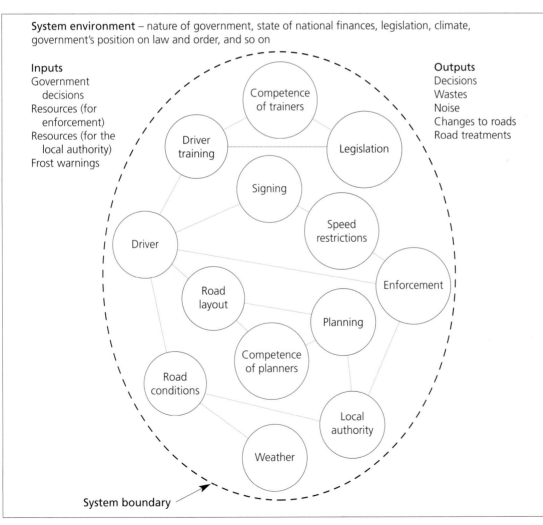

Figure 21.7
Richer system diagram for a road traffic accident

System environment – nature of government, state of national finances, legislation, climate, government's position on law and order, and so on

Inputs
Government
  decisions
Resources (for
  enforcement)
Resources (for the
  local authority)
Frost warnings

Outputs
Decisions
Wastes
Noise
Changes to roads
Road treatments

Competence of trainers
Driver training
Legislation
Signing
Speed restrictions
Driver
Road layout
Enforcement
Planning
Competence of planners
Road conditions
Local authority
Weather

System boundary

Table 21.4

Examples of inputs and unintended outputs

Single input

Input	Unintended output
Toxic chemical	Poisoning
Electricity	Electric shock

Combined inputs

Input	Unintended output
Vibration, reverberation	Noise
Two chemicals	Toxic chemical, fire or explosion

Lack of inputs

Input	Unintended output
Hard hats not available	Head injury
Ladder not available	Fall

Systems failures analysis

Systems failures analysis is a collection of techniques based on systems and systems thinking which can be used to analyse the causes of failures. It is particularly useful in risk management, since it takes into consideration both technical failures, such as failures in the hardware elements of a system, and failures arising from human errors. This means that systems failures analysis is an ideal approach when undesired emergent properties or undesired outputs have multiple causes.

Systems failures analysis begins by asking two fundamental questions:

1. was there a failure?
2. was there a system?

However, for our current purposes, we can assume that injuries, damage and other losses are, by definition, failures, and that where they occur, a system exists which is either inappropriate or dysfunctional. These assumptions need only be questioned when, for example, a loss arises from sabotage, where one system (the sabotage system) has been successful and the 'system' which has sustained the loss was never designed to combat sabotage – in other words, there was no sabotage prevention system.

The usual approach to systems failures analysis is to compare the system which has 'failed' with an ideal system and establish the causes and consequences of any discrepancies identified. The initial stage might, for example, involve comparing the complete system being analysed with a formal system paradigm of the type shown in Figure 21.8.

This comparison would allow the identification of the point, or points, at which there were departures from the ideal, and the next stage would be to investigate these in more detail. To assist with this investigation, there is a range of more detailed paradigms against which we can compare the system under analysis. For example, there are detailed paradigms for the decision and control subsystem and the performance monitoring subsystem identified in Figure 21.8.

Waring[78] has provided a detailed description of how systems failures analysis can be applied, together with a range of case studies, including the Littlebrook D hoist failure.

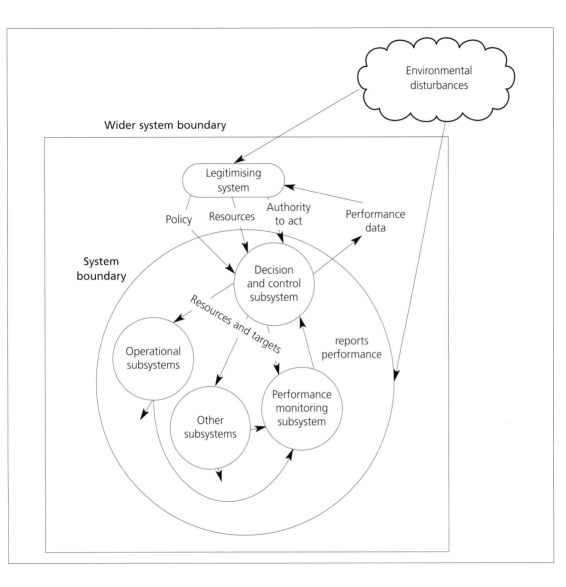

Figure 21.8
Formal system paradigm[78]

Acknowledgment: Formal system paradigm, Figure 5.2 in *Practical systems thinking*, © 1996 Alan Waring. Reprinted with the permission of Thomson Learning. (Original sources: adapted from Checkland P, *Systems thinking, systems practice*, John Wiley and Sons, 1981; and Watson L, *Systems paradigms*, Open University T301 course material, 1984, revised 1993.)

Safety culture*

There is a growing perception, at least in the UK, that safety culture is an important issue in risk management. This is reflected in quotations such as:

> The prevailing health and safety culture within an organisation is a major influence on the health and safety related behaviour of people at work. Developing a positive health and safety culture is important if high standards of health and safety are to be achieved and maintained.[79]

* 'Safety culture' is the term most frequently used for this concept and is the term used in this book. However, the concept is also known as 'health and safety culture', 'occupational safety and health culture' and 'safety climate'.

Arguably, a focus on safety culture can be seen as the next stage in the evolution of safety management from its beginnings as a purely reactive procedure (wait for something to go wrong and then fix it), to the more recent emphasis on risk assessment (find out what could go wrong and fix it before it does). The HSE appears to accept this argument:

> There is a limit to the health and safety performance an organisation can achieve without addressing the contribution which human factors have to play in eliminating occupational accidents and ill health.
>
> Many organisations have now reached a point where they feel ready to tackle the complex area of health and safety culture.[79]

However, as the HSE points out, safety culture is a complex area and the aim of this section is to clarify some of the issues involved in safety culture by providing information on the following topics:
- what we mean by safety culture
- why safety culture is important
- how safety culture can be measured
- how safety culture can be influenced.

Within each topic, we give brief details of the relevant issues, together with references, as necessary, to sources of more detailed information.

What do we mean by safety culture?

The word 'culture' has a wide variety of meanings from the attributes of religious and ethnic groups to a collection of bacteria in a Petri dish. In this section our discussion will be restricted to organisational culture, which we will define in terms of what people think, say and do in the context of the organisation in which they work.

In order to have an identifiable culture, there must be a recognisable range of issues involved, and to have a safety culture it must be possible to identify relevant safety issues. This is normally done by having a definition; there are various definitions of safety culture which attempt to encapsulate the issues involved, including the following:

> [Safety culture consists of] shared values (what is important) and beliefs (how things work) that interact with an organisation's structure and control systems to produce behavioural norms (the way we do things around here).[80]

> ... the product of individual and group values, attitudes, competencies, and patterns of behaviour that determine the commitment to, and the style and proficiency of, an organisation's health and safety programmes. Organisations with a positive safety culture are characterised by communications founded on mutual trust, by shared perceptions of the importance of safety, and by confidence in the efficacy of preventive measures.[81]

The latter definition is also quoted in HSG65.[1]

General definitions of this type have the following advantages:
- They highlight the issues which are important in safety culture, and which differentiate safety culture from other measures of health and safety performance. In particular, they emphasise the role of people's attitudes and beliefs.
- They emphasise risk and risk control as being central to good safety management rather than, for example, just compliance with legislation.

However, these definitions are of limited practical value since they do not provide a direct means by which safety culture can be measured and influenced.

To some extent, this has been addressed by certain authors. For example, the ACSNI report[81] quoted earlier provides a 'safety culture prompt list', which subdivides safety culture issues into more detailed, and more measurable, elements. This prompt list will be discussed later in this section.

Why safety culture is important

The rationale for concentrating on safety culture is the belief that improving it will result in a reduction in losses. That is, organisations with a 'good' safety culture will have fewer injuries, less ill health, and fewer incidents which result in damage to assets or the environment. This means there will be a negative correlation between safety culture and losses. As safety culture improves (a measurement increasing), losses will be decreasing. However, the statistical techniques for dealing with negative correlations are well worked out, so this should not create practical difficulties. Where it does, one or other of the measures can be inverted, for example, the extent of the reduction in losses can be used instead of the amount of losses.

If this correlation between safety culture and losses is to be tested, we obviously need to be able to measure both safety culture and losses. In essence, we need to measure at both ends of the causation spectrum described in Chapter 19. Figure 19.1 is reproduced as Figure 21.9 by way of a reminder.

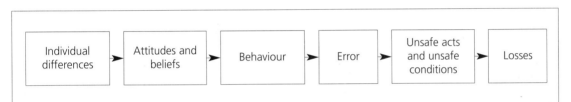

Figure 21.9
An illustrative causation continuum

We dealt with measuring losses in detail in Chapter 8, and we will deal with measuring safety culture later in this section. However, measuring safety culture involves measuring people's attitudes and beliefs as well as what people do, and assessing how attitudes link with behaviour. These are dealt with as separate topics in their own right in Chapter 26 but, for the purposes of this section, we will use a simplified terminology.

It is possible to identify three broad categories into which most of the relevant safety culture terminology will fit:
- **think**, including attitudes, beliefs, motivation and so on
- **say**, including stated intent, verbal behaviour such as body language, and written statements
- **do**, including all other behaviour and physical responses.

These three categories are linked in the way illustrated in Figure 21.10,* which shows the normal assumption that what people think determines what they say and do. If this assumption holds, it means that if people 'think safety', they will 'do safely'.

Where the links shown in the diagram exist, the condition is very stable and it is this type of stability, combined with 'thinking safety', which is required if safe and healthy behaviour is to be self-sustaining, rather than imposed by, for example, strict supervision. However, these links may be absent, and this is illustrated by example in Table 21.5, together with examples of when the links are present.

---

* It is possible to include another element in this diagram, namely how people feel, or the 'affect' element of attitudes. However, this is unnecessary in the present discussion. See Chapter 26 for more information on the affect element.

Figure 21.10
Links in 'think, say
and do'

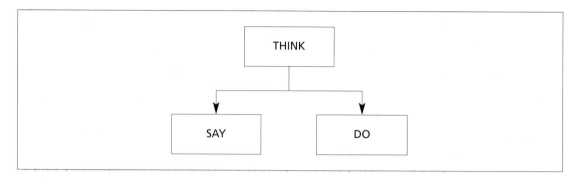

Table 21.5
Examples of
inappropriate and
appropriate links

	Positive behaviour	Negative behaviour
Positive thought	Wearing eye protection is a good idea. I wear my eye protection.	Wearing eye protection is a good idea. I do not wear my eye protection.
Negative thought	Wearing eye protection is a bad idea. I wear my eye protection.	Wearing eye protection is a bad idea. I do not wear my eye protection.

The first step in using attitudinal measures is to identify that there is a mismatch between thought and behaviour. There can be a variety of reasons why this mismatch exists. For example, people may wear eye protection because it is an enforced safety rule rather than because they believe their eyes require protection, and people may not wear eye protection, despite thinking it is a good idea, because of peer pressure. Although wearing eye protection is used as an illustration, the problem can apply to all health and safety activities, including those normally associated with management. These include chairing health and safety meetings, planning for health and safety and organising health and safety funding.

Unfortunately, the links between thinking, saying and doing are not as simple as those shown in Figure 21.10. There are feedback loops involved, as illustrated in Figure 21.11.

Figure 21.11
Feedback loops in
'think, say and do'

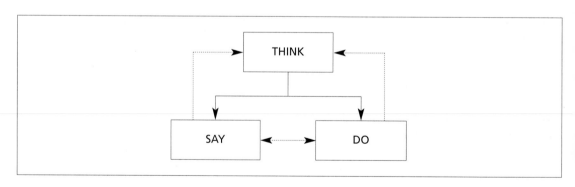

Figure 21.11 shows that what people do influences what they think and say. For example, people's experience of new foods, new technologies and new techniques changes how they think about these things, and what they subsequently say about them.

Similarly, what people say influences what they do and what they think. For example, people make promises in haste or to solve a short-term problem and this later influences what they do (fulfil the promise) and think (promises are not a good idea!).

'Think, say and do' can be considered on a corporate as well as an individual level. There are many examples of organisations committing themselves to a course of action and later regretting it.

The complexity of these feedback loops is often summed up in the phrase: 'Attitudes influence behaviour and behaviour influences attitudes.' These links and feedback loops are important since, if the aim is safe behaviour that is self-sustaining, the links will have to have the following characteristics:

- 'Think, say and do' will be consistent, with 'think' determining 'say' and 'do'. This consistency will be in the top left cell of Table 21.5 – positive thought and positive behaviour.
- The feedback loops will have beneficial results. For example, there will be no rewards for undesired behaviour, and no penalties for desired behaviour.

Having dealt with the links between attitudes and behaviour, it is now possible to return to why safety culture is important. As we will see later, in the section on measuring safety culture, there are many indicators which are important in safety culture. However, three of the main ones will be discussed in this section:

1. self-monitoring
2. consensus
3. range of issues.

Self-monitoring

It is known from research findings that 80 to 90 per cent of accidents arise as a result of human activities, and if this number is to be reduced, there will have to be some method of reducing the numbers of errors and rule violations.

In the past, rewards and sanctions have been used as the main methods, combined with supervision. However, these methods last only as long as the rewards, sanctions and supervision are effective (see Chapter 29 for a detailed discussion of this issue).

For long-term success, what is required is self-monitoring and, where appropriate, monitoring of colleagues. People must do 'the right things' because they believe in them. That is, they must have the integrated 'think, say and do' shown in the top left cell of Table 21.5.

Consensus

It is known from research findings that perception of safety performance differs at different management levels in an organisation. For example, when managers at different levels in an organisation (from board level down to supervisors) are asked to answer the same set of questions about their organisation's health and safety performance, the percentage scores correlate well with the management level. The higher the management level, the higher the scores.

When the same questions are answered by independent auditors, it results in scores which match more closely those from the lowest management level. In other words, senior management have an over-optimistic view of their organisation's health and safety performance.

Only when there is a consensus of views at all management levels and these views are an accurate reflection of the real state of affairs, will it be possible to have a good, strong safety culture.

Range of issues

It is known from the continuing high rates of accidents and ill health that conventional management techniques are inadequate. One response is to use more of these conventional techniques (more supervision, more machinery guarding, more PPE and so on), but this, too, is likely to be ineffective. Newer approaches to health and safety management based on risk assessment and the techniques of quality management are an improvement on the conventional techniques, but not the complete solution. What we need is health and safety management that includes the best of both conventional techniques and the new techniques but, in addition, embraces other issues.

There are many issues which fall into this category, but they can be illustrated by considering four:
1. belief in the efficacy of risk control
2. democracy
3. care and concern
4. personal problems.

Belief in the efficacy of risk control

The belief that 'accidents are bound to happen' can be a self-fulfilling prophecy. If people believe that nothing can be done to prevent accidents, then they will do nothing to prevent them. If a good health and safety performance is to be maintained, people must believe that risk control will work, at least most of the time.

Democracy

Typically, the people exposed to the highest risks have the least say in how these risks are controlled. If good health and safety behaviour is to be maintained, people must have a significant influence in assessing the risks to which they are exposed and in deciding on the risk control measures they will use.

Care and concern

Injuries and ill health are only special cases of a more general lack of wellbeing, but they are given particular attention because of their salience, seriousness, or legal implications. Organisations which have a good health and safety performance will usually be committed to more general wellbeing, including, for example, ways of reducing stress at work.*

Personal problems

This is similar to care and concern but extends the concern to non-work problems. It is important because of the causal links between personal problems and injuries and ill health, with factors such as stress and lack of attention making people more likely to have accidents and show symptoms of ill health.

There is still debate about the full range of issues which should be covered by the term 'safety culture', as will become clear in the next section.

Measuring safety culture

Measuring an organisation's safety culture will require:
* separate measurements at each management level so that differences in perception can be identified at these different levels
* the inclusion of issues such as belief in risk control, consensus and wellbeing away from work
* special measurement techniques to measure 'thought'.

The most obvious way to meet these criteria is by using some form of attitude survey, since this can be administered at different levels in the organisation and the results from the different levels compared.

The statements or questions to be used in such a survey can be derived from the ACSNI report,[81] which contains an extensive and detailed list of issues in the form of a 'safety culture prompt list'. Some of the items

---

* Note that BS OHSAS 18001 'is not intended to address other health and safety areas such as employee wellbeing/wellness programmes' but it could be extended to cover these issues.

in this list are described next, together with the sorts of difficulty that may be encountered in designing a safety culture attitude survey.

The first item in the 'safety culture prompt list' is concerned with general organisational culture and asks:

> Has the organisation evidence to demonstrate that communications at all levels are founded on mutual trust?

A priori, this would appear to be an excellent question about an organisation's culture and, for the present purposes, it can be assumed that it is. What evidence, therefore, would an organisation be expected to produce to demonstrate the current level of the safety culture with respect to this issue? There are several practical problems with this:

- **What is 'trust' in relation to communications?** Obvious dimensions are truth (I trust that what you tell me is true), comprehensiveness (I trust that you will tell me everything I need to know) and timeliness (I trust that you will tell me what I need to know when I need to know it). While a simple 'yes' or 'no' answer to 'Is there mutual trust?' may provide an initial measure of an organisation's safety culture, if effective remedial action is to be taken, the answers to more detailed questions about the nature of any lack of trust will also be needed.
- **How is 'mutual' to be assessed?** The principle is fairly straightforward, as a sample of communications can be chosen and each participant asked about their level of trust in the other participants' communications. There are three possible outcomes: mutual trust, mutual lack of trust, or lack of trust in one direction only. The measure of 'mutuality' could, therefore, be the relative percentages of communications falling into each category. However, this straightforward principle is likely to be extremely difficult to implement because of the practical difficulties involved in identifying suitable samples of communications.*
- **What is meant by 'at all levels'?** The implication is that the concern is with communications between people who are all at the same level and that the analyses of the results would produce, for example, a figure for 'percentage trust' at each level. This, however, is clearly inadequate since there are at least two other important issues with respect to level. First, there is communication between levels, so trust in 'downward' and 'upward' communications must be measured separately. Second, there is communication between different functions in the organisation, irrespective of the levels at which these communications take place. Poor communication arising from lack of trust between, for example, purchasing and production, or quality and health and safety, can have detrimental effects on safety culture.†

These points are not, of course, a criticism of the ACSNI question, since it comes from a set of questions clearly labelled 'prompt list'. Rather, these points are made to illustrate the practical problems which still have to be resolved before it can be said that there is an effective safety culture attitude survey.

To reinforce this, we will now consider the first question from the ACSNI prompt list that deals specifically with safety culture. This is an apparently straightforward question:

> Has the organisation evidence to demonstrate that the chief executive takes a personal and informed interest in safety?

There are practical problems with this question:

---

* See Chapter 29 for a discussion of how interpersonal relationships can be measured.

† A discussion of the kinds of question needed for detailed analysis can be found in Chapter 22.

- **Who is the chief executive?** The identification of the chief executive in the UK nuclear industry may not be a problem but this is less likely to be the case in other industries. Ignoring the problem of the job title, what are the job characteristics which would lead to the identification of a particular individual as the chief executive for a given safety culture attitude survey? In larger organisations, it could be the case that the chief executive is not based at the site being surveyed, or even in the country in which the survey is taking place. What we need is a specification of the criteria to be met for a person to be considered the chief executive for the purposes of a particular safety culture attitude survey.
- **What sort of evidence demonstrates taking a 'personal' interest in safety?** The best evidence is likely to be specified in terms of the activities the chief executive should carry out in order to demonstrate a personal interest, along with, if appropriate, the amounts of time to be spent on these activities. If the evidence is restricted to 'think' and 'say' items, then it is possible that the chief executive can be accused of hypocrisy or paying 'lip service' to safety; he or she claims or states a commitment, but does not demonstrate the claimed or stated commitment.
- **What sort of evidence demonstrates taking an 'informed' interest in safety?** In theory, the safety competences required of a chief executive could be specified and individual chief executives assessed against these competences. While it would not be too difficult to specify the competences, it may be more difficult to get individual chief executives to agree to be assessed. However, the authors of the report use 'informed' rather than competent, which suggests less rigorous criteria, although what these are is not clear.

All of these sorts of problem will have to be resolved in order to create an effective safety culture attitude survey which produces data that allow us to describe accurately the nature and strength of an organisation's safety culture.

Practical approaches to measurement

There are two basic practical approaches to beginning the measurement of safety culture:
1. **Devise a set of questions or statements.** Developing a complete measure of all aspects of safety culture would be a significant task, but major benefits can be achieved by devising a set of questions or statements which suits a particular organisation.
2. **Adopt a set of questions or statements prepared by someone else.** This is an attractive option, since it cuts out the work needed to devise questions or statements. However, care is needed in choosing the set of questions. As we have seen, attitude measurement is not straightforward and poorly designed questions can do more harm than good.

Sets of safety culture measurement questions and statements are available from a number of sources, including learned journals and commercial health and safety organisations.

The UK's Health and Safety Laboratory has also produced a Safety Climate Tool,[40] which consists of safety culture statements and software for recording and analysing the survey results. The safety culture statements deal with the following:
- organisational commitment
- health and safety oriented behaviours
- health and safety trust
- usability of procedures
- engagement in health and safety
- peer group attitude

- resources for health and safety
- accident and near miss reporting

Irrespective of the source of the questions or statements, sampling will be critically important, since a balance will have to be struck between collecting all the relevant data from everyone in the organisation (ideal but impractical) and collecting a sufficient sample of data to enable valid inferences to be drawn about the organisation, or part of the organisation, being studied.

Analysis of safety culture data

Even when the relevant measures are available and the data have been collected, the data will require special analyses and these analytical techniques are described next.

The analytical techniques used for a safety culture attitude survey will include all the techniques used to analyse the results of a conventional survey – for example, the percentages of people answering particular questions, the range and variance of answers and so on. However, these analyses will have to be supplemented by detailed cross-referencing, as follows:

- answers from different levels of management will have to be compared in order to identify, for example, differences in perception at the different management levels
- various issues will have to be cross-referenced – for example, stress management in the organisation is likely to be addressed under a number of headings and all the 'stress' questions will have to be drawn together and analysed as a unit
- the 'think, say and do' questions on a particular topic will have to be analysed together to arrive at an estimate of their cohesiveness.

Because the data analyses are complex and time-consuming, even when only a sample is being analysed, they are best done on a computer using specially designed software. Special types of software may also be needed at the research and development stage to help with exploratory techniques such as repertory grid, item analysis and analysis of variance.

There are several commercially available survey recording and analysis software packages which you could use to develop your own set of safety culture questions (see Chapter 23). There are also online survey services, some of which are free for small-scale surveys.

Symptoms of a poor safety culture

The previous discussion was concerned with the systematic measurement of an organisation's safety culture, which is what will be required for long-term assessment of improvement or deterioration. However, it is possible to detect particular symptoms of a poor safety culture in the course of other risk management work. Some examples are:

- **The interviews carried out in the course of incident investigations, risk assessments and active monitoring can produce evidence of issues that are relevant to safety culture.** Typical issues are belief in the efficacy of risk control measures, the lack of democracy in selecting risk control measures, and a belief that risk control measures are for management's benefit, not for the benefit of those exposed to the risk.
- **The behaviour of individuals with respect to risk management activities can give an indication of their commitment to risk management.** Typical evidence includes the amount of time spent on risk management tasks, the time and trouble taken to achieve risk management competences, and the extent to which allocated risk management duties are effectively performed.

- **Observation of communication on risk-related topics can provide evidence of two sorts.** First, it shows whether the necessary communication is taking place. This ranges from informal discussions between supervisors and workers about whether risk control methods are effective, to whether the health and safety committee meets at the specified frequency. Second, it shows whether the content of the communications is what would be expected if the safety culture were good. Again, this ranges from the degree of trust between supervisors and workers to the extent to which the health and safety committee is adversarial (management versus workers).

Once an organisation has raised awareness of the issues relevant to safety culture, it will be able to identify evidence on these safety culture issues in many aspects of the operation of the HSMS.

Causes of a poor safety culture

The wide range of issues which has to be taken into account when measuring safety culture suggests that there will be a correspondingly wide range of causes of a poor safety culture. Identifying causes of a poor safety culture is possible only by first determining which issues are causing problems. For example, if people do not believe that risk control measures are effective, this can be for any one of a number of reasons:

- they have not had a say in selecting the risk control measures
- they do not, in general, believe what they are told
- they believe the risk control measures benefit only management
- they have not had an explanation of how the risk control measure is supposed to work
- using the risk control measure conflicts with other management priorities.

In general, any of the organisational factors discussed in Chapter 29 can be a cause of one or more aspects of a poor safety culture – see that chapter for a more detailed discussion.

Influencing safety culture

Safety culture can be influenced in a large number of ways but it is always necessary to start with the nature of the problem identified. If some form of survey is used, this will probably identify a number of problems which should be addressed. The solutions to these problems can then be formulated as aims, and the planning methods described in Chapter 23 can be applied.

However, in drawing up a plan for influencing safety culture, bear in mind that:

- **it is essential to address senior management issues first** – health and safety performance can be changed at lower levels in the organisation, but this change will not be permanent unless it is continuously supported by senior management
- **many of the issues to be addressed may not be seen as being part of risk management** – for example, a lack of trust on risk management issues may be part of a general problem of lack of trust on all issues; liaising with other departments is likely to be needed to solve these kinds of problem
- **people's attitudes and beliefs must be changed, and this is a notoriously sensitive and difficult task** – to be successful, it must be taken slowly and with well-thought-out steps; making radical changes or changing things too fast is likely to cause resistance.

Many of the problems associated with a poor safety culture can be influenced by conventional management techniques (training, information, rewards and so on), but others require special techniques.

Useful techniques have evolved from quality management, including safety circles and empowerment. Safety circles are analogous to quality circles and are particularly useful for promoting consensus and benefiting from peer pressure. Empowerment involves providing people with the competences and authority to manage their own work risks by, for example, carrying out risk assessments, implementing their own risk control measures, and being able to stop the job on health or safety grounds.

Another technique is behavioural modification, used in various behavioural safety techniques. This is based on the fact that what people do influences what they think, and it works by getting people to 'do the right thing' by any reasonable means. Once people understand the benefit of 'doing the right thing', it is argued, the behaviour will become self-sustaining.

Behavioural safety begins by identifying what it is that people are doing, or not doing, which could contribute to an incident or accident. Simple examples include:

- **doing** – riding on the forks of a fork-lift truck, standing on swivel chairs, sitting with a poor posture, lifting heavy weights unassisted
- **not doing** – not wearing PPE, not holding the hand rail when walking down stairs, and not checking tyre pressures on vehicles.

When the behaviours have been identified, any effective and reasonable means are used to change the behaviour. What will be effective in changing the behaviour, and what constitutes reasonable means, will vary according to the circumstances, but commonly used methods include:

- **Continuous supervision.** People tend to do what they are supposed to do when they know they are being supervised or observed by their colleagues. This method is strengthened if it is combined with either, or both, of the next two methods.
- **Sanctions for nonconformity.** Sanctions range from verbal criticism, through more severe forms of disciplinary action, to dismissal. The last has the effect of 'encouraging the others', since it demonstrates managers' intentions very clearly.
- **Rewards for conformity.** These can vary from verbal commendation, through financial rewards, to early promotion.

If, by whatever means, the required behaviour can be enforced for long enough, the behaviour becomes internalised – that is, the person now considers that the behaviour is natural and normal and behaves in the desired way from choice. When this has happened, we say that the behaviour is self-sustaining.

An example of how this works can be seen in the requirement to wear seat belts in vehicles in the UK. When cars were first fitted with seat belts, very few people wore them. However, when it became a legal offence not to wear a seat belt, people wore them to avoid being fined. Having worn seat belts for this reason for a number of years, people now feel 'uncomfortable' without a seat belt, and it is likely that most people would continue to do so even if they could no longer be fined for not wearing one.

A good safety culture requires both appropriate behaviour and appropriate attitudes. Behavioural safety tries to change behaviour in the hope that attitudes will also change, while the more conventional safety culture approach is to change attitudes in the hope that behaviour will change. The two techniques can, therefore, be used in parallel when necessary.

In addition to these direct techniques for influencing safety culture, it is also possible to influence it indirectly. In general, this can be achieved by any method which demonstrates that effective risk management has benefits for those involved. For example, effective risk assessment techniques which involve the people exposed to the risks will have a 'spin-off' benefit on safety culture by increasing people's belief in the efficacy of risk assessment.

Finally, changing safety culture should be seen as a process of continual improvement, not a once-and-for-all solution. There will never be 'perfect' safety attitudes, but good organisations, implementing a well-thought-out plan for influencing safety culture, should have 'better' safety attitudes each time they measure them.

Summary

This chapter began with a description of the ways in which the risk reduction effects of risk control measures can be calculated or estimated. This was followed by a description of how two techniques – brainstorming and systems thinking – can be used to generate a wider range of ideas for risk control measures. The chapter ended with a discussion of what is arguably the next major step in risk control – improving an organisation's safety culture.

22: Emergency planning

Introduction

For the purposes of this chapter, we will consider an emergency as an event which necessitates a rapid and more or less complex response in order to minimise losses. On this basis, emergency planning requires us to identify:
- the events which could lead to an emergency
- the activities required in response to the emergency and the timescales for these activities.

However, there are particular problems with emergency planning in that it is not possible to test the efficacy of an emergency plan* by creating a real emergency, or allowing one to occur. Thus, we must devise and use alternative methods of testing emergency plans or aspects of them.

In addition, the fact that an emergency is not a continuous process can also cause problems. Where a process is continuous, there can be arrangements for measuring outputs and taking remedial action which ensure that the requirements of the continuous process are met. With emergencies, there have to be special arrangements to ensure that the resources required for an effective response continue to be available. In order to deal with this difference, the following terminology will be used:
- **The implementation of the emergency plan** – this term is used to describe the work needed to put in place everything needed to respond appropriately in the event of an emergency.
- **The activation of the emergency plan** – this term is used to describe the work needed to respond in the event of an emergency.

For the most part, meeting the requirements of emergency responses in terms of implementation can be satisfied by using techniques such as active monitoring and review, which have been described elsewhere in this book. However, the one major omission so far has been the planning process itself – so this chapter deals in detail with the techniques of planning.

Identifying events

Events which constitute an emergency have one or more of the following characteristics:
- **They require the rapid deployment of resources which would not normally be needed as part of the core activities of the organisation concerned.** These resources can range from such things as spillage kits and emergency showers, through the resources required to fight a major fire, to complete backup systems for the essential elements of an organisation's business activities. Since rapid deployment is required, special arrangements have to be made to ensure that the required resources are in place and available at all times.
- **They require competences which would not normally be needed as part of the core activities of the organisation concerned.** These competences can range from knowledge of evacuation procedures, through the skills needed to deal with a major spillage, to experience of dealing with the media and local communities. Since these competences will be required in a timescale which precludes them being obtained during an emergency, special arrangements have to be made to ensure that they are in place and available at all times.
- **They require a rapid series of concerted actions which it would not be realistic to work out within the timescales involved.** These actions will include such things as evacuation of personnel, minimising damage

---

* In BS 18001 and BS 18004, 'programme' has the same meaning as 'plan' and the two words are used interchangeably in this chapter.

to assets and the environment, and keeping relevant people informed. Since the timescales during an emergency will not allow proper consideration of the actions required, special arrangements have to be made to ensure that they have been planned in advance.

These requirements for resources, competences and concerted action mean that there has to be an appropriate level of emergency preparedness in all organisations. The first step in this preparedness is to identify those events which could lead to an emergency and this is usually done in two stages:

1. **Identify outcomes that would need an emergency response.** Typical outcomes include fires, explosions and losses of containment of chemicals, substances or radiations. This identification of outcomes is, in essence, a risk assessment exercise and can, therefore, be carried out using an appropriate range of risk assessment techniques.

2. **Identify the events which would have to occur to produce a particular outcome.** As appropriate, events which may occur accidentally and events which may occur as the result of arson, vandalism or terrorism will have to be identified separately. Some events may contribute to a number of different outcomes; for example, spillages of certain substances can result in ill health, fire and damage to the environment. However, for any given outcome, all relevant events should be identified, since the relationships between the events may vary for different outcomes. An effective technique for identifying these events is Fault Tree Analysis, with the relevant outcome as the top event.

Sometimes this stage of the emergency planning process identifies the need for additional or alternative risk control measures, since it is, essentially, a risk assessment procedure. However, the rest of this chapter will concentrate on planning the response to an emergency, not the planning required to prevent an emergency. In other words, we are aiming to minimise the consequences of an emergency which has occurred by the timely and effective deployment of emergency procedures.* This minimising of consequences is also referred to as mitigation. In addition, the emphasis will be on those emergencies which have serious adverse consequences for the organisation concerned. These are usually referred to as disasters, and planning for minimising their consequences is described as disaster planning. But since the planning techniques to be described in this chapter are equally applicable to all levels of emergency planning, we will continue to use the term emergency planning.

Once we have identified the events which could lead to specific outcomes, we will know how likely specific outcomes will be, how serious they will be, and where they are likely to occur. This knowledge can then be used to draw up priorities for emergency planning, where such plans do not exist, or for the review of emergency plans where they are already in place. For the rest of this chapter, we will assume that this preliminary work has been carried out, so it is appropriate to move on to the next phase of preparing an emergency plan – that is, identifying the actions required when an emergency happens.

Identifying required actions

Identifying the actions needed in response to an emergency is the first stage in the planning process proper, since the earlier work was primarily risk assessment.

There are many planning techniques which can be used but, for the purposes of this chapter, we will use the planning process described in Annex D of BS 18004[2] with slight modifications to suit emergency planning requirements.

---

* Preventing emergencies relies on effective risk control measures; how risk can be controlled effectively has been dealt with elsewhere in the book.

It is easier to illustrate the use of the planning process by using an example and, since fire is arguably the most common outcome in the context of emergencies, we will use this. If you are interested in how this planning process can be used in non-emergency situations, see Annex D of BS 18004.

The identification of the required actions should be carried out in two stages:

1. specify the objectives to be achieved
2. for each objective, specify the actions required to achieve it.

These two stages are dealt with separately in the next two subsections.

Specifying what is to be achieved

For practical reasons, specifying what has to be achieved is better done in two steps. The first step is an exercise in creativity intended to produce a list of aims to be achieved. Typical aims in the case of response to a fire could be:

- prevention of injury and ill health to personnel on site
- prevention of injury and ill health to people off site
- minimisation of damage to the organisation's assets
- minimisation of damage to other people's assets
- minimisation of damage to the environment
- minimisation of business interruption.

Each of these aims will have to be dealt with separately and in order of priority, since they may conflict. For example, the actions required to prevent injury and ill health may reduce the options available for minimising business interruption. In general, minimising business interruption is qualitatively different from the other aims listed above since it is primarily concerned with business risk and not pure risk. For this reason, planning that mainly addresses minimising business interruption is often referred to as 'contingency planning' rather then emergency planning.

Once an aim has been identified, the second stage is to translate it into a form that can be measured – that is, the aim must be expressed in a quantified form. A quantified aim will be referred to from now on as an objective; some sample aims and objectives are given next.

Aim	Prevention of injury and ill health to off-site personnel as a result of a fire.
Objective	No-one off site will be injured or sustain ill health as a result of a fire on site.
Aim	Minimise business interruption.
Objective	Normal business will be resumed within 24 hours of a fire on site.

The reason for working through the 'aims then objectives' sequence is that it is easier to generate aims in 'brainstorming' mode, since there is no need to consider the practicalities of quantification. However, quantifying the aims is an important step since it focuses attention on the measurable aspects of what is to be achieved and this, as will be described later in the chapter, is essential if the success of an emergency plan is to be measured. Quantifying aims is not necessarily an easy process and BS 18004 gives the following guidance:

> Objectives to increase or reduce something should specify a numerical figure... and a date for the achievement of the objective;

> Objectives to introduce or eliminate something should be achieved by a specified date;

> Objectives to maintain or continue something should specify the existing level of activity.

As an illustration for the rest of this chapter, we will use prevention of injury and ill health to on-site personnel as the main example since this is likely to be the highest priority aim in most circumstances. The relevant objective will be no injuries or ill health to on-site personnel in the event of a fire.

Specifying the actions required

In order to achieve the overall aim of preventing injury and ill health to on-site personnel in the event of a fire, certain actions will be required and these can be considered as subsidiary aims. Identifying these actions is again an exercise in creative thought. Any suitable creativity-enhancing techniques will help with this process, but BS 18004 suggests the use of 'guide words' to stimulate thinking, the analogy being the guide words used in HAZOP. The guide words given in BS 18004 are shown in Table 22.1, but other guide words can be added or substituted to suit particular planning exercises. Table 22.1 also gives examples of the actions the guide words might suggest in the context of preventing injury and ill health in the event of a fire. Note, however, that in BS 18004 other examples are used.

Table 22.1
Guide words for helping to generate aims

Guide words	Examples
Increase Improve	Increase awareness of evacuation procedures Improve emergency lighting
Maintain Continue	Maintain effective fire detection systems Continue fire induction training
Reduce	Reduce the number of obstructions in emergency exit routes
Introduce	Introduce refresher courses in evacuation procedures
Eliminate	Eliminate 'dead' areas in the fire alarm system

As with the 'what has to be achieved' aims described earlier, these 'action aims' will also have to be quantified, and typical objectives for some of the aims listed in Table 22.1 are given next.

Aim	Increase awareness of evacuation procedure.
Objective	All personnel on site can describe accurately the evacuation procedure for all parts of the site they may occupy.
Aim	Continue fire induction training.
Objective	All the organisation's personnel on site at any given time will have received basic fire induction training during their first day on site.*
Aim	Eliminate 'dead' areas in the fire alarm system.
Objective	The fire alarm will be audible at all parts of the site.

Note that there are now two sets of objectives: those relating to what is to be achieved, and those relating to the actions required for this achievement. It is essential to keep these two sets of objectives separate since, if they are allowed to become interlinked, there will be no method of measuring why a plan is a success or a failure. We will look at this point in more detail later in the chapter.

Once we have drawn up an initial list of required activities in the form of objectives, we can move on to the next stage of the planning process – that is, specifying targets. Planning is usually an iterative process or,

---

* Other aims and objectives will, of course, be required for contractors working on site and for visitors to the site.

as it is put in BS 18004, 'The various steps in the process might need to be re-visited several times before the primary objectives, the programme itself, or the performance indicators are finalized.'

Specifying targets

Each activity objective of the type described above will only be achieved if things are done and we will refer to these things as targets. Targets should meet the requirements for standards set out in Chapter 10, that is, they should specify **who** is to do **what**, by **when** and with what **result**.

It is only by specifying and implementing an adequate range of targets that particular activities will be carried out. We will deal with preparing targets in more detail later, using as an example the training of managers in fire response procedures. However, it should be apparent at this stage that a large number of targets will be required in order to describe fully the detailed response required in an emergency, and that recording all of these requirements will not be a trivial task. The next section deals, therefore, with recording the plan.

Recording the plan

The most common method of recording plans is straightforward narrative description. However, there are more precise ways of recording, one of which is to record the plan in the form of a structured set of questions, and we will consider this in this section, beginning with a summary of why questions should be used in this way and then moving on to the procedures for developing a structured set of questions.

There are several advantages to developing and using a structured set of questions:

- To design a structure for the questions, it is necessary to think systematically about the steps required to achieve the objective or objectives. This should be done in all planning, but having to produce a structured set of questions provides a discipline which is often useful.
- In order to write individual questions, it is necessary to think through the practicalities of what has to be done, and this usually results in a better thought-out plan.
- Once a structured set of questions has been produced, it can be passed to the relevant people who will then have a detailed 'blueprint' of what is required. They can then use this both as an aid when first implementing the plan and for periodically monitoring the implementation of the plan.
- A structured set of questions can be used as an auditing tool (see Chapter 23).
- As we saw in Chapter 10, if a suitable format is used for the questions – for example, they all require 'yes', 'no' or percentage answers – it is possible to quantify the extent to which the plan is being implemented. This enables comparisons over time and between different parts of an organisation.

As we have already mentioned, there will be a large number of targets, and hence questions, required for any but the most trivial plan. Therefore, an essential first step is to decide on a structure for the proposed questions. This will be, in effect, a set of headings and subheadings which can be used to group related issues. Producing a detailed and suitable set of headings has the following advantages:

- **At the planning stage, it enables issues to be addressed one at a time and in a systematic manner.** Many planning exercises get 'bogged down' because the participants attempt to create a single list of requirements and, once this list gets beyond about 50 items, no-one can easily remember what has and has not been included.
- **At the monitoring and audit stage, it enables rapid identification of areas of weakness.** However, this aspect of plan use is dealt with in more detail in Chapter 23, where advanced auditing techniques are considered.

The appropriate structure for a set of questions will depend on the type of emergency being considered, the nature of the organisation, and the range of personnel available to take action in the event of an emergency. However, a generally useful structure is a 'tree' or 'directory' structure with main headings and various levels of subheadings. An example of this sort of structure is shown in Table 22.2. Note that details of the lower levels are given only for training in fire procedures, which we will use as an illustration later in the chapter.

Table 22.2
Partial structure for a generalised emergency plan

Level 1	Level 2	Level 3	Level 4	Level 5
Procedures	Preparation Review and revision			

Competences	Competence needs
	  analysis
	Competence provision
	Competence
	  assessment
		Fire response
		procedures
			Managers	Population
				Course content
				Delivery and so on

Equipment	Specification Provision Testing and maintenance			
Communication	Internal External	Community Emergency services Media		
Business contingency	Backup services Customer liaison			
Insurances	Personnel Assets Business interruption			

Notice that there are no items in Table 22.2 relating to the testing of the plan or its effectiveness in practice. This is because, for reasons which will be dealt with later in the chapter, it is important to keep data on how well the plan is being implemented separate from data on how effective the plan is in practice.

Once a provisional overall structure for the plan has been established, the preparation of individual targets in the form of questions can begin. However, the detailed work involved in question preparation usually identifies issues which make it desirable to amend the original structure. That is, the advanced planning process, like the advanced risk assessment process, is, as has already been mentioned, an iterative one.

At appropriate levels for each of the headings in Table 22.2, and for the many more headings required in a full emergency plan, clear objectives will have to be established and numerous targets set. In this section, the question-writing process is illustrated using as an example the training of managers in the requirements of fire response procedures. The aim and the objective will be:

Aim   To have managers trained in the procedures to be followed in the event of a fire.
Objective   To have all managers attend a fire response procedures course every two years.

Note the wording of this objective. People who use this planning technique for the first time often fall into the trap of writing objectives in the form 'All managers to be trained in fire response procedures by [a specified date]'. While this sort of objective may be required as an interim measure, it is not adequate for a complete emergency plan since it does not take into account issues such as recruitment and placement. Sometimes objectives with a natural end are referred to as 'project objectives', while those that have to be maintained indefinitely are called 'system objectives'.

Once a clear objective has been established, the next step is to determine a measure of whether or not the objective is being achieved. This is done by defining 'outcome indicators' – that is, the measures by which success will be judged. It is also necessary to ensure that a system can be put in place to collect the data for these outcome indicators, referred to as outcome data.

The recording system is then used to establish the baseline, that is where the organisation is now. Obviously, if the start point is not known, it will not be possible to judge whether things are getting better or worse.

These main stages are shown in Figure 22.1, using the example of training managers in fire response procedures.

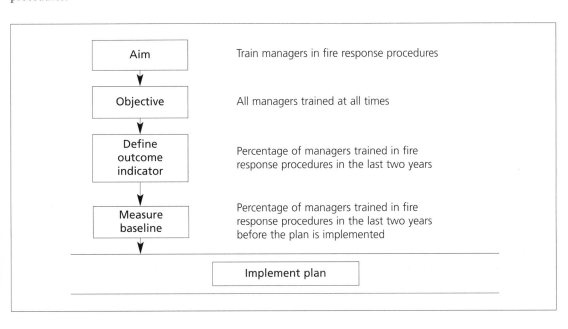

Figure 22.1
Planning the outcome

As far as planning the outcome is concerned, the steps in Figure 22.1 are all that are required. However, for completeness, Figure 22.2 shows the actions required after this part of the emergency plan has been implemented.

Figure 22.2 shows the ongoing procedure for measuring the success of this part of the plan. The nature of the corrective action required will, of course, depend on why the objective is not being achieved – we will deal with how this can be established later in this section.

Returning now to plan preparation, if we have a clear objective, we can then work out what has to be done, or what has to be in place, in order to achieve that objective. This is best done in three stages:
1. Brainstorm the issues to be addressed so that a structure for this part of the plan can be developed. Effectively, this is the same procedure as was used to develop the overall structure of the plan. It is simply being applied at a lower level in the overall structure (see Table 22.2). Continuing with the example of training managers in fire response procedures, the sorts of issues to be addressed would include:

Figure 22.2

Action following the implementation of a plan

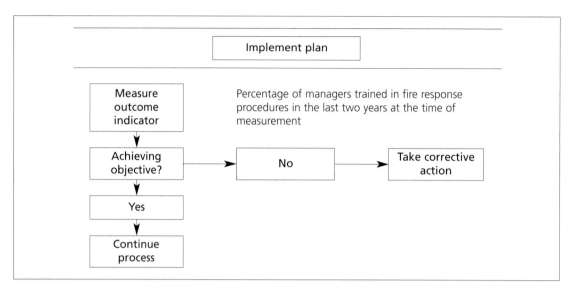

- defining and keeping up-to-date details of the population to be trained
- drawing up the course content, preparing course materials, and reviewing the course content regularly
- implementing and monitoring a system for setting up courses (choosing dates, booking tutors, arranging accommodation and so on).

2. Work out the details required within each main issue. For example, in the first issue listed above, these more detailed matters would include:
- the definitions of 'managers' for the purposes of this training
- generating and keeping up to date a list of these managers, and so on.

3. Convert each detailed issue into a question, or questions, encapsulating the standards to be met or the targets to be achieved.

To illustrate how questions are written, we will consider in detail the first of the issues identified in the example – the population to be trained. Let us assume for this illustration that we need to define 'manager' for the purposes of this training and that it may be necessary to change this definition from time to time.*

If these requirements are to be met, in the future there will have to be a check that the most up-to-date version of the definition is being used throughout the organisation. The check on these requirements could be expressed in the form of a single question, for example: 'Is the most up-to-date version of the definition of a "manager" in use?'

However, this type of question is of limited value for analysis, for several reasons. If someone answers 'no' to the question above, it may be for one of the following reasons:
- there is no definition of a 'manager' available
- a definition is available but it is not up to date
- an up-to-date definition is available but is not being used.

---

* This is a common problem with competence provision. It arises because not all the people with 'manager' in their job title will need the training to be provided, and it is also likely that some people who are not officially designated as 'managers' will need the training.

To increase the diagnostic value of questions it is, therefore, usual to restrict each question to a single aspect of the issue so that the composite question above would be written as three separate questions.

1. Is a definition of 'manager' available?
2. Is the available definition up-to-date?
3. Is the up-to-date definition being used?

When written in this way, the pattern of answers allows us to identify what in particular is going wrong.

Figure 22.3 provides a summary of the procedure for preparing a plan, or part of a plan, in the form of a structured set of questions.

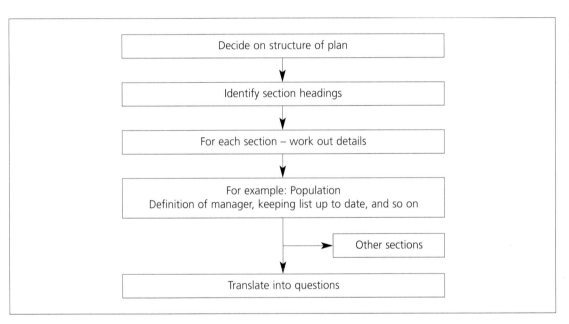

Figure 22.3
Summary of plan preparation procedure

The output of the planning process can be a more or less extensive list of questions, organised in some appropriate structure. Readers who are familiar with any of the commercially available audit packages, such as ISRS, QSA or CHASE, will know that these, too, are structured sets of questions. Although these products are sold as audit tools, in practice they are specifications for HSMSs, since they describe what, in the opinion of the authors of the audit system, an organisation should have in place. The fact that the specification is recorded in the form of questions is primarily for the convenience of the auditors. These audit tools are dealt with in detail in Chapter 23.

An emergency plan produced in the way described in this section will have similar uses as an audit tool, but it will also be useful to individual managers. These managers will use it initially as a specification of what they have to do, and periodically thereafter as a conformity monitoring tool to help them to check that what should be in place is in fact in place. This use of structured sets of questions is dealt with in more detail in the next section.

Using structured sets of questions

Figure 22.3 shows the complete process for producing structured sets of questions, and Figure 22.4 shows how these questions are used for monitoring purposes after the plan has been implemented.

Figure 22.4
Use of question sets
for monitoring

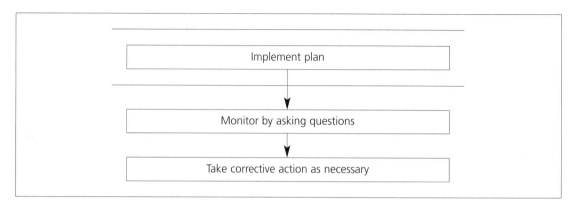

The monitoring shown in Figure 22.4 is conformity monitoring, designed to check whether the plan is being implemented (see Chapter 10). This should not be confused with the monitoring of the outcome data, which is designed to check whether the plan is having the desired effect. The links between the two types of monitoring data are shown in Figure 22.5.

Figure 22.5
Links between
outcome and
monitoring data

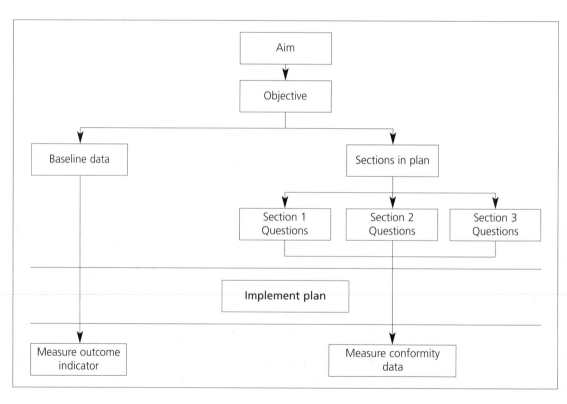

When the outcome data and conformity monitoring data are available, they can be compared in a variety of ways. We will explain why these comparisons are necessary in the following section on testing the plan.

Testing the plan

All plans, including emergency plans, can be tested in a variety of ways, and this section deals with the various tests which should be carried out.

Testing before implementation

Plans are usually prepared by a small group of individuals who, although they should be knowledgeable about the circumstances in which the plan will be used, are unlikely to be omniscient. For this reason, it is advisable to make such tests of the plan as are possible before implementing it. These tests fall into two broad categories:

1. **Tests of individual elements of the plan.** If, as we recommended above, the plan is in the form of a structured set of questions, it is possible to test individual elements. This can be done by, for example, inspecting to find out whether required physical resources are in place. Relevant sections of the plan can also be sent to those individuals and groups who will have to implement them, to obtain their opinions on the practicality of the targets. The overall aim of this part of pre-implementation testing is to check that there are no weak links in the chain.

2. **Tests of the likely effectiveness of the total plan.** The overall aim of this part of pre-implementation testing is to check whether the chain, even if it has no weak links, will serve the intended purpose. That is, if the plan is activated, will it produce the desired results? There are various options for this type of test and these options are dealt with next.

There are two broad categories of methods for testing the activation of emergency plans, which, for convenience of description, we will call 'desktop tests' and simulations. The latter category is best used for testing plans after implementation and will, therefore, be dealt with in the next subsection. However, desktop tests can be used at any time and are particularly useful for testing before implementation.

These desktop tests can take a variety of forms but the examples below, which are in order of increasing rigour, illustrate the main techniques.

1. The complete plan can be sent out to a number of informed individuals with a request for comments on whether, in their view, it would be successful when activated. This approach means that a number of 'fresh eyes', with their associated broader range of experience, have an opportunity to consider the plan, and gaps and overlaps missed by the planners may be identified.

2. A relevant group of individuals can be given copies of the plan to read and then be brought together to discuss it, guided by an appropriate chairperson. This has the advantage of the cross-fertilisation of ideas produced by any well-led group discussion.

3. The sort of group discussion just described can be enhanced in a number of ways. For example, the group could inspect relevant locations, work with scale models of the site, or call in relevant experts or site personnel to obtain detailed information on particular aspects of the plan.

4. An expert in the sorts of advanced risk assessment techniques described in Chapter 20 can lead a group discussion and the plan can be the subject of the sorts of techniques described in the previous two paragraphs. The rationale for this approach is that the plan can be considered as a system and, for example, HAZOP techniques can be used to identify likely causes of system failure and how these can be avoided.

It is useful to test an emergency plan before implementing it, since it is usually easier to change a plan than to change arrangements which are already in place. (Similarly, it is usually easier to change the design for a machine rather than the machine itself.) However, once the plan has been implemented, further tests are necessary and these are discussed next.

Testing after implementation

Three levels of test are required after implementation:

1. Is the plan being implemented effectively?
2. Is the implementation meeting the specified objectives?
3. Does the plan work when activated?

The first two levels of test can be carried out as part of the organisation's normal monitoring activities since they involve collecting relevant outcome data and conformity data. What is required is best illustrated by example, so we will continue with the training of managers in fire response procedures.

Remember that the objective in this example was to have all managers trained in fire response procedures every two years and that a section of the full emergency plan was devoted to the actions required to achieve this. This part of the plan was in the form of a structured set of questions covering all of the relevant targets. Once this part of the plan has been implemented for a suitable period of time, it will be possible to collect the following data:

- **outcome data**, that is the percentage of managers who have been trained in fire response procedures in the last two years
- **conformity data**, that is the extent to which the plan is effectively implemented – for example, the proportion of particular standards being met and the number of individuals who are achieving particular targets.

We mentioned earlier that it is necessary to keep these two types of data separate. This is because separation enables comparisons of the type shown in Table 22.3. From this table, we can see that having separate data sets enables conclusions to be drawn which would not be possible if the two types of data were linked. In simple terms, it is possible to work out why the plan is or is not working, not just whether the plan is working.

These sorts of comparison should be carried out periodically for all of the relevant objectives in the plan, so that there can be a level of assurance that, were the plan to be activated, all of the resources previously identified as being necessary will be in place and able to be used.

Table 22.3
Comparison of outcome data and conformity data

	Outcome indicator showing improvement	Outcome indicator not showing improvement
Plan is being implemented effectively	Plan is working. Maintain activities	Plan is not working. Make alternative plan
Plan is not being implemented effectively	Plan is not relevant to this objective. Resources spent on implementation may be being wasted	Effectiveness of plan cannot be assessed. Implement plan as intended and continue to monitor

The final testing of the plan involves establishing whether the plan is effective when activated, and this is the ultimate test. However, it is also a test which no-one concerned would ever like to see made! What is needed, therefore, is a substitute in the form of a simulation – that is, a 'pretend' emergency.

Most people are familiar with simulations which take the form of fire drills or fire evacuation exercises, where it is assumed that, for the purposes of the simulation, a fire has occurred at a particular location and the evacuation procedure part of the emergency plan has to be activated. This basic idea can be extended to any type of emergency and any part of the plan. For example, explosions, gas releases, floods and business interruptions can all be simulated and, as appropriate, communications, the response of the internal and external emergency services, and the availability of physical resources can all be tested.

The effectiveness of these simulations as tests will, of course, depend on the extent to which they match the circumstances which would occur in a real emergency, but even relatively unrealistic simulations can provide valuable additional data. This section ends with a description of one such simulation to illustrate the use of this technique.

Following the fire at King's Cross Underground station in London in 1987,[82] I was involved in a number of strands of work including fire safety training and the emergency evacuation of stations, particularly subsurface stations.* A range of desktop work was undertaken, but I was not certain that this had identified all of the relevant issues and requested a simulation. This, in brief, took the following form:

1. A subsurface station which was normally closed to the public on Sundays was made available, fully staffed, for the whole of one Sunday.
2. Students from a local university were recruited as 'passengers'.
3. Senior members of the relevant emergency services were briefed and agreed to having their personnel called out without being informed that this was a simulation.
4. Relevant controllers were briefed so that the underground system would not be brought to a halt by 'fires' and people in a station which should have been closed.
5. Various fire and health and safety professionals, as well as senior emergency service personnel, were recruited to act as observers during the simulations and to take part in the debriefings and the analysis of the results.

With these matters in place, several simulations were run, beginning with the station locked so that no-one would enter inadvertently. For this initial simulation, the student 'passengers' wore large numbers front and back (like marathon runners) so that the observers could record who was where at particular times. This enabled a picture of 'passenger' movements to be built up during subsequent analysis. For the later simulations, which were run with the station open, and involved the emergency services, these numbers were removed to increase the verisimilitude of the exercise.

It will be apparent from this brief description that this simulation was extremely resource-intensive and, hence, expensive. However, the data obtained were extremely valuable for making decisions about how to handle future emergencies, and they could not have been obtained through desktop exercises.

The general conclusions were that the more realistic the simulation, the more valuable the data obtained, but that increasing realism, in general, required more resources. For example, opening the station gates increased realism but required additional personnel to ensure that people who entered the station were dealt with appropriately. However, where the emergency plan is dealing with outcomes that could be catastrophic, or where the lessons learned have widespread applications, the resources spent on adequate simulation for test purposes are likely to be justifiable.

UK legislative requirements

In the UK, there are several statutory requirements for emergency planning and a selection is given in the *UK legislation – emergency planning* box overleaf.

---

* Not all of the London Underground network is underground and only some of the stations are underground. Those that are are usually referred to as subsurface stations.

UK legislation – emergency planning

Statutory requirements for emergency planning can be found in:

- the Regulatory Reform (Fire Safety) Order 2005[83]
- the Confined Spaces Regulations 1997,[84] regulation 5
- the Control of Major Accident Hazards Regulations 1999 (as amended),[85] part 4

Summary

This chapter has described the requirements for an emergency plan and the planning process contained in BS 18004. The use of the BS 18004 planning process was illustrated in the context of the development of an emergency plan and the chapter ended with a description of the ways in which plans can be tested.

23: Advanced audit and review

Introduction

This chapter deals with the more advanced techniques of audit required by ISO 19011 and describes in detail the range of techniques used for management reviews. The chapter ends with a brief look at using computers in health and safety.

Overview of the audit process

The principles of auditing described in this section are those set out in BS EN ISO 19011:2011 *Guidelines for auditing management systems*.[34] It is important to make a distinction between HSMS auditing and typical traditional 'safety' audits, which have the following characteristics:

- They are a check on compliance with health and safety legislation. This means that key management activities, such as monitoring and review, are not audited, unless the organisation is operating under a safety case regime.*
- They are carried out by health and safety professionals who have not been formally trained in audit techniques.
- They are restricted to inspections with, perhaps, some questioning of people encountered during inspections. This means that there are no formal interviews and no formal checks on documents and records.

HSMS audits go beyond these typical traditional safety audits because they include the management areas not covered by health and safety legislation and use a wider range of techniques for collecting audit evidence.

For the majority of health and safety professionals, HSMS auditing is likely to be an unfamiliar process that will require time and effort to master. However, since there is an International Standard on management system auditing, it is important that health and safety professionals are able to comply with it. If they do not, they will be falling behind their colleagues in the quality and environmental professions.

The section begins with an overview of the ISO 19011 audit process in the form of a flowchart and the rest of this section deals with each stage in more detail. Note the following points about the flowchart:

- The quotations are from ISO 19011 unless otherwise stated.
- In ISO 19011 terminology, audits are initiated by the audit client, who is the 'organization or person requesting an audit'.
- Audits can be conducted by a single auditor or a team of auditors. Where a team of auditors is involved, there must be an audit team leader. Audit team leaders are allocated certain functions in ISO 19011 and an auditor working alone has to fulfil these functions. However, some functions, such as allocating tasks to members of the audit team, are not relevant to auditors working alone. Since this section is intended for auditors working alone, the flowchart gives audit team leader functions for completeness but, where appropriate, these are accompanied by a note to the effect that they are not dealt with further in this section.

These days, auditors commonly use specially designed audit software to facilitate their work. The final part of this section deals with audit software.

---

* The term 'safety case regime' is used to describe organisations that are legally required to have an HSMS, for example those operating under the COMAH Regulations.[85]

Figure 23.1

Flowchart of the ISO 19011 process

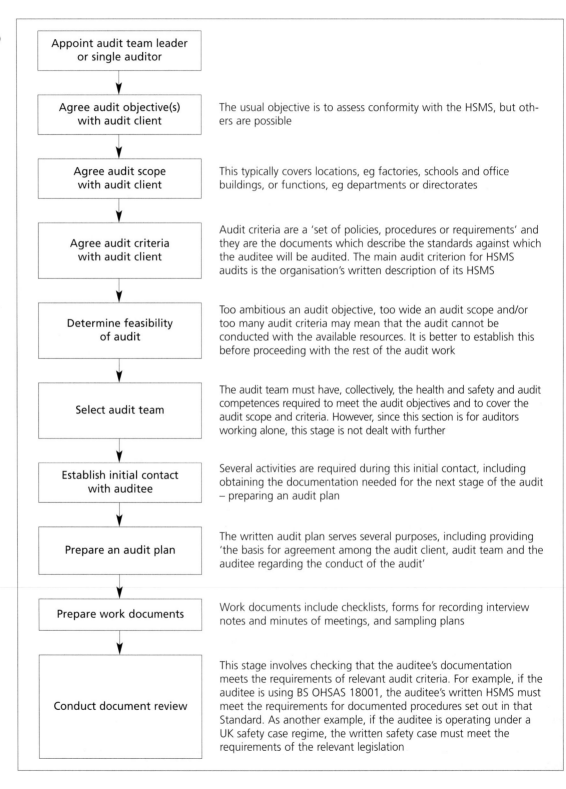

Appoint audit team leader or single auditor	
Agree audit objective(s) with audit client	The usual objective is to assess conformity with the HSMS, but others are possible
Agree audit scope with audit client	This typically covers locations, eg factories, schools and office buildings, or functions, eg departments or directorates
Agree audit criteria with audit client	Audit criteria are a 'set of policies, procedures or requirements' and they are the documents which describe the standards against which the auditee will be audited. The main audit criterion for HSMS audits is the organisation's written description of its HSMS
Determine feasibility of audit	Too ambitious an audit objective, too wide an audit scope and/or too many audit criteria may mean that the audit cannot be conducted with the available resources. It is better to establish this before proceeding with the rest of the audit work
Select audit team	The audit team must have, collectively, the health and safety and audit competences required to meet the audit objectives and to cover the audit scope and criteria. However, since this section is for auditors working alone, this stage is not dealt with further
Establish initial contact with auditee	Several activities are required during this initial contact, including obtaining the documentation needed for the next stage of the audit – preparing an audit plan
Prepare an audit plan	The written audit plan serves several purposes, including providing 'the basis for agreement among the audit client, audit team and the auditee regarding the conduct of the audit'
Prepare work documents	Work documents include checklists, forms for recording interview notes and minutes of meetings, and sampling plans
Conduct document review	This stage involves checking that the auditee's documentation meets the requirements of relevant audit criteria. For example, if the auditee is using BS OHSAS 18001, the auditee's written HSMS must meet the requirements for documented procedures set out in that Standard. As another example, if the auditee is operating under a UK safety case regime, the written safety case must meet the requirements of the relevant legislation

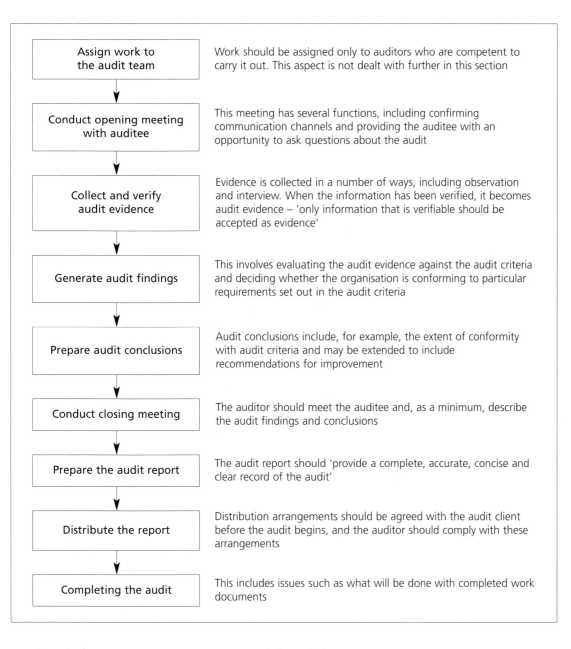

Figure 23.1
continued

Assign work to the audit team	Work should be assigned only to auditors who are competent to carry it out. This aspect is not dealt with further in this section
Conduct opening meeting with auditee	This meeting has several functions, including confirming communication channels and providing the auditee with an opportunity to ask questions about the audit
Collect and verify audit evidence	Evidence is collected in a number of ways, including observation and interview. When the information has been verified, it becomes audit evidence – 'only information that is verifiable should be accepted as evidence'
Generate audit findings	This involves evaluating the audit evidence against the audit criteria and deciding whether the organisation is conforming to particular requirements set out in the audit criteria
Prepare audit conclusions	Audit conclusions include, for example, the extent of conformity with audit criteria and may be extended to include recommendations for improvement
Conduct closing meeting	The auditor should meet the auditee and, as a minimum, describe the audit findings and conclusions
Prepare the audit report	The audit report should 'provide a complete, accurate, concise and clear record of the audit'
Distribute the report	Distribution arrangements should be agreed with the audit client before the audit begins, and the auditor should comply with these arrangements
Completing the audit	This includes issues such as what will be done with completed work documents

Audit objectives, scope, criteria and feasibility

Taken together, the audit objectives, scope and criteria define what has to be done to complete the audit. The audit objectives are normally suggested by the audit client, and they, along with the scope and criteria, are agreed between the client and the auditor. Objectives, scope and criteria must all be documented and any changes must also be agreed and documented.

A clear, documented description of the audit objectives, scope and criteria enables decisions on the following:

- **requirements for the audit team,** both in terms of numbers and competences
- **other resources required,** including personnel such as technical experts and guides, and other resources such as offices and PPE
- **the feasibility of the audit** – if the audit as specified is not feasible, and cannot be called off, changes have to be made to one of more of the specifications in the objectives, scope or criteria.

Audit objectives

The audit objectives define what the audit is supposed to accomplish. A wide range of objectives is possible, including:

- conformity with one or more pieces of health and safety legislation
- conformity with one or more health and safety Approved Codes of Practice (ACoPs)
- conformity with one or more British Standards relevant to health and safety
- conformity with one or more voluntarily adopted health and safety standards
- conformity with one or more internal health and safety procedures
- conformity with the requirements of a published HSMS (for example, BS OHSAS 18001, ILO 2001 or HSG65)
- conformity with the requirements of an in-house HSMS
- conformity with contractual requirements (are we meeting our customers' health and safety requirements?)
- conformity with contractual requirements (are contractors meeting the employing organisation's health and safety requirements?)
- conformity with contractual requirements (are suppliers meeting the organisation's health and safety requirements?)
- effectiveness of the HSMS in meeting health and safety objectives
- checking for continual improvement in the HSMS
- identify ways of improving the HSMS.

The widest of these objectives is conformity with BS OHSAS 18001, since this covers all relevant health and safety legislation, other health and safety requirements such as contractual requirements, and many management issues not covered by UK health and safety legislation. Where examples are given later in the section, they will be based on audits with the objective of checking conformity with the requirements of BS OHSAS 18001.

Audit scope

The audit scope describes the extent and boundaries of the audit. Typical issues taken into account in defining scope include:

- **physical locations** – usually documented with a map, sketch map or site plan
- **organisational units** – usually documented with organisation charts, and supplemented with a listing of the job titles and numbers of the various categories of people making up the organisational unit
- **activities** – these have to be identified so that relevant audit criteria can be selected; the activities can be documented simply by listing them
- **processes** – these are a linked list of activities, and can be a convenient shorthand way of recording activities
- **the time period covered by the audit.**

Audit criteria

The audit criteria are the 'reference against which audit evidence is compared'. Because they involve different types of work by auditors, it is useful to divide audit criteria into external and internal types.

External audit criteria

These include:

- **international and British Standards, guidance and specifications**, such as HSMSs, (for example HSG65, BS 18004 and BS OHSAS 18001), standards for documentation (for example ISO/TR 10013) and, for some audits, hardware standards such as those for machinery guarding
- **health and safety legislation** – what constitutes applicable legislation will, obviously, depend on the audit objectives and scope
- **industry codes of conduct** – these are normally voluntary and it is necessary to establish whether the auditee has agreed to conform to the code of conduct concerned.

Because external audit criteria do not vary from organisation to organisation, it is possible to prepare checklists which are applicable to all organisations. This has led to health and safety consultancies developing and selling checklists for auditing conformity with health and safety legislation and HSMSs. These checklists are dealt with in more detail later in the section.

Internal audit criteria

These are documents describing the standards the organisation sets for itself. As was mentioned earlier, HSMS audits are not like traditional safety audits; instead, they check whether an organisation is doing what it said it would do.

In order to carry out this type of audit, it is necessary to know what the organisation said it would do and, realistically, this is restricted to what the organisation has put in writing. The organisation's written procedures are, therefore, the internal audit criteria.

The internal audit criteria have to be the starting point for an HSMS audit. If an organisation does not have a written HSMS, it cannot be audited. Other types of audit, such as audits of compliance with health and safety legislation or with British Standards, are possible because these are based on written external criteria.

A typical finding when starting an HSMS audit is that the organisation has written guidance on how it will comply with various items of health and safety legislation but no written procedures on, for example, audit, monitoring, measuring or review.

Determining feasibility

The audit objectives, scope and criteria determine what has to be done and from these it is possible to estimate the resources required. The main resource requirements are usually auditors' time and travel and accommodation expenses. However, in some circumstances it may be necessary to take into account auditees' time.

Where the estimates suggest that the audit is not feasible, and the audit is not to be called off, one or more of the following can be done:

- Agree a less ambitious objective with the audit client. For example, audit only certain elements of the HSMS rather than the complete management system.
- Agree a smaller scope with the audit client. For example, visit fewer sites or departments.
- Agree a smaller range of audit criteria with the audit client. For example, audit only selected items of health and safety legislation.

It is important to remember that all auditing is done on a sample basis, and that all of the above examples are simply reducing the sample for the audit. Effectively, successively smaller samples are drawn until a size is reached which can be audited with the resources available. It is important, however, to ensure that the sample is representative, and this is dealt with later in the section.

Establishing initial contact with the auditee

ISO 19011 allows the initial contact with the auditee to be informal or formal. The purposes of the initial contact are as follows:

- **To establish communication channels with the auditee's representatives.** These days, this is usually done by swapping mobile phone numbers.
- **To confirm the authority to conduct the audit.** This is usually done by informing the auditee of the identity of the audit client.
- **To provide information on the audit objectives, scope, methods and audit team composition, including technical experts.**
- **To request access to relevant documents and records for planning purposes.** It is particularly important to get a copy of the auditee's written HSMS since this is usually reviewed before on-site audit activities begin.
- **To determine applicable legal and contractual requirements and other requirements relevant to the activities and products of the auditee.**
- **To confirm the agreement with the auditee regarding the extent of the disclosure and the treatment of confidential information.**
- **To make arrangements for the audit, including scheduling the dates.**
- **To determine any location-specific requirements for access, security, health and safety or other.**
- **To agree on the attendance of observers and the need for guides for the audit team.** Observers are people the auditee nominates to observe what the auditors do. Neither observers nor guides should take part in the auditing work. For auditors working in their own organisation, these arrangements may not be relevant.
- **To determine any areas of interest or concern to the auditee in relation to the specific audit.**

In practice, the initial contact with the auditee is usually via a series of phone calls and emails, which continue until the auditor has all the required information and the auditee has had all his or her queries dealt with.

Maintaining an audit plan

ISO 19011 requires that auditors prepare, and keep up to date, an audit plan. In this section, this process is referred to as maintaining an audit plan.

The information required to maintain an audit plan has to be collected from different people, at different places and at different times, but a draft audit plan has to be available before the opening meeting for the site work so that its contents can be discussed with the auditee.

Contents of an audit plan

ISO 19011 gives the following as the contents of an audit plan:

- the audit objectives
- the audit scope, including identification of the organizational and functional units, as well as processes to be audited
- the audit criteria and any reference documents
- the locations, dates, expected time and duration of audit activities to be conducted, including meetings with the auditee's management
- the audit method to be used including the extent to which audit sampling is needed to obtain sufficient audit evidence and the design of the sampling plan, if applicable
- the roles and responsibilities of the audit team members, as well as guides and observers
- the allocation of appropriate resources to critical areas of the audit.

The following should also be included, as appropriate:
- the identification of the auditee's representative for the audit
- the working and reporting language of the audit where this is different from the language of the auditor or the auditee, or both
- the audit report topics
- logistics and communications arrangements, including specific arrangements for the locations to be audited
- any specific measures taken to address the effect of uncertainty on achieving the audit objectives
- matters related to confidentiality and information security
- any follow-up actions, for example, from a previous audit
- any follow-up activities to the planned audit
- co-ordination with other audit activities, in the case of a joint audit.

Keeping the audit plan up to date

Auditing is a highly iterative process. That is, the work carried out at any given stage may make it necessary to go back to an earlier stage and alter what had been decided at that earlier stage. For example:
- discussions with the auditee during the opening meeting may show that the sample of personnel selected for interview is no longer appropriate
- initial observation of the site may result in a change of route for the site inspections
- document examination may take longer than expected, leading to a decision to inspect fewer records so that the audit can be completed in the allocated time.

Irrespective of the reason for, or the nature of, the change, the changes should be agreed with the audit client and/or the auditee as appropriate and the audit plan should be amended in a timely manner to ensure that it is up to date and gives an accurate description of what is currently planned.

Preparing work documents

The term 'work document' is used in ISO 19011 to describe the checklists and forms used for reference and to record audit proceedings, and the sampling plans used to demonstrate that appropriate samples have been chosen.

Various types of work document are required during an audit and the following are dealt with in this section:
- checklists for reviewing documentation
- checklists for auditing the implementation of procedures
- checklists for physical conditions
- forms to accompany checklists
- forms for recording interview results
- forms for recording meetings
- sampling plans.

Some of these work documents are available commercially and some can be prepared as soon as the audit criteria are agreed. However, others can only be prepared when details of the site to be audited are known. These details should be collected during the initial contact with the auditee. Nevertheless, for convenience, all of these types of work document are dealt with together.

Checklists for reviewing documentation

While it is possible to review documentation directly from an audit criterion, it is more effective to convert the contents of the audit criterion into a checklist. Using a checklist makes it less likely that individual requirements will be overlooked and it provides the opportunity for numeric results from audits.

For a full audit of conformity with BS OHSAS 18001 documentation requirements, the following checklists will be required:

- a checklist setting out the BS OHSAS 18001 requirements for documentation
- a checklist setting out the health and safety legislative requirements for documentation – for example, the written arrangements for protective and preventive measures required by the Management of Health and Safety at Work Regulations 1999[13]
- a checklist setting out any internal requirements for documentation.

Checklists for auditing implementation of procedures

BS OHSAS 18001 requires organisations to have procedures, and in most organisations with a proper HSMS there will a written description of how they carry out such activities as internal audit, monitoring and review. When preparing checklists for auditing the implementation of procedures, auditors should go through the relevant written procedures, identify all the standards which can be audited, and record these standards as questions as described in Chapter 22. The complete set of questions can then be used as a checklist to audit implementation of the procedure.

In organisations that are not already using BS OHSAS 18001, their written procedures are likely to contain a mixture of standards, which can be audited, and guidance and other material, which cannot.

Checklists for physical conditions

These checklists are primarily used during site inspections. Most health and safety professionals are familiar with these checklists, since they are used widely to check, for example, fire safety arrangements.

Physical conditions checklists are commercially available for legal requirements, such as those contained in the Health and Safety (Display Screen Equipment) Regulations[17] and the Workplace (Health, Safety and Welfare) Regulations.[70] However, auditors have to prepare their own checklists when the organisation sets its own standards for physical conditions.

Forms to accompany checklists

It is useful to have a form as a preliminary to checklists to remind auditors to record relevant information and as a convenient way to record the information.

For checklists for auditing documentation, the form should contain spaces for such things as the title, version, issue date and so on of the document which was audited and the date of audit.

For checklists for implementation of procedures, the form should contain spaces for the name of the procedure being audited and its version number and date of issue, details of inspections, checks and interviews used to collect audit evidence, and the dates and times of collection of audit evidence.

For checklists for physical conditions, the form should include spaces for the name of the area, its physical characteristics and the activities taking place, the date, time and duration of the inspection, and the names of any technical experts or guides involved.

All these types of form should include suitable identification of the auditor, audit client and auditee.

Forms for recording interviews

Interviews are a major source of information during audits and it is important that the results are recorded. Using an appropriate form makes it more likely that the required questions are asked and that the results are recorded.

A typical form would include spaces for:

- suitable identification of the audit client and auditee
- date, time and duration of the interview
- name of the interviewer
- name and job title of the interviewee.

There would then be a list of questions to be asked, with a suitable space for recording the answers.

It is important that the questions to be asked are properly phrased since closed questions (questions which can be answered with 'yes' or 'no') are of limited value. It is better to convert the closed questions used in checklists to open questions for interviews. This is illustrated in Table 23.1 for two questions on the health and safety policy.

Closed question for checklist	Open question for interview
Have the contents of the health and safety been communicated to all employees?	What are the contents of the health and safety policy?
Are all employees aware of their individual health and safety obligations?	What are your individual health and safety obligations?

Table 23.1
Closed and open questions

In a full audit of conformity with BS OHSAS 18001, interview forms are likely to be required for:
- different levels of employee, from operatives to top management
- people with specific health and safety roles – for example, the management appointee and the health and safety adviser
- people with specific health and safety responsibilities – for example, risk assessment and monitoring
- people exposed to specific health and safety risks.

Forms for recording meetings

Minutes should be taken at meetings and an appropriate form helps to ensure that the required information is collected.

ISO 19011 has requirements for meetings and the layout for forms for meetings should provide a reminder of what these requirements are, together with sufficient space to record the relevant information. The ISO 19011 requirements for opening and closing meetings are given later.

Sampling plans

ISO 19011 points out that audit evidence is:

> ... based on samples of the information available, since an audit is conducted during a finite period of time and with finite resources. An appropriate use of sampling should be applied, since this is closely related to the confidence that can be placed in the audit conclusions.

Sampling techniques should be used for:
- interviews – for example, a sample of managers at various levels, supervisors and operatives within all relevant functions
- documents – while all HSMS documents may have to be audited, the use of documents on site will be audited on a sample basis
- records – a sample of records such as incident reports and risk assessments should be audited
- inspections – a sample of sites and activities should be visited.

Brief guidance on sampling is given in Annex C of ISO 19011, but the sampling techniques which can be used are discussed in more detail in what follows. When a sample is to be used, it must be adequate if the audit is to be sufficiently searching. Using the information obtained during the discussions required to decide on the audit objectives, scope and criteria, and during the initial contact with the auditee, auditors should draw up a provisional sampling scheme of, for example, what is to be inspected and who is to be interviewed.

The single most important part of any sampling technique, and the part which usually takes the most time, is determining the population to be sampled. Unless the auditor has comprehensively defined the population, he or she has no way of demonstrating that the sample drawn is an appropriate one. Key areas to be addressed in drawing up this definition include:

- **who** – employees are usually relatively easy to define but others who may have to be included will be contractors and visitors
- **where** – fixed sites are usually relatively easy to define but other areas which may cause problems include temporary accommodation, locations which do not fit within core activities (playing fields, social clubs, off-site storage) and sites owned by other companies but visited by the organisation's employees for work activities
- **what** – this should include all major plant, equipment, machinery and vehicles. It should also include all the major activities carried out by the organisation
- **when** – when is particularly important for organisations which operate shift systems, since all shifts will have to be represented in the sample.

For all of the people, places, things, activities and times the population description should include the appropriate management infrastructure. It should be possible, for example, to identify, for a particular machine, who is responsible for its safe operation and maintenance.

Sampling criteria are the bases on which populations are divided, and typical sampling criteria include the following:

- **Management responsibility.** Samples should be drawn which include all the main lines of responsibility from the senior person in the area being audited down to the 'shop floor'. As the audit is, primarily, a systems audit, this is normally the most important factor to consider.
- **Core activities and their timings.** Each of the major activities should be represented in the sample and a check should be made that all levels of risk are represented. When timings vary, such as when shifts are used, timing is also a criterion.
- **Geographical area.** All relevant geographical areas should be represented.
- **Previous health and safety performance.** Part of the audit is to assess the current health and safety performance of the area being audited. However, during the planning stages, it is necessary to consider past health and safety performance. The required information should be obtained from previous audits or from other information if audit results are not available.

Note the correspondence between population descriptors and sample criteria. In general, any way in which the population can be divided is a potential sampling criterion.

The ideal sample is a random one, since the data from this type of sample are appropriate for statistical techniques such as determining confidence limits. However, random samples are only appropriate where there are relatively large, homogeneous populations. Some examples of the use of random samples are given here.

- **Conformity with PPE requirements – for example hard hats on construction sites.** If we know that there are 75 people on site, we can observe a random sample of 15 and draw conclusions as to the rate of compliance of the whole population.
- **Knowledge of the health and safety policy.** If we know that there are 200 people who should be familiar with the health and safety policy, we can question a random sample of 20 and draw conclusions as to the knowledge in the whole population.

- **Understanding of hazards associated with a chemical.** If we know that there are 20 people on site who should understand the hazards associated with a particular chemical, we can ask a random sample of 10 about these hazards and draw conclusions as to the level of understanding in the whole population.

There are various methods for choosing random samples, including tossing coins or throwing dice, drawing names written on slips of paper from a suitable container, and selecting every 10th (or other appropriate number) name from a list.

In practice, the use of random sampling alone is rarely adequate and probably the most used sampling method for the majority of audit applications is stratified random sampling. This procedure can be illustrated by the following example.

Stratified random sampling – example

The first step is to list the population in a way which reflects the sampling criteria, which in this example are 'manager', 'activity' and 'age of site'. These criteria are then arranged in order of importance, starting from the left. This is the stratification part of the process and the results are shown in Table 23.2 where, for illustration, it is assumed that there are three managers (1, 2 and 3) each with 10 sites under their control, and that these sites carry out three different activities (A, B and C) and can be new or old sites.

Manager	Activity	Sites	Sample
Manager 1	Activity A	New – 1 site	
		Old – 2 sites	
	Activity B	New – 1 site	
	Activity C	New – 2 sites	
		Old – 4 sites	
Manager 2	Activity A	New – 2 sites	
		Old – 1 site	
	Activity B	New – 1 site	
	Activity C	New – 2 sites	
		Old – 4 sites	
Manager 3	Activity A	New – 1 site	
	Activity B	New – 4 sites	
		Old – 1 site	
	Activity C	Old – 4 sites	

Table 23.2
Table for preparing a stratified random sample

When the population has been stratified in this way, a decision is then made on the size of the sample – say a 50 per cent sample for illustrative purposes. A 50 per cent sample would be 15 sites; since there are 14 different categories in the stratification, the obvious approach is to choose one site from each category. Where there is more than one site in a category, the site to be audited should be chosen randomly; using a die would be a convenient way of doing this since the numbers are small. The 15th site to be audited can also be chosen randomly from the sites not already selected.

When auditors have drawn their sample in this way, they should look at the overall sample and check that it matches the overall criteria. For example, there are more sites carrying out activity C than the other activities and this should be reflected in the overall sample. If, for any reason, the sample is not representative, all, or relevant parts, of the sample should be reselected.

Document review

The auditee's documented HSMS should be reviewed to determine whether it conforms with the requirements of the relevant audit criteria. For example, if the auditee is claiming conformity with BS OHSAS 18001, the auditee's documentation should be reviewed against the relevant requirements of BS OHSAS 18001.

Document review is normally done off site because this is usually more convenient. However, it may require a preliminary site visit to establish an overview of, and collect copies of, the available information. Many document reviews these days are, however, done using electronic copies of the documents emailed to the auditor by the auditee as part of the initial contact.

Document review is done early in the audit sequence so that the audit client can be told about nonconforming documentation. This is particularly relevant for certification auditing where clients may not wish to pay for site work when they know that they cannot achieve certification because of nonconforming documentation. Advising the clients about such documentation before the site work enables them to make a decision on whether or not to continue with the rest of the audit.

Document review is most conveniently done in two stages:
* checking whether all of the required documents are available
* checking whether the available documents meet the relevant requirements.

Each of these steps is dealt with below in more detail.

Checking required documents

Management systems such as BS OHSAS 18001 require organisations to have documentation of various types. The first check in document review is on whether all the required documents exist. The full list of documentation required by BS OHSAS 18001 is:
* the OH&S policy and objectives
* description of the scope of the OH&S management system
* description of the main elements of the OH&S management system and their interaction, and reference to related documents
* documents, including records, required by this OHSAS Standard
* documents, including records, determined by the organization to be necessary to ensure the effective planning, operation and control of processes that relate to the management of its OH&S risks.

Except in health and safety legislation that requires a safety case, there are very few requirements for written procedures in UK health and safety legislation. Although the Management of Health and Safety at Work Regulations 1999[13] require employers to carry out risk assessments (regulation 3) there is no requirement to have a written risk assessment procedure. In contrast, regulation 5, for employers with five or more employees, does require the recording of arrangements, but only in regard to 'preventive and protective measures'.

The output from this part of the document review is a list of the required documents with an indication of which of them the auditee has and does not have. A checklist is usually the most convenient means of recording this output.

Checking adequacy of documents

Where an audit criterion states that a procedure is required, it usually goes on to specify minimum contents. Here are two examples to illustrate this:

- BS OHSAS 18001[3] requires that organisations have an internal audit procedure and that this procedure must address 'a) the responsibilities, competencies, and requirements for planning and conducting audits, reporting results and retaining associated records; and b) the determination of audit criteria, scope, frequency, and methods.'
- The requirement to record arrangements in regard to 'preventive and protective measures' in the Management of Health and Safety at Work Regulations must cover 'planning, organisation, control, monitoring and review'.

When checking whether a document meets the requirements of an audit criterion, it is easier to work from checklists – we dealt with preparing suitable checklists earlier.

The output from this part of the document review is one or more completed checklists showing the extent to which the auditee's documentation conforms with the requirements of the relevant audit criteria. Although the term is not used in ISO 19011, document review is sometimes referred to as document gap analysis.

Opening meeting

ISO 19011 requires that an opening meeting is held with 'the auditee management and, where appropriate, those responsible for the functions or processes to be audited'. The relevant material from ISO 19011 is given below.

The purpose of the opening meeting is to confirm the agreement of all parties (eg auditee, audit team) to the audit plan, introduce the audit team, and ensure that all planned audit activities can be performed.

In many instances, for example internal audits in a small organisation, the opening meeting may simply consist of communicating that an audit is being conducted and explaining the nature of the audit.

For other audit situations, the meeting may be formal and records of the attendance should be kept. The meeting should be chaired by the audit team leader, and the following items should be considered, as appropriate:

- introduction of the participants, including observers and guides, and an outline of their roles
- confirmation of the audit objectives, scope and criteria
- confirmation of the audit plan and other relevant arrangements with the auditee, such as the date and time for the closing meeting, any interim meetings between the audit team and the auditee's management, and any late changes
- presentation of the methods to be used to conduct the audit, including advising the auditee that the audit evidence will be based on a sample of the information available
- introduction of methods to manage risks to the organisation which may result from the presence of the audit team members
- confirmation of formal communication channels between the audit team and the auditee
- confirmation of the language to be used during the audit
- confirmation that, during the audit, the auditee will be kept informed of audit progress
- confirmation that the resources and facilities needed by the audit team are available

- confirmation of matters relating to confidentiality and information security
- confirmation of relevant health and safety, emergency and security procedures for the audit team
- information on the method of reporting audit findings including any grading, if any
- information about conditions under which the audit may be terminated
- information about the closing meeting
- information about how to deal with possible findings during the audit
- information about any system for feedback from the auditee on the findings or conclusions of the audit, including complaints or appeals.

Note that in some circumstances, for example internal audits in a small organisation, the opening meeting may simply consist of communicating that an audit is being conducted and explaining the nature of the audit.

Collecting and verifying audit evidence

Audit evidence includes 'records, statements of fact or other information, which are relevant to the audit criteria'. It can be qualitative or quantitative but it must be verifiable. Audit evidence can be collected using a number of techniques but the main techniques are by interviewing people, by checking documents and records and by observing locations and activities. All of these techniques were dealt with in Part 1.1.

Preparing audit findings and conclusions

Audit findings are the 'results of the evaluation of the collected audit evidence against audit criteria', while the audit conclusion is the 'outcome of an audit, after consideration of the audit objectives and all audit findings'. Each of these topics is dealt with separately below.

Audit findings
If checklists containing the requirements set out in audit criteria are being used, the completion of these checklists generates the audit findings since 'yes' answers indicate a conformity and 'no' answers a nonconformity. The audit findings recorded in this way can also be used to meet the following ISO 19011 requirements for audit findings:
- they can be used to summarise conformity with audit criteria
- they can be used as a list of the nonconformities which will have to be reviewed with the auditee to obtain acknowledgement that the audit evidence is accurate and that the nonconformities are understood. Every attempt should be made to resolve any diverging opinions concerning the audit evidence and/or findings, and unresolved points should be recorded.

Although it is an optional requirement in ISO 19011, the checklists and any associated guidance can also be used to indicate to the auditee opportunities for improvement.

Audit conclusion
ISO 19011 makes the following points with regard to audit conclusions:

> Audit conclusions can address issues such as
> - the extent of conformity with the audit criteria and robustness of the management system, including the effectiveness of the management system in meeting the stated objectives;
> - the effective implementation, maintenance and improvement of the management system

- the capability of the management review process to ensure the continuing suitability, adequacy, effectiveness and improvement of the management system;
- achievement of audit objectives, coverage of audit scope, and fulfilment of audit criteria;
- root causes of findings, if included in the audit plan;
- similar findings made in different areas that were audited for the purpose of identifying trends.

If specified by the audit objectives, audit conclusions can lead to recommendations for improvements or future auditing activities.

Closing meeting

The closing meeting is held to present the audit findings and audit conclusions in a way that can be understood and acknowledged by the auditee. If appropriate, the closing meeting is also used to agree a time frame for the auditee to present a corrective and preventive action plan.

The closing meeting should be chaired by the audit team leader and attended by the auditee and, if appropriate, the audit client and other parties. The meeting should be formal and minutes, including records of attendance, should be kept. In addition to presenting the audit findings and audit conclusions, the following should be explained, as appropriate:

- advising that the audit evidence collected was based on a sample of the information available
- the method of reporting
- the process of handling of audit findings and possible consequences
- presentation of the audit findings in such a manner that they are understood and acknowledged by the auditee's management
- any related post-audit activities (eg implementation of corrective actions, audit complaint handling, appeal process).

Any diverging opinions regarding the audit findings or conclusions between the audit team and the auditee should be discussed and, if possible, resolved. If not resolved, this should be recorded.

If specified by the audit objectives, recommendations for improvements may be presented but if they are it should be emphasised that they are not binding.

Note that in many instances, for example internal audits in small organisations, the closing meeting may consist of just communicating the audit findings and conclusions.

Preparing the audit report and completing the audit

The audit team leader is responsible for the preparation of the audit report and the content of this report and its approval and distribution are dealt with next, followed by the management of audit documents once the audit has been completed.

Content of audit report

The audit report should provide a complete, accurate, concise and clear record of the audit. The following should be included, or referred to:

- the audit objectives
- the audit scope, particularly identification of the organisational and functional units or processes audited

- identification of the audit client
- identification of audit team and auditee's participants in the audit
- the dates and locations where the audit activities were conducted
- the audit criteria
- the audit findings and related evidence
- the audit conclusions
- a statement on the degree to which the audit criteria have been fulfilled.

In addition, the following may be included or referred to, if appropriate:

- the audit plan, including time schedule
- a summary of the audit process, including any obstacles encountered that may decrease the reliability of the audit conclusions
- confirmation if the audit objectives have been achieved within the audit scope, in accordance with the audit plan
- any areas within the audit scope not covered
- a summary covering the audit conclusions and the main audit findings that support them
- any unresolved diverging opinions between the audit team and the auditee
- opportunities for improvement, if specified in the audit plan
- good practices identified
- agreed follow-up action plans, if any
- a statement of the confidential nature of the contents
- any implications for the audit programme or subsequent audits
- the distribution list for the audit report.

Distribution

The audit report should be issued within the agreed time period. However, if this is not possible for any reason, the audit client should be informed of the reasons for the delay. The audit report should be dated and, where review and approval arrangements have been agreed, these arrangements should be followed. Copies of the report should be sent to the recipients, as defined in the audit plan.

Completing the audit

The audit is completed when all activities described in the audit plan have been carried out and the approved audit report has been distributed. However, the documents used and generated during the audit have to be managed. ISO 19011 specifies the following:

> Documents pertaining to the audit should be retained or destroyed by agreement between the participating parties and in accordance with audit programme procedures and applicable requirements.
>
> Unless required by law, the audit team and the person managing the audit programme should not disclose the contents of documents, any other information obtained during the audit, or the audit report, to any other party without the explicit approval of the audit client and, where appropriate, the approval of the auditee. If disclosure of the contents of an audit document is required, the audit client and auditee should be informed as soon as possible.
>
> Lessons learned from the audit should be entered into the continual improvement process of the management system of the audited organisation.

Audit software

A number of organisations sell computer software for use by auditors. Using appropriate computer software helps with the following:

- **Preparing and maintaining checklists.** The software is designed to deal with information in the form of structured sets of questions and has many facilities not available in, for example, word processors or spreadsheets.
- **Recording and analysing results.** Although it is usually more convenient to use paper recording during inspections and interviews, the results of document reviews and audits can be entered directly into the software. In addition, the software is designed to analyse the results in a variety of ways.
- **Providing numerical results.** The software can produce a range of numerical results which it would be prohibitively time-consuming to produce manually.
- **Preparing reports.** The software has a number of features which will help in the preparation of reports.

Review

Types of review

By examining carefully the main HSMSs' treatment of the term 'review', we can identify several types of review:

- The initial status review. This appears as a formal element only in the BS 18004 model but, for many organisations, it would be a useful exercise. However, since, by definition, it only happens once, it is not properly part of a sustained HSMS and this type of review will not be discussed further.
- Management reviews, which take on different terminology in the different HSMSs. It is possible to subdivide these reviews into two main types:
  - Reactive reviews – in other words, reviews that are conducted in response to something going wrong. These occur only in HSG65 and, as it is not clear how they differ from a detailed investigation with recommendations for remedial action, they will not be considered further during this chapter.
  - Proactive reviews, which are planned and carried out to a schedule.

The proactive reviews can be further subdivided by the level of management involved. BS OHSAS 18001 is the only HSMS which restricts review to top management, but since this is arguably the most important review in an organisation, this would appear to be acceptable. In practice, the techniques of proactive review will be the same irrespective of the management level involved. It is these management proactive reviews which will be considered in the remainder of this chapter, and they will be referred to simply as management reviews.

Management reviews

In Part 1.1, various functions of review were identified as follows:

1. checks that the individual elements of the HSMS (such as incident investigation and risk assessment) are functioning as intended
2. checks that the HSMS as a whole is functioning as intended – for example that data are being passed from one element to another
3. data analysis that takes into account the output information from all elements of the HSMS, such as comparing loss data and conformity monitoring data to check that risk control measures are having the intended effect
4. looking for ways in which what is being done now can be done more efficiently or more effectively

5. checking that what is being done now is still appropriate (that is nothing is being done which is not required) and adequate (that is everything that is required is being done).

Each of these functions is dealt with in more detail in the remainder of this section.

Checking individual HSMS elements

As we mentioned earlier, most processes and systems deteriorate with time and the processes making up the individual elements of the HSMS are no exception to this. For this reason, a necessary function of review is to check that each element of the HSMS is functioning as intended. It is worthwhile making this the first step in the review process, since any failures identified at this stage are likely to have implications for the other functions of review. For example, if it is found that some loss data are not being reported, or not being recorded adequately, this will have to be taken into account in interpreting the results of any comparative analyses undertaken later in the review.

The checks required will, of course, depend on the detailed workings of each individual element of the HSMS, but some of the more general checks required are listed in Table 23.3.

Table 23.3
General checks on individual elements of the risk management system

Loss monitoring	Are all incidents for all relevant key losses being reported? Are all incidents for all relevant key losses being recorded adequately? Have all reported incidents been adequately investigated by a competent person? Have all relevant personnel been informed of the results of incident investigations? Have adequate trend and epidemiological analyses been carried out by competent people?
Risk assessments and risk control	Have all required risk assessments been carried out and adequately documented? Are the arrangements for carrying out risk assessments before new tasks are started still effective? Are all risk assessments being carried out by competent personnel? Are the results of risk assessments being used to design, implement and maintain adequate risk control measures? Are risk control measures being designed, implemented and maintained by competent personnel? Are risk assessments being reviewed at the required frequency?
Conformity monitoring	Is all required conformity monitoring being carried out, including 'monitoring the monitors'? Are the results of conformity monitoring being adequately recorded? Is all conformity monitoring being carried out by competent personnel? Is conformity monitoring resulting in corrective action where necessary? Have trend and epidemiological analyses of conformity monitoring data been carried out by competent personnel?

We have suggested how these checks can be made in the relevant chapters in Part 1.1. However, in those chapters, the responsibility for the checks is allocated to the personnel who are operating the relevant procedure. At the review stage, the checks are made by personnel other than those responsible for the procedures and the approach should be to carry out a random sample of checks to establish the extent to which the procedure is operating as intended.

Note that key losses and the review process itself will have to be reviewed, but these are special cases. We will look at the details of how to carry out these aspects of the review later in the section.

Checking the HSMS as a whole

Even if the individual elements of the HSMS are operating as intended, this is not necessarily a guarantee that the system as a whole is operating as intended. For this reason, it is necessary to make some additional checks, preferably before moving on to data analysis.

As with the checks on the individual elements of the HSMS, the details of the checks required will vary from organisation to organisation but, in general, they are concerned with the flow of information around the system. Figure 23.1 shows a part of the HSMS and the related information flows.

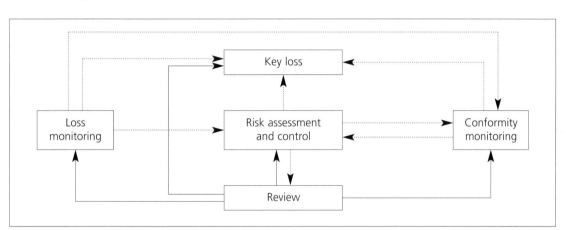

Figure 23.1
Information flow in the risk management system

Notice that there are two types of line shown in the diagram: each represents one of two types of primary function for the information flow.

Arrows shown as ┈┈┈┈┈┈┈➤
The primary function of this type of information flow is to provide ongoing updates to ensure that different elements in the risk management system 'learn' from what is happening in the other elements. For example, loss monitoring, risk assessment or conformity monitoring could each identify the need to record additional types of key loss. Similarly, loss monitoring and risk assessment could identify new requirements for conformity monitoring.

Arrows shown as ────────➤
The primary function of this type of information flow is feedback. Although there are some elements of feedback in the dotted line links, the main source of feedback is, as has been seen, the review process.

The method of checking on the effectiveness of the dotted line links is to select from each element a sample of data where the nature of the data is such that it could be expected that changes would be evident. These can then be 'traced through' the system to check that the appropriate changes have, in fact, been made. The solid line links are checked during the review of reviews, which is dealt with later in this section.

Some examples of the sorts of checks which can be made are:
1. **Loss and risk assessment data.** Where an accident has occurred, the relevant risk assessments should be revised as necessary, and the review is an appropriate time to check that this is happening. The fact

that an accident has happened is not, *a priori*, grounds for altering the likelihood or severity estimates of relevant risk assessments. However, where data on a number of accidents are available, they can be used to review and, if appropriate, revise likelihood and severity estimates.[*]

2. **Risk assessment and conformity monitoring data.** The end point of risk assessment is recommendations on risk control measures and these recommendations should take into account the reliability of the risk control measures. Conformity monitoring data, if appropriately analysed, can provide information on how reliable particular risk control measures are in 'real life' circumstances. The review process should check that appropriate analyses are being carried out on conformity monitoring data and that amendments to reliability estimates are being made as necessary. However, as with loss data, the review process should have many more data available, so that more options are possible.

3. **All three data types (loss, conformity and risk assessment).** This aspect of review provides an opportunity for overall checks. For example, a particular task may be chosen and the relevant data of the three types already discussed can be checked for that task.

Improving effectiveness and efficiency

The review stages described so far have all been intended to check whether the HSMS is doing what it is supposed to do. However, a full review goes beyond this, in that it also attempts to find ways of improving the system. This is done in two ways. First, it reviews its effectiveness and efficiency, which will be dealt with in the present section; second, it reviews its scope, which will be dealt with in the next section.

The review of effectiveness and efficiency begins with two statements and two questions:
- This is what the organisation wants to achieve.
- This is how the organisation achieves it now.
- Can it be achieved more effectively?
- Can it be achieved more efficiently?

For our current purposes, the terms 'effectiveness' and 'efficiency' are used as follows:
1. Improving effectiveness means raising the standards to be achieved – for example, more effective noise reduction or more effective use of PPE.
2. Improving efficiency means achieving current standards more cheaply or by using fewer resources.[†]

The two need not necessarily be in conflict. With the rapid development in risk control technologies, an improvement in standards can sometimes be achieved at a lower cost.

The two are also linked in that an organisation's management can take a policy decision on whether they would rather cut costs or raise standards, but this will depend on factors not related to the HSMS, especially the current and likely future profitability of the organisation. However, neither option is possible unless improved risk control measures can be identified, and this usually requires an element of creative thought on the part of the reviewers. This creative thought process can be assisted by suggestion schemes, which, although

---

[*] If it is not obvious why this is so, consider the case of tossing a coin. The fact that the coin comes down heads on a particular throw does not justify changing the estimated probability of the coin coming down heads on the next throw. However, if a coin has been tossed 20 times and has come down heads on 19 of these tosses, there are good grounds for revising the estimate. See Chapter 19 for more discussion of this point.

[†] As mentioned earlier, effectiveness is defined in ISO 9000:2005 as the 'extent to which planned activities are realized and planned results achieved' and efficiency is defined as the 'relationship between the result achieved and the resources used'. However, the less formal definitions given above suit our purposes better.

they are more commonly associated with productivity improvements, can be used to great effect in collecting ideas for improved risk control measures.

Like any creative process, there is no simple formula which can be used to ensure the success of this aspect of the review, but experience shows that groups of managers who systematically focus their attention on ways of improving risk control measures can, as with other aspects of their management, devise improved methods and systems.

The important thing is to undertake the review with a clear understanding of what this aspect of the review process entails. This is also true of the next aspect of review – checking on continuing appropriateness.

Checking on continuing appropriateness

There is a story, probably apocryphal, about a psychologist who was asked to conduct an efficiency review of army gun crews. He was very impressed by their efficiency but at the end of the review he asked why the most junior member of the gun crew always stood 100 metres away from the gun and took no part in the loading or firing. His hypothesis was that this was to protect the hearing of the young soldier but, after much enquiry among the senior army personnel, it transpired that he was there to hold the horses. During the transfer from horses to mechanised transport, someone had forgotten to cancel the need for a horse-holder, and this need had not been questioned since. If this sounds a bit too fanciful, let us look at some more up-to-date examples:

- An organisation had a very efficient and complex local exhaust ventilation (LEV) system which was used to remove rather nasty fumes from the workplace. Despite the fact that the process creating the fumes was no longer run, the organisation still spent several thousand pounds per year on testing, maintaining and repairing the LEV system.
- An organisation, as part of a long-term contract with its employees, provided high-quality safety footwear and continued to do so even when the employees moved to jobs where this safety footwear was no longer required.
- An organisation had a safe system of work which required two people to carry out certain tasks, and the sole function of the second person was to take a message back to the depot if anything went wrong. No-one had considered the relative costs of two people and one person with a mobile phone operating under a 'lone worker' system.

A high proportion of long-established organisations are likely to have this type of redundant control measure, and savings can be made by reviewing the appropriateness of safety systems.

The other side of the appropriateness coin is to check whether the HSMS is still covering all of the issues considered relevant. This part of the review becomes more important as the management system matures and the losses it was originally set up to reduce come under control, so new methods of measuring performance are required, or new losses have to be considered. In more detail, these two options could take the following forms.

New methods of measurement

If injury accidents are the key loss indicator, then this aspect of review can be summarised in the question: 'What is done if there are no more injury accidents?' Obviously, it would not be appropriate for the organisation to stop doing all the things it has previously been doing to reduce injuries to zero, but if it continues to use numbers of injuries as its only measure of success, then it cannot get better! There can be no improvement on no accidents.

In order to drive continual improvement, some other measure is needed and the most obvious candidates are numbers of near misses or numbers of nonconformities identified during regular inspections. Each of these will present problems of reporting, recording and analysis, but if the HSMS has evolved to the level where it is preventing all injuries, it is likely that it is capable of solving these problems.

New losses

An effective HSMS focusing on, for example, injury and damage accidents can readily be expanded to cover other losses. This can either be existing types of loss (for example, injury) but extended in scope to cover areas such as accidents at home or road traffic accidents, or it can be entirely new types of loss, such as environmental damage or product safety. Even if the HSMS is not at the stage where this level of review is appropriate, the techniques can still usefully be practised by identifying suitable measures for losses where there are currently no loss data (fire is usually a good example in this category) or new losses where there is likely to be a short-term benefit (days off sick is usually a good example in this category).

More information on review

If you need more information on safety reviews, study the review sections of BS OHSAS 18001 and OHSAS 18002,[86] since this is the best-developed review process in all of the published HSMSs.

Review of reviews

Review is a safety management activity and as such should itself be subject to review. It should not be assumed that a review procedure is effective; evidence should be collected which demonstrates that it is. However, since the people who carry out the normal reviews are the people who review the review process, there is always the difficulty of ensuring sufficient objectivity. For this reason, it is often preferable to leave the review of reviews to 'outsiders' and to incorporate it into the audit process.

Using computers

Audit software was mentioned earlier in the chapter, but this section deals with using computers in risk management more generally.

Hardware and system software

There are many types of computer (usually referred to as hardware), but the most frequently encountered is the personal computer (PC), either in its desktop form, or as a portable or laptop. The discussion in this section will, therefore, be restricted to software available for PCs.

Before any program can be run on a PC, it has to be equipped with system software. This software does a number of things but essentially it is an interface between the hardware and any other program to be run on that hardware. The major practical value of system software is that it means that people who write programs for specific purposes (for example, for recording risk assessment data) do not have to produce a different version for each different type of hardware. Instead, they write a program for a particular type of system software. The most common system software is Microsoft Windows in its various versions, including Vista. The discussion in this section will be limited to software packages which run on Windows, but many of the points made will also apply to other system software and many of the software packages are also available in versions that will run on other operating systems.

The nature of programs

The sorts of program we will discuss in this section all operate in essentially the same way. Each one provides a framework, or shell, into which people can enter data and, for the present purposes, the programs can be classified according to the types of data they accept. The main categories are:

- **Free format text, diagrams, pictures, tables and so on.** These data types are all accepted by programs such as word processors, desktop publishing packages, photo editing software and presentation packages.
- **Structured alphanumeric data.** These data types consist of mixed letters and numbers in a highly structured format of records and fields. Database programs accept these data types, including the specialised database programs used for specific purposes such as recording risk assessments and accidents.
- **Questions and answers.** This is a subcategory of the structured alphanumeric data but it has special relevance in risk management because of its use in auditing, conformity monitoring and safety culture measurement.
- **Numeric data.** Spreadsheets are the most common programs for numeric data, but these data types are also used by specialised statistical packages.

There is usually an overlap between programs: for example, word processors will do elementary calculations. However, all programs are designed to deal primarily with a single data type. These specific programs will be considered after some points about programs in general.

General points on programs
In theory, there could be one computer program which did everything, but, in practice, the more a computer program does, the more difficult it is to learn and use. For this reason, program authors compromise in two main ways:
1. **They reduce functions.** This involves limiting the number of things the program will allow people to do, such as the sorts of calculation people can do using a word processor, or the level of word processing people can do using a spreadsheet.
2. **They reduce flexibility.** This involves limiting the data the program will accept, or the number of things which can be done with these data. For example, any database program can be used for recording risk assessment data, but database programs are difficult to learn. A program designed solely for risk assessment data, although it is less flexible, should be much easier to learn and use.

The links between functionality and flexibility and between speed of learning and ease of use depend on the skill of the software designer. Some very limited programs are badly designed and are difficult to learn and use, while some very powerful programs are relatively easy to learn and use.

Programs for text
The main programs in this category include word processors, desktop publishing packages and presentation packages. As far as risk management is concerned, the primary uses for word processors are:
- for preparing and maintaining risk management documentation, including policies, procedures and work instructions
- for designing and preparing forms and checklists for use in risk management activities such as risk assessment and active monitoring
- for preparing reports on, for example, accidents, risk assessments and audits.

In addition, where data handling requirements are modest, word processors can also be used to record inventories and other data more usually kept in a database.

A key point to consider when selecting suitable software of this type is whether it will accept data directly from the other packages which have to be used. Having to retype text or data is tedious and error-prone, and it is preferable to have a word processor and presentation package which will read data directly from the output of the other packages in use.

The best strategy is to select well-known packages such as Microsoft Word (word processor) and

PowerPoint (presentation package), since authors of other software are likely to ensure that the output from their programs will be compatible.

Alphanumeric data

The main programs in this category include general databases and databases designed for use with specific types of data, such as risk assessment and accident records. General database programs such as Microsoft Access have a very wide range of functions and are very flexible. However, they are difficult to use without some programming experience or extensive time devoted to learning how to use them. There are two separate stages in the use of general databases:

1. **Setting up the database so that it will do the recording and analysis required.** If, for example, people wanted to use a general database to record and analyse risk assessment data, they would have to set up the fields for recording such things as the name of the person who carried out the assessment, the date and time of the assessment, and the results of the risk calculation. This is specialised work requiring a high level of skill.

2. **Entering data into the framework created in step 1.** This requires a lower level of skill but, unless step 1 has been carried out properly, it will be highly error-prone. For example, step 1 should include building in automatic checks on the data being entered, with appropriate error messages when incorrect data are entered.

Because of the high levels of skill required to set up general databases for specific uses, it is usually not worthwhile for health and safety professionals to learn the necessary skills. What normally happens is that the health and safety professional specifies what is required and then hands over the work of setting up the database to IT professionals. They then produce a program which looks like a specific database when it is being used for data input and analysis.

Specific databases

Specific databases are available for a wide range of uses, including recording and analysing data on accidents, risk assessments and various test results, such as audiometry and LEV tests. There are usually several different versions of each database type on the market and they differ in functionality, flexibility and price. The key selection strategies for these types of database involves two main elements:

1. **Be clear about what data you need to record and what analyses you need to carry out.** Software suppliers will try to convince potential buyers that what their program does is what the potential buyer should be doing, but this is not always the case. Instead, take the opportunity to review what is being done in the organisation in which the software is to be used, because there is little point in, for example, computerising a poor paper system.

2. **Look to the long term.** Many program demonstrations are carried out with just a few records on a high-powered computer and they appear fast and easy to use. Ask to see demonstrations involving the sort of computer that will be used in the organisation, with the numbers of records there will be in the system in two to three years' time. Some programs may be so slow as to be unusable.

A particularly important issue with these databases is integration, and this is considered next.

Integrated databases

As we saw in the discussion of review earlier in this chapter, integrating data is particularly important for detailed analyses and, for a full range of analyses, a number of different data types are required. There are databases on the market which claim to enable two or more types of data to be integrated for analysis purposes. However, it is necessary to check that the package will allow the types of integration required by the organisation which will use it. In particular, check that an 'integrated' package is not simply a number of different 'stand-alone' modules, linked together for marketing purposes, which do not allow transfer of data from one module to another.

The relevant trade press carries advertisements for these types of specific database, both stand-alone and integrated, and it is easy to get more information by contacting the suppliers.

Questions and answers

The main uses for programs of this type are for recording and analysing monitoring data, audit data, and data from surveys such as attitude surveys or safety culture surveys. The strategy for choosing these programs includes the points already made about specific databases, plus:

- **The flexibility of the question set.** Some programs are supplied with a set of questions which the purchaser cannot alter, while others can be supplied in a form which allows questions to be entered 'from scratch', or questions provided with the program to be tailored. As we discussed earlier, fixed questions are fine as long as they exactly meet an organisation's requirements, but this is not often the case.
- **The use of more than one question set.** Some programs allow the use of only one set of questions (fixed or tailored) for all analyses, while others allow as many different sets of questions as the user requires. The latter type of program is to be preferred when, for example, an organisation has a wide range of risks and wants to create sets of questions matching specific ranges of risk. This is done, for example, to avoid asking particular managers a lot of questions that do not apply to them.
- **The analysis options.** Some programs have very limited analysis options while others provide a range of alternatives. An important point to note is the extent to which the program allows 'labelling' of the answers to a particular set of questions. For programs designed for auditing, it may be adequate to have a limited range of labels for each set of questions – for example, the locations which were audited and the audit date. For attitude surveys, however, a range of labels will be required, including, for example, department, level in the management hierarchy, name, age, sex and job title.

The software required for attitude survey data is usually advertised for use with conventional attitude surveys and does not normally appear in the health and safety trade press.

Numeric data

Programs for numeric data are similar to databases, in that they are split into general programs – that is, spreadsheets – and programs which are designed to do specific things with numeric data, such as statistical packages. The principles for choosing and using spreadsheets are the same as for general databases, although people in general tend to be more familiar with spreadsheets.

Spreadsheets are particularly useful in risk assessment for the construction of severity distributions and the creation of graphs based on these distributions. In addition, where data handling requirements are modest, spreadsheets can be used, as can word processors, for inventories and recording basic details of risk assessment results.

There is also a range of specialised statistical packages available, ranging from cheap and easy-to-use packages that will do most of the basic statistical tests to expensive, 'heavyweight' packages suitable only for the professional statistician. Health and safety professionals would normally only require a package of this type if they were carrying out advanced work on trend or epidemiological analysis.

Specialised packages

A whole range of packages is available for specialised applications. Some that may be of use in particular aspects of risk management are:

- **Image manipulation software.** The availability of cheap colour printers, digital cameras and scanners means that photographic images can now be inserted relatively easily into documents and presentations. However, for the best results, some kind of image manipulation software is usually required.
- **Software for use with advanced risk assessment techniques.** Specialised programs for use with Fault Tree Analysis are available and graphics packages can be used for Event Tree Analysis. These types of program are advertised on the internet.
- **Software to help with specific tasks, such as preparing organisation charts and flow diagrams.**

Choosing appropriate software

To end this section, the notes below provide a summary of the main steps in choosing appropriate software of any type.

- Know the hardware and system software you are using, since this will restrict your choice of programs.
- Know exactly what you want to achieve by using the software. But always take the opportunity to review what you are currently doing in the organisation, since the software available may make it possible to do more than you are currently doing.
- Check what relevant software is on the market. This is probably best done by reading the appropriate trade press or one of the many computer magazines, or browsing on the internet.
- Get a demonstration of the software under conditions which match the hardware and system software available in your organisation and which correspond to your longer-term data analysis requirements. Many software publishers supply 'demonstration versions' which you can try out on your organisation's computer system before you commit to buying them.
- Do a cost–benefit analysis on the options available. It is unlikely that any package will exactly meet all your requirements, but remember that having a program written specially for your organisation is likely to be several orders of magnitude more expensive than buying one 'off the shelf'. You must make a decision on whether being able to do exactly what you require is worth the extra cost.

Summary

This chapter dealt with some of the more advanced uses of review and audit, and ended with a brief discussion of the use of computer software in risk management.

24: Financial issues

Introduction

It could be argued that the best, in the sense of most effective, motivation for managing risk well is that such management will, in the long term, improve the financial performance of the organisation concerned. However, it is notoriously difficult to demonstrate that effective risk management improves an organisation's financial performance. In the current jargon, it is difficult to assess the effect of risk management on an organisation's 'bottom line'. For this reason, the chapter begins with a discussion of typical problems associated with the financial aspects of risk management.

However, as we pointed out in Chapter 18, there has been a concerted effort to deal with the financial implications of poor quality management and the estimation of the costs which arise from the failure to meet quality standards. This work is directly analogous with the costs of promoting health and safety and the second part of this chapter deals with a number of ways in which the cost of promoting health and safety, and the losses arising from not promoting health and safety, can be used to assess the effectiveness of risk management. To provide a structure for this material, the chapter has five main sections:

1. cost–benefit analysis
2. breakeven analysis
3. costing risk control measures
4. types of insurance
5. risk management tools.

Some of the terminology used in this chapter has already been introduced, but there are several new terms. For ease of reference, the main terms to be used are summarised here.

- **Pure risk and business risk.** These terms are used to differentiate between the sorts of risks dealt with in risk management (pure risks) and other sorts of risks such as investing in new products or services (business risks, also known as speculative risks). In general, business risks are characterised by the opportunity to gain as a result of an investment, whereas with pure risk, any 'investment' is intended to prevent loss, with no opportunity for any other gain. However, as will be seen later in the chapter, the two categories can overlap. Unless otherwise qualified, risk will be used for the remainder of this chapter to mean pure risk.
- **Risk avoidance.** This is a conscious decision not to import risk into an organisation by deciding, for example, not to manufacture a particular product or provide a particular service.
- **Residual risk.** This is the risk which remains even when specified risk control measures are being used and are operating effectively.
- **Risk acceptance.** This is the point at which no more is done to reduce risk. A decision not to do any more about a risk can be taken with knowledge of the risk, for example because it is too expensive to reduce the risk further. However, much risk acceptance is without knowledge of the risk. These two forms of risk acceptance can also be referred to as conscious and unconscious risk acceptance, and the word 'retention' can replace 'acceptance'.
- **Risk transfer.** This involves moving the source of a risk from one place to another, usually from one organisation to another. This can be done ethically by, for example, employing competent contractors to remove asbestos, or unethically by, for example, having high risk processes carried out in developing countries where proper risk control measures are not enforced.
- **Transfer of the financial aspects of risk.** This involves taking out one or more of the many types of insurance available. Note that this is often, incorrectly, referred to as risk transfer. The last section of this chapter discusses types of insurance in more detail.

Difficulties with financial issues

There are several reasons why it is difficult to deal with health and safety in terms of pure finance and the main ones are described here.

- **The cost of promoting health and safety are relatively easy to quantify and the relevant expenditure is certain.** For example, expenditure on safety personnel, risk control measures and safety training is easy to quantify and is incurred if the relevant action is taken. In contrast, the losses arising from lack of health and safety management are difficult to estimate (see below) and whether or not these costs will be incurred is uncertain. Not all organisations that spend little on health and safety have accidents and not all organisations that spend large amounts on health and safety are accident free.
- **It is difficult to calculate accurately the costs of accidents since there are so many factors involved.** Because accident costing is so difficult, many organisations carry out no accident costing at all. This results in health and safety expenditure being seen only as a drain on resources and not as an investment. This can partially be overcome by estimating accident costs, rather than calculating them in detail, and the techniques for this will be discussed later in the chapter.
- **Few organisations have the accounting procedures necessary to balance the costs of health and safety with the costs of ill health and injury.** For example, insurance premiums are usually paid centrally in an organisation so that individual managers have little incentive to reduce these premiums. This leads on to a final point.
- **Individual managers can see that safety costs them money while ill health and injury accident costs are met by a central budget.** For example, expenditure on noise reduction measures may have to be met out of a manager's budget, while deafness claims are paid from a central fund. In these circumstances, a 'financially rational' manager with an eye on the 'bottom line' will allow people to work in conditions which will make them deaf in the long term. While some organisations have attempted to make managers 'pay' for lack of health and safety, this does not appear to be widespread.

Improvements in risk management driven by financial considerations will require appropriate accounting systems and it is worthy of note that one of the major recent influences on risk management was the Turnbull Report,[103] which was an accountancy report, not a health and safety report. The remainder of this chapter is based on the assumption that effective accounting systems can be put in place and describes the sorts of finance-based techniques which can be used in risk management.

Cost–benefit analysis

The techniques of cost–benefit analysis can be used at various levels in risk management. For example, they can be used to determine which supplier of training courses should be employed or whether in-house or contracted resources should be used for audit work. A particular use of cost–benefit analysis is in deciding between risk control measures, but since this is a special case, it will be dealt with in a separate section later in this chapter.

The most generalised use of cost–benefit analysis is to determine whether expenditure of resources on risk management throughout an organisation is producing benefits. For this reason, the remainder of this section deals with cost–benefit analysis at the organisational level. The first, and obvious, steps in cost–benefit analysis are the calculation of the costs and the benefits, although these calculations are not always easy. The next two subsections deal, respectively, with the calculation of costs and benefits.

Calculating costs

If the cost of risk management throughout an organisation is to be calculated, a large number of individual costs will have to be considered. Some of these will be reasonably easy to calculate and others will be extremely difficult, but using a three-stage approach will make the procedure more systematic:

1. **Identify the sources of costs.** These may be financial, such as fees paid to trainers or payments for equipment, or they may be time spent on risk assessment or conformity monitoring. For effective identification of all relevant sources, some structure will be required – possible structures are discussed later in this subsection.

2. **Quantify the sources of costs.** This stage involves quantification in whatever units are appropriate to the source of cost being considered. This may be money (for fees and payments), number of hours or days (for time spent on risk management activities), or any other convenient unit. This stage is often necessary because it is not always possible to move to the third stage.

3. **Calculate monetary values.** Some elements at stage 2 may already have a monetary value, but this is usually only for a part of the cost. For example, the price of a training course may be known, but it will still be necessary to work out how much it cost to send people on the course. Similarly, the purchase price of software is usually easy to ascertain, but it may not be easy to determine how much it cost to select and install this software.

We have suggested this three-stage approach because there may be a tendency to assume that, because an exact monetary value cannot be calculated, it is not worthwhile to collect data at all. However, if some information is available, this is better than nothing, and estimates can be made when actual data cannot be obtained. We will now consider each of the three stages listed above in more detail.

Identification of sources of costs

As was mentioned above, the identification of all relevant sources of costs will be more systematic if some structure is used for classifying the sources. Any appropriate structure could be used, and a good approach is to look at each element of the HSMS in turn and estimate the costs of maintaining that element. This can be done irrespective of which HSMS you are using.

Using this approach, a full costing of risk management in an organisation could be a time-consuming exercise. However, it can be done over a period of time, with items being recorded in a spreadsheet as they are identified.

Quantification

As was mentioned above, this stage involves quantification in any units which are convenient for the source of cost being considered. The most commonly required units are time, in hours or days, and the number of items which have to be purchased. However, a range of other costs may be incurred in particular circumstances, including travel expenses and photocopying.

Quantification is an interim step to full costing but it is important to record these elements since it is not always possible to obtain information on actual costs. For example, it may be relatively easy to determine how much time the board of directors spends on discussing the organisation's health and safety policy, but it will be much more difficult to find out what this costs in terms of their salaries and overheads. As with sources of costs, the units used for quantification can be recorded using a simple spreadsheet.

Particular problems may arise with items which fall into the category of opportunity costs. These are items where resources spent on risk management mean that resources cannot be used in other areas. Typical examples include:

- managers spending time on risk management activities which means that they are not spending their time on more 'profitable' activities, such as ensuring the timely delivery of products or services

- resources spent on risk control measures which means that these resources are not available for business risk investments. For example, a replacement machine has to be purchased because the current machine is too noisy when these resources could be used to purchase an additional machine which would produce new products.

However, as we mentioned in the discussion of quality management in Chapter 18, it is important to devise some method of estimating even these sorts of costs. Expenditure on quality management has opportunity costs in the same way that expenditure on risk management has opportunity costs. Accurate calculation of these costs is usually impractical, but it is possible for the interested parties to agree a value.

Deriving monetary values

The last stage in the costing process is deriving monetary values for the sources and quantities already identified. In theory, this should be a fairly straightforward process but complications arise for various reasons, such as:

- **Time of personnel.** It may be possible to determine an hourly or daily rate for particular individuals but extremely difficult to estimate their opportunity costs.
- **Units purchased.** It is usually relatively easy to determine how much was paid for an item of software or 1,000 earplugs but it is much more difficult to work out how much it cost to purchase these items. For example, how much time was spent on their selection? How much did the purchasing process itself cost?

A particular problem in quantification is how to deal with infrastructure costs, often referred to as 'overheads'. It may, for example, be possible to calculate quite easily a person's hourly rate of pay, but some method will be required for calculating, or estimating, such things as the cost of the person's office accommodation, the fuel used to heat it, and so on.

However, it is always possible to arrive at agreed estimates of costs that cannot be calculated accurately and, if a spreadsheet is being used to record costs, some convention can be used to differentiate between calculated costs and estimates. For example, a bold or italic typeface can be used for estimated values and a normal typeface for calculated values.

From the above discussion, it can be seen that the calculation of the costs of risk management is not a trivial task. But it is essential if the value of risk management is to be demonstrated on business grounds.

The important thing to remember is that the best should not be allowed to become the enemy of the good. In other words, because it is known that accurate, detailed costings (the best) are unlikely to be achieved in practice, this should not be used as a reason for not making agreed estimates of likely costs (the good).

Calculating benefits

The costs incurred in risk management should, in theory and practice, produce benefits for the organisation and there is a strong argument for discontinuing any risk management activity which does not produce benefits. There are practical problems associated with identifying these benefits, quantifying them, and converting them into monetary values, but if these problems are not solved, cost–benefit analysis cannot be carried out.

The main benefits arising from risk management are:

- **The reduction of known losses – in other words, the prevention of recurrence of losses which have occurred in the past.** Losses which do occur can be identified and costed, and risk management techniques can then be used to reduce the extent of these losses in the future. The costing of known losses is dealt with later in this subsection.
- **The prevention of possible losses.** It can be argued that one of the primary functions of risk management activities is to prevent losses and that the cost of such 'prevented losses' should be included on the benefit

side of the cost–benefit equation. Since these losses are probabilistic in nature, and there may be few previous data, estimation techniques are required to arrive at benefits, and these are described later in this subsection.

- **'Spin-off' or opportunity benefits.** These are benefits which do not involve loss reduction but nevertheless can be calculated or estimated. For example, the replacement of a machine in order to reduce noise in the workplace may produce other benefits: the new machine may be faster or more economical to run, or allow the manufacture of a wider range of products or products of a higher quality. These spin-off benefits are also dealt with separately later in this subsection.

Known losses

The calculation of losses is a similar procedure to that just described for the calculation of costs. The first stage is to identify which losses are going to be considered (see Chapter 8) and then, for each loss, work through the three stages already described.

1. Identify the sources of loss. For an injury accident, for example, these would include the time lost from work by the injured person, time lost from work by other people involved (rescuers, witnesses, those who 'cleared up', accident investigators, supervisors, managers and so on), the costs of medical treatment, and the costs of repairing any associated damage to assets.
2. Quantify the sources of loss in convenient units.
3. Allocate monetary values as appropriate.

This costing work will be more methodical if some structure is used for the identification of sources of loss. Ideas for suitable structures can be found on the accident costing pages of the HSE's website (www.hse.gov.uk/costs/accidentcost_calc/accident_costs_intro.asp).

As with costs, a simple spreadsheet table can be used to record the data collected. However, some of the commercially available accident recording software includes facilities for structuring and recording accident costs.

Using accurate calculations of losses which occur, or best estimates of these losses, enables trend analysis of loss data and any reduction in losses can be compared with the costs of the HSMS.

Possible losses

Where there are no historical data on losses, some estimate should be made of the losses which would be sustained if the HSMS did not exist. This may, at first sight, appear to be an unfamiliar concept, but consider the case of losses resulting from fires.

Relatively few organisations have direct experience of losses from fires, either because there has never been a fire, or because fires which have occurred have been effectively controlled. Nevertheless, most organisations spend money on, and devote other resources to, a range of fire prevention and fire control measures, and also to those measures required for an effective response in the event of a fire. In the UK, the motivation for this may, in part, be to meet legislative requirements, but there are probably sufficiently frequent reports of organisations going out of business or suffering severe business interruptions as a result of fires to keep fire prevention on management agendas. Thus, fire prevention is seen as a benefit and this benefit can, in theory, be allocated a monetary value. This can be done in practice by expressing the fire risk in financial terms, but since this can also be done with most risks, including fire risks, the procedure will be described in a wider context later in this subsection.

The allocation of monetary values to risk reduction is important since, unless it can be done, it will not be possible to decide how much it is worth spending on risk management activities. To continue with the fire example, unless a financial value for the risk can be estimated, there is no rational basis on which to make a decision on such matters as the effectiveness (and hence expense) of the fire detection and control systems which should be installed.

What is needed, therefore, is some method of calculating risk in financial terms and one such method was described in Chapter 5, where the probability of a hazardous event was multiplied by the estimated cost of the most likely outcome if the hazardous event occurred.

This use of monetary values allows qualitatively different types of risk to be taken into account, for example:

- A minor injury may be less than £50, but if it is associated with a spillage of chemicals affecting the environment, the overall severity may be £3,000.
- The repair of accidental damage to plant may be £300, but the stoppage caused by this damage may move the overall cost to £7,500.
- A chemical leakage may cost only £10 as far as injury or environmental damage are concerned, but possible spoilage of products or raw materials may put the overall cost up to £30,000.

Of course, agreement will have to be reached among the interested parties as to the costs to be used, but at least this procedure gives a rational basis for discussion. Using this procedure, the possible losses for the organisation as a whole can be calculated by summing the results from individual risk assessments. Since these risk assessments should already be carried out for other reasons, the amount of additional work required to arrive at a costing of possible losses should be minimal.

It is also possible to make an estimate of possible losses using the following procedure:

1. classify accidents into various categories according to the nature and severity of losses
2. allocate a mean monetary value to the accidents in each category, such as £1,000,000 for a fatality and £50 for a minor injury
3. estimate the expected number of accidents in each category which would be prevented by risk management activities
4. multiply the estimated number of accidents in each category by the mean value for an accident in that category
5. sum the resulting figures over all categories.

Spin-offs

As was mentioned above, there are benefits other than loss reduction which arise from expenditure on risk management. These spin-off benefits are many and various and some examples are given below.

- **Training in risk management.** Good risk management training is based on sound management principles which can be applied to most aspects of a manager's work. Many managers have received little formal training in management techniques so that the benefit of their risk management training is generalised to the management of other activities.
- **Conformity monitoring.** Although conformity monitoring for risk management purposes has a limited number of aims, it is unlikely that only these aims will be met during conformity monitoring. Conformity monitoring is a key element of other management systems (although, as has been noted, it is referred to by a variety of different terms) and managers who are making best use of their time will combine a range of monitoring requirements.
- **Safe working procedures.** Sometimes the work required to devise safe working procedures identifies aspects of tasks which, although they are not unsafe, are very inefficient. Thus, the resources used to devise safe working procedures may also result in improvements in productivity.

While it may be difficult to allocate accurate monetary values to these sorts of spin-off, there is no good reason why estimates cannot be made and agreed with the interested parties.

These spin-offs are different in nature from the benefits arising from loss reduction, in that they are not probabilistic. Managers are either receiving a general benefit from risk management training or they are not;

managers are either combining risk management monitoring with other monitoring requirements or they are not; and so on. The monetary value may be uncertain and have to be estimated, but the benefits themselves either accrue or they do not.

For this reason, it is easier, in accounting terms, to treat the monetary value of spin-offs as reductions in the cost of risk management rather than increases in the benefits. This has a neutral effect on the results of the cost–benefit analysis but it makes calculation easier. Two examples are given here:

- If a risk management training course is to cost £10,000 but it is estimated that 10 per cent of the course will be of general value, or devoted to matters other than risk management, then the cost of the course in the risk management cost–benefit analysis is £9,000.
- If it took five days to devise a safe working procedure, but two of these days were spent dealing with inefficiencies in the existing working procedure, then only three days are allocated to the costs of risk management in the cost–benefit analysis.

This technique is used in other types of cost–benefit analysis. For example, in the cost–benefit analysis for a new machine, any scrap value of the old machine can be deducted from the cost of the new machine rather than being counted as a benefit of the new machine. However, these are accounting conventions used to make life easier at the calculation stage and any suitable conventions can be agreed among the interested parties.

Comparing costs and benefits

At the end of the work described in the previous three subsections, the following data should be available:

- an agreed overall cost for the risk management activities over a specified period
- an agreed overall value for the losses being prevented over the same period.

Note the use of the word 'agreed' in both of these items. In the previous discussion it has frequently been mentioned that estimates are likely to be required and that this may arise either because data are not available, or because the resources needed to collect the required data could be prohibitive (essentially, it is not cost-effective to collect the cost data!). However, where there are gaps in the data, agreed estimates can be used, and arriving at such estimates will be a standard negotiating process. The risk managers will argue that risk management activities cost less than they really do, and that the losses being prevented are much higher than they really are. The senior management may take the opposite views! Somewhere between the two negotiating positions, there will be a reasonable estimate of true costs and benefits.

Although the work required for adequate cost–benefit analysis is not trivial, it is important to carry out the work if risk management is to be justified on cost–benefit grounds. In addition, there is a breakeven point in expenditure on risk management as shown in Figure 24.1. What Figure 24.1 shows is that increasing expenditure on risk management will always result in a further reduction in losses. However, beyond a certain level of expenditure – the breakeven point – there are diminishing returns. That is, the amount of loss reduction is less than the expenditure on risk management required to produce that reduction. Organisations can, of course, choose to devote resources to risk management beyond the breakeven point but, unless the sorts of cost and benefit data just described are available, this choice will not be an informed one.

This type of breakeven analysis is, however, only a special case of more general types of breakeven analysis, and more information is given in the next section.

Breakeven analysis

Breakeven analysis is used to determine at what point costs and benefits are equal – that is the breakeven point illustrated in Figure 24.1. Where, as in Figure 24.1, both costs and benefits are curvilinear, or where they are

Figure 24.1
Relationship
between
expenditure on risk
management and
losses

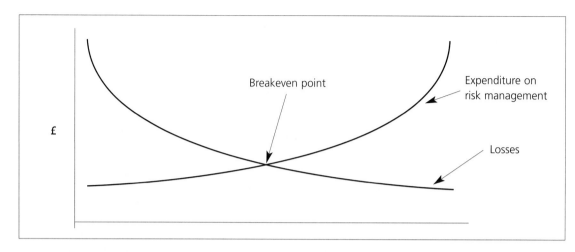

Breakeven point

Expenditure on
risk management

£

Losses

linear, the breakeven point can be determined simply by establishing the crossover point. However, where this is not the case, more detailed techniques are required.

A common reason for lack of linearity in costs is that the costs have a fixed element and a variable element. How breakeven analyses are carried out in these circumstances is described next, using the provision of a training course as an example.

The financial benefits of a training course may be difficult to establish but, as we suggested earlier, they can be estimated. However, what is relatively straightforward is that there will be a linear relationship between the overall financial benefit and the number of people on the course. That is, for the organisation concerned, there will be 10 times more benefit if there are 10 people on the course than if there is only one person on the course.* However, the cost of providing a course has two main elements:

1. **The fixed costs.** These are the costs which are incurred irrespective of how many people there are on the course. They include the tutor's costs for delivering the course, the cost of the course venue, the costs of any training equipment, and the costs of publicising the course.
2. **The variable costs.** These are the costs which are incurred per student. They include such things as course handouts, the various costs incurred by students, such as subsistence and accommodation, the cost of having delegates' normal jobs done by someone else, and the tutor's costs for assessment (for example examination marking).

Once the relevant data have been collected, a table of the type shown in Table 24.1 can be drawn up and the breakeven point can then be established by inspection. Table 24.1 is based on the following cost and benefit figures:

Fixed costs	£2,000
Variable costs (per student)	£100
Benefits (per student)	£400

From Table 24.1 it can be seen that the breakeven point is not reached until there are seven students on the course.

---

* This is an oversimplification. In addition, there are limits. Depending on the type of course, there can be too many people for effective training to take place.

Number of students	Cumulative fixed and variable costs	Cumulative benefits
1	£2,100	£400
2	£2,200	£800
3	£2,300	£1,200
4	£2,400	£1,600
5	£2,500	£2,000
6	£2,600	£2,400
7	£2,700	£2,800
8	£2,800	£3,200

Table 24.1
Breakeven point data for a training course

Costing risk control measures

In Chapter 6 there was a description of basic cost–benefit analysis for risk control measures and we showed that it is possible to calculate a cost per unit of risk reduction. However, in Chapter 6, the costs of risk control measures were assumed and there was no detailed discussion of how they were calculated. These calculations are the subject of the present section.

When calculating the cost of a specific risk control measure, three categories of costs have to be taken into account:

1. **Implementation costs.** These are the costs of setting up the options – for example, designing appropriate risk control measures, buying PPE, buying and installing machinery guarding, and training. There will also be the costs of planning for appropriate installation and maintenance.
2. **Maintenance costs.** These are the costs of maintaining the options – for example, replacing PPE, testing and maintaining guards, and refresher training. This category will also include the organisational and operational costs associated with the option.
3. **Opportunity costs.** These are the costs associated with any reduction in productivity, or with other effects the changes may have on the system.

The collection of the required data may not be straightforward but, as with the costing exercises discussed earlier in the chapter, estimates are better than no information at all. Whether actual figures or estimates are being used, the following points should be taken into account:

- **Risk control measures often have benefits other than risk reduction, that is the sorts of spin-off described earlier in this chapter.** Where these exist, their value (calculated or estimated) should be deducted from the cost of the risk control measure to give a 'true' cost.
- **The 'life' of the options is important, that is the time period over which the options will be used.** For example, hearing protection has low implementation costs but relatively high maintenance costs whereas acoustic guarding has a high initial cost but low maintenance costs. This means, in effect, that there are fixed and variable costs for risk control measures so that a breakeven analysis is likely to be required to compare the true costs over different time periods.

The calculation of the costs of risk control measures may be done as an exercise in its own right, that is simply to find out how much they cost. However, it is more usual to do these costings as an aid to decision making and there are two frameworks within which decisions have to be made:

1. The risk has to be reduced to a specified level and a choice has to be made between options that will produce the required risk reduction. This framework can be referred to as cost-effectiveness analysis. This is the framework required by organisations which adopt the approach of setting a level of acceptable risk.
2. The risk has to be reduced to a level as low as reasonably practicable, so there is no compulsion to implement a risk control measure where the costs of doing so would outweigh the benefits. This framework can be referred to as cost–benefit analysis. This is the framework required by organisations working under UK legislation and organisations using business cases for risk control measures.

The costing work required for these two frameworks is the same since, as can be seen from their descriptions, they differ only at the decision point: which risk control measure to implement or whether to implement a risk control measure.

Once the necessary data have been collected, or appropriate estimates made, the calculations are relatively simple. The steps are:

1. Decide on the relevant life for the calculation. This may be easy if, for example, it is known that the process will run for four years, but where the risk control measure will have to be in place for an indeterminate or indefinite period of time, it will be necessary to select an arbitrary period such as three or five years.
2. Work out, for each risk control measure, the 'true' cost over the chosen time period, where the true cost is all costs minus all spin-off benefits.
3. Select the option with the lowest true cost (for cost-effectiveness analysis) or an option where the benefits outweigh the costs (cost–benefit analysis).

The main complication in cost–benefit analysis of risk control measures is that the extent of risk reduction has to be taken into account. The most straightforward method of doing this is to calculate a true cost per unit risk reduction (described in Chapter 6).

All of the techniques described above can have problems associated with them.

- **It may be difficult to collect the cost and benefit data.** For example, PPE may slow down production but it may be difficult to estimate how much this will cost.
- **It may be necessary to allocate arbitrary monetary values to certain costs and benefits.** For example, not having to wear PPE is likely to have a 'comfort' benefit, but this will be difficult to measure objectively.
- **It may be necessary to make subjective estimates of risk.** All risk estimates are subjective in that they depend on the risk assessor's knowledge at the time. However, estimates of possible risk reduction often have to be made on the basis of very limited information, so they can be even more subjective.

However, despite these problems, the methods described above provide one rational way of deciding between risk control options.

Insurance

As was noted earlier, insurance transfers only the financial implications of risk; it does nothing to reduce the risk *per se*. However, the insurance companies may require certain action and the size of premiums may make it cost-effective to implement risk control measures, but these are side effects, not an integral part of the insurance.

A wide range of types of insurance is available and some of the commonly used ones are:

- **employers' liability** – it is compulsory in the UK for employers to insure against their liability to pay damages for bodily injury or disease sustained by their employees in the course of their employment

- **public liability** – this type of insurance is cover against liability for injury or illness to third parties, or damage to the property of third parties, where these arise from accidental causes
- **professional indemnity** – this covers mainly liability for damage and so on arising from professional advice given in the course of employment
- **product liability** – this covers mainly liability for damage or other losses arising from defects or other failures in an organisation's products.
- **other losses** – there is a wide range of other insurances; some of these are concerned with specific sources of loss, such as theft and fire, while others are associated with particular types of asset, such as information, vehicles and personnel.

The costs associated with losses can be divided into insured and uninsured costs, and examples of insured to uninsured ratios for injury accidents are shown in Table 24.2.

Type of industry	Insured : uninsured ratio
Construction site	£1 : £11
Creamery	£1 : £36
Transport company	£1 : £8
Oil production platform	£1 : £11

Table 24.2
Insured and uninsured cost ratios[4]

From the ratios shown in Table 24.2, it can be seen that insurance is of limited value in protecting an organisation against losses. However, insurances do have a place in a balanced 'portfolio' of risk management tools, the main one being cover for very low likelihood but extremely high severity risks. The portfolio of tools concept is discussed in more detail in the next section.

Risk management tools

As we have seen at various points throughout this book, risks are many and various and there is a wide range of ways of dealing with them. In addition, the nature and status of a particular organisation may influence both what risks it wishes to accept and the way it manages those risks it has. For these reasons, there is no general 'best' way of managing risk which will apply to all risks and all organisations. It is up to the risk management professional in a particular organisation to decide, for each type of risk, how best to approach its management. This can best be done by the systematic application of four key management tools: risk assessment, risk control, costing techniques and use of insurance.

Risk assessment

If risk assessment is not used, or is not used effectively, the following are possible:
- the ratio of known to unknown risk in the organisation will be heavily biased towards unknown risk, and unknown risks cannot be managed effectively
- it will not be possible to practise effective risk avoidance, since the necessary data for deciding whether particular activities present unacceptable risks will not be available.

Risk control

If a detailed knowledge of risk control methods is not available, or applied ineffectively, then the following are possible:

- there will be no accurate discrimination between controlled and uncontrolled risk or, to put it another way, the magnitude of the residual risk will be unknown
- it will not be possible to make accurate judgments on when risk transfer should be used, since the effectiveness of in-house risk control measures cannot be estimated accurately.

Costing techniques

The sorts of costing techniques described earlier in this chapter have to be implemented if it is to be possible to make rational decisions on, for example, choosing risk control measures.

Use of insurance

A knowledge of required and available insurances is needed so that, in appropriate circumstances, at least some of the consequences of failures in risk controls can be insured against.

Summary

This chapter dealt with a number of financial issues relevant to risk management. The chapter began with an overview of the difficulties associated with these financial issues and then moved on to various types of financial analysis including cost–benefit analysis, breakeven analysis and the costing of risk control measures. The chapter ended with the role of insurance and a brief review of risk management tools.

Part 2.2: Human factors – advanced

25: Part 2.2 – overview

Introduction

Part 2.2 covers the more advanced human factors material required for the NEBOSH Diploma syllabus and the knowledge requirements for certain aspects of the NVQ/SVQ (see Appendices 1 and 2). It also deals with topics covered in some university health and safety syllabuses and the relevant IOSH requirements (see Appendix 3). The present chapter gives a brief description of the human factors material covered in this part.

As far as Part 2.2 is concerned, human factors are primarily confined to individual reliability – that is people avoiding doing things which may cause harm to themselves or others and doing those things which are necessary to prevent harm to themselves and others. Other sources of harm – for example, being injured when no action was possible to avoid the injury – are not dealt with here.

Human reliability can be defined informally as doing the right things at the right times and not doing wrong things. Therefore we need to consider what might cause people to fail to do the right things at the right time, and what might cause them to do things that are wrong. These causes are many and various but, for ease of presentation, they will be dealt with in separate chapters under the following headings:

- **Chapter 26: Individual differences.** It has been emphasised earlier in this book (see Chapter 15) that individuals differ and that variations in factors such as personality and attitudes influence all aspects of an individual's behaviour. These factors also affect human reliability, and Chapter 26 deals with the main sources of individual differences. This will make it possible, for the remainder of Part 2.2, to proceed without too many caveats about the importance of individual differences.
- **Chapter 27: Human error.** Everyone, at some stage, learns by 'trial and error', so dealing with human error is not a straightforward removal of all errors of all types. Rather, it involves identifying those errors which could result in harm, and the causes of these types of error. With accurate information on causes, it is then possible to design and implement remedial measures. This chapter ends with a discussion of how human error has contributed to major disasters.
- **Chapter 28: Perception and decision making.** Accurate perception of hazards and risks is central to avoiding risk, since, if people cannot perceive that there is a risk, they can take no action to avoid it. Decision making is also important, since it is necessary to choose the right avoiding action at the right time.
- **Chapter 29: External influences on human error.** The physical and social environments in which individuals have to operate influence the likelihood of them making errors. This chapter describes the main external factors which influence human error rates.

The four chapters just described cover the various types and causes of human errors but only in passing do they deal with ways of reducing errors or, putting it another way, ways of increasing human reliability. The last chapter in Part 2.2 – **Chapter 30: Improving human reliability** – deals more systematically with the ways of improving human reliability, including ergonomics and environmental factors.

26: Individual differences

Introduction

This chapter deals with a number of related sources of individual differences: motivation, intelligence, personality and attitudes. All of these sources were introduced in Part 1.2 but in this chapter we will consider them in more detail and describe some of the major theories associated with each.

Motivation

As with many of the concepts dealt with in this book, there is no generally agreed definition of what 'motivation' means. However, for the purposes of this chapter, it is sufficient to define motivation informally as being those factors which influence an individual to behave in certain ways.

A major goal of the study of motivation is to identify accurately the relationships between a particular factor – for example, pay or the opportunity to socialise – and the effect this has on a person's behaviour. If these relationships were clearly understood, we would be able to manipulate the factors in ways which produced the behaviours required. This raises obvious ethical issues since, in effect, people would be manipulated, but in the present state of knowledge on motivation, this is rarely a practical problem. There are few aspects of motivation which are sufficiently well understood to enable accurate predictions of cause and effect.

There are various reasons for this lack of accuracy, but the overriding one is the usual problem of individual differences. The factors which motivate one person need not necessarily have the same motivating effects on another person, and what motivates a person today may not motivate him or her tomorrow.

Despite, or perhaps because of, these difficulties, there has been extensive research on motivation over the years, with various researchers putting forward a wide variety of theories. Some of the more important theories are introduced next.

Maslow's five classes of human need

Maslow[87] put forward the idea that humans have five classes of need, each of which is motivating, in that individuals behave in ways which enable them to satisfy these needs. The five needs are:

1. physiological – the basic biological needs for food, drink and sex
2. safety – the need for an environment which poses no threat to individuals' mental or physical wellbeing
3. belongingness – the need for a sense of attachment to other individuals, either singly or in groups
4. esteem – the need to feel valued by people the individual considers important, as well as the need for 'self-value' or 'self-esteem'
5. self-actualisation – a complex category of needs which can be summarised as the needs individuals have to fulfil their potential.

Maslow argued that the five classes of need have to be satisfied in ascending order. For example, safety needs only become motivating if, and when, physiological needs are satisfied, and self-actualisation only becomes motivating if all of the four lower classes of need have been satisfied.

There are numerous problems, both theoretical and practical, with Maslow's theory but it has stimulated a great deal of further research. Various authors have put forward alternative theories based on the idea that motivation is about fulfilling needs of different sorts, for example the need for achievement.

The primary focus of Maslow's theory (and other theories based on need fulfilment) is why people do anything at all, and it is not particularly successful at dealing with alternative actions which could satisfy the same need. To take a simple example, need theories are good at explaining why people eat, but not at explaining why people eat one type of food rather than another. Attempts to explain this type of motivation include an influential theory by Vroom, which is described next.

Vroom's expectancy theory

Vroom[88] argued that when individuals are faced with a number of possible actions, they will make their choice after considering three factors:

1. expectancy – the individual's estimate of how likely it is that he or she will be able to complete the action successfully
2. instrumentality – the individual's estimate of whether or not performing the action would lead to an identifiable outcome
3. valence – the extent to which the individual values the outcome identified in 2.

Although the theory is described as an expectancy theory, it is also known as instrumentality theory and as a VIE theory (from Valence, Instrumentality and Expectancy).

For any given action, the strength of the motivation to carry it out is, according to Vroom, the product of the three factors listed above. That is:

$$\text{Motivation} = \text{Expectancy} \times \text{Instrumentality} \times \text{Valence}$$

Vroom's theory has a number of important practical applications:

- If people do not believe they have the competence to carry out a task, the expectancy for the task will be zero. The implication for safety is, therefore, that if people do not have the competences required to work safely, expectancy for the safe actions will be zero.
- If people cannot clearly see the results of a particular action, the instrumentality of that action will be zero. The implication for safety is that if people cannot see clearly the links between particular actions and the protection these actions produce, the instrumentality of these actions will be zero.
- If people do not value the outcomes of the actions, the valence for these actions will be zero. In safety terms, if no value is put on personal safety, the valence of safe actions will be zero.
- If the value of V, I or E in the motivation equation is zero, the motivation will be zero, because V, I and E are multiplied together to arrive at the value for motivation.

These points may appear obvious with our current state of knowledge, but they were not in 1964 when Vroom put them forward. It is still a useful exercise when considering any existing or proposed safety actions to consider carefully their likely V, I and E for the people who will have to carry out the activities.

While there are problems with Vroom's theory, which are dealt with below, there seems little doubt that the three factors he identified do have an impact on motivation.

The main problems with Vroom's theory, and the other theories which are variants of Vroom's theory, are that:

- it is not clear that individuals really go through the complex cognitive processes required to estimate the values of V, I and E for all of the possible alternatives
- it may be that V, I and E, when they are calculated, are added together to arrive at a value for motivation, and not multiplied as Vroom suggested. This is an important issue since, as was described earlier, if V, I and E are multiplied then a zero value for any one of them makes the motivation zero. This is not the case if V, I and E are added
- it does not work when the outcomes have a negative valence, that is the person sees the outcomes as undesirable. This creates obvious difficulties for safety, where there are many undesirable outcomes such as injury and ill health.

The types of theory put forward by Vroom and Maslow are attempts to describe all forms of motivation and they have limited success in doing this. Theories of motivation which are more specific tend to be more successful and goal-setting theories of motivation, which can be very specific indeed, are considered next.

Goal-setting theories

A number of authors have contributed to the range of goal-setting theories and they have included people working on practical management issues as well as academics working on theoretical issues. The former group has been involved in the development of techniques such as management by objectives (MBO), while the latter group has been concerned with such issues as what types of goals produce the strongest motivation and why goals should be motivating at all.

There is a wealth of detailed research results on goals and their motivational effects but not all of it is easy to interpret. However, the following points appear to have been clearly established.

- Goals are only motivating to the extent to which people commit themselves to achieving them. This commitment may arise because people have decided on a goal for themselves, or because they have had a say in setting the goals. Where goals are simply imposed on people without prior consultation and agreement, they are rarely motivating. The remainder of the points below assume that the goals have been accepted by the people concerned.
- Goals which are difficult to achieve are more motivating than goals which are easy to achieve.
- Specific goals are more motivating than general goals.
- Feedback is essential if goals are to continue to motivate. This feedback usually takes the form of knowledge of results of particular actions and the extent to which the actions have contributed to the achievement of the goals.

Various texts, including Annex D of BS 18004,[2] refer to these general principles, which are often summed up with the acronym SMART. This encapsulates the requirements for goals that will motivate effectively:

- Specific
- Measurable
- Achievable
- Relevant
- Timely.

The main problem with goals as motivators is getting people to commit themselves to the goals in the first place. As was mentioned earlier, unless this happens the goals are not motivating for the people concerned. However, there is a variety of ways of getting people to commit themselves to goals, including participative decision making during goal setting and a 'tell and sell' management style.

Relevance of theories of motivation

There is no single theory of motivation which enables an accurate prediction of the effect on an individual's behaviour that will result from the manipulation of factors considered likely to affect motivation. This is partly due to the weaknesses in the theories themselves, but it is mainly due to the wide range of individual differences. As we pointed out earlier, what motivates one person may not motivate another, and what motivates a person today may not motivate that person tomorrow. These individual differences will be considered in the remainder of this chapter but, to end this section on motivation, there are two points in favour of having a working knowledge of motivation theories.

1. Although it may not be possible to predict an individual's motivation at a given time, it is possible to make general statements about the motivation of numbers of people over an extended period of time. If it is not clear why this is so, think about the simple example of throwing a coin. It is not possible to predict

whether a coin will come down heads or tails on a given throw but over, say 1,000 throws, it is possible to predict that there will be roughly 500 heads and 500 tails.

2. A knowledge of the motivation theories provide an awareness of the sorts of issues which are relevant when considering how to motivate people to take, or not to take, certain actions. By taking into account these issues when reaching decisions on how to influence people's behaviour, a more systematic approach should be possible and successful intervention will be more likely.

As we have already mentioned, a major problem with theories of motivation is the way their predictions are confounded by individual differences. A number of aspects of individual differences have this effect, and we will now consider three of the most important of these: intelligence, personality and attitudes.

Intelligence

As long ago as 1923, intelligence was defined by Boring[89] as 'what intelligence tests measure', which was a neat way of avoiding the question of what intelligence actually is. It will probably not be surprising to learn that in the succeeding years there has still been no agreement on the definition of intelligence. However, there are some general findings on intelligence about which there is agreement, even if there is no agreement about what the findings mean. This section deals with some of the more important of these general findings, starting with the concept of Intelligence Quotient.

Intelligence Quotient

Most people are familiar with 'intelligence tests' consisting of a number of questions of various levels of difficulty. Since the first of these tests was developed by two French psychologists (Binet and Simon[90]) early in the 20th century, thousands of variants have been created in different countries and for different purposes. However, they are all based on the same premise: the 'better' a person does in the test, the more 'intelligent' that person is.

If a number of people take the same test, under the same conditions, it is possible to make comparisons between these people and, if appropriate, to use the results to select people for schools, universities or jobs. Indeed, this was the motivation behind Binet and Simon's original tests: they wanted to identify 'retarded' children who would have special educational needs. However, it was not adequate to identify children as retarded simply because they were in, say, the bottom 5 per cent of a sample of children tested. While the bottom 5 per cent of the whole population of children may have been defined by Binet and Simon as retarded, they had no way of knowing whether a particular sample was representative of the population. What was required, therefore, was a way of describing how a particular child's performance compared with the general population of children.

The solution adopted by Binet and Simon, and which is still in use, was the Intelligence Quotient (IQ), which is calculated as follows:

1. A large number of children of different ages were tested using the same range of tests and the average* scores for five-year-old children, six-year-old children, and so on, were calculated.

2. The score for an individual child could then be compared with these average scores until a match was found between that child's score and the average score for an age group.

3. The child's IQ was then calculated using the following equation:

$$\frac{\text{Age represented by the child's test score}}{\text{Actual age of child}} \times 100$$

---

* More accurately, the mean. See Chapter 19.

Note the following points about the IQ:

- **Children whose performance on the tests is commensurate with their age will have an IQ of 100, so that an IQ of 100 is often referred to as 'the average IQ'.** Children whose performance is higher than would be expected for their age will have an IQ greater than 100, and children whose performance is lower will have an IQ of less than 100.
- **Various methods are used to increase the sensitivity of the IQ score but the majority are based on the principle that the normal scores vary from age group to age group.** For example, one simple method of increasing sensitivity is to measure age in years and months rather than just in whole years.
- **In theory, it does not matter at what age IQ is measured since intelligence is assumed to be a characteristic of an individual which does not vary over time.** In practice, this is not the case, since the majority of IQ tests require reading, writing and numerical skills, all of which can be taught. However, a distinction has to be made between improving a child's performance on IQ tests, which can be done by appropriate teaching, and improving underlying intelligence. This distinction will be dealt with in more detail later under the heading of the nature/nurture controversy.

Although the original work on IQ (and much subsequent work) has been with children, there is also a need to measure the IQs of adults. In these cases, age-based IQ measures are not appropriate. This is because the performance of adults in IQ tests does not normally change beyond the age of about 18. For adults we need a measure of the general spread of IQ scores in the population as a whole, against which the score of a particular individual can be compared. These population scores are referred to as 'norms' and it is possible to look up an individual's score in a table of norms for the test he or she has taken. It is then possible to make such statements as 'This person is in the top 5 per cent of the population (on IQ scores)', or 'This person has an average IQ'.

The main problem with IQ is the one referred to at the beginning of this section: the only certainty about intelligence is that 'intelligence is what is measured by intelligence tests'. Although this has practical benefits in, for example, selection, it is unsatisfactory on a theoretical level since it contributes little to an understanding of what intelligence actually is. For this reason, we need to look briefly at the nature of intelligence.

The nature of intelligence

The simplest way of considering intelligence is probably to treat it as an individual's ability to process information or, putting it another way, an individual's ability to think, referred to by psychologists as cognitive ability.

IQ tests measure cognitive ability in a standardised and very structured manner and not all psychologists agree that this is the sort of intelligence which is used, or is useful, in real life. The dissenting psychologists refer to this 'real life' intelligence as 'practical' intelligence, which differs from intelligence as measured by IQ tests in a number of ways, including these:

- In IQ tests, the problem to be solved is clearly set out, while in real life there can be major difficulties in identifying accurately the nature of the problem. People with high levels of practical intelligence are good at putting problems into a form which makes them more easily solved.
- In IQ tests, everything which can be standardised is standardised, including the instructions given, the time available to complete the test, and the correct answer. In real life, intelligence has to be applied in non-standard settings and there may be more than one 'correct' answer.

These and other differences between the two types of intelligence may mean that practical intelligence is more desirable when deciding on the intelligence requirements for a job or when specifying job competences. Irrespective of the type of intelligence, there are many unresolved issues which are the subject of ongoing research and debate. Two of these – the nature/nurture controversy and the structure of intelligence – are described next.

The nature/nurture controversy

The two extremes of this controversy can be summarised as follows:

1. **Nature.** People are born with a certain level of intelligence which is genetically determined and nothing that happens in their life can increase this inborn level of intelligence.
2. **Nurture.** People's intelligence is totally determined by what happens during their life, with factors such as a visually rich environment in early life and good teaching increasing their intelligence.

There have been numerous attempts to determine which of these extremes is the correct one, including studies of identical twins. The rationale for these studies is that the genetic make-up of identical twins is, by definition, the same, so that any differences in intelligence must be due to 'nurture' factors. However, since the nurture factors are likely to be similar for twins brought up together, the studies have involved twins who have been separated at, or soon after, birth.

For a variety of reasons, even these studies of separated identical twins did not provide conclusive proof, and for most practical purposes it is now assumed that neither of the extremes is correct. The simplest way to deal with the nature/nurture controversy is to assume that a person's potential intelligence is genetically determined (nature) but that the actual level of intelligence reached will depend on factors such as environment and education (nurture), especially early in life.

Before leaving the nature/nurture controversy, however, it is necessary to look at one area where it has important practical implications – that is, in racial and other forms of discrimination. There are people who argue that because some races do less well in intelligence tests than others, it is because they are genetically less intelligent. This is not true for the following reasons:

- IQ tests are designed for a particular group and use vocabulary and problem-solving methods common to that group
- the norms for a particular test are calculated for a specific group.

Thus, if a test designed for white UK males is used on a different group, this group may do badly for 'nurture' reasons (education, vocabulary, etc) irrespective of any genetic differences.

This poses problems when intelligence tests are used in selection procedures which involve candidates from different races, religions or ethnic groups. Some candidates may be rejected on the basis of low scores which are not a true measure of their intelligence (because an inappropriate test has been used) and, in the UK at least, this can be construed as an illegal form of discrimination.

The structure of intelligence

Although intelligence has been described as cognitive ability, it is easy to observe that people differ not only in their overall ability but also in particular types of ability. For example, some people have high levels of ability in dealing with numbers, while others are particularly able with words. People can also show high levels of the sort of intelligence measured by intelligence tests without demonstrating a similarly high level of ability on practical intelligence.

These and similar observations have led psychologists to suggest that there is a structure to intelligence of the type illustrated in Figure 26.1.

General intelligence is thought to play a part in all cognitive performance, so people with high general intelligence are, at least potentially, able to perform at relatively high levels in all of the subdivisions of intelligence. The fact that they may not do so is usually attributed to, for example, lack of motivation to perform or lack of opportunity to learn.

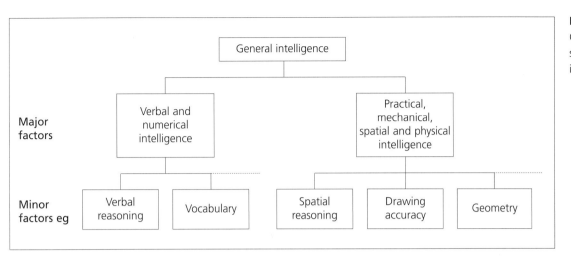

Figure 26.1
Generalised
structure of
intelligence

The main subdivision below general intelligence can be thought of as the division between the abstract world and the concrete world, and it is a common observation that individuals tend to perform better in one rather than the other, perhaps due to the motivational and learning factors just mentioned. Some theorists argue that there should be a third subdivision for emotional intelligence, but this is not, as yet, generally accepted.

Below the level of this main subdivision, there are further subdivisions into minor factors; some examples of these are given in Figure 26.1. These minor factors are separated out for a number of reasons, including the fact that particular individuals can have intelligence levels on particular factors that are much higher than their level of general intelligence would lead an observer to expect. This has two important practical implications:

1. **If a job requires a particular type of cognitive ability, this must be measured directly.** It is unlikely that a measure of general intelligence will give sufficiently accurate results, since even people with high levels of general intelligence may be low on a specific ability.
2. **People with high levels of verbal intelligence can give the impression that they also have a high level of general intelligence.** This can be of concern, for example, in job interviews where a person with a high level of verbal intelligence will 'interview well' and may, deliberately or otherwise, hide a lack of intelligence of the type required for the job for which he or she is applying.

In some variations of the generalised structure of intelligence shown in Figure 26.1, a further level is added, variously described as specific abilities or aptitudes. This level is used to describe either further subdivisions of the minor factors or to include aptitudes, such as musical aptitude, which are not normally included in the overall structure of intelligence.

To conclude this brief discussion of intelligence, it is worth repeating that levels of general intelligence are not good predictors of levels of specific types of intelligence. This means that job requirements and competences have to be specified in terms of the types of intelligence required, and these types of intelligence have to be measured if intelligence testing is to be used successfully in selecting people for particular tasks.

Personality

The psychology of personality is probably one of the most complex and most researched areas of psychology. This is understandable if we remember that, for most psychologists, personality is what defines the individual. In essence, having a particular personality is what makes people behave in particular ways.

However, as with other aspects of psychology, there is no general agreement about how personality should be defined, nor about the links between particular types of personality and the behaviours associated with them. This section will, therefore, contain a brief overview of some of the major approaches to personality theory, their main weaknesses and, where appropriate, their practical applications.

Psychoanalytic theories

The most famous psychoanalytic theory is the one developed by Freud, but alternative psychoanalytic theories have been put forward by such people as Jung, Erikson and Adler.

In outline, Freud's theory was that personality was determined by instinctive forces, collectively known as the psyche, and that different personality types were created by different balances among these psychic forces. Freud identified three main psychic forces:

1. **The id** is the expression of people's instinctual energies, such as sex and aggression. It is the 'primitive' facet of the individual and it seeks immediate gratification of what it leads the individual to desire – that is, it operates on the pleasure principle. The id, operating unchecked, has no inhibitions and cannot distinguish between fantasy and reality.

2. **The ego**, unlike the id, operates on the reality principle: it can tolerate delays in gratification and can distinguish between fantasy and reality. The role of the ego is to direct the impulses of the id so that they are expressed in socially acceptable ways and at appropriate times. However, the ego cannot eliminate or block the impulses from the id, it can only direct them.

3. **The superego**, unlike the id and ego which are present from birth, only develops during childhood. Standards set by a child's parents are internalised by the superego which becomes the child's conscience and governs the child's morality. The superego operates on the perfection principle: it tries to get the individual to conform to the internalised ideal standards received from the parents.

An individual's personality, according to Freud, is determined by the current balance in the struggle between the three facets of the psyche, most of which takes place subconsciously. Freud was primarily interested in personality disorders which, he said, were due to inappropriate balances between the three facets, for example a too dominant id. Since the causes of this imbalance were likely to be subconscious or unconscious, Freud had to use special techniques to identify them. These techniques included the study of dreams and slips of the tongue (Freudian slips) where people 'accidentally' say what they really mean rather than what it would be socially acceptable to say.

Freud believed that no behaviour was truly accidental, but that people could not explain the reasons for their behaviour. In its extreme expression, this would mean that at least a proportion of the people injured in accidents 'wanted to get hurt' but cannot explain why this is what they wanted. As with many aspects of psychology, this will be true in some cases. There are people who indulge in self-mutilation for a variety of reasons and getting hurt at work may be seen as a more 'socially acceptable' form of self-mutilation. Some cases of stress-related ill health at work can also be explained in Freudian terms; for example, people with an overdeveloped superego (which works on the perfection principle) can succumb to stress if they have insufficient time to work to the high standards their superego imposes on them.

Many aspects of Freudian theory are no longer considered valid or useful, but a number of his ideas have passed into everyday use (for example, Freudian slips) and others have been adopted as useful explanations of particular types of behaviour. An example of the latter category is what Freud referred to as defence mechanisms which, he argued, were methods people used to avoid unpleasant anxiety. Typical defence mechanisms include:

- **Denial.** People in an extreme form of denial simply pretend that things are not really as they are. For example, people in extreme denial over being made redundant simply refuse to accept that they no longer have a job and may genuinely believe that they are on extended sick leave, despite all evidence to the contrary. Less extreme forms of denial lead to ignoring any facts which are causing the anxiety, such as letters from bank managers about overdrafts.

- **Projection.** People using projection attribute their own faults to other people and then get indignant about these supposed faults and, not infrequently, take action to try to correct them. For these people, dealing with other people's attributed faults is used as a substitute for dealing with their own faults.
- **Reaction formation.** People using reaction formation deal with their socially unacceptable impulses by making strong protestations of the opposite view. For example, someone expressing strong anti-pornography views may be doing so as a way of dealing with their own true sexual desires.

Freud argued that having to use defence mechanisms detracted from people's capacity to live a full life and that it was important, therefore, to identify the reasons why people used such mechanisms and remove the need for them.

Psychoanalysis in various forms is still used in the treatment of mental illness but it is a time-consuming and, hence, expensive approach.

An increasingly popular form of therapy is cognitive behaviour (or behavioural) therapy (CBT), which focuses on the thoughts, actions and feelings created by particular situations. The aim of the therapy is to change the way people think and feel about situations so that they move from negative to positive thoughts and feelings. In parallel, the therapy also aims to change negative behaviours to positive ones. By working on thoughts, actions and feelings together, it is possible to break the vicious circle in which negative thoughts and feelings produce negative actions, which in turn create additional and/or stronger negative thoughts and feelings.

Trait theories

There are two basic variants of the trait theory approach:
1. a fixed number of personality types into one of which each individual fits
2. a number of dimensions of personality, each a continuum, and an individual takes a characteristic place on the continuum for each dimension.

The first approach has a long history going back to Hippocrates in the late 5th century BC (he of the Hippocratic Oath), who put individuals into one of the following four categories:
1. **phlegmatic** – these people were calm, unexcitable and unaggressive
2. **choleric** – these people were irritable and quick-tempered
3. **sanguine** – these people were cheerful and optimistic
4. **melancholic** – these people were sad and depressed.

This simple classification by Hippocrates can be used to illustrate a number of important factors about personality:
- Although individuals may be classed as, for example, phlegmatic, they are not calm all of the time. While modern evidence suggests there is a genetic element in personality, which helps to explain why personality is relatively stable, the majority of variation in personality (about two-thirds) is due to non-genetic factors.
- It appears, *a priori*, that the population as a whole contains more than four different types of personality. There have been attempts to produce classifications with greater numbers of categories, including the signs of the Zodiac. However, they all suffer from the same problem of a limited number of categories.
- If someone is classed as phlegmatic, it provides information about one aspect of his or her personality, but gives no information on how optimistic he or she is. This can be used to illustrate why a dimensional approach to personality is to be preferred. Hippocrates's four categories are really two dimensions of personality which could be described as shown in Figure 26.2.

Figure 26.2
Dimensions of
personality

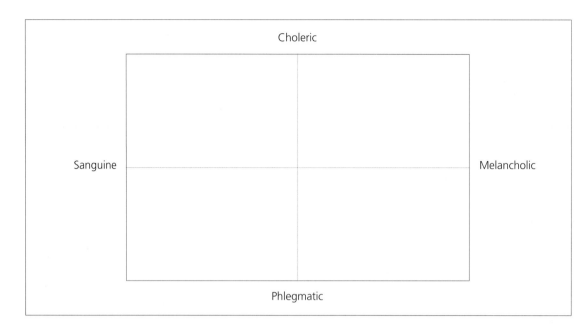

Using Figure 26.2, it would be possible to place an individual anywhere in the two-dimensional space defined by the sanguine–melancholic and phlegmatic–choleric dimensions. This is, in essence, the approach used by the trait theorists, although they differ in the number of dimensions (traits) they use and what they call these traits.

An individual's personality is usually measured by using some form of personality questionnaire which consists of a number of questions. On the basis of someone's answers, his or her position on each trait is determined and the combination of trait positions describes the person's personality or, as it is more usually called, his or her personality profile.

As will be illustrated below, different researchers have identified different numbers and types of trait, but there appears to be an emerging consensus that the following five traits are likely to be the most important:

1. **extroversion–introversion:** extroverts are lively, sociable, excitable people while introverts have the opposite characteristics
2. **neuroticism:** neurotics are characterised by high levels of anxiety and tension
3. **conscientiousness:** conscientious individuals are well organised and focus on targets, goals and deadlines
4. **agreeableness:** agreeable individuals are good-natured, eager to co-operate with others, and concerned to avoid conflict
5. **openness to experience:*** people who are open to experience are influenced by any new experience they encounter and show high levels of curiosity about what goes on around them.

The assumption underlying trait theory is that a person occupying a particular location on a trait (for example, either of the extremes or somewhere near the middle) will behave in certain ways. Taking this to its logical conclusion, therefore, if we want people to behave in certain ways at work, then we should select people who have the personality traits which are linked with the behaviour we want. This is only possible if there are ways of measuring personality and there is a small industry based on the development and sale of personality tests, usually in the form of personality questionnaires.

---

* This trait is also described as 'intellect' because it is related to, but not the same as, the sort of intelligence measured by intelligence tests.

There are many different personality questionnaires but well-known ones used in the UK include:

- the **Eysenck Personality Questionnaire** (EPQ), which measures only three traits: extroversion, neuroticism and psychoticism
- the **Cattell 16 Personality Factor Questionnaire** (16 PF) which, as its name suggests, measures 16 traits including submissiveness, self-assurance and tendermindedness; a full list of the traits is given in Table 26.1
- the **Saville and Holdsworth Occupational Personality Questionnaire** (OPQ32), which measures 32 traits considered to be relevant in the context of work. The traits are grouped into 'relationships with people' (R1–R10), 'thinking style' (TS1–TS12), 'feelings and emotions' (FE1–FE10), and 'social desirability' (SDE). The full list of these traits is given in Table 26.2.

Factor	Low score characteristics	High score characteristics
A – Warmth	Reserved, detached, critical	Outgoing, warm-hearted
B – Reasoning	Less intelligent, concrete thinking	More intelligent, abstract thinking
C – Emotional stability	Affected by feelings, easily upset	Emotionally stable, faces reality
D – Dominance	Humble, mild, accommodating	Assertive, aggressive, stubborn
F – Liveliness	Sober, prudent, serious	Happy-go-lucky, impulsive, lively
G – Rule consciousness	Expedient, disregards rules	Conscientious, persevering
H – Social boldness	Shy, restrained, timid	Venturesome, socially bold
I – Sensitivity	Tough-minded, self-reliant	Tender-minded, clinging
L – Vigilance	Trusting, adaptable	Suspicious, self-opinionated
M – Abstractedness	Practical, careful	Imaginative
N – Privateness	Forthright, natural	Shrewd, calculating
O – Apprehension	Self-assured, confident	Apprehensive, self-reproaching
Q1 – Openness to change	Conservative	Experimenting, liberal
Q2 – Self reliance	Group-dependent	Self-sufficient
Q3 – Perfectionism	Undisciplined, self-conflict	Controlled, socially precise
Q4 – Tension	Relaxed, tranquil	Tense, frustrated

Table 26.1
Cattell's 16
personality factors

These personality questionnaires are powerful tools which, if used incorrectly, can have adverse effects on the individuals 'tested'. For this reason, they are normally only available to relevant professionals, such as chartered psychologists, or to those who have been trained in their use.

While personality questionnaires have been used successfully, they have to be treated with care because of the following potential problems:

- **The link between personality, as measured by a questionnaire, and behaviour may be situation-specific.** Links will normally have been identified during some form of research which will have been conducted in

Table 26.2
Occupational personality questionnaire traits

Trait name	High score characteristics
Persuasive (RP1)	Enjoys selling, comfortable using negotiation, likes to change other people's views
Controlling (RP2)	Likes to be in charge, takes the lead, tells others what to do, takes control
Outspoken (RP3)	Freely expresses opinions, makes disagreement clear, prepared to criticise others
Independent minded (RP4)	Prefers to follow own approach, prepared to disregard majority decisions
Outgoing (RP5)	Lively and animated in groups, talkative, enjoys attention
Affiliative (RP6)	Enjoys others' company, likes to be around people, can miss the company of others
Socially confident (RP7)	Feels comfortable when first meeting people, at ease in formal situations
Modest (RP8)	Dislikes discussing achievements, keeps quiet about personal success
Democratic (RP9)	Consults widely, involves others in decision making, less likely to make decisions alone
Caring (RP10)	Sympathetic and considerate towards others, helpful and supportive, gets involved in others' problems
Data rational (TS1)	Likes working with numbers, enjoys analysing statistical information, bases decisions on facts and figures
Evaluative (TS2)	Critically evaluates information, looks for potential limitations, focuses on errors
Behavioural (TS3)	Tries to understand motives and behaviour, enjoys analysing people
Conventional (TS4)	Prefers well-established methods, favours a more conventional approach
Conceptual (TS5)	Interested in theories, enjoys discussing abstract concepts
Innovative (TS6)	Generates new ideas, enjoys being creative, thinks of original solutions
Variety seeking (TS7)	Prefers variety, tries out new things, likes changes to regular routine, can become bored by repetitive work
Adaptable (TS8)	Changes behaviour to suit the situation, adapts approach to different people
Forward thinking (TS9)	Takes a long-term view, sets goals for the future, more likely to take a strategic perspective
Detail conscious (TS10)	Focuses on detail, likes to be methodical, organised and systematic, may become preoccupied with detail
Conscientious (TS11)	Focuses on getting things finished, persists until the job is done
Rule following (TS12)	Follows rules and regulations, prefers clear guidelines, finds it difficult to break rules
Relaxed (FE1)	Finds it easy to relax, rarely feels tense
Worrying (FE2)	Feels nervous before important occasions, worries about things going wrong
Tough minded (FE3)	Not easily offended, can ignore insults, may be insensitive to personal criticism

Optimistic (FE4)	Expects things will turn out well, looks to the positive aspects of a situation, has an optimistic view of the future
Trusting (FE5)	Trusts people, sees others as reliable and honest, believes what others say
Emotionally controlled (FE6)	Can conceal feelings from others, rarely displays emotion
Vigorous (FE7)	Thrives on activity, likes to be busy, enjoys having a lot to do
Competitive (FE8)	Has a need to win, enjoys competitive activities, dislikes losing
Achieving (FE9)	Ambitious and career-centred, likes to work to demanding goals and targets
Decisive (FE10)	Makes fast decisions, reaches conclusions quickly, less cautious
Social desirability (SDE)	Has been less self-critical in responses, is more concerned to make a good impression

Table 26.2
continued

a (usually small) number of situations. There is no guarantee that the link will be as strong, or exist at all, in a wider range of situations.

- **The questionnaires rely on self-reporting, since people answer questions about how they have behaved or would behave.** There are methods of detecting if people lie when answering the questions, and most questionnaires have lie scores incorporated in them. However, as we pointed out during the discussion on psychoanalytic theories of personality, people are very poor at describing their own behaviour and the reasons for it. A much better assessment of personality would be obtained by watching what people actually do (rather than asking them what they have done or would do), and some personality research has been carried out using these techniques. However, their use in selection and other practical applications is precluded because they are too time-consuming and, hence, too expensive.

To end this section on personality, there is a final caveat which can be set out as follows:
- people who have a certain personality trait behave in certain ways – for example, extroverts spend a lot of time with other people
- people who indulge in certain types of behaviour are said to have a certain personality – for example, people who spend a lot of time with other people are said to be extroverts
- no-one has so far established whether, or how, a personality trait determines behaviour or, more fundamentally, what causes some people to have one type of personality (or exhibit a particular type of behaviour) rather than another.

Attitudes

This section begins with a discussion of what is meant by attitudes and the relationship between attitudes and behaviour. It continues with a description of how attitudes are measured, paying particular attention to measuring attitudes towards safety. The section ends with an overview of the factors which influence attitude change.

Definitions of attitudes
As with practically every other concept in psychology, there are various definitions of attitude. If, however, the various definitions are considered together, the following three components of attitudes emerge:
1. **a cognitive component** – how a person perceives the object of the attitude and what the person thinks and believes about the object of the attitude

2. **an affective component** – how a person feels about the object of the attitude
3. **a behavioural component** – how a person acts towards the object of the attitude.

The phrase 'object of the attitude' has been used for convenience in these descriptions, but people can have attitudes about any physical aspect of their environment, abstract concepts (such as safety), or people and animals. In effect, people can have an attitude about anything at all.

In most discussions of attitudes, it is the cognitive component which is considered to be the most important and it is the one which is most frequently researched. It will also be the one which we deal with in most detail in this section. However, there are certain attitudes which are strongly influenced by the affective component; phobias (irrational fears) strongly influence people's attitudes towards, for example, spiders and snakes (even harmless ones) and high places.

The inclusion of the behavioural component is more contentious, as many people argue that behaviours are an outcome of attitudes, not part of the attitudes. In other words, because people have a certain attitude towards an object, they behave in a particular way towards that object. The counter-argument is that true attitudes can only be inferred from behaviour, since what people think about the object of an attitude and how they behave towards it may not match up. This has already been illustrated in Part 1.2 (see Table 15.2).

In practical terms, it could be argued that the study of attitudes is only useful to the extent to which attitudes influence behaviour, and since this is a major assumption underlying work on safety culture, it was dealt with as a separate topic in Chapter 21. However, measuring attitudes to safety is only a special case of attitude measurement, and what follows is a description of how attitudes in general are measured.

Measuring attitudes

All three of the components of attitudes listed earlier can be measured in direct ways:

1. **The cognitive component of attitudes can be measured by any of the techniques which enable discrimination between true and false statements.** Examples of these include drugs (the so-called 'truth drugs'), hypnosis and the polygraph, which is commonly known as the 'lie detector'.
2. **The affective component of attitudes can be measured using a variety of physiological concomitants of affect.** This is rather a grandiose way of saying that feelings such as fear or pleasure influence such things as pulse rate, pupil dilation, blood pressure and rate of sweating. The last of these can be determined by measuring the electrical conductivity of the skin, the so-called galvanic skin response (GSR). The more people sweat, the better their skin conducts electricity, and people tend to sweat more when they are under stress. The GSR is often used as one part of the lie detector test. On a more complex level, the affective component of attitudes can also be investigated using electroencephalograms (EEGs), which are a record of the electrical activity in the brain.
3. **The behavioural component can be measured by observation.** Watch what people do and infer their attitudes from the behaviour identified. However, as we mentioned earlier, this may not be particularly accurate; people may wear hard hats because of situational factors (a strict supervisor) rather than any attitude they may have towards hard hats (see Table 15.2).

However, all of the techniques for direct measurement of attitudes are time-consuming and some require high levels of expertise. For these reasons, it is more common to measure attitudes indirectly by using self-report – by asking people how they think, feel and behave towards the object of the attitude. Self-reported attitude measurement will now be considered in more detail.

The most common method used for attitude measurement is some form of questionnaire. The development of adequate questionnaires is a highly skilled and time-consuming activity and not simply a matter of collecting together a set of likely-looking questions and asking an easily accessible group of people to answer them. An extensive review of the techniques for developing attitude survey questionnaires are beyond the scope of this book, but two common techniques are as follows:

1. Instead of being asked questions requiring a 'yes' or 'no' answer, the people completing the questionnaire are given a list of statements and asked to what extent they agree or disagree with them.

2. Instead of being asked what they think or feel themselves, people are asked what they think other people think or feel. This so-called attributional technique is used because it is believed that people will attribute their true opinion to others when they are not willing to express it as their own. For example, if 10 people working for the same boss were asked what they thought of their boss, it is likely that they would give a positive opinion, especially if they thought that the survey results might get back to their boss. However, if they were asked what the other nine thought of the boss, the response might be 'I think the boss is OK but the others think he's terrible'. This attribution allows people to express their true opinion without taking responsibility for that opinion.

Table 26.3 shows examples of attributional and non-attributional forms of statements relevant to health and safety.

Attributional form	Non-attributional form
People here are not clear about their health and safety responsibilities	I am not clear about my health and safety responsibilities
People here take risks to get the job done	I take risks to get the job done
People here pay no attention to health and safety	I pay no attention to health and safety
People here do not understand the health and safety risks	I do not understand the health and safety risks

Table 26.3
Examples of attributional and non-attributional forms of statements

The problem with all self-report questionnaires is that there is not always an adequate method of ensuring the truth of the answers. Various techniques, including appropriate attributional items, can be used to measure the likelihood of truthful answers and, in some circumstances, it is possible to include a lie score which measures the likely truthfulness of answers. However, the fundamental problem of level of truthfulness is always present.

This can have major practical implications in, for example, a survey of attitudes to safety where people know the 'acceptable' answer (hard hats are a good thing) and give this answer, rather then the one which truly reflects what they think, feel or do.

Another problem with self-report questionnaires is that they usually measure only the extent to which people agree or disagree with particular statements. They do not go on to measure how strongly people hold these attitudes, how important they are to them, or how much they know about the object of the attitude. It has been argued by some psychologists that these factors are the important ones in determining the extent to which attitudes influence behaviour. For example, people are more likely to behave in ways which are consistent with their attitudes if these attitudes, or the objects of these attitudes, are important to the people concerned.

This distinction can be illustrated by examples:

• If 100 people were asked whether they disliked spiders, 20 of them might be in the 'strongly agree' category at the top of the scale. However, the lives of the majority of these 20 are unlikely to be affected in any major way by this dislike, while for a few, the extreme phobics, their lives can be seriously disrupted by the presence of a spider.

- If 100 people were asked whether they approved of smoking, 80 might disagree – that is, they are anti-smoking. However, it is likely that some of these would themselves be smokers and, of the non-smokers, some would take no action on the basis of their expressed attitude while others might be active members of an anti-smoking organisation.

These examples show that even when a person's attitude is known and there is some measure of the strength of that attitude, this does not, *per se*, provide information on the behaviour which will follow. Table 26.4 provides a summary of the components of attitudes and the ways in which they can be measured.

Table 26.4
The components of attitudes and their measurement

	Components of attitudes		
	Cognitive	Affect	Behaviour
Self-report questions – questions based on the stems...	What do you think about...? What would you think about...?	How do you feel about...? What would you feel about...?	What do you do about...? What would you do about...?
Direct measures	Hypnosis Truth drugs Lie detectors	Galvanic skin response Electroencephalogram	Observation

Two basic question stems are used for self-report measures. The first, 'What do you...' is used to measure attitudes about existing circumstances, while the second, 'What would you...' is used to measure attitudes to hypothetical circumstances. The latter form would be used, for example, to determine attitudes to a proposed new product, shift pattern, risk control measure or payment structure.

Changing attitudes

Before discussing how attitudes can be changed, it is important to be clear about what we are aiming to change, since, as we noted at the beginning of this section, there are three components of attitudes – cognitive, affect and behaviour – and two ways of measuring each of these – indirect (self-report) and direct.

It could be argued that the only important combination is direct measurement of behaviour, since there is little practical value in changing attitudes unless there is also a change in behaviour. On a less extreme level, it could be argued that only direct measures are important since there are only limited ways of ensuring the veracity of self-report measures. However, it is much easier to use self-report measures for both research and consultancy. For example, on the most basic level, it is possible to get the results from an attitude survey questionnaire for 10, or 100, or 1,000 people in an hour, depending on the size of the available lecture hall or canteen. Alternatively, large numbers of questionnaires can be sent out electronically with very little effort. It is then possible to get the results scored and analysed by a research assistant or secretary. In contrast, the amount of time required to observe individuals' behaviour is proportional to the number of people to be observed, and the observations have to be carried out by people competent in this work, both of which requirements make this technique time-consuming and expensive.

Because of the large differences in the resources required, much research and consultancy work on attitude change focuses on changes in self-reported attitudes with no check on the extent to which these are correlated with changes in behaviour. What is required, however, is more evidence linking attitude change with stable changes in behaviour.

Having dealt with these caveats, we can now list some of the factors which make it more likely that attitudes will be changed:

- The higher the credibility of the communicators, the more likely it is that they will change people's attitudes. Credible communicators are those who can demonstrate that they are experts in the relevant subject and who have a record of honesty, or appear to be honest by, for example, arguing against their own interests.
- Communicators who are perceived to be attractive by those whose attitudes they are trying to change are more likely to be successful in changing attitudes. Note, however, that there are major variations in what people find attractive and that a communicator that some members of an audience find attractive may not be attractive to other members of that audience.
- Communicators who present both sides of an argument, but clearly state why they think one side of the argument is to be preferred, are more likely to change attitudes than communicators who present only their preferred side of the argument. However, this may not be the case when the people whose attitudes are to be changed are of limited cognitive ability, or are unfamiliar with the issues involved. In these cases, presenting only the preferred side of the argument is more likely to be successful.
- Communicators who present information to groups of people and then allow the groups to discuss the issues involved are more likely to change attitudes than communicators who only present the information.*

In safety, a technique of attitude change which is particularly relevant is the use of fear – for example, fear of injury or ill health, or fear of prosecution for contravening health and safety legislation. While fear has been used to change attitudes, the relationship between the amount of fear induced and the degree of attitude change is not a straightforward one. There is evidence to suggest that when the level of fear arousal is fairly low, the attitude change is greater than when high levels of fear arousal are used. This may be because people do not like being frightened and ignore messages they find too frightening.

Even when appropriate levels of fear arousal are used, changes in attitudes are only likely to be successfully achieved if the following conditions are met:
- the threat is a serious one
- the threat may affect the person whose attitude is to be changed
- there is some specific action which the person can take to avoid the threat
- the person is capable of performing the action required to avoid the threat.

These are the sorts of conditions which will have to be met if, for example, poster campaigns illustrated with injured people or campaigns to change managers' attitudes based on fear of prosecution are to be successful. Finally, remember that, although the change factors listed above apply to all three components of attitude, what is important in practical safety terms is behaviour change.

So far in this chapter the emphasis has been on the dimensions of individual differences – motivation, personality, and so on – with a brief mention of how differences in nature and nurture might produce these differences. The final section of the chapter deals more generally with the causes of individual differences under two headings: genetic factors and cultural factors.

Genetic factors

Many individual differences are genetically determined. These include obvious ones such as height and hair colour as well as less obvious ones such as intelligence and speed of reaction. In general, individual differences which are genetically determined are fixed and have limited or no scope for modification, and what happens from conception onwards determines only whether individuals reach their full genetic potential. For example, if

---

* This has obvious practical implications for designing training courses. These should, ideally, consist of briefings (to present information) followed by syndicate group exercises during which the delegates can discuss the issues.

the full genetic potential is for a height of six feet, this will only be reached if there is an adequate supply of nutrients and vitamins from conception to the end of the growth phase. Similarly, if the genetic potential is for an IQ of 140, this will only be reached if there is adequate nutrition and a sufficiently rich learning environment.

Within each factor which is genetically determined, there can be a wide range of variability, as illustrated by height and IQ. There are also many of these genetically determined factors, so the combination of intra-factor variability and the large numbers of factors involved make every individual genetically unique (with, of course, the exception of identical twins).

In addition, there is the variability in other individual differences such as attitudes which are not thought to be genetically determined but produced by the culture in which the individual lives.

Cultural factors

Everyone lives in a culture and the culture in which they live determines a large number of individual differences. Some examples of individual differences which are culturally determined are:

- **Language.** It is obvious that different cultures use different languages but what is less obvious is that different languages impose different structures of thinking and contain concepts which are not used in other languages. There is a temptation to assume that it is always possible to make direct translations between one language and another, but this is not necessarily the case. For example, the Chinese concept translated as 'face' is so complex and so interlinked with other aspects of Chinese culture that it is unlikely that anyone not brought up in the Chinese culture can fully understand what it means.
- **Modes of thought.** Partly these differences are a function of language but differences in modes of thought are more fundamental than the differences inherent in different languages. Concepts such as 'personal property' and 'truth' vary in meaning from one culture to another, although people brought up in one culture find it difficult to appreciate that the definition they use is not the only possible one.
- **Interpersonal relationships.** The culture determines the appropriate ways of behaving towards other people in that culture. Again, there are wide variations between cultures, perhaps most markedly in the behaviour expected between the sexes.
- **Body language.** What gestures and body postures mean are specific to a culture and can have very precise meaning not understood by those outside that culture. In the UK, for example, the meaning of raising the first two fingers in a 'V' sign depends on whether the palm is turned towards or away from the audience, and this subtle distinction is not always appreciated by members of other cultures.

The examples given above apply to broad cultures such as nations and ethnic groups, but within these broad cultures there will be subcultures each with its own influence. For example, within a language there will be differences in accents and dialects.

Irrespective of the nature of the culture, it appears to have its influence on individuals in the following main ways:

- **Providing a model for imitation.** Children have an apparently innate tendency to behave in ways which are similar to the ways in which the people around them behave, so, for example, children will imitate their parents. This imitation is not necessarily conscious but it is a very powerful formative influence in infancy and childhood. It explains, for example, why children adopt the local accent and the local preferences in food and drink.
- **Conditioning.** Behaviour which is rewarded, or is found to be rewarding, tends to be repeated while behaviour which is punished, or found not to be rewarding, tends not to be repeated. Many aspects of training, education and experience provide information on new behaviours and their likely results and individuals can then try these behaviours for themselves (or avoid them, in the case of behaviours with clearly unpleasant results).

Operating through either or both of these mechanisms, the psychological make-up of the people with whom individuals interact, the nature of the society in which the individuals live and the different subcultures to which individuals are exposed all influence how individuals think and behave. Since every individual will have had a unique combination of experiences with respect to these psychological, sociological and anthropological factors, the result is that each individual is unique. Thus, these factors have a major role in the creation of individual differences.

Summary

This chapter dealt with four important aspects of individual differences – motivation, intelligence, personality and attitudes – and pointed out that inter-individual and intra-individual variations make the prediction or control of human behaviour extremely difficult. The chapter ended with a discussion of the main sources of individual differences – that is, genetic factors and cultural factors.

27: Human error

Introduction

It is a natural human characteristic to want to achieve certain aims and to indulge in a wide variety of behaviours in order to achieve these aims. Some of these behaviours will be successful and will tend to be repeated, while others will be unsuccessful and will tend not to be repeated. In general terms, behaviours which do not achieve the intended aim can be thought of as errors. This is the familiar process of 'learning by our mistakes'.

In a wide variety of circumstances, this sort of learning is perfectly acceptable since it results in no harm to anyone. It is one of the main ways in which people learn skills such as writing, painting and playing musical instruments. However, this trial and error learning is not an acceptable strategy where errors could result in severe harm. For example, it would not be considered appropriate to allow people to learn how to fly an aeroplane, use a parachute or defuse bombs solely via trial and error.

It is important to bear in mind this general role of errors in human learning since the errors people make teach them a lot about their physical and social environments and how they interact with them. This 'useful' role of errors makes it more difficult to study and eliminate those errors which may result in harm, since it is not always possible to predict which behaviours will result in useful learning experiences and which will result in harm.

This chapter on human error covers the following topics:
1. classification of human errors – various authors have attempted to classify human errors in ways which make them easier to study
2. error reduction strategies – there are various general strategies for error reduction and evidence suggests that certain strategies are more effective for particular types of error
3. human error in disasters – reports of major disasters have identified the important role played by human error in the causation of these disasters.

Classification of human errors

Various authors have produced classifications of human errors based on research in different industries and with different sorts of tasks and a summary of this research can be found in Glendon et al.[91] A typical error classification quoted by Glendon et al. is that devised by Miller and Swain and this is reproduced in Table 27.1.

Error type	Description
Commission	Adding or including something that should not be there
Omission	Missing something out
Selection	Incorrect choice from a range of options
Sequence	Incorrect serial position of actions or events
Time	Too late or too early with an action
Qualitative	Not performing an action properly

Table 27.1
Miller and Swain's classification of errors

A particularly influential classification of errors is that produced by Rasmussen.[92] This classification is described next, followed by descriptions of the way in which Reason[93] extended the classification into generic error modelling systems and how Hale and Glendon[94] used it as a basis for their model of individual behaviour in the face of danger.

Rasmussen identified three basic types of error:

1. **Skill based.** These are errors which occur during routine repetitive tasks which can be carried out 'automatically', such as typing. The skill-based errors are subdivided into slips, which are failures in attention, and lapses, which are failures in memory.
2. **Rule based.** These are errors which occur during tasks which require a sequence of steps to be carried out in the correct order, such as isolating electrical switchgear.
3. **Knowledge based.** These are errors which occur during tasks which require working from first principles, such as planning and problem solving.

This classification of error types is important since, for example, it allows more accurate focusing of training (see Glendon *et al.*) and, as was mentioned earlier, it has been extended by Reason into generic error modelling systems (GEMS). The error types identified by Reason are illustrated in Figure 27.1 and explanatory notes are given opposite.

Figure 27.1
Human error types[93]

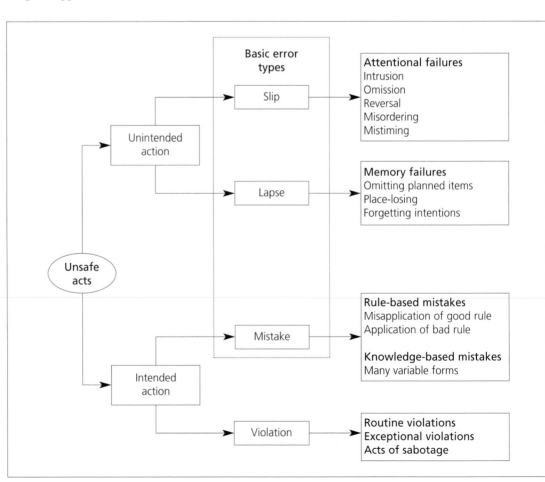

Reason divides unsafe acts into unintended and intended actions. Unintended actions which arise from slips or lapses usually occur at the execution stage of a task and they are, therefore, relatively easy to detect by observation. Since they are relatively easy to detect, it is also relatively easy to provide suitable remedial or compensatory actions. Unsafe acts which arise from mistakes usually occur at the planning stage of a task and the consequences of such mistakes may not materialise until some time after the mistake has been made. Mistakes in the design of buildings which create risks during the construction or maintenance of the building would fall into this category, as would design errors which make machinery difficult to operate without risk. Because these errors remain in the system until circumstances occur which trigger their effects, they are often referred to as 'latent' errors. Many of the inappropriate decisions by management on safety matters fall into the category of latent errors. Reason refers to these latent errors as 'pathogens', the analogy being a medical one. Pathogens are, for example, bacteria and viruses normally present in the body but kept under control. When circumstances change – for example a person becomes unusually fatigued or stressed – control over the pathogen is lost and the person becomes diseased.

Violations are described by Reason as deliberate, but not necessarily reprehensible, deviations from those practices deemed necessary (by designers, managers and regulatory agencies) to maintain the safe operation of a potentially hazardous system. Remember that there are cases where safety rules are so complex and inappropriate that systems can only operate if there are frequent rule violations condoned by managers. In the UK, 'working to rule', that is accurately following all safety rules, has been used by workforces as an alternative to strikes as a means of putting pressure on managers during pay negotiations. Where this is possible, it is legitimate to doubt the appropriateness of the rule base being used; the criteria for good safety rules were discussed in Chapter 7.

The second major application of Rasmussen's work is the Hale and Glendon[94] model of individual behaviour in the face of danger. This model is reproduced in Figure 27.2 and explanatory notes are given below.

The Hale and Glendon model should be 'read' from the bottom of the model upwards. It begins with objective danger, which might also be described as objective risk, and this can be increasing, static or decreasing. In order to keep the danger under control, individuals have to make the appropriate responses and this requires the danger to be brought to their attention by signals or warnings.

Where signals or warnings are available and responded to appropriately, the danger is brought (or stays) under control. Where there are no signals or warnings, individuals have to seek information on hazards for themselves, but in order to do this successfully they need knowledge about appropriate tests for danger.* Once the danger has been accurately identified, it is then necessary to recognise that action is required and accept, or allocate, responsibility for the action. However, successful implementation of each stage in the plan–procedure–response sequence is also required if the danger is to be brought or maintained under control.

The major implication of the Hale and Glendon model is that a danger is more likely to be controlled successfully at the skills level since there are far fewer opportunities for things to go wrong.

A clear danger signal with a matching programmed response is likely to be the most successful way of dealing with danger. Where controlling danger relies on action at the knowledge level, there are many stages at which failure can occur, as we will show later in the chapter in the discussion of the role of human error in major disasters. However, before dealing with these major disasters, there is a brief note on error reduction strategies.

---

* The Hale and Glendon distinction between hazard seeking and tests for danger appears to be analogous to other authors' distinctions between hazard identification and risk rating, and it suffers from the same conceptual difficulties. See Chapter 20 for a discussion of these difficulties.

Figure 27.2
Individual behaviour
in the face of
danger model[94]

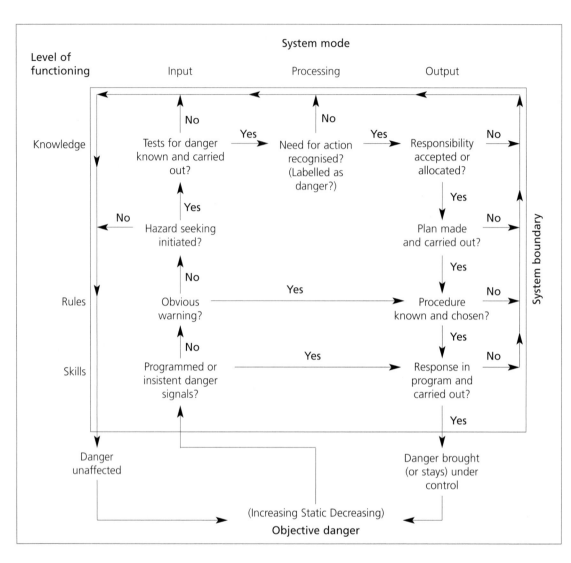

Figure 27.2
Individual behaviour in the face of danger model[94]

Error reduction strategies

Error reduction strategies are intended to reduce the risk arising from human error. There are two broad approaches:

1. **Strategies designed to reduce the likelihood of an error.** These can be intended either to eliminate the possibility of the error completely (reduce the likelihood to zero), or to reduce the likelihood to a greater or lesser degree.
2. **Strategies designed to reduce the severity of the consequences if an error occurs.** Strictly, these are not 'error reduction' strategies but they are mentioned here for completeness.

Within the strategies designed to reduce the likelihood of an error, the most effective methods involve design or redesign of machines, equipment and tasks. As we have seen, relying on human behaviour for error reduction is not a good strategy and 'designing out' the potential for errors is always preferable.

Where errors cannot be designed out, it is essential to identify accurately the sorts of errors which can occur since, as we have seen, there are different types of error and each requires a different approach, or combination of approaches, to prevention. Given that the likely types of error have been identified, possible approaches to error reduction include:

- labelling, colour coding, warning signs and other reminders of hazards
- selection and placement – in other words, ensuring, so far as is practicable, that individuals have the physical and mental capacities required to avoid errors
- training and information – in other words, ensuring that people have the necessary knowledge and skills to enable them to avoid errors.

Various authors have attempted to establish the most effective methods of reducing the different types of error and Table 27.2 (reproduced from Mason and quoted in Glendon *et al.*) illustrates one such author's results. However, a detailed consideration of these strategies will be left until Chapter 30, which deals with ways of improving human reliability. The remainder of this chapter is a description of the role played by human error in disasters.

Error type	Error reduction methods
Slips and lapses	Design improvement, training
Potential for mistakes	Training (team and individual), overlearning, refreshers; duplication of information, clear labelling, colour coding
Knowledge-based errors	Hazard awareness programmes, supervision, work plan checks, post-training testing
Violations	Motivation, correcting underestimation of risk, balancing perceived risks and benefits, supervision, group norms, management commitment

Table 27.2
Error reduction
strategies[91]

Human error in disasters

Over the years, there has been a large number of disasters or catastrophes where a single incident has resulted in the loss of a large number of lives and/or significant damage to assets or the environment. Each of these disasters could, quite legitimately, be referred to as 'a catalogue of errors' and the study of any of them provides many examples of how human error contributed to the final outcome.

We can see from a reading of even a small selection of the reports on disasters that the final outcome is a result of a number of causal factors. Whether the result is an explosion and/or fire as at Flixborough and on Piper Alpha, a train crash as at Moorgate or Clapham, or an aeroplane crash as at Kegworth, multicausality is involved. The outcome results from a concatenation of circumstances which, had any one of them been different, would have resulted in a near miss or a non-disastrous outcome. Here are some examples:

- the difference between an 'air miss' and a midair collision can be a few hundred metres
- the difference between a signal passed at danger (SPAD) and a train crash can be what happens to be on the line beyond the signal
- the difference between a leak of a flammable gas and an explosion can be whether there happens to be an ignition source in the vicinity
- the difference between a small fire and a large fire can be whether the appropriate resources for extinguishing a small fire are to hand.

When reading about disasters, it is important to remember that it is only the final outcomes and the specific combination of circumstances which produced this final outcome that are rare. The individual causal factors are usually present in the relevant system all the time and, since these systems are dynamic, there is a constantly changing pattern of causes. It follows from this that in any system with the potential for a major disaster, it is only a matter of time before a pattern of causes occurs that could produce a disastrous outcome.

Since, by definition, these patterns of causes are rare, it can be argued that attempting to predict specific patterns is of doubtful value. However, removing a single causal factor from a system, or reducing its frequency or prevalence, will reduce the number of possible patterns of causes which can result in a disaster. More generally, the greater the number of causal factors removed from a system, the fewer disaster-causing combinations there will be.

When reading reports of disasters, therefore, it is of limited value to think in terms of whether the specific risks being described occur in a particular organisation or whether the particular combination of circumstances could arise in a particular organisation. It is more important to consider whether the individual weaknesses identified in the reports exist, or could exist, in a particular organisation. Poorly designed display and control systems create risk, whether they are in an aircraft cockpit or on the shop floor; failure to comply with permit-to-work procedures is a problem not only on offshore installations; and failure to check the quality of work adequately is not confined to the installation of railway signalling.

Summary

This chapter dealt with human error, including the various ways in which human error can be classified and the strategies that can be adopted for reducing the number of errors people make.

28: Perception and decision making

Introduction

Perception in humans is a complex process and we must appreciate this complexity in order to understand the ways in which failures in perception can lead to human error. The naïve view of perception is that the eyes operate like a camera, registering exactly what is in the visual environment and that the other sense organs operate in analogous ways. However, as we pointed out in Chapter 14, this is not the case. Rather, the eyes (for example) are part of a complex visual system, which includes several information processing regions of the brain. It is, therefore, more accurate to think of human perception as an information processing task, rather than a mechanistic registering of what is in the environment.

For ease of description, this chapter deals separately with the following aspects of the information processing required for perception:

- **The attention mechanism.** Perception is only possible if people are attending, consciously or subconsciously, to the relevant information. Lack of attention, or inappropriately focused attention, can result in human errors.
- **Long-term memory.** Accurate interpretation of the environment is necessary if errors are to be avoided and the information needed for this accurate interpretation is usually held in the long-term memory. Failures in retrieving information from the long-term memory can, therefore, result in human errors.
- **Short-term memory.** The short-term memory is the brain's working memory. It has a limited capacity, so human error can occur because of 'information overload' – there is simply not enough room in the short-term memory for all the essential pieces of information.*
- **Expectancies.** Individuals have to have an overview of how the world works – for example, they have to have an understanding of cause and effect. This overview of the world is referred to as an individual's expectancies, and these expectancies have to be learned. If the learning is incomplete or inaccurate, then errors can result.
- **Decision making.** Making the correct decisions is important if errors are to be avoided and although decision making is only one aspect of human information processing, it is dealt with as a separate topic to emphasise its importance.

This chapter is intended only as a general introduction to human perception and decision making and the ways in which failures in perception and decision making can lead to human error. When reading this chapter, it is important to remember the role of individual differences.

Certain machines, such as computers, are designed specifically to process information. Each type of computer has design limits which restrict the types of information it can process, and the speed at which it can process it. In these respects, human beings are similar to computers. There is a restricted range of types of information they can process, and the rate at which they can do this is also subject to an upper limit. As with other things about human beings, there is a great deal of variability from one person to another as to where these limits are set, and information processing speed and information processing capacity are important determinants of cognitive ability.

---

* It may be helpful to think of the long-term memory as the brain's equivalent of the hard disk of a computer, which stores information permanently, and short-term memory as the brain's equivalent of RAM (random access memory), where information is stored temporarily during processing and gets overwritten as new information comes in.

The attention mechanism

If you are reading this book 'with attention', you will be aware mainly of the book, the words being read and their meaning. However, you can voluntarily change the focus of your attention. For example, you can make yourself aware of the feel of your clothes and the noises around you simply by making the effort to do so. When a person 'attends' to a particular aspect of the environment, he or she gives it priority in information processing terms. Other aspects of the environment considered irrelevant for the time being are 'filtered out' by the attention mechanism.

The most useful way of thinking of the brain's attention mechanism is as a system which sets a filter so that it allows through certain sorts of information but blocks other sorts. This is shown diagrammatically in Figure 28.1.

Figure 28.1
The attention mechanism

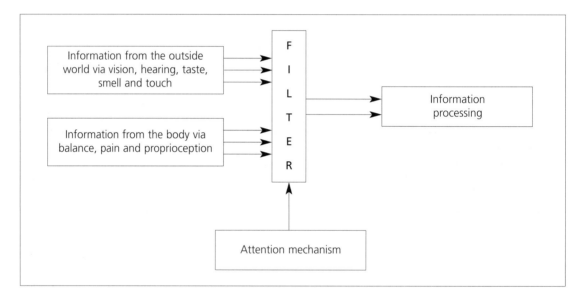

The important feature of the filter is that it can be set to allow different types of information to pass through. This can be done by conscious effort, as is the case when you deliberately think about the feel of your clothes or the noises around you. This has obvious implications for hazard identification, since if the filter is not set to let through information about the relevant features of a hazard, then that hazard will not be perceived.

The filter can also be changed by activity in the outside world, such as a telephone ringing. When this happens in a way which changes the filter setting from 'appropriate' to 'inappropriate' in a given circumstance, then this is distraction. For example, if someone has his or her filter set in a way appropriate to driving on a motorway and their mobile phone rings, then the filter settings change to those appropriate to answering a telephone call. If the driving task is continued, this change in the filter settings will be a distraction and a conscious effort will be required to maintain the appropriate driving settings.

Conscious changes in the filter and distraction are probably the most familiar aspects of the attention mechanism, although you may not have thought of them in terms of a filter. In order to ensure human reliability in situations containing hazards, the filter has to remain open to the relevant information. The various ways in which the filter can be closed, or set inappropriately so that the relevant information does not get through, will be considered later in the chapter.

Consciously setting the filter produces a short-term change and unless a continuous effort is made to maintain the change, the filter reverts to its long-term settings. Before looking at these long-term settings, it is necessary to consider two changes in the filter which are produced by the information in the environment:

1. **If information remains the same over a period of time, the filter will no longer let it through.** This is one of the mechanisms underlying habituation and helps to explain why, for example, people are not aware of the feel of their clothes unless they make a conscious effort. Note that it is not just touch which is involved; if the visual or auditory environments are unchanging, they too will cease to be perceived.

2. **If information changes very slowly, the filter will not let through details of the change.** As with habituation, the inability of the senses to perceive slow changes has been dealt with in an earlier chapter. For example, a leak of domestic gas will remain undetected by people in the vicinity if the leak is small enough to let the gas concentration build up slowly. Again, the operation of the filter over all senses explains why this inability to detect slow changes is common to all senses.

Both limitations in the way the filter operates have advantages and disadvantages. The main advantages are as follows:

- Habituation means that the information processing system is not overloaded with information which is not relevant. As we mentioned earlier, the human information processing system has a limit and if processing capacity is taken up with, for example, registering the feel of clothes or irrelevant noises, then it is not available for other tasks.

- In evolutionary terms, things which changed very slowly were rarely, if ever, a threat, so that there was evolutionary pressure to keep information processing capacity free to register things which changed quickly since they were, or could be, a threat. However, as with other aspects of human evolution, this adaptation is not necessarily appropriate to modern living. See, for example, the 'fight or flight' response discussed in the section on stress.

The disadvantages of these aspects of the filter mechanism are that they may, as indicated above, be inappropriate in modern life. Two examples are:

1. Car drivers do not notice the slowly decreasing efficiency of a car's braking system. When they brake for the first time after having had the car serviced, they are usually surprised by the effect.

2. People working with edged tools rarely notice the slow decline in the sharpness of the tool and this decline may continue until the tools are inefficient and perhaps dangerous.

In order to cope with examples like this, some artificial system has to be adopted to overcome the human limitations, such as car servicing at fixed intervals and frequent sharpening of edged tools.

Habituation and the inability to detect slow changes are 'built-in' features of the filter and are common to everyone. However, how an individual's filter is set at any given time depends on many other factors. For example, the filter will let through less information if the individual is tired. The notes which follow give a brief overview of the main influences on the filter arising from within the individual.

General closing down of the filter

Fatigue and boredom will usually result in the filter letting less information through. Typically, there is little selectivity and information of all types is filtered out. Two other factors – stress and drugs – can also have this effect but they can also result in increased selectivity. For example, people under high levels of stress may focus their attention exclusively on aspects of the environment related to their stressors and exclude information on such things as personal relationships.

Long-term filter setting

People can 'learn to perceive', that is, training or experience can influence the long-term settings of the filter. For example, once a trainee driver has consciously learned to attend to important hazards and filter out information not relevant to the driving task, these filter settings become automatic and they come into operation without conscious thought whenever they are required. Learning, however, is not the only thing which can have this effect. People's interests and motivation, their attitudes and personality all affect how the filter is set. If a person is interested in photography, that person is much more likely to perceive and remember a camera shop during a walk down a high street than a person who is not interested in photography taking the same walk.

The state of the filter at any given time, for a particular individual, will result from an amalgam of all of these personal factors. The photographer who would normally perceive the camera shop may not do so if he or she is tired or under stress. If an individual is in pain, the 'pain information' may block the filter to all other types of information. Pain also shows the power of consciously switching attention since, if individuals are in pain, they find it difficult to concentrate on anything else; if, by conscious effort, they can switch their attention, the pain can be 'forgotten', at least temporarily. This conscious switch of attention is recognised in the phrase 'taking your mind off it'.

To recapitulate, hazards will only be perceived if the filter is set in a way which allows through the relevant information. Since perception is a necessary first step if an individual is to take avoiding action, the role of the attention mechanism is central to hazard avoidance. However, the attention mechanism is subject to many influences which can alter its setting over time.

Environments with low information content

For most of the time, there is too much information in the environment and some of it has to be filtered out. Humans are therefore used to an information-rich environment, and when people are placed in an environment which has too little information, the brain takes action.

There are two typical actions the brain takes in response to information-poor environments. The first is to send the body to sleep, since a constant input of information is not required during sleep. Indeed, for most of us, a drastic reduction in information input is required before we can go to sleep at all. The second action the brain can take is to generate its own information flow so that the required information level is sustained. This generation of information usually takes the form of daydreaming and it is possible to induce daydreaming deliberately to relieve boredom or stress – that is, to fantasise. However, daydreaming can occur involuntarily in response to inadequate supplies of information from the outside world.

The onset of sleep or daydreaming have similar effects – that is, blocking the filter to external information. During a task such as motorway driving, this can induce an extremely dangerous vicious circle. The low information content of the task results in drowsiness or daydreaming which blocks the filter, reducing even further the input of information from outside. This, in turn, results in increased drowsiness or more intense daydreaming, further blocking the filter. This vicious circle can develop on any task with a low information content, not just motorway driving.

Human beings need a minimum amount of information input in order to operate reliably. Short periods without sufficient information input result in daydreaming or sleep; longer periods without sufficient information input have more serious effects. One of the first steps in brainwashing is to deny people information for long periods of time by keeping them in dark, soundproof rooms.

Any task designed for human beings which has to be carried out for more than a few minutes at a time must, if errors are to be avoided, take this need for information content into account. Typical methods of countering low information content include:

- increasing the information content of the task itself, such as with 'job enrichment'
- providing other sources of information, such as radio programmes

- doing the task only for short periods of time with frequent breaks for activities with a higher information content
- removing all hazards from the task so that drowsiness or daydreaming will have no adverse effects.

The filter is never completely closed

In terms of human reliability, the important features of the filter are those things which close the filter to information about hazards. However, there is one final point to make about how the filter works.

Except in extreme cases, for example traumatic or drug-induced loss of consciousness, the filter is never completely closed. Even when a person is asleep, they can be woken by a noise. However, the filter is still operating selectively and the person is less likely to be woken by a familiar noise, even if it is quite loud. This, of course, partly explains why it is more difficult to sleep in a strange environment. In contrast to filtering out information, the filter will also let through information of personal significance to the sleeper. People are more likely to be woken by their own name than by any other word, even if it is said at the same volume of sound. Parents are woken by noises of distress from their children even when these sounds are relatively quiet.

Even when daydreaming, quite complex actions are possible. Most experienced drivers will at one time or another have 'come to' after a spell of daydreaming and remember nothing about the last few miles. All the necessary perception, information processing and actions have taken place completely automatically. The fact that this can happen shows that the filter is open to relevant information such as bends in the road, but this 'automatic' driving is inherently dangerous because the filter may not be open to the more general information needed to perceive unexpected hazards. Tasks with a low information content are particularly prone to being carried out in a daydream, and even complex tasks, such as driving, can be done in this 'automatic' manner if the person doing them is highly experienced.

Long-term memory

When a human being is processing information, for example in the course of making a decision, information from the outside world, filtered by the attention mechanism, is only one of the sources of information used. The other main source of information is the person's long-term memory. All normal adults have, stored in the long-term memory areas of their brain, a vast quantity of information. To get an idea of how much information is stored, consider this:

- How long would it take you to draw a road map of the area around your home, putting in as much detail as you can remember?
- How long would it take you to do a similar exercise for the area around which you work?
- How long would it take you to repeat the exercise for all your previous homes and places of work?
- More generally, how long would it take you to draw maps of all the areas you know, including journeys you have made, and towns you have visited at home and abroad?

For most readers, doing this exercise would be very time-consuming, and maps are only one aspect of the information stored in long-term memory. How much longer would it take if you added everything you could remember about houses, shops, pubs, parks and street furniture? If you added your store of information about the people associated with all of these things, how much extra time would be required?

And this is only 'real life'. What if you added everything you could remember about the books you have read, the films and television programmes you have seen, and the radio programmes and music you have listened to?

And the final question: how much time would it take to write down everything you have in your long-term memory?

It is quite likely that most adults could spend the rest of their life, working seven hours a day, five days per week, writing down just what they knew before they started the task. If the reader is not really convinced that this is so, spend 10 minutes making a list of the headings of things which could be written about if the time were available. It is worth doing this because it is important to appreciate at this stage the extent of the information being dealt with when long-term memory is under discussion.

Despite the amount and diversity of information in the long-term memory, it is organised in a manner which enables particular items of information to be recalled very quickly. Selecting particular items of information for recall is referred to as information retrieval and the speed of information retrieval from the human brain is usually very fast, much faster than can be achieved by computers.

Consider a contestant in a general knowledge quiz. In these sorts of quiz, the contestants search the information in their long-term memory and may come up with an answer in a fraction of a second. Even if it were possible to put the amount of information necessary for a general knowledge quiz into a computer, it would take even the biggest and fastest computer several minutes to produce the correct answer.

Considering the vast amounts of information in the long-term memory and the speed of retrieval, the surprising thing about human memory is not that it fails occasionally but that it operates at all. However, from the point of view of human reliability, it is the occasional failures which are important.

If a particular item of information is required in order to deal with a hazard, there are two main ways in which a failure might arise:

1. the required information may not be in the long-term memory
2. it may not be possible to retrieve the required item fast enough for the circumstances.

There are several possible reasons for each type of failure. We will deal first with why the required information might not be in long-term memory.

Information not in long-term memory

The fact that a person has been exposed to a particular item of information does not ensure that it is retained in long-term memory. There is a school of thought which argues that everything is retained in long-term memory and it is only the retrieval system which fails. Evidence in support of this includes rare individuals who 'can forget nothing' and the retrieval of information under hypnosis which was not available to the conscious mind. However, for practical purposes, it is preferable to treat absence of information and failure of retrieval as separate topics. For example, people look up telephone numbers, remember them for long enough to dial them, and then forget them. Often the retention period is so short that if the number is engaged or has to be redialled for some other reason, the person has to look the number up again. Similarly, people learn things for examinations and as soon as the examination is over, they 'forget' most of the information.

This obviously has implications for those circumstances where we have to 'know' something in order to avoid hazards, and if it is important that an item of information should be transferred to long-term memory and remain there, then there are steps which can be taken to make this more likely. Three of these – rehearsal, high significance and association – are described below.

Rehearsal

Making use of an item in long-term memory by retrieving it is one way of making sure it stays in long-term memory. Most people can remember how to multiply and divide because they use the rules for these operations fairly often. Very few people can remember how to multiply and divide fractions because they probably have not used these operations since school. More generally, any item of information which is not rehearsed will tend to 'fade' from long-term memory until eventually it cannot be retrieved at all.

The effectiveness of rehearsal can be demonstrated by fire drills. If you ask all the people in an organisation where their fire drill assembly points are, you would get different results three months after a fire drill compared to the day after a fire drill. This is only one aspect of fire drills, and not necessarily the most important, but fire drills are one of the few forms of organised rehearsal in the safety world.

Fire drills are carried out because real fires are too infrequent to allow 'natural rehearsal' – some form of artificial rehearsal is required in order to keep the necessary information in long-term memory. There are, however, many analogous situations where arrangements could be made for artificial rehearsal.

- A skid in a car is, for most drivers, a rare event and even drivers who have received tuition in how to cope with a skid may have forgotten the necessary action. Despite the relative ease with which rehearsal facilities could be provided, they rarely are.
- For events like tyre blowouts and windscreens shattering while the car is moving, the provision of natural rehearsal facilities is unlikely to be practical. There is, however, no reason why artificial rehearsal facilities should not be provided.

It should be noted that in the commercial flight industry, where the results of accidents can be very serious, extensive rehearsal facilities are provided via flight simulators.

There are many industrial tasks which require particular actions in response to infrequent hazards, but it is rare for adequate facilities to be provided for rehearsal.

High significance

Items of information which are important to an individual are less likely to fade from long-term memory, although this is partly because items of high significance are likely to be rehearsed more often. What determines whether or not an item is significant varies widely from individual to individual, but everyone's memory contains items of information which they remember because they are significant to them but which would fade rapidly from other people's long-term memory.

In the safety field, hazards take on a high significance if a person has had an accident as a result of that hazard. The significance of a hazard can also be increased by seeing someone else having an accident relating to that hazard, or hearing about someone having such an accident.

In order to maximise human reliability, the hazards an individual considers significant have to correspond to the actual hazards in the environment. Certain people, for example, refuse to wear seat belts in case they get trapped in a car fire or they drive into a canal. The fact that car fires and driving into canals are comparatively rare events is ignored. For these people, their high significance hazards are inappropriate to their environment.

For any task, the important hazards have to be established and action taken to increase their significance for the people carrying out the task.

Association

A particular item of information is less likely to fade from long-term memory if it is associated with other items in long-term memory. This is why people find it easier to learn more about a familiar subject than to start learning a new subject. There are more 'hooks' available with the familiar subject for making links to the new information.

The effect of association on remembering information about hazards can be illustrated by tasks having a number of hazards associated with one aspect of the work, and a number of isolated hazards. For example, a job which involves drilling different components would have various hazards associated with the drilling operation itself. These hazards, because they form an associated group, are likely to be well remembered. However, there may be other hazards, for example, some components may be brittle (risk of eye damage), others may have sharp edges (risk of cuts to the hands), while others may produce harmful dusts (risk of long-term health damage). These individual hazards, because they are not directly associated with the central task of drilling, are much more likely to fade from long-term memory.

The implications for human reliability of this aspect of human memory is that special attention should be paid to isolated hazards – that is, hazards that are not readily associated in a person's memory with the other hazards of the job being done.

Speed of retrieval

So far, we have discussed three ways of improving the chances of an item of information remaining in long-term memory. However, the fact that an item is in long-term memory does not guarantee that it will be available when required. Everyone has had the experience of having an item of information 'on the tip of their tongue', when they know that the required item is in their memory but they cannot retrieve it immediately. Unfortunately, with most hazards, time is not usually available for retrieval, as the appropriate information must be available immediately.

In general, there are three factors which affect the speed of retrieval: recency, association and overlearning.

Recency

Other things being equal, the shorter the time since the last retrieval, the faster the next retrieval will be. Recency is the retrieval equivalent of rehearsal as a means of ensuring that items remain in long-term memory; the two are linked because retrieval is one form of rehearsal.

Association

In addition to helping an item remain in long-term memory, highly associated items of information can also be retrieved more quickly. It is likely that this is because a large number of associations provide alternative routes to the required item of information. This is shown diagramatically in Figure 28.2.

Figure 28.2
Low and high association memory items

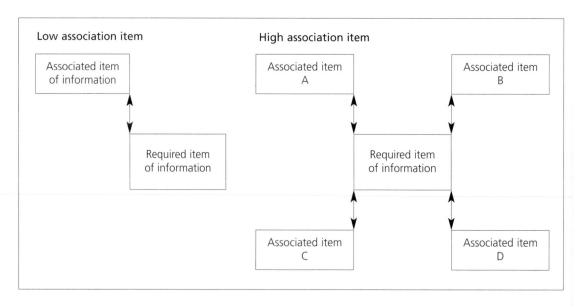

With the low association item, the person has to produce the required item of information directly, or by thinking first of the single associated item. Thus, there are only two possible routes for the retrieval of the required item. In contrast, the high association item provides five possible routes for retrieval and, in real life, the highly associated items may have thousands of links.

It is possible to demonstrate how important the organisation of information in the brain can be for retrieval time by noting how long it takes to answer the following two questions:
1. What is the ninth letter of the alphabet?
2. What is the letter after H?

For most people, the retrieval time for the answer to the second question is much shorter than for the first. This is because of the way the alphabet is 'arranged' in most people's brains as a sequence from A to Z. Another example is that very few people can say the alphabet backwards in anything like as smooth a manner, or at anything like the speed, as they can say it going forwards. This is because the 'backwards' association has never been developed. Most people can 'count down' with numbers as fast as they can 'count up' because the necessary backward associations have been built up for numbers.

Overlearning

Reference to the alphabet and numbers introduces a final factor which influences retention in long-term memory and speed of retrieval. This factor is known as overlearning and it describes what happens when particular items of information, like the alphabet and numbers, become so familiar that it is almost impossible to imagine anyone forgetting them in non-traumatic circumstances.

Overlearned items of information can be extremely valuable in human reliability because they maintain themselves in memory and can be retrieved rapidly. However, they can present problems.

These problems arise when circumstances change in such a way that the information itself, or the way it is organised, is no longer appropriate. For example, the overlearned response of depressing the clutch when changing gear in a car with a manual gearbox becomes inappropriate, and potentially dangerous, when the driver switches to a car with an automatic gearbox. In effect, the main advantages of overlearning become disadvantages if circumstances change. Many readers will have had the experience of moving to a new house or office and, after a few days of turning up at the correct location, they go to the 'old' location. This arises because the route to the old location has been overlearned and, if a conscious effort is not made to override this, there is a reversion to the now inappropriate old learning. This sort of reversion can have important safety implications when, for example, new machinery or processes are introduced.

Long-term memory and the filter

So far, we have considered long-term memory and retrieval as though they were independent of other types of information. This was to make description easier but, in practice, the long-term memory is linked with a number of other sources of information and is subject to attentional filtering in the same way as information from the outside world. Figure 28.1 is redrawn as Figure 28.3 to show how this operates.

The filter, as it affects long-term memory, can be set consciously or unconsciously and in the short term and long term, just as it could for information from other sources. For example, when people start to drive a car, they set their long-term memory filter to let through more readily those items of information relevant to car driving. General blocking of the filter by, for example, fatigue or stress, will also adversely affect retrieval from long-term memory.

Short-term memory

In Figures 28.1 and 28.3, the information processing function was indicated by a single label, but this part of the chapter considers how information processing is carried out and the implications of this information processing for human reliability.

Figure 28.3

The attention mechanism and long-term memory

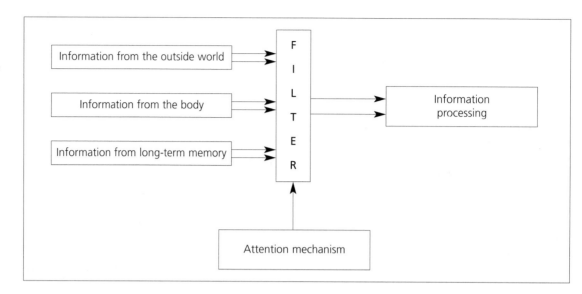

The majority of human information processing is carried out by a special type of memory known as the short-term memory, although in some books it is referred to as the 'working memory' in recognition of its active role in information processing.

There are two features of the short-term memory which make it different from the long-term memory described earlier:

1. it has a very limited capacity
2. information in short-term memory fades very quickly.

It is short-term memory which is used for tasks like looking up a phone number and keeping it in memory long enough to dial the digits. This task illustrates both limitations of short-term memory. It is extremely difficult to keep a number of more than seven digits in memory and any delay between looking up the number and dialling will probably result in the number being lost from short-term memory altogether.

To illustrate the limited capacity of the short-term memory, try the following exercise. Read out aloud the following numbers, then look up and say them again without looking at the page:

5, 3, 7, 9, 6

Now repeat the exercise with the following numbers:

6, 3, 0, 5, 1, 7, 9, 2, 4, 8

For most people, the first of these tasks is extremely easy while the second task is almost impossible. This is because there is a limit of around seven items in short-term memory which means that the early numbers in the sequence get 'pushed out' by the numbers arriving later. In a routine task, such as number reading, they are lost completely although it is possible to learn a long sequence where this is necessary. This is done by transferring the sequence to long-term memory and while this prevents the items being 'lost', it requires a certain amount of effort.

It might seem that a restriction to just seven items for processing at any one time could be a severe handicap. However, since everyone has this restriction, and everyone processes information more or less continuously, it must be, in general, adequate. The solution to this apparent paradox lies in the nature of an item of information.

Try the memory for numbers test with the following sequence:

1, 0, 6, 6, 1, 9, 1, 4, 1, 9, 1, 8, 1, 8, 1, 2

Now try the same sequence with the numbers arranged like this:

1066, 1914, 1918, 1812

There should now be no difficulty in remembering a complete sequence of sixteen digits which, if they had no special meaning, could not be retained in short-term memory. Note that there are two memory aids in operation here. The first is the grouping of the digits into dates, and the second is the link between the second and third dates (the start and end of the First World War). Both rely on our previous experience and knowledge for their effectiveness.

This simple example illustrates how 'serious' information processing can be carried out with just a few items of information. The answer is to merge pieces of information so that they become a single item (1, 0, 6, 6 becomes 1066). If there are still too many items to deal with, then the process of combination is repeated (1914 and 1918 are linked in many people's memory and one will 'trigger' the other).

With this background information on the operation of the short-term memory, it is now possible to consider some examples of the implications for task design.

- If a person is being given instructions, it will be impossible for them to keep more than five or six instructions in short-term memory.
- In tasks which involve a sequence of operations, it will not be possible for people to keep the required sequence in short-term memory for more than a few minutes.
- Tasks which involve the monitoring of displayed information have to be designed so that no more than five or six of the displayed items are significant at any one time. If more are involved, the person doing the monitoring will be unable to hold all of the significant items in short-term memory at the same time.

The limited capacity of the short-term memory is overcome by 'chunking' items of information during prior learning so that there are never more than five or six chunks required at any one time. Think of a straightforward chunked instruction like 'drive to Carlisle'. Now think of the number of instructions that would be required if the person being instructed had never heard of Carlisle and had no idea where it was. Then think of the number of instructions that would be required if the person being instructed did not know how to read a map, or did not know how to drive a car. People who are dealing effectively with long strings of instructions, or complex instructions, are chunking them together in some way so that they are, in practice, dealing with shorter, simpler sequences.

The limited capacity of the short-term memory is only one of the aspects of memory which influence human reliability. The other is the fact that items fade from short-term memory very quickly. For example, it is extremely unlikely that you will remember the first sequence of numbers given earlier, since items in short-term memory typically remain there for only a few seconds.

There is a simple trick for keeping items in short-term memory which involves creating a short-term memory loop by rehearsing the item or items to be remembered. Most people have done this by, for example, saying a phone number over and over between the time of looking it up and the time of dialling. Getting the engaged tone, or any other interruption, will break the loop and result in the phone number being lost from short-term memory.

The limitations of short-term memory have obvious implications for human reliability in that if tasks rely for their safe completion on people remembering items of information for more than a few seconds, one of the following strategies must be adopted:
- the item of information must be kept in short-term memory by rehearsal
- the item of information must be transferred to long-term memory in such a way that it is available for retrieval when required
- some artificial memory aid, such as a written checklist, must be used
- the memory aspects of the task must be allocated to a machine or computer.

These descriptions of the nature of long-term memory and short-term memory demonstrate that both have an important role to play in human reliability and that the limitations of each type of memory must be taken into account when designing tasks. However, a more detailed discussion of this aspect of safety will be found in the chapter which deals with ergonomics (Chapter 30).

Figure 28.4 shows the place of short-term memory in the information processing sequence. This figure also includes a box labelled 'expectancies', which is the subject of the next section of this chapter.

Figure 28.4
The place of short-term memory and expectancies in information processing

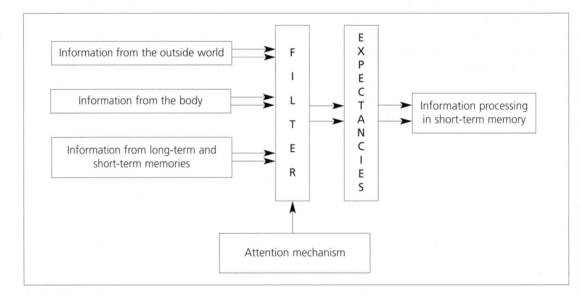

Expectancies

At various points in this section, we have pointed out that there is usually too much information available for it all to be processed and that a number of techniques are used by humans to avoid information overload. The idea of expectancies ties together a number of these techniques under a single heading.

Everyone has within their brain a representation of how the world 'works', which can be thought of as a model world, although it is not, of course, a physical model. It is this model world which people use when making predictions about, for example, cause and effect.
- People playing pool use their model world to predict what will happen if one ball strikes another in a certain way.
- People drinking from a glass of water use their model world to predict what will happen when they tilt the glass through varying degrees.

This model world is extremely important as it enables people to predict the effect of the actions they intend to take. One of the main aspects of children's learning is to build up an accurate model of the world, usually by trial and error. As most parents know to their cost, the only way a young child can find out what will happen when a glass of water is tipped is to tip a glass of water!

As people get older, their model of the world takes into account many more things than simple physical properties and, at its most complex, it takes into account the behaviour of other people. In order to operate effectively and safely, people need a very rich and accurate model of the world and this model forms their 'expectancies' – that is, how they expect the world to behave.

In the context of human reliability, there are two main ways in which expectancies can result in increased probability of error:
1. a person's expectancies may not correspond with the real world
2. two people's expectancies may not match.

People expect that something which looks like a chair will bear their weight and they are willing to sit down without testing it first. People expect that the chair they have decided to sit on will still be there when they have turned their back on it in order to sit down. These two examples involving chairs provide simple illustrations of how people can sustain injuries. If the expectancy is not fulfilled in either case, the person ends up on the floor.

More seriously, there are many circumstances where a mismatch between expectancy and the real world can create high risk situations. Examples include:
- crossing a road in a country where the traffic drives on the left when you are used to traffic driving on the right, or *vice versa*
- moving around in a familiar environment soon after it has been changed without your knowledge
- picking up an object which is much lighter or much heavier than expected.

This potential for creating hazardous situations is increased when people interact and there is the possibility of two or more people operating with slightly different models of the world. The safe movement of road traffic depends on all drivers having the same world model. For example, it is expected that everyone will:
- stay on their side of the road
- stop when traffic lights are on red
- give way at main roads, and so on.

Unless these expectancies are shared by everyone, and are mostly fulfilled, all road traffic would grind to a halt, and this would be particularly noticeable where traffic densities are high, such as in cities. Hazardous situations arise whenever one of these expectancies is not met – for example, an oncoming car turns right without warning, someone jumps a red light, or a driver takes a right of way to which he or she is not entitled.

Certain driving behaviours are, at present, ambiguous and present hazards for everyone. For example, flashing headlights can mean 'I am here, take care' or 'I have seen you and I am giving way' or 'I have seen you but I am not giving way'. If the two drivers concerned have the same expectancy about what the signal means, then no harm will result. If the driver flashing his or her lights is giving a warning and the second driver takes this as an invitation to proceed, then a high risk situation is created.

Away from driving, conflicting expectancies can result in hazards at work. If a person expects that a piece of electrical equipment has been isolated, and it has not, that person could receive an electric shock or be injured in other ways when the machine starts up.

If human reliability is to be maintained, the relevant expectancies have to be a true reflection of the real world. Establishing the correct expectancies will normally be a function of training and will require three stages:

1. establishing the relevant real world expectancies
2. establishing the trainee's existing expectancies
3. modifying existing expectancies where necessary.

One final point on expectancies. There are some tasks which have to be carried out with a minimum of expectancies if they are to be carried out safely. These include tasks which have a wide range of hazards, tasks which are new to the person carrying them out, and tasks which have unknown or unpredictable hazards. These sorts of task should be carried out with the minimum number of expectancies or, to put it in another way, little reliance should be placed on the model world. Going back to the example of the chair, tests should be made that the chair will take a person's weight, and that it is still there, before anyone sits on it.

Stereotyping

Stereotyping is one of the procedures by which individuals form expectancies and it operates as follows:
1. A category of objects – animals, people, and so on – is identified. For example, a category might be light switches, or large dogs, or police officers.
2. On the basis of experience with a sample of the category, certain characteristics are identified. For example, 'all light switches in the sample use down for on', 'all large dogs in the sample are aggressive', or 'all the police officers in the sample are helpful'.
3. The identified characteristics for the sample are attributed to all members of the category. For example, 'all light switches use down for on', 'all large dogs are aggressive' and 'all police officers are helpful'.

Stereotypes can be formed on the basis of a sample of one or on the basis of information received from others about the category but, whatever the basis of their formation, they can be accurate or inaccurate.

Where stereotypes are accurate, they are extremely useful in that they remove the need to investigate separately each member of a category. However, except in very restricted circumstances, stereotypes are rarely sufficiently accurate to be relied on in safety management. Even if 99 per cent of light switches use down for on, the 1 per cent which do not may be enough to cause problems. More importantly, stereotypes which are based on people's behaviour are notoriously inaccurate.

Decision making

So far, the various influences on information have been shown to lead to a box labelled 'information processing'. This label is rather vague, since human beings carry out all sorts of information processing. From the human reliability point of view, the most important type of information processing is probably decision making, which is the subject of this section.

It is only by making and acting on appropriate decisions that hazards can be avoided consistently, and decision making in hazardous situations will be considered later in this section. This is preceded by a description of decision making in a more familiar setting.

Suppose that a couple were considering buying a new house. The process they would have to go through would be roughly as follows:
1. decide what they wanted from the house in terms of geographical location, size and extra features such as a garage or a greenhouse
2. decide what they could afford to pay for a new house
3. find out what is on the market and establish how closely each possible house meets their requirements
4. decide which of the available houses they will buy.

This is, of course, an oversimplified scheme of what is usually a very complex (and harrowing) procedure. However, it does illustrate a number of important points about human decision making. These points are described below, followed by a discussion of how they relate to decision making in hazardous situations.

- **People differ in what they want.** Different people will require different things from a new house and the same people will require different things at different times in their life.
- **Not all of a person's requirements will be easily measurable.** Some features of a house, such as its size and geographical location, will be easy to determine. However, people usually want other things too, such as a nice area, a pleasant atmosphere or an attractive façade, and these kinds of thing are extremely difficult, if not impossible, to quantify.
- **There will be short-term and long-term factors to be taken into account.** For example, it is possible to buy a cheap house in a 'poor' area in the hope that the area will improve. It is also possible to buy a house which is expensive in the short term, on the assumption that your earning power will increase.
- **What is available may put severe constraints on choice.** The sort of house required may not exist in the appropriate area, or it may be there but not be for sale.
- **Deciding on one particular house when more than one satisfies all requirements will involve a decision on the basis of a cost–benefit ratio.** The idea behind cost–benefit ratios is quite simple. If there are three houses which meet the requirements, the one chosen will not necessarily be the cheapest or the one which gives you the most benefits. Rather, it will be the one which gives the most benefits per unit cost. In common parlance, this is the one which is 'best value for the money'. This type of cost–benefit ratio is the type referred to in the Hale and Hale model (see Chapter 9) and it takes into account factors in addition to the financial costs and benefits.

Although buying a house may be traumatic and involve a good deal of stress, it is not usually thought of as a hazardous activity. Yet the various aspects of the decision making process outlined above also apply during the performance of tasks that might be hazardous. The notes which follow show how they apply to the hazardous task of driving.

- **People differ in what they want.** One person's overall aim in driving might be to make each journey as safely as possible, another person might want to make each journey as quickly as possible, while another person might want to make each journey as fuel-efficiently as possible. Each of these requirements would have implications for the way in which the person drove and the extent to which he or she took risks of damage or injury. Note also that the same person might adopt different strategies for different journeys or even swap strategies part way through a journey. It is possible to start off intending to drive as safely as possible and then switch to driving as fast as possible because traffic jams have made you late for an appointment.
- **Not all of a person's requirements are easily measurable.** Things like time taken and amount of fuel used on a journey are, at least in theory, easily measurable. However, things like 'amount of enjoyment from driving' or the status conferred by being thought of as a 'good driver' are much more difficult to measure. Nevertheless, they can be important factors in determining how a person drives.
- **The way people drive will have long-term and short-term costs and long-term and short-term benefits.** These affect both the driver (short-term benefits of getting to a meeting on time, long-term costs in terms of coronary and digestive damage) and the car (wear and tear on tyres, clutch and engine). Short-term and long-term are, of course, relative terms. Short can mean anything from instant to the time until the next service or change of car. Long can mean anything from tomorrow to the end of a person's life. In safety management, the problem is often to find ways of countering short-term benefits which have long-term costs. For example, stopping people putting themselves under stress to arrive at meetings on time (short-term benefit) in order to reduce the chances of heart attacks and gastric disorders (long-term costs).

- **When considering hazards, short-term and long-term are further complicated by the probability of occurrence.** Not everyone who consistently drives fast gets stomach ulcers, and not everyone who consistently drives carefully avoids a crash. This affects the value people place on long-term costs – the 'it won't happen to me' syndrome. When people make cost–benefit decisions, it will be their own estimate of the costs and benefits they will use and this may or may not be an accurate reflection of the true costs and benefits.
- **Lack of appropriate actions.** In order to deal with hazardous situations, at least one of the possible actions considered must allow for the hazard being dealt with in a safe manner. For example, if a driver has never learned how to behave when a car skids, the appropriate action will not be available as one of the options in the cost–benefit decision. Note the distinction between the lack of availability of the item (not knowing how to behave in a skid) and knowing the appropriate action but not selecting it, perhaps because of panic or because there was insufficient time to retrieve the relevant information from long-term memory.

Although these considerations about decision making have been illustrated using car driving as an example, they are applicable to all hazardous tasks. When designing tasks, or training programmes for those who have to carry out the tasks, pay attention to these various aspects of decision making and, in particular, to the costs and benefits associated with particular actions. The next section deals with cost–benefit decisions in more detail.

Cost–benefit decisions

As we mentioned above, it will be the ratio of costs to benefits that will determine which action is chosen. Remember too that costs and benefits are not restricted to financial costs and benefits. In the context of human reliability, in hazardous situations at least one action must be available which will enable the person to deal with the hazard safely, and a safe action must be chosen.

Assuming that a safe action is available, there are two options for increasing the chances of its being chosen: increase the benefits associated with choosing the action or reduce the costs associated with choosing the action.

Influencing the choice of a safe action is not, however, always possible and, in these circumstances, it is necessary to influence the unsafe actions so that they are less likely to be chosen. Again there are two approaches: the costs of the unsafe actions can be increased or the benefits of the unsafe actions can be decreased.

Whether or not to wear PPE is a decision which is subject to this sort of cost–benefit analysis. In Table 28.1, the four possible approaches to increasing the likelihood of a person wearing PPE are listed. Opposite each of the approaches is one example of how it might be applied.

Table 28.1
Influences on wearing PPE

Approach	Example
Increase benefits of wearing	Praise or reward people who wear PPE
Reduce costs of wearing	Make PPE more comfortable
Increase costs of not wearing	Disciplinary action if caught not wearing PPE
Reduce benefits of not wearing	Ensure that wearing PPE does not slow down work

In order to ensure human reliability in hazardous situations, therefore, one of the things which we must do is to ensure that the cost–benefit ratios of the possible actions consistently favour safe actions. Remember, too, the importance of individual differences and variations. The important cost–benefit ratio is not the one the person setting up the task would use; it is the one used by a particular individual when carrying out that task at a particular time.

Examples of cost–benefit ratios which may operate directly against safe behaviour include:

- **Most forms of piecework, where people are paid more for producing more output.** The financial benefits from working quickly are sure and obvious. The costs of reduced safety are unsure, both in terms of the likelihood of an accident and the seriousness of its outcome.
- **Pedestrian subways and bridges.** The costs in terms of increased time and the effort of walking the additional distance are obvious and certainly incurred. The costs in terms of accidents when crossing on the carriageway are not.
- **Hearing protection which is uncomfortable and reduces the possibilities for social contact.** Immediate costs (discomfort) are incurred and immediate benefits (social contact) are lost, in return for a benefit (limitation of hearing damage) at a time far in the future.
- **Most forms of the work practice known as 'job and finish'**, which encourage speed at the expense of safety.

Although it may not be possible to predict accurately what costs and benefits a particular individual will take into account at a particular time, it is possible to predict the aggregate effects of the costs and benefits associated with a particular way of performing a task. In designing tasks and the ways of carrying them out, it is necessary, therefore, to consider this aspect of decision making and do what is possible to implement the sorts of approaches shown in Table 28.1.

Online and offline processing

At various points in this chapter, we have mentioned that people process information both consciously and subconsciously. For example, people can consciously 'think through' a problem or they can collect all the available data and 'forget' about the problem until a solution occurs to them. The majority of mental activity takes place at a subconscious level and, taking an analogy from computing, this can be referred to as 'offline processing'. The alternative form of computer processing, 'online processing', is an analogy for conscious mental processes. This analogy was more meaningful in the days when computing resources were very limited and data were submitted on punched cards and paper tape for processing when computer capacity was available. These days, with the ubiquitous personal computer, comparatively little processing is done offline.

There is, however, another interpretation of on-line and off-line processing. Stranks[95] defines the two terms as follows:

1. **On-line processing:**

   ... the spur of the moment decision making that an individual has to take in order to survive.

   It is not clear how individuals can 'take' decision making, but the sense would appear to be that on-line processing involves rapid decision making.

2. **Off-line processing:**

   ... process whereby people actually simulate in their own minds the outcomes of a different course of action prior to making any final decision as to which course of action to take.

The rationale for this distinction is not clear since, as we pointed out in the discussion on the Hale and Hale model, individuals can 'simulate in their minds' very rapidly indeed. The speed at which processing is carried out is not necessarily related to whether the processing is carried out consciously or subconsciously.

Summary

This chapter dealt with the complex processes of perception and decision making in humans. It attempted to describe the wide range of issues involved and the implications of these issues for the reliability of performance of humans.

29: External influences on human error

Introduction

This chapter deals with a range of issues which influence human error and, hence, human reliability. It would be possible to adopt a number of different frameworks to provide a structure for discussion; however, for the purposes of this chapter, we will assume that there are three main requirements if human error is to be avoided and human reliability maintained.

1. **The individuals concerned must be capable of avoiding errors.** The factors in this category include such things as physical ability and competence, and we will deal with the category under the general heading of competence in avoiding errors.
2. **The individuals concerned must be willing to avoid errors.** This group of factors is primarily concerned with motivation and will be dealt with under that heading.
3. **The individuals concerned must be enabled or empowered to avoid errors.** This group of factors deals with those things which prevent individuals behaving in ways in which they are able and willing to behave. We will use the general heading of organisational factors for this group.

Within each category, the relevant external factors will be described, with notes on how manipulating these factors influences human reliability. For the sake of simplicity during these discussions, we will assume that increasing human reliability is equivalent to reducing human error rates.

Note that one important factor – ergonomics – is excluded from this chapter as it is covered in more depth in Chapter 30.

Competence

General competences
The subject of individual competence was dealt with in the section on training in Chapter 12, but there are various factors which influence general competence in the context of reliability.

Recruitment and placement*
It is obvious that errors will occur if the individuals selected to fill posts do not have the competences required to avoid the sorts of errors associated with their post. However, few organisations take this into account in their selection procedure, and there are two main weaknesses. First, organisations do not take enough care over task analysis, so the identification of possible errors is inadequate. Second, even when adequate error data are available, it is unlikely that the selectors, using procedures based on typical 'job interviews', would be able to tell whether the candidate had the relevant competences or not. Where placement is involved, there are additional opportunities for testing for particular competences (such as observation of work, and reports from supervisors and subordinates) but these are not always used. In the case of recruitment, testing for competences is usually limited to interview data with, perhaps, one or more psychometric tests.

Using recruitment and placement as a means of improving human reliability will, therefore, require:
* accurate identification of relevant likely or possible errors using some form of task analysis

---

* These terms are used to differentiate between filling a post from outside the organisation (recruitment) and from within (placement). Where both options are relevant, we have used the term 'selection'.

- accurate identification of the competences required to avoid these errors
- identification or development of suitable instruments for measuring these competences (in this context, instruments could be anything from observation of task performance to psychometric testing)
- the use of the measuring instruments as part of the selection procedure.

Many organisations expend substantial resources attempting to select people who can carry out specific tasks, and the sequence described above is used to identify people who can do these tasks. It is unfortunate that similar levels of resources are not devoted to identifying those people who can carry out specific tasks most reliably.

The safety competences required by people at work in the UK have been dealt with in detail by Proskills. The list of standards published by Proskills[96] is:

PROHSS1   Make sure your own actions reduce risks to health and safety
PROHSS2   Develop procedures to safely control work operations
PROHSS3   Safely control work operations
PROHSS5   Investigate and evaluate health and safety incidents and complaints at work
PROHSS6   Conduct a health and safety risk assessment of a workplace
PROHSS8   Review health and safety procedures at workplaces other than your own
PROHSS9   Supervise the health, safety and welfare of an individual at work
PROHSK1   Basic hazard awareness (knowledge unit).

Training

As we saw in Chapter 12, training – whether it is on or off the job – is only one method of providing competences. It is, though, probably the one which is most used. As far as human error is concerned, the main weakness with training is that, like selection, it usually focuses on how to get the job done 'properly' and not enough attention is paid to likely errors and how to avoid or recover from them. It is, for example, possible to debate whether training should include information on 'short cuts' which speed up the job but increase the likelihood of error. One side of this debate says that these short cuts, and the likely consequent errors, should not be demonstrated since it makes it more likely that trainees will use them in the future. The other side of the debate says that they should be demonstrated, and the results of likely errors pointed out, in order to discourage trainees from using them.

Both sides of the debate miss two critical points:

1. **Where there are known likely errors, these should be addressed before training.** There is an unfortunate tendency to see training as a way of compensating for over-complex work systems. Instead of simplifying the system, so that likely errors are eliminated or reduced, the training is used to enable individuals to carry out tasks which are inherently error-prone.
2. **Whether or not individuals take short cuts after training is unlikely to be a function of the training received.** It is much more likely to be influenced by, for example, motivational factors, peer pressure and the effectiveness of supervision. Unless these sorts of issue are addressed, training will, at best, have a temporary influence on reliability. These sorts of issue are discussed later in the chapter.

Thus, training provision, like recruitment and placement, requires more emphasis on likely errors and the competences required to avoid them. The actions required to improve matters involve the same sequence as we described above for recruitment and placement, with the substitution of training for selection in the final stage.

Assessing competences

Whether organisations are obtaining competences by selection or by training, the emphasis during competence assessment is, as we have pointed out, usually on whether the individuals being assessed can carry out a particular task. There is rarely any systematic assessment of individuals' competence in error avoidance and recovery.* Partly this is because too little is known (or documented) about likely errors and their recovery procedures, but mainly it is because too few people see this as a relevant and necessary part of competence assessment. This failure to assess error avoidance and recovery skills means that little is done to improve these skills and this, in turn, means that errors continue to happen.

Although the importance of this sort of competence assessment in the context of selection and training has already been emphasised, these are only special cases. There should be a role for ongoing competence assessment, since any competence is likely to degrade. Periodic assessments should be used to check that this has not happened. The 'competency stages for the individual' described in Chapter 12 demonstrate why this continuing assessment is required.

Specific competences

As we have seen in previous chapters, humans tend to operate in particular ways in, for example, problem solving because of how their mental activities are organised. In particular circumstances, these characteristic ways of operating can result in increased error rates and, where it is not possible to alter the circumstances, special competences are required. This subsection deals with the more important of these specific competences.

Problem solving

As we saw in Chapter 21 (Figure 21.1), the normal procedure for problem solving involves implementing the first solution which appears capable of solving the problem. However, even when we have identified the problem correctly (see below), this is not the best strategy and it will only be by chance that we identify the optimum solution. As we showed in Chapter 21, a better strategy is to generate possible solutions and then evaluate all of them, and people can be taught to use this strategy. With appropriate training, people can also be taught to use this strategy even in situations involving imminent danger.

Problem identification

Finding the optimum solution to a problem presupposes that we have correctly assessed the nature of the problem in the first place. Studying the reports of major disasters (see Chapter 27) suggests that responses to critical situations can be inappropriate because the people involved were trying to solve the wrong problem. On a less dramatic level, it is likely that many minor accidents result from errors in problem diagnosis, often as a result of incorrect expectancies about the cause of a fault. People often operate on the basis of the most likely cause rather than beginning with a conscious attempt to identify the actual cause. As with problem-solving strategies, people can be taught the analytical techniques required for effective problem identification.

Assessing implications of solutions

Even when we have correctly identified a problem and have chosen an appropriate solution, there is still a potential for error. This arises because the action required to implement the chosen solution may have implications for other parts of the system. In other words, one problem is solved but another is created. It can be argued that a solution which has adverse effects elsewhere in the system cannot, by definition, be the best solution, and this is probably true. However, the competences we need to determine the ancillary effects of a solution are different from those we need to determine its effects on the problem. The former require creative thought in the 'what if..?' mode and are, therefore, more akin to risk assessment techniques than to the evaluation techniques required to choose between solutions.

---

* There are, of course, notable exceptions to this, such as the competence assessment for airline pilots and the specialised training in skid avoidance and recovery provided for drivers.

We must separate these three stages in problem solving and related issues since, as the discussion above has indicated, they require different competences. However, all of these competences are in the 'cognitive learning' category – they are mental abilities rather than physical abilities. All of these abilities can be taught, but in putting them into use there may often be time constraints that have a direct effect on human reliability.

The problem identification, problem solving and assessment of implications sequence can be applied in circumstances where there are no critical time constraints. This would be the case, for example, if we were using the sequence as part of an advanced risk assessment exercise. There will also be circumstances during normal work activities where the sequence can be applied at a speed determined by the competence of the personnel involved and the complexity of the problem being addressed. The difficulty arises when decisions have to be made in timescales which are too short for considered decision making. When this is the case, stimulus-response sequences are required, and these are dealt with next.

Stimulus–response sequences

In circumstances requiring a rapid response, one possible solution is to inculcate appropriate stimulus–response sequences which remove the need for considered decision making. The ballistic movements described in Chapter 14 are specific examples of this type of stimulus–response sequence and, more generally, the first two levels in the Hale and Glendon model[94] (see Figure 27.2) can also be stimulus-response sequences. That is, there is a danger signal (stimulus) which evokes a programmed response at the skills level, or there is an obvious warning (stimulus) which evokes a known procedure (response) at the rules level. However, the effectiveness of these stimulus–response links in error prevention depends on a number of factors:

*   there needs to be a unique danger signal or warning – that is, it must be possible to devise signals and warnings which refer to a specific problem, or a category of problems that have the same solution
*   there needs to be a response or procedure which will solve the problem without creating adverse effects for other parts of the system
*   the response or procedure must be capable of being taught to the people who will need it, and there must be a way of ensuring that they will be able to retrieve and execute it in the time available
*   there needs to be a way of maintaining a one-to-one relationship between stimulus and response – there must not be circumstances in which the same stimulus could require a number of possible responses.

Where these criteria can be met, stimulus–response sequences can be an effective method of reducing the likelihood of errors. However, they can create their own problems, as we saw in the discussion of overlearning in Chapter 28. In order to be effective in short time periods, stimulus–response links have to be very well learned (overlearned), which means that if circumstances change, they can result in errors. In other words, what was an appropriate response to a stimulus is no longer appropriate and results in an error.

This discussion should have made it clear that anything which reduces someone's competence to avoid errors will result in an increased likelihood of errors. The corollary of this is that anything which increases someone's competence to avoid errors increases their reliability. These statements may be truisms, but few organisations have procedures in place to test error-avoiding competences during recruitment and placement, to provide training in error-avoiding skills during training, or to assess individuals' on-the-job error-avoiding competences. In other words, few organisations have a competence management system of the type described in Chapter 12.

Motivation

The fact that people are sufficiently competent to avoid errors is not, in itself, enough to ensure that they do not make errors. Individuals must also be willing to apply their competences in the correct way and at the relevant times. In effect, individuals must be motivated to avoid errors.

In Chapter 26, we looked at the complexities of human motivation and the wide inter-individual and intra-individual variations in human motivation, and the discussion here of the motivation to avoid errors should be read in the context of the points made in that chapter. Remember throughout that the descriptions are broad generalisations and that it is not usually worthwhile to try to predict a particular person's motivation in a given set of circumstances.

For the purposes of this discussion, it is useful to divide attempts to influence motivation into two broad categories:

1. **Attempts to influence job performance directly through such things as payment systems and reward schemes.** It can be argued that the first modern systematic application of these techniques was the 'scientific management' system put forward by F W Taylor in his book *Principles of scientific management*, published in 1911.[47] This broad category will be dealt with under the general heading of 'scientific management'.

2. **Attempts to influence motivation through such things as peer pressure and job satisfaction.** These broader motivational factors were identified partly through the weaknesses in scientific management, and they will be dealt with under the general heading of 'other motivation theories'.

Scientific management

There seems little doubt that, for many people, financial reward can be a motivating factor. However, in most organisations, payment schemes are focused on increasing productivity, not reliability. The often quoted piecework system can be used to illustrate the distinction. In simple terms, piecework payment systems are those where the amount of pay received is a function of the amount of work done (originally, the number of 'pieces' of work produced). Under piecework systems, the motivation is to work quickly, not to avoid errors.

Piecework illustrates the fundamental problem faced by risk management: balancing a sure and measurable benefit (amount of pay) against an uncertain and hard-to-measure cost (injury). A further complication arises where the likely cost is ill health, since the benefit is short-term (pay at the end of the week or month) while the cost (hearing loss, lung disease or cancer) may be 20 years in the future.

The concept of piecework, and related pay systems, was formalised by Taylor in his description of scientific management.* Taylor's system involved the application of a number of principles and the key ones, as far as the present discussion is concerned, are described below. However, in the years since Taylor put forward his system, a number of weaknesses have been identified. These weaknesses are dealt with at the relevant points in the notes which follow.

Efficiency

Taylor believed that the majority of the work he observed could be done more efficiently. He studied how particular tasks were carried out and was able to identify ways in which they could be done more efficiently by, for example, using more appropriate tools, re-ordering the work sequence or redesigning the complete task.

These exercises were the beginnings of the 'work study' approach. While this type of work study did, and still does, identify ways in which productivity could be increased, it was in large measure reliant for its effectiveness on treating people rather like machines. That is, people were expected to use a limited range of movements and repeat these movements a large number of times during the working day without error. As we have already seen (for example, in the discussion on the Hale and Hale model), this is not a realistic expectation. Where a limited range of movements is to be repeated a large number of times over an extended period, machines are a better alternative as far as reliability is concerned, and this is discussed in more detail in the next chapter.

---

* In some texts, this management system is called 'Taylorism'.

Links with performance

Taylor believed that individuals were motivated more strongly if rewards and sanctions were linked directly to task performance. He assumed that individuals would be motivated if they could see a clear link between their task performance and, for example, the level of pay they received and their rate of promotion. He also assumed that this motivation would be reinforced if there was the possibility of dismissal for poor performance, or poor performance led to individuals being transferred to less desirable tasks.

These assumptions were given a rationale in a psychological theory of motivation known as behaviourism, first propounded by Skinner in the 1970s[97] and hence also known as Skinnerism. In essence, this theory states that behaviours which are rewarded, or are not punished, will tend to be repeated, while behaviours which are not rewarded, or are punished, will tend not to be repeated. While this may generally be true, the motivation of human beings is too complex for it to be applied in the direct and simple manner implied in scientific management. For example, some individuals will find it more rewarding to 'buck the system' from time to time rather than to continue with the actions they have already carried out numerous times at work, and may have to carry out for the foreseeable future.

Another problem with this aspect of scientific management is that it may be difficult to establish an accurate measure of performance. Taylor began with tasks where productivity could be measured readily in terms of physical output, such as the amount of coal a person moved, but there are many tasks where measurement of performance is more difficult – for example, supervisory and managerial tasks.

Financial rewards

Taylor believed that the primary motivating factor for individuals was financial reward. He assumed that people would do things which brought financial reward and avoid doing things that either brought no financial reward or resulted in a financial penalty. This view of human motivation was summed up in the idea of 'rational man' – people who maximised or optimised financial rewards. However, as we have seen at various points in this book, human motivation is complex and to base a management theory too heavily on assumptions about the importance of one particular motivating factor is inadequate.

This inadequacy has been demonstrated in a number of ways since Taylor first put forward his ideas, but the problems were highlighted in an early series of studies which became known as the Hawthorne experiments (because they were carried out at the Hawthorne Western Electric Works near Chicago), whose main findings were published in 1939.[98] These experiments involved manipulating a range of variables, such as the level of lighting in the workplace, to determine the effect on productivity. The experimenters began with the not unreasonable assumption that improving these variables (for example, increasing the levels of light) would improve productivity. They successfully demonstrated that this was the case but were surprised to find that, in a continuation of the study, reducing the lighting back to its original level produced a further increase in productivity.

Their conclusion from this, subsequently confirmed by other researchers, was that the motivating factors were not, for example, the level of lighting, but the fact the someone (in this case the experimenters) was taking an interest in the workforce and what they were doing. This effect, which became known as the 'Hawthorne Effect', had important implications for management theories, since it could no longer be assumed that individuals would be motivated solely by financial rewards and sanctions. This point will be discussed in more detail when we deal with organisational factors.

The individual

As we mentioned above, Taylor was primarily concerned with individual performance, and he attempted to devise ways in which this could be measured and linked to rewards. His argument was that people who produced more should receive higher (financial) rewards than those who produced less. While this may appear to be a reasonable argument, it breaks down in practice. Numerous studies, including the Hawthorne experiments just mentioned, have demonstrated that groups of workers involved in the same task will establish informal group norms for productivity so that the amounts earned by the members of the group tend to level out. In strongly established groups, this can extend to the stronger or more skilled group members 'contributing' part of their output to 'help out' less able members, or deliberately slowing down their rate of output to reduce the discrepancy between their output and the output of the least able.

These sorts of behaviours indicate that there are social motivational factors in addition to financial ones. In other words, for some people at least, it is more important to remain on good terms with the work group than to maximise their earnings. These behaviours also indicate that, again for some individuals, altruism (helping others at a cost to oneself) is a motivating factor which can override financial motivation.

Other aims of scientific management

Increasing productivity by motivating individuals was not the only aim of scientific management. However, the other two main aims – increasing co-operation between the management and the workforce, and reducing conflict at work – will be dealt with in the 'organisational factors' section later in this chapter. The present subsection ends with a review of the implications of scientific management for human reliability.

Scientific management and human reliability

Scientific management, as far as motivation is concerned, can be summed up as an attempt to increase productivity by making people work more efficiently and by linking individual performance directly to financial rewards and sanctions. While scientific management as a complete management system has largely been discredited, certain elements of it – such as piecework and financial bonuses for individuals – remain as elements in other management systems. In addition, certain parts of scientific management continue to be applied in risk management and it is these elements that we will consider next.

Taylor was concerned with finding ways of getting people to exhibit certain types of behaviour and to maintain them over extended periods of time. His concern was with behaviours which resulted in high productivity, but more modern practitioners have applied his ideas to health and safety-related behaviours. The rationale is as follows:

- Behaviours can be identified which will increase reliability or counter the effects of errors. Examples include rigorous adherence to safe working procedures and consistently wearing relevant PPE.
- Rewards, including financial rewards, can be devised which will reinforce the identified behaviours. In addition, or as an alternative, sanctions, including financial sanctions, can be devised which will discourage non-compliance.
- The rewards and sanctions are applied consistently at the level of the individual so that particular individuals comply with the required behaviours.

It will be noted that this rationale is based on the behaviourism approach described earlier and, indeed, it is often referred to as 'the behavioural approach to safety management', or simply 'behavioural safety'. Using these techniques has been shown in some studies to produce improvements in the short term, but there is some doubt as to the adequacy of the controls for the Hawthorne Effect in these studies. That is, we cannot be sure whether the improvements arise as a result of the rewards and sanctions rather than as a result of the experimenters or consultants taking an interest in the work being done. In addition, there is little evidence to demonstrate that changes in behaviour produced in this way are sustained in the longer term. Behaviour changes will be sustained only if they are supported by parallel changes in attitudes, and attitudes and behaviour remain consistent. If the behaviour is supported only by extrinsic rewards, then once these are removed the behaviour will tend to decline.

See also Chapter 21 for a discussion of attitudes and behaviour in the context of safety culture.

Other motivation theories

The theories of motivation discussed in Chapter 26 have, among other things, had an influence on theories of management, and this subsection deals with the effects of this influence as far as human reliability is concerned. The following subheadings are used to structure the discussion:
1. group motivation
2. rewarding reliability
3. providing feedback on reliability
4. higher level motivations.

Group motivation

The limitations of concentrating on motivational issues at the individual level were discussed in the previous subsection. We pointed out that one of the reasons for the failure of this approach is the influence of groups. The present subsection begins, therefore, with a brief outline of how interpersonal relationships can be studied. There is then a discussion of the advantages of group motivation.

Interpersonal relationship studies

As we have seen in the discussions on groups (see, for example, Chapter 16), intra-group influences are varied and complex. However, it is known that certain group members have more influence on group behaviour than others and that these influential group members need not be the 'official' group leader. For this reason, it is important to study interpersonal relationships in order to identify the people with influence in a group. It may not be possible, for organisational reasons, to implement attempts to motivate other than through official group leaders but, unless we establish the real influences within groups, it will be impossible to determine why attempts to motivate succeed or fail.

One way of establishing where the influences flow in a group is to collect the data necessary to construct a sociogram of the type shown in Figure 29.1.

Figure 29.1
A simple sociogram for a nine-person group

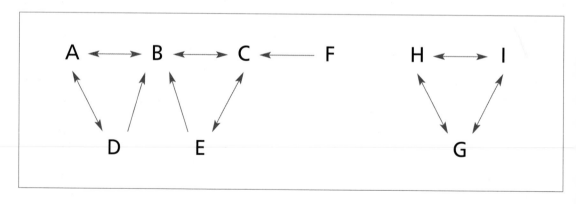

Sociograms are one form of social structure diagram which are used to represent the relationships or communications between individuals or groups. The organisation charts illustrated in Chapter 16 and 'family trees' are common examples of social structure diagrams.

The sociogram may be a less familiar example, but it has a number of uses when investigating relationships within groups. The simple sociogram shown in Figure 29.1 was constructed in the following way:

- A group consisting of nine people (labelled A to I) were asked to name the two other people in the group they would most like to have as friends.*
- Each of these nominations is then represented by an arrow to produce the sorts of results shown in Figure 29.1.

This simple sociogram shows a number of things:
- there are a lot of mutual friendship votes, as might be expected
- F is unpopular, and voted for only one other member of the group
- B is very popular
- A–B–D and B–C–E are 'cliques', linked only by B's general popularity
- the 'group' A to I is, in practice, two distinct groups: A to F and G to I.

These facts about groups are the sorts of information you will need if you want to try to influence group behaviour, including influencing motivation with respect to reliability. However, as we will see in the next subsection, group motivation is potentially very powerful – it can therefore be worthwhile to make a detailed study of the dynamics within a group before trying to influence its motivation.

The advantages of group motivation

The major advantage of group motivation as far as reliability is concerned is that it is much more likely than individual motivation to be sustained in the long term. The main reasons for this are:
- Behaving in the desired way may in itself be rewarding to individuals, but this reward is supplemented by the sorts of social rewards (acceptance by the group and altruism) mentioned earlier in the chapter. Thus, there is no conflict between motivations as may happen when the emphasis is on individual rewards.
- The group 'polices' compliance with the desired behaviour. Members of the group reward compliance with praise and ongoing acceptance and provide sanctions against non-compliance in the form of criticism and, in extreme cases, ostracism.

In terms of reliability, motivational schemes that focus on the group have the benefits listed above but, in addition, there is an increased error recovery potential arising from group work. Highly cohesive groups informally monitor each other's health and safety performance as well as those aspects of the group's behaviour directly related to productivity.

Rewarding reliability

It is a truism that an organisation's motivational efforts should be directed at the objective the organisation wishes to achieve. However, although the majority of organisations would probably list high levels of reliability among personnel as one of the things they wish to achieve, it is rare that motivational efforts are targeted directly at reliability or absence of error. This is a misdirection which organisations could solve – this has been demonstrated by changes in quality management, and this change is outlined below.

The majority of organisations have always been concerned to ensure the quality of their products and services in that they had a desire to ensure that products and services were produced to meet a pre-defined specification. However, these organisations were also keen for levels of productivity to be high. The conventional method of meeting these two, often contradictory, aims was to pay the producers of the products or services according to their level of productivity and to employ teams of quality inspectors and teams of people who repaired faults. Although this worked, in terms of ensuring product and service quality, it was rarely a totally satisfactory arrangement since, even when quality was maintained at the required level, it was

---

* Friendship is one way of identifying possible influence. Other ways include asking people in the groups who they most respect, who they are most likely to ask for advice, and who they consider most suitable for promotion.

an expensive exercise because of the amounts of inspection and reworking required. A major breakthrough came when it was realised that a better approach would be to identify and eliminate the causes of the failure to meet the pre-defined standards. This change, which seems blindingly obvious now that it has occurred, revolutionised quality management, although it required extensive rethinking of production and delivery systems to enable the required measurements to be collected and used.

It can be argued that an analogous revolution is required in risk management. People should be paid for doing their job without errors which could lead to losses, as people are now paid to avoid errors which could lead to nonconformities. In theory, this is an attainable goal, since it has proved possible in quality management. However, in practice, it requires organisations to devote the necessary resources to rethinking how work is done and measured, and the ways in which people are paid to do this work. Whether organisations are willing to do this is likely to be a commercial decision. The revolution in quality management was driven by financial considerations and it is likely that any equivalent revolution in reliability management will only occur if there is similar financial pressure. This point is discussed in more detail in Chapter 24, on financial issues.

It follows from the preceding discussion that, where the aim is to reduce the sorts of error which result in losses, the rewards and sanctions should be targeted at reducing these types of error. However, the reward systems which have been used in the past, variously referred to as payment systems, reward schemes and incentive schemes, have usually failed to do this. Examples of this type of misdirection are given here:

- **Rewards based on productivity where productivity is defined solely in terms of numbers of units produced.** These include the piecework systems already mentioned, but any system of reward where the reward received is commensurate with output, irrespective of other measures such as quality or error rates, will fall into this category. However, this type of reward can be used successfully if the required output is redefined, as has been done with modern quality management systems.
- **Rewards based on numbers of accidents.** There have been attempts in the past to reward individuals or groups on the basis of the number of accidents sustained in a given period with, for example, the groups having the lowest number of accidents receiving the highest rewards. These sorts of schemes are, however, largely discredited since the evidence is that what they do is suppress reporting of accidents, rather than reduce the number of accidents that happen.
- **Rewards focused on particular risk reduction activities.** These include such things as rewarding people for wearing PPE or for conducting regular 'safety inspections'. While these reward systems may produce the required behaviour while the reward is available, the people concerned are usually behaving in the required way to receive the reward, not to reduce risk. As has already been noted, this means that the required behaviour ceases when the reward is no longer available, or is no longer rewarding.
- **Rewards focused on speed of work.** There are certain patterns of employment which encourage people to work quickly, rather than to work without errors. For example, the work patterns known as 'job and finish' fall into this category. In these ways of working, individuals are paid a fixed rate for a fixed amount of work, irrespective of how long it takes to complete the tasks. Typical examples include unsupervised working away from a fixed base, for example refuse collection and local deliveries. In these cases, the motivation to complete tasks quickly may be financial, in that finishing quickly enables the individuals concerned to take another paid job, or it may be that finishing quickly provides the individual with valued free time. Other tasks which are subject to these sorts of motivation include driving tasks, particularly over long distances, and unsupervised 'home working' on, for example, word processing tasks where adequate breaks might not be taken.
- **Rewards based on numbers of hours worked.** For most tasks, error rates increase the longer the task is carried out continuously.* Despite this, a number of reward systems and patterns of employment

---

* This is a complex area with conflicting evidence provided by different studies. However, the broad generalisation made here is adequate for our purposes.

encourage extended working hours. In particular, all of the reward systems based on higher levels of pay for overtime at the ends of shifts, or for working on days which would normally be rest days, build in this sort of bias. For the 'rational man' described earlier, it makes financial sense to limit productivity as much as possible during the normal working hours so that employers are forced to extend the working hours, at higher rates of pay, in order to achieve the required output.

- **Rewards based on 'unnatural' work patterns.** As we saw in Chapter 15, humans have a diurnal cycle which, taking into account the wide range of individual differences common to all human characteristics, means that there is a 'natural' cycle of activities. For most individuals, for most of the time, fatigue is at is lowest and alertness at its highest during normal daytime hours. However, patterns of work such as shift work reward individuals for working counter to their natural cycle. There is some evidence to suggest that those who work 'permanent nights' are able to compensate for this, which might reduce the effects, but this is not possible for patterns of employment which involve rotating shifts.

The list of examples given above is not intended to be exhaustive; it illustrates the general point that many systems of reward and patterns of employment militate against rewarding reliability. You may, no doubt, be able to think of other examples from your own experience of how this principle manifests itself in particular organisations.

The lesson to be drawn from this subsection is that if human reliability is to be improved, then it must be measured so that the extent of the problem is known, and reward systems and patterns of employment have to be devised so that they reward reliability. Subsequent measurements can then be carried out to establish the effectiveness of particular reward systems or patterns of employment with respect to increasing reliability, and adjustments made as appropriate.

Providing feedback on reliability

Earlier in this chapter we pointed out that avoiding errors is a competence and, like all competences, it requires knowledge and skills. It has long been established that the most effective way to teach and maintain skills is to provide feedback on performance. It follows from this that the most effective way to maintain the skills required for reliability is to provide feedback on reliability performance.

It can be argued that the work system itself provides a certain amount of feedback on reliability, in that errors are 'punished' by injury, ill health and other losses. However, even if this is true, it is not the ideal type of feedback since it is being provided by incidents of the type higher reliability is intended to prevent. What is required, therefore, is feedback on reliability of performance at earlier stages, before errors which result in losses.

This type of feedback will, of course, require the skills needed for reliability to have been identified correctly (see the earlier section on competence) but, assuming that they have, providing the necessary feedback follows the general principles of all skills maintenance feedback. However, when current feedback systems are considered, it is unusual to find systematic feedback on reliability skills. The average appraisal system, for example, has a range of criteria against which performance is assessed, but these criteria usually focus on the same range of issues as the reward systems already discussed – namely productivity, extent of working hours and speed of work. These criteria may include some related to accuracy of performance, which is one form of reliability, but rarely do they include reference to the sorts of reliability relevant to risk.

As with other reliability issues discussed in this section, long-term improvements in human reliability will require accurate identification of the skills required and appraisal systems which provide adequate feedback on individuals' performance with respect to these skills.

Higher level motivations

The emphasis in this section so far has been on financial rewards, although the influence of other rewards such as group acceptance and altruism has been mentioned. However, remember the description of Maslow's

Hierarchy of Needs (see Chapter 26), which argued that financial reward is relatively low in the hierarchy of motivations. Maslow's theory predicts that once people have satisfied their basic financial requirements, financial rewards are no longer motivating. This is in contrast to the 'rational man' prediction that individuals are motivated to earn greater and greater sums of money. As might be expected, given the nature of human beings, neither of these predictions is universally accurate, but it is true that most individuals are not motivated solely by financial rewards.

While there are the inevitable inter-individual and intra-individual differences, it is possible to make some general comments on these higher motivations and their relevance to reliability.

- **Motivation arising from social interaction.** For those people who are motivated to work in order to satisfy social needs, the important aspects of their jobs are what management might consider to be peripheral issues concerned with groups and their interaction. While attempts to pay these people to work more reliably may be successful, the cause and effect relationship is likely to be complex. People with this sort of motivation are more likely to work reliably if they can see that doing so will make them more acceptable to the group or result in an enhancement of the social benefits they receive from being at work. For example, emphasising that reliable working is likely to contribute to the wellbeing of other group members is more likely to be an effective motivator than an emphasis on the personal financial benefits of reliable working.
- **Motivation arising from altruism.** For those people who are motivated by a desire to help others, financial rewards are, in extreme cases, irrelevant. Where such individuals have to work to obtain money to support themselves and their dependants, the motivating effects of the financial rewards is likely to cease once they have earned enough to meet their perceived needs. Increasing reliability among individuals motivated by altruism is more likely to be achieved by appeals to avoid harming others than by increasing personal financial rewards.
- **Motivations arising from self-actualisation.** This, remember, is Maslow's highest level of motivation, which can be extremely complex.

There have, however, been attempts to produce theories which integrate various levels of motivation at work. A particularly influential one was produced by Herzberg in 1966.[99]

Herzberg argued that the observed complexities with motivation at work were, at least in part, due to the fact that motivation was dealt with as a single dimension, for example the one-dimensional hierarchy described by Maslow. In response, Herzberg produced what is referred to as a 'two factor' theory but is, more accurately, a theory based on two groups of factors. Herzberg divided the factors which had previously been identified as motivational into two separate groups:

1. factors which prevent people becoming dissatisfied with their work, which are referred to as hygiene factors or maintenance factors
2. factors which motivate people, referred to as motivating factors.

Table 29.1 provides a summary of where Herzberg fitted various aspects of work and the working environment into his two factor theory. Notice in Table 29.1 that the hygiene factors are largely to do with the working environment while the motivating factors relate to what a person does at work.

Herzberg's work is linked with the idea of job satisfaction in that, if his theory is correct, such things as lack of financial reward or poor supervision will lead to dissatisfaction with the job, but will not necessarily be demotivating. Conversely, people who are satisfied with their job in terms of hygiene factors need not necessarily be motivated. For high levels of job satisfaction, combined with high levels of motivation, attention to both sets of factors is required.

Herzberg's theory has been criticised on a number of grounds, including that it does not take into account the observation that some people are genuinely motivated by money. However, an awareness of Herzberg's ideas should make people who are attempting to set up a motivational approach to increasing human reliability less inclined to make assumptions about what will, and will not, be motivating.

Hygiene factors	Motivating factors
Money	Sense of achievement
Status	Recognition
Relationship with boss	Enjoyment of job
Company politics and administration	Possibility of promotion
Work rules	Responsibility
Working conditions	Chance for growth
Supervision	The work itself
Relationship with peers	Advancement

Table 29.1
Herzberg's classification of factors

Summary of motivation

Since motivation and human reliability are both complex issues, it is worth attempting to distil key points from the previous discussion:

1. **Human reliability will best be improved if attempts to motivate people are targeted directly at reliability.** This will require direct, preferably quantified, measures of performance reliability so that the effects of attempts to motivate can be assessed.

2. **An individual's reliability will best be influenced if the motivation is one which produces an outcome valued by the individual concerned** (see the discussion of Vroom's theory in Chapter 26). Applying the same type of motivation to everyone may be successful for some people but it is likely to have a negative effect for others.

3. **Individuals' motivation changes from time to time so that what will be motivating will also change.** If reliability is to be maintained through motivation it will, therefore, be necessary to monitor reliability of performance and take appropriate action if reliability diminishes.

Organisational factors

In the previous two sections of this chapter, we have discussed people's ability to be reliable (competence) and their willingness to be reliable (motivation). In this final section of the chapter, the emphasis will be on empowerment to be reliable and, in particular, the organisational factors which militate against individual reliability.

To a large extent, the competence and motivational factors already discussed are organisational factors in that they are under the control of the organisation. However, there are several other factors that are under organisational control but which, at first sight, do not appear to have any relevance to individual reliability. In practice they do, although their effect is usually mediated by competence or motivation, or both. Since this idea of mediation may not be a familiar one, this sections begins with a simple example of mediation; this is followed by a description of the main organisational factors relevant to human reliability.

The mediation process

The senior management in an organisation sometimes implements a new procedure without adequate prior consultation and preparation. If this happens, the change can have any or all of the following effects:

- The individuals concerned are unfamiliar with the new procedure so they make mistakes, some of which may have implications in terms of risk. This is an organisational factor influencing reliability mediated by competence or, more accurately, lack of competence.

- The lack of consultation is likely to have a variety of effects in itself. For example, it may make the individuals concerned feel that their opinions are not valued, or even that they themselves are not valued.

These feelings can then be translated into lower morale and lower motivation. This is an organisational factor influencing reliability mediated by morale or motivation.

- The fact that individuals are being expected to carry out new tasks with which they are unfamiliar may be a stressful experience and the level of stress may be such that reliability is adversely affected. This is an organisational factor influencing reliability mediated by stress.
- The activities required to move from an old procedure to a new procedure may mean that, during the changeover, people have to work more intensively or for longer hours, or both. This could result in greater than normal levels of fatigue which, in turn, could have an adverse effect on reliability. This is an organisational factor influencing reliability mediated by fatigue.
- Since the same change can result in any or all of these effects, organisational factors can have more than one effect, mediated through more than one channel.

How change is managed is an important organisational factor as far as reliability is concerned and it will be dealt with in more detail in the next subsection. However, there are other relevant organisational factors and these are described in the separate subsections which make up the remainder of this chapter.

Managing change

As we saw in the example above, change in an organisation can have many effects that are peripheral to the change being introduced. Nevertheless, since these peripheral effects can influence reliability, it is important that the change management process includes the identification of possible effects on reliability.

There is an aphorism that 'no-one likes change', but this is only true if change is very carefully defined. Another aphorism is that 'one man's meat is another man's poison', and this illustrates the point. One person's change is another person's improvement, and *vice versa*. Consider a domestic example, where one partner wants to move to a new house and the other does not. For the first partner, the move is an improvement ('a change for the better') while for the second partner the change is one of the ones that no-one likes.

In general terms, if a person is initiating a change, or approves of what the change is intended to achieve, the actions required to bring about the change will be less stressful and less fatiguing and the motivation to bring about the change will be high. In these circumstances, the change process will have the minimum of effects on reliability. However, this is not the case where the change is being imposed on the person, or the person disapproves of what the change is intended to achieve.

It follows from this informal description that change, *per se*, need not have a detrimental effect on reliability, but that the way in which the change process is managed may result in such effects. In essence, good change management involves allowing people to initiate change themselves (or creating circumstances that make them think they are doing the initiation) and/or ensuring that people approve of what the change is intended to achieve. A detailed consideration of the techniques of change management is outside the scope of this book, but Handy[50] provides a useful summary of the 'psychological progression' which should be adopted when a change in strategy is required:

1. Create an awareness of the need for change (preferably not by argument or rationale but by exposure to objective fact).
2. Select an appropriate initiating person or group ('appropriate' in this context refers to sources of power as perceived by the recipients of the strategy).
3. Be prepared to allow the recipients to adapt the final strategy. (That which one adapts one can more easily call one's own. Ownership equals internalisation, ie self-maintaining.)
4. Accept the fact that, like the good psychoanalyst, the successful doctor gets no credit but must let the patient boast of his sound condition. Good managers live vicariously.

5. Be prepared to accept a less than optimum strategy in the interests of achieving something rather than nothing. Compromise has its own morality.

Managing conflict

As with managing change, detailed consideration of managing conflict is outside the scope of this book. However, Handy deals with conflict and its management in detail and this section is a selection of his key points with supplementary notes from the present author on the implications for reliability and risk management.

Handy begins his discussion of conflict with a description of the symptoms of conflict:

- **Poor lateral and vertical communication.** Lack of communication can affect reliability in a number of ways, including failure to pass on information required to identify hazards or risk control measures. In addition, poor communication can affect reliability through stress or low morale.
- **Intergroup hostility and jealousy.** These can result in poor communication, with the problems identified above, but they can also distract attention from reliability issues and remove the advantages of groups 'looking out for each other's welfare'.
- **Interpersonal friction.** This has similar consequences to those described for group hostility but at the level of the individual.
- **Escalation of arbitration.** This symptom of conflict results in the problem which is the original source of the conflict being referred to higher and higher levels of management for arbitration instead of being dealt with at the level at which it occurs, or at the most, the level above. In health and safety management, a symptom of conflict can be the referral of inappropriate health and safety issues to the health and safety committee and the role of the health and safety committee is dealt with later in this section.
- **Proliferation of rules and regulations.** This symptom usually arises more from attempts to resolve conflicts than from the conflicts themselves. For example, the senior management may try to clarify what is, and what is not, acceptable behaviour by encapsulating these in the form of rules.
- **Low morale.** This can arise directly or be caused by a mixture of the other symptoms listed above. We discussed the effects of low morale on reliability earlier in this section.

These symptoms of conflict can arise for a number of reasons, but a major source of conflict is differences between goals, which can arise at a number of different levels. However, for the purposes of this chapter, these are divided into intra-individual and inter-individual conflicts.

Intra-individual conflicts

These are conflicts which arise within an individual and there are two common reasons for this:

1. The organisation imposes conflicting goals on the individual, such as a high productivity goal and a high reliability goal.
2. The organisation imposes a goal on an individual which conflicts with that individual's personal goals. For example, organisational goals that require levels of performance which can only be met by extended working hours are likely to conflict with personal goals related to family life and hobbies.

These sorts of goal conflict within the individual are likely to create stress and other negative psychological states such as anxiety and depression, and individuals suffering these sorts of psychological states are likely to be less reliable. The resolution of these conflicts involves the application of general principles of conflict management and these are dealt with later in the section.

Another source of intra-individual conflict arises where there is a mismatch between two or more of the following aspects of someone's work:

- **accountability** – the application of rewards and sanctions to an individual on the basis of certain targets being met

- **responsibility** – the allocation of particular tasks to an individual
- **authority** – the allocation of control of resources to an individual.

Where accountability, responsibility and authority match, there is no conflict. But mismatches can occur, for example:

- an individual is rewarded (or sanctioned) on the basis of the collective performance of a group of people (full accountability) but some of the group report to another person who directs their work (partial authority)
- an individual is rewarded (or sanctioned) on the basis of the number of units he or she produces (full accountability) but parts of the operation required for production have to be carried out by someone else (partial responsibility)
- a person is allocated a task (full responsibility) but is given insufficient resources to carry out the task (partial authority).

The types of conflict created by these sorts of mismatch are similar to those created by conflicting goals, and they have to be resolved using similar general principles.

Inter-individual conflict

Inter-individual conflict is used here as a shorthand for the various types of conflict which involve more than one person. These include conflicts between individuals, conflicts between individuals and a group, and conflicts between groups.

Handy gives a detailed list of the causes of such conflicts and a selection of these is reproduced here, with supplementary notes.

- **Formal objectives diverge.** Where organisations set formal objectives, these may result in conflict. For example, objectives concerned with high productivity can come into conflict with objectives based on low accident rates. Even when the objectives are not formal, but expressed as less formal goals, the same potential for conflict exists.
- **Role definitions diverge.** For example, the role of the safety function (to reduce losses) diverges from the role of the production function.
- **The contractual relationship is unclear.** For example, the debate about whether the health and safety professional's role is to assist production to work more safely, or to 'police' production activities on behalf of senior management.
- **Roles are simultaneous.** See the example above; the health and safety professional may have to be both an adviser and a police officer.

As can be seen from these last two examples, the various causes of conflict overlap and the results they produce (low morale, low motivation and so on) also overlap. However, the ways in which conflict should be managed are common to the various causes and these conflict management techniques are dealt with next.

Conflict management techniques

Handy identifies two main approaches to conflict management:
1. turning the conflict into fruitful competition or purposeful argument
2. controlling the conflict.

He suggests that the second approach should be used only if the first is not successful, although certain techniques used for controlling conflict may be useful as short-term measures while arrangements are made to turn the conflict into fruitful competition or purposeful argument.

While Handy points out that there is no sure way to turn conflict into fruitful competition or purposeful argument, he suggests that such transformations will be more likely if the following features are in place (all quotations below are from Handy[50]):

- **There is a clear and shared purpose for the group or organisation.**

  Without a common goal individuals are, in effect, licensed to do the best for themselves within the rules. Management then has to start regulating the competition by adjusting the rules.

The importance of this aspect of conflict management can be judged from the number of times conflict between goals, objectives and roles were mentioned in the earlier discussion.

- **Information is available on progress towards the goal.**

  It is little use to preach 'service to the customer', 'higher standards' or 'profitability' if the information system is all geared to controls rather than results, to time-keeping, costs and budgets instead of achievements. Individuals are realistic, they judge the real objects of the exercise by the type of feedback that comes their way.

Although Handy does not mention reliability, the importance of collecting and making available data on reliability has been mentioned several times in this chapter.

- **The system does not punish failure.**

  To do one's best is all that can be expected, even though some people's best may be better than others'. Fruitful competition will encourage risk taking and accept failure if it is the result of honest endeavour. The only fault is not to try at all. Where no-one has to lose there is more likely to be trust, collaboration and mutual help.

This is also an important issue in reliability and the negative effects of a 'blame culture' have been discussed elsewhere in this book (see Chapters 8 and 9). However, as a reminder, one of the main detrimental effects is that punishing people for errors means that they are less likely to report errors, so that the data necessary to prevent recurrence cannot be collected.

Handy points out that to achieve the states just described can take some time so that conflict control tactics may be required as interim measures. He lists a number of these tactics, including:

Arbitration. The use of the lowest cross-over point in the organization to resolve conflicts. Only useful when the conflict is apparent and specific. Not very valuable in episodic or continual conflict.

Rules and procedures. These are often arrived at by negotiation. They can easily become one of the bargaining elements and are seen as not a solution, but another constraint. They are useful when the conflict is recurrent and predictable, but should not be regarded as a permanent solution.

Co-ordinating device. A position is created on the organization chart to resolve the issue in conflict. The position often carries the name of the problem as its title, eg 'sales/production liaison', or 'new product co-ordination'. This may, however, complicate communication and increase conflict since problem-solving groups may have a vested interest in maintaining the problem they were set up to resolve. This device, therefore, is not good for episodic or occasional conflict, but it is useful where the conflicting pressures are going to be continual but are not so predictable that they can be handled by rules and regulations.

Confrontation. A technique much favoured by those who believe in 'openness' in organizational communications. This strategy will be effective if the issues can be clearly defined and it is not a symptom of more underlying differences. It is an approach analogous to arbitration but to be preferred to it in that the solutions will be 'owned' and 'internalized' by both groups rather than imposed on them by the arbitrator.

Separation. If interaction increases the depth of sentiments, separation should cool them. This strategy can work if interdependence is not anyway necessary because of the task. If it is, then the interaction will have to be managed somehow, either by a co-ordinating device, or rules and regulations. The strategy of separation works best if two groups are coincidentally interacting and conflicting, ie because they happen to be lodged in the same location. It also works if the true cause of the conflict is two incompatible individuals – incompatible in personality, or more often with relative status inappropriate to the situation. To separate the individuals, ie by transferring one, may then be a successful strategy. But interpersonal friction is more often the result of conflict than the cause of it. If this be so, then separation of the individuals will merely bring about a temporary lull in the situation, and two more conflicting personalities will soon arise.

In safety management, there is an additional opportunity for conflict resolution: the health and safety committee. This is dealt with next.

The health and safety committee
Health and safety committees can take many forms and they can have a range of different functions,* including monitoring the organisation's health and safety performance, reviewing this performance and, in the better committees, planning health and safety strategy. However, health and safety committees can have positive and negative influences on the sorts of issues being discussed in this section and some key points are:

- **Where the health and safety committee is seen as bipartisan (managers *versus* workers), the shared goals required for conflict avoidance and resolution are absent.** The goal of managers is seen as minimising the resources spent on health and safety while the goal of the workers is seen as getting as much as possible spent on health and safety. The committee will only be really useful when there is a shared goal such as 'cost-effective loss reduction'.
- **Where the health and safety committee is unrepresentative of the workforce, then those parts of the workforce which are not represented will see themselves as disenfranchised, with possible negative effects.** 'We don't take any interest in health and safety because the health and safety committee isn't interested in us.' Some method is required to ensure that all parts of the workforce are represented on the health and safety committee and that the method used is transparent to all.
- **Where the health and safety committee is seen as a conflict resolution body, delays can occur in resolving health and safety problems.** It can be the case that referral to the health and safety committee is seen as a way for managers to avoid having to make decisions on health and safety matters. This is, in effect, the sort of 'escalation of arbitration' described earlier in the subsection on the symptoms of conflict. The terms of reference of the health and safety committee should be clear about what action is to be taken on items like these which are inappropriately referred to the committee.

---

* In the UK, health and safety committees, often abbreviated to safety committees, are the subject of legislation which specifies a (minimum) range of functions. However, the present discussion is not restricted to the functions set out in UK legislation.

- **When there is no feedback of the results of the health and safety committee's deliberations to the people represented on the health and safety committee, those represented cannot see whether the work of the committee is relevant to them.** It should be made clear to all health and safety committee members that one of their primary duties is to pass on information to all those they represent.

An effective health and safety committee, with properly constituted terms of reference, can make a significant contribution to safety management in an organisation. However, it can be argued that a poor health and safety committee is worse than none at all because of its potential for negative effects.

Summary

This chapter dealt with three broad topics that have implications for the reliability of human behaviour: competence, motivation and organisational factors. Within each topic, we explored the key issues, together with their potential effects on human reliability.

30: Improving human reliability

Introduction

As we saw in previous chapters, humans are complex and prone to error in a wide variety of ways. It has long been recognised that the variability in performance between individuals (inter-individual variability) and by the same individual at different times (intra-individual variability) mean that tasks which have to be carried out by people are, in the long term, likely to result in an error.

There are ways of minimising the effects of inter-individual performance (for example, by selecting the most suitable individuals) and intra-individual performance (for example, refresher training). However, real advances in improving human reliability can usually only be gained by treating the human as part of a system, with the other elements in the system being the tasks to be carried out, the machinery and equipment, and the environment in which the tasks have to be carried out.

This systems thinking approach to human reliability means that those aspects of performance where human error rates are likely to be high can be identified and action taken which will reduce the likelihood of these errors. This approach to increasing human reliability is variously referred to as the ergonomic approach or the human factors approach. Some authors see ergonomics and human factors as synonymous but, in this book, the term ergonomics is used to describe the approach to human reliability (dealt with in this chapter) while human factors has been used as a more general term for all aspects of human psychology and physiology of relevance to risk management.

The chapter is set out in the following sections:
1. ergonomics as a discipline
2. identifying error-prone tasks
3. the 'man-machine' interface
4. environmental factors
5. application of ergonomics.

Ergonomics as a discipline

The roles played by ergonomists were described briefly in Chapter 13 but a more detailed description of the application of ergonomics has been given by Sanders and McCormick[100] and this is paraphrased below.

The application of ergonomics seeks to:
- design equipment and the work environment to match human capabilities with the aims of ensuring effective operation of equipment and optimising working and living conditions
- enhance the effectiveness and efficiency of work and other activities
- design work systems so that the requirements for human physical and mental wellbeing are met
- enhance safety, reduce fatigue and stress, and increase such things as comfort, job satisfaction and the quality of life.

There are two important points to be made about this summing up of the aims of applied ergonomics (and other, similar, summaries).
1. **Ergonomics is not just about hardware, it also deals with work systems.** Similarly, ergonomics is not just about errors and safety; it encompasses work efficiency and effectiveness, and physical and psychological wellbeing. However, because this chapter is dealing with ergonomics in the context of risk management, only a subset of all of the applications of ergonomics will be discussed.

2. **Ergonomics improves reliability (and the other factors described above) by changing such things as equipment, work systems and the work environment.** Ergonomics recognises that humans are inherently unreliable (even when the sorts of issue discussed in the previous chapter have been dealt with) and seeks to change things (other than human behaviour) in order to minimise this inherent unreliability or reduce its consequences.

The successful application of ergonomic principles should result in individuals carrying out only those tasks which pose no threat to their long-term physical and mental wellbeing and which allow recovery, in risk terms, when the inevitable unreliability occurs. However, it should be remembered that these principles will only be effective within their defined system. An ergonomically designed chair is likely to result in less fatigue and strain during keyboard work and contribute to the reliability of this work. However, an ergonomically designed chair is still not a good platform for retrieving objects from high shelves!

Since ergonomics is systems-based, it is important to define the system. This is usually done by considering the tasks which have to be carried out. Once the tasks have been identified, it is then possible to consider how prone to error each task might be. These two topics are the subjects of the next section.

Identifying error-prone tasks

The preliminary stages for specifying a system involving tasks have already been described in Chapter 7 during the introduction to task analysis, and the same sorts of task analysis procedures can be used for the first stages in specifying the task elements of systems in the context of ergonomics. The other elements in the system will include equipment and the work environment, but these will be dealt with in separate sections later in the chapter.

The basic principles of Hierarchical Task Analysis (HTA) were described in Chapter 7 and HTA is a good first step in the analysis required for the purpose of ergonomics. However, where the primary concern is with reliability, it is preferable to analyse the error-proneness of tasks in more detail.

This analysis can be done at a very general level since there are various classifications of what humans are 'good at doing' and 'not good at doing'. These classifications can be used as a general guide to whether particular tasks should be allocated to humans or to machines and two of these classifications, reproduced from Glendon et al.,[91] are given in Tables 30.1 and 30.2.

However, the general guidelines summarised in Tables 30.1 and 30.2 only provide a starting point. What is required for fully effective ergonomic design of tasks is a more detailed analysis of the likely human errors associated with a system. Various techniques are available for this type of analysis and two of these, Task Analysis for Error Identification (TAFEI) and Predictive Human Error Analysis (PHEA), are described by Glendon et al. Brief details of these two techniques are given next.

Task Analysis for Error Identification (TAFEI)

This method of analysis is used in those circumstances where an individual is, or will be, operating a machine or using equipment which has a finite number of states, all of which can be defined. This would be the case, for example, with switching using high voltage switchgear. The technique would not be appropriate for analysing, for example, driving tasks, where there will be an infinite number of states and not all of these states will be capable of accurate definition.

Within its limitations, TAFEI is extremely useful for identifying possible errors which are, or would be, critical to the safe and efficient operation of the system. Having identified these errors, appropriate action can be taken to reduce the likelihood of their occurrence, preferably to zero. We will deal with possible methods for reduction later in the chapter.

Property	Human capacity	Machine capacity
Speed	Inferior	Superior
Power	Two horsepower for 10 seconds	Consistent at any level
Consistency	Not reliable, needs to learn, subject to fatigue	Ideal for routine, repetition and precision
Complex activity	Single channel, low throughput	Multi-channel
Memory	Large store, multiple access, best for principles and strategy	Best for literal reproduction and short-term storage
Reasoning*	Good inductive, easy to re-program	Good deductive, tedious to reprogram
Computation	Slow, subject to error, good at error correction	Fast, accurate, poor at error correction
Input	Wide range and variety of stimuli dealt with by one unit (eg eye); affected by heat, cold, noise, vibration etc; good at pattern detection, good at detecting very low signals and signals in noise	Some outside human senses (eg radioactivity); insensitive to extraneous stimuli; poor pattern detection
Overload	Graceful degradation	Sudden breakdown
Intelligence	Can adapt, anticipate, deal with the unpredicted and the unpredictable	None, cannot switch goals or strategies without direction
Dexterity	Great versatility and mobility	Specific

Table 30.1

Comparison of human and machine capabilities for allocation of function (Fitts' list, amended by Singleton[101])

Predictive Human Error Analysis (PHEA)

The initial stages of Predictive Human Error Analysis (PHEA) are similar to those of other techniques of this type – that is, the HTA approach. However, the output from the HTA is then used as a basis for determining types of error. The error classification used is given in Table 30.3 but, in use, this classification is expanded by giving more detailed descriptions of the form each error will take. For each error identified, four parameters are estimated:

1. **Consequence.** That is, what would happen if the error occurred.
2. **Probability.** That is, how likely is it that the error will occur. This is estimated as low, medium or high.
3. **Criticality.** That is, whether the combination of consequence and probability is such that action is required to prevent this error. This parameter is binary, that is a 'yes' or 'no' decision as to whether action is required. (Note that the three parameters correspond to the sorts of risk calculations described in earlier chapters but also incorporate an assessment of whether or not the risk is acceptable since the criticality parameter is binary.)
4. **Recovery.** That is, whether there is any possibility of recovery from the error at a later stage (or stages) in the procedure.

The combination of criticality and recovery is used in the final decision on whether a 'remedy' is required. In others words, whether remedial action needs to be taken with respect to this error.

---

* Inductive reasoning is the inference of general laws from particular instances; deductive reasoning is inference of particular circumstances from a general law.

Humans are generally better at:
Sensing very low levels of stimuli, eg visual, auditory, tactile, olfactory, taste
Detecting stimuli against high noise level backgrounds, eg blips on DSE displays with poor reception
Recognising complex patterns of stimuli which may vary between presentations, eg objects in aerial photographs and speech sounds
Remembering (storing) large amounts of information over long periods – better for remembering principles and strategies than a lot of detailed information
Retrieving relevant information from memory (recall); frequently retrieving many related items (but low recall reliability)
Drawing on varied experience in making decisions and adapting decisions to situational requirements; acting in emergencies (do not require reprogramming)
Selecting alternative modes of operation if certain modes fail
Reasoning inductively and generalising from observations
Applying principles to solutions of varied problems
Making subjective estimates and evaluations
Developing entirely new solutions
Concentrating on the most important activities when required to by overload conditions
Adapting physical responses to variations in operational requirements (within reasonable limits)

Machines are generally better at:
Sensing stimuli outside the normal range of human sensitivity, eg X-rays, radar wavelengths, ultrasonic vibrations
Applying deductive reasoning, eg recognising stimuli as belonging to a general class
Monitoring for prescribed events, especially infrequent ones
Storing encoded information quickly and in substantial quantity
Retrieving coded information quickly and accurately
Processing quantitative information when so programmed
Performing repetitive actions reliably
Exerting considerable physical force in a highly controlled manner
Maintaining performance over extended periods
Counting or measuring physical quantities
Performing several programmed activities simultaneously
Maintaining efficient operations under conditions of heavy load and distractions

Although PHEA is described here in the context of ergonomics, it can also be used as a risk assessment tool (the similarities between criticality and risk were pointed out earlier) and as an aid when designing remedial action following an accident investigation.

However, a limitation of PHEA is that it is qualitative with, for example, probability being specified as low, medium or high, and risk (criticality) being specified as yes or no. While this may be adequate in a number of cases, there are circumstances in which it is preferable to make quantitative estimates and, in these circumstances, Quantified Human Reliability Assessment techniques are likely to be more appropriate.

Quantified Human Reliability Assessment (QHRA)

Quantified Human Reliability Assessment (QHRA) is analogous with the reliability assessments used for hardware described in earlier chapters. Basically, instead of, for example, estimating whether a human error is of low, medium or high likelihood, these techniques attempt to allocate a more exact probability to the error. Obviously, where this can be done with any degree of precision, it allows accurate calculation of any given outcome by using, for example, Fault Tree Analysis techniques. However, the practical problem with QHRA is obtaining the error probabilities and this is usually done in one of two ways.

1. **By collecting empirical data.** For example, extended observations can be made of people carrying out particular tasks and the errors they make during these tasks. Since this provides relevant data, an empirical

		Table 30.3
Planning errors Associated with performing a sequence of actions	Plan preconditions ignored Incorrect plan executed Correct but inappropriate plan executed Correct plan executed too soon or too late Correct plan executed in wrong order	Error categories for Predictive Human Error Analysis
Action errors Associated with performing observable actions	Operation too long or too short Operation in wrong direction Operation too little or too much Misalignment Right operation on wrong object Wrong operation on right object Wrong operation on wrong object Operation mistimed Operation incomplete	
Checking errors Associated with performance checks	Check omitted Check incomplete Right check on wrong object Wrong check on right object Wrong check on wrong object Check mistimed	
Retrieval errors Associated with the retrieval of information from memory, paper or screen	Information not obtained Wrong information obtained Information retrieval incomplete	
Information communication errors Associated with communicating information to, or receiving information from, other parties	Information not communicated Wrong information communicated Information communication incomplete	
Selection errors Associated with selection from alternatives	Selection omitted Wrong selection made	

probability can be calculated by dividing the number of errors made by the number of times the error could have been made. A typical set of data of this type was given in Table 21.3.

2. **Using expert judgment.** For example, a panel of people experienced in the relevant type of work can draw up an agreed table of error rates for particular tasks.

For each of the two methods, the probabilities (or rates) derived are likely to be accurate only for the specific tasks and circumstances for which the data were collected. This is because of the influence of a wide range of environmental and other factors, which will be described later in the chapter. Whichever method, or methods, are used to identify the likelihood and consequences of human errors, when an error requiring action is identified, there are two broad approaches.

1. **Remove the need for the task.** This usually involves redesigning the system so that the error-prone task is no longer required. A simple example of this in the UK is the requirement that electrical equipment is sold with an integral plug, thus removing the need for the error-prone task of wiring plugs for newly purchased electrical equipment.

2. **Allocating the task to machines which are not subject to the human error.** As has already been illustrated in Tables 30.1 and 30.2, machines are good at certain tasks and human error can be eliminated, or at least reduced, by mechanising appropriate tasks. However, this approach may be unsuccessful if due consideration is not given to the interaction between the machines and the people who have to operate them and this 'interface design' is the subject of the next section.

The 'man–machine' interface*

A person operating a machine and the machine being operated can be thought of as a single system, or as two separate systems which interact. Machine designers who adopt the latter mode of thought usually design machines in which the design decisions are taken mainly on the basis of engineering considerations with little, if any, consideration of how the machine will be operated and maintained. Perhaps the most famous example of this is the lathe design which resulted in the production of 'Cranfield Man', introduced in Chapter 16 and reproduced as Figure 30.1.

Figure 30.1
Cranfield Man

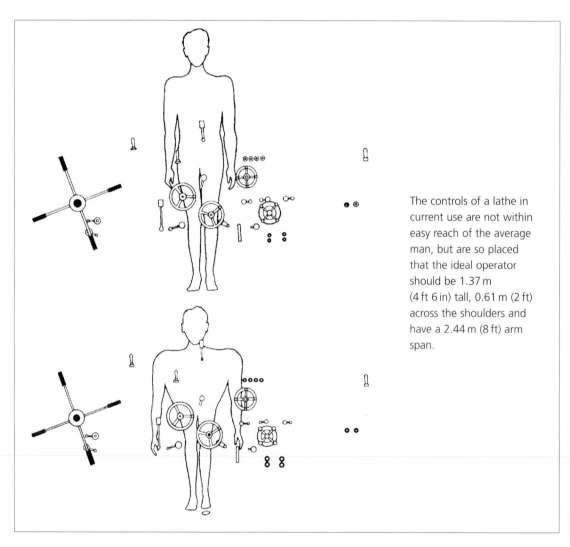

The controls of a lathe in current use are not within easy reach of the average man, but are so placed that the ideal operator should be 1.37 m (4 ft 6 in) tall, 0.61 m (2 ft) across the shoulders and have a 2.44 m (8 ft) arm span.

However, if designers consider the machine they are designing as part of a system of the form shown in Figure 30.2, then the approach to the design of the machine has to be rather different.

---

* Early ergonomists, like UK legislators, seem to recognise only one sex. The term 'man–machine interface' has been used as the section title since it is the most commonly used phrase to describe the subject of the section.

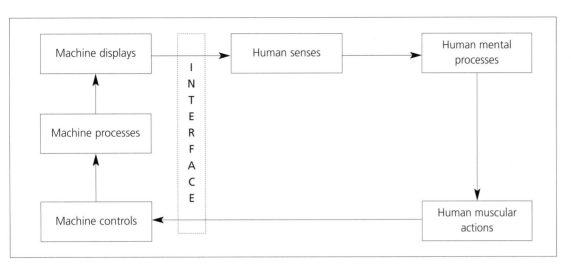

Figure 30.2
Human–machine
system, and the
human–machine
interface

When machine design is done using the system shown in Figure 30.2, the machine designers have to be aware of, and take into account, considerations other than engineering ones. Examples of the sorts of factors which should be taken into account include:

- **The limitations of the human sense organs.** These limitations were described in some detail in Part 1.2 and it is clear from this material that the human senses have their own limits which machine designers must respect if long-term damage is to be avoided.
- **The limitations of human mental processes.** Limitations of, for example, short-term memory, long-term memory and decision-making capacity have to be taken into account.
- **Anthropometrics.** Humans vary in physical dimensions but only within known ranges. These ranges should be taken into account in machine design.
- **Posture.** Humans can take up a wide variety of postures but only a subset of these can be maintained for long periods without long-term damage. Machines should be designed such that they do not require damaging postures.
- **Physical effort.** Humans have limitations on the absolute physical effort they can make without damage, and on the periods of time over which they can sustain a given level of physical effort without damage. Both types of limitation should be taken into account in machine design.
- **Speed of movement.** Humans have a 'natural' rate of working and movements which are too fast or too slow will induce fatigue or stress or both.
- **Environmental factors.** Humans have such high tolerance levels for such factors as temperature, humidity, lighting and noise that they can tolerate levels which are damaging in the long term. Machine designers should work to limits for these factors which are within those known not to inflict long-term harm. They should take account of the influence of the machine itself on environmental factors (for example noise and heat creation) and also the effects of the environment in which the machine is likely to be used.

For routine machine operations, the machine designers will have to consider each of the human elements shown in Figure 30.2 and consider the implications of the factors listed above, where relevant. In a complete study, this should be repeated for such things as maintenance, repair and physical relocation of the machine.

The detailed ergonomics of each of the elements in Figure 30.2 is a field of study in itself and outside the scope of this book. However, the two subsections which follow give key points to be considered in the ergonomics of displays and controls.

Displays

There are several factors which have to be considered in designing displays which are optimal so far as the human elements in the system are concerned and the more important of these are listed below.

- **Location.** Displays which have to be used during operation should be visible from the normal operating position and similar considerations apply to other tasks such as maintenance.
- **Modality.** Displays are usually visual or auditory, but tactile displays can also be used, such as vibrating pagers and mobile phones. The choice of an appropriate modality will depend on a number of factors, including the location of the operator and the likely focus of attention of the operator. For example, displays combined with audible warnings are to be preferred when it is necessary to redirect the operator's attention before the display will be read.
- **Digital *versus* analogue displays.** The characteristics of these displays mean that they are better suited to some tasks than others. For example, if the important information to be displayed is the rate of change of the process, analogue displays are the preferred option. However, if the important information is an exact value, digital displays are to be preferred.
- **The extent to which the display accurately reflects the real world.** The ultimate real world display is the car windscreen, but simplified representations of the real world include flight simulators, mimic diagrams and maps. The effectiveness of these representations of the real world depends critically on the extent to which they encode the required information. For example, a mimic diagram used in the direction of high voltage electrical switching from a control room is usually set out to show the sequence of switches and the links between them. However, what may be more important during switching, and particularly switching during emergencies, is the travel time between switches since only one engineer may be available for switching operations.

Controls

A number of factors have to be considered in designing controls which are optimal as far as the human elements in the system are concerned and the more important of these are listed below.

- **Location.** The requirements here are similar to the requirements for displays, the difference being that the critical factor for controls is access.
- **Effort required for operation.** Controls which require too much effort cause fatigue in long-term use and may not be suitable for operation in an emergency, for example when an operator's normal functions are impaired. However, controls which require too little effort for their operation can be as bad as controls which require too much effort since there is the potential for inadvertent operation (see 'Feedback' below) or 'over operation', for example car steering systems which are 'too light'.
- **Feedback.** The operation of a control should provide appropriate feedback to the operator so that the operator can be sure whether or not the control has functioned. This feedback can be provided by the control itself, for example, it moves positively from one position to another, or by a linked display which shows the effect on the system. The latter approach is essential for those controls which provide no feedback of their own, for example touch-sensitive panels and screens.
- **Ease of identification.** Controls are also displays in that they have to 'display themselves'. This display is required partly so that they can be located, but good controls should also display their function. Imagine a word processor keyboard where all of the keys were blank! Various conventions are used to aid ease of identification including labels (as on a keyboard), colour codes, shape codes and texture codes. The last two methods of coding are used extensively in those circumstances where visual information is not available (for example photographic darkrooms) or cannot be used because the operator's vision is required for other aspects of the task (for example operating mechanical diggers).

Ergonomists also need to consider the links between controls and displays. This has already been identified in the case of controls which give no feedback (for example, touch-sensitive panels) but, more generally, operators should be able to identify accurately and in an appropriate timescale the effects of using particular controls.

Environmental factors

At various points in this chapter, we have pointed out that ergonomists have to take environmental factors into account, and factors such as humidity and noise were mentioned. The present section deals with these environmental factors in more detail, beginning with some general points about their effect on reliability.

In general, humans can operate reliably for short periods even in extremes of environmental conditions but, for sustained reliability, environmental conditions have to be within certain limits. However, there are variations between individuals in preferences for environmental conditions. If there are four people travelling in the same car, there will be four different views on which windows should be opened, and by how much, and four views on the settings for the heating controls. These individual differences should be borne in mind when reading about optimal levels or ranges for environmental factors.

Note also that environmental factors can have direct effects on reliability, for example, high noise levels may mask the audible information required to avoid an error. However, inappropriate environmental conditions can also lead to fatigue or stress and the effects of this is dealt with in a separate section below.

There are four main categories of environmental conditions – thermal comfort, lighting, air quality and noise – and each of these is dealt with briefly below. For a more detailed treatment of these and other environmental factors, see Hartley.[102]

Thermal comfort
This term is used to describe the interrelated effects of ambient temperature, heat sources, humidity and air movement. All of these are taken into account, albeit subconsciously, in an individual's perception of thermal comfort and the issue is further complicated by the fact that, at least most of the time, humans wear clothes and, in general, are in control of which clothes they wear.

Notwithstanding these various complications, there are sources of general guidance on, for example, temperature and humidity levels appropriate to various tasks and, in general, staying within these guidelines will reduce the likelihood of human error. Lack of thermal comfort can, however, result in errors for a variety of reasons and some examples are given below.
- If it is too cold, people may wear inappropriate gloves which reduce their ability to use controls accurately.
- If air movements are too fast, people may wear hoods or other head coverings which obscure their view of information required to avoid errors.
- General low levels of thermal comfort, for whatever reason, may distract attention from the core activity and so make errors more likely.

Lighting
The increase in error potential arising from lighting levels which are too low is obvious. However, there are other aspects of lighting which can also influence error rates and these are listed below.
- Light levels which are so high that they result in glare which obscures information required for error avoidance.
- Rapid changes in light levels, either from bright to dark or *vice versa*.
- Light which is of an inappropriate colour balance so that, for example, colour codes cannot be interpreted properly.

These sorts of problem were dealt with in more detail in Chapter 14.

Air quality

Air quality is used as a general term to describe the extent to which the ambient atmosphere is contaminated with substances other than those found in 'pure' air. These contaminants can have effects because they are in themselves harmful, because they displace too much oxygen in the atmosphere, or because the have some other adverse effect. A selection of the possible effects on errors of poor air quality is given below.

- **Contaminants may directly affect human performance by being, for example, narcotic.** Any substance which has a detrimental effect on a human's ability to operate is likely to result in an increased likelihood of errors.
- **Contaminants whose effect is to deplete the oxygen supply will produce a general degradation of performance and a consequent increased likelihood of error.**
- **Contaminants such as dust and fumes which reduce visibility can have detrimental effects by obscuring the information required to avoid errors.** This can be directly because there is so much dust or fumes that visibility is impaired, or indirectly because people take action to avoid the dust getting into their eyes, or because the dust makes their 'eyes water'. These indirect effects can also be produced by substances other than dusts, for example irritant vapours.
- **Contaminants which cause peripheral effects such as coughing, sneezing or nausea in circumstances where such effects influence the ability to carry out tasks in an error-free manner.**

Noise

The main direct effect of noise is, as was mentioned earlier, the masking of auditory information necessary for error avoidance. However, high levels of noise induce fatigue and stress and these are dealt with in the next subsection.

Fatigue and stress as mediators

Any of the environmental factors just described are, if they are outside the levels an individual finds comfortable for any length of time, likely to induce fatigue or stress or both in that individual. As has been seen in Chapter 15, the effects of fatigue and stress are many and various but it is known that, however produced, they have a detrimental effect on error rates.

Application of ergonomics

It is possible to identify the application of ergonomics in most aspects of life and some examples are given below. The examples are divided, for convenience of description, into three categories:

- The application of **classical ergonomics**. This term is used to describe the sorts of techniques which were applied at the beginning of work on ergonomics and which are still in use today.
- The application of **systems ergonomics**. This term is used to describe the sorts of ergonomics work, described earlier in this chapter, where ergonomics is applied to a complete system.
- The application of **error ergonomics**. This term is used to describe the application of classical and systems ergonomics specifically for the purposes of error reduction, rather than, for example, reduction in fatigue, muscular damage or stress.

Classical ergonomics

Many aspects of the tools and equipment used at work, and away from work, have benefited from the application of classical ergonomics and some examples related to motor cars are given below.

- **The design of displays in motor cars.** These are being continually improved, both in terms of layout and readability. Years ago, displays of information for car drivers were arranged rather haphazardly over the dashboard with, for example, the speedometer in the centre of the dashboard, that is between the driver and the passenger. Not only are these weaknesses in the display layout in individual makes of car being improved, but the differences between cars of different marque and manufacturer are being reduced so that changing between cars is less likely to result in errors. The readability of displays is also being improved with, for example, changes in the contrast between the dial and the indicator needle, and better back lighting for improved readability in low levels of ambient light.

- **The design of controls in motor cars.** The controls for motor cars have undergone a similar range of improvements to those just described for displays. Only vintage car enthusiasts are now exposed to cars where the familiar clutch, brake and accelerator pedal layout is not standard, and only older readers will remember idiosyncratic control layouts such as indicators being operated from the centre of the dashboard and headlight dipping being via a foot-operated floor pedal. These days, the majority of new cars have all the relevant controls within easy reach (improved layout) and the layout in different cars is similar, enabling the benefits of stereotyping.
- **The design of seating in motor cars.** There have been two main areas of advance in the design of seating in motor cars. First, non-adjustable seats have been improved in terms of support for 'the average person' and, second, the range of adjustments possible on adjustable seats is becoming much wider. The availability of adjustable seats is also increasing, but this may be due to market pressure, since it has long been known that such seats were desirable for ergonomic reasons.

We have used cars to illustrate the application of classical ergonomics because they are likely to be familiar to most readers. However, similar principles have been applied to work equipment and a comparison of most 'old' items of work equipment with their most recent equivalent is likely to result in a list of the sorts of changes described above for motor cars.

Systems ergonomics

Systems ergonomics is probably the most generally useful approach and two examples are given below to illustrate the effects it has had in recent years.

- **Handling loads.** It has been known for many years that back injuries and other tissue damage resulting from the manual handling of loads is a common cause of loss in many organisations. The application of systems ergonomics to the 'load handling system' is used to identify opportunities to eliminate manual handling by, for example, re-ordering the work or introducing mechanical handling. However, where it is not possible to eliminate manual handling, systems ergonomics is applied to identify, and suggest remedies for, manual handling activities which would result in damage to those handling the loads. This work is clearly systems-based since it takes into account the complete range of factors making up the system, including the load itself, the direction, speed and distance of the movements required, the number of times movements are required and their periodicity, and the environment in which the movements have to take place.
- **Using computer workstations.** Computer workstations are relatively complex systems including, as they do, displays (screen and audible signals), controls (keyboard and mouse), software, the supports for the equipment and the user (usually desks and chairs), and the environment in which the work is undertaken. The application of systems ergonomics has resulted in recommendations for improvements in all of these over the years. These improvements have been implemented either through changes in design (detachable keyboards, adjustable seats, more 'user-friendly' software) or by making widely available guidance on ergonomic best practice (adjustment of workstation layouts, posture and physical environment).

It can be difficult to discriminate between the application of classical ergonomics and systems ergonomics since, if necessary, systems ergonomics will include the redesign of equipment, controls and displays which is usually thought of as the province of classical ergonomics.

Error ergonomics

As we mentioned above, error ergonomics is a term used to describe the application of the ergonomic principles just illustrated specifically for the purpose of reducing errors. As has already been seen, the application of ergonomic principles may reduce error rates indirectly since they reduce fatigue and stress but many applications also have a more direct effect. The changes to the layout and design of the displays and controls used in motor cars illustrate these error effects. For example, improvements in 'readability' of displays not only reduces the number of errors in reading the display concerned, it means that the display can be read more quickly, thus enabling the driver's attention to return to the road more quickly. Improvements in control layouts have similar effects; not only are the controls easier to find and operate, which reduces errors in the use of these controls, but the controls can, for example, be used without drivers having to take their hand away from the steering wheel.

On occasions, ergonomic principles are applied specifically to error reduction, usually in cases where errors could have serious consequences. Examples include the display and control elements of air traffic control, high voltage electrical switching from control rooms, and flying aircraft.

Summary

This chapter has described how the application of ergonomics can be used to increase human reliability.

Appendices

1: NEBOSH syllabus

Introduction

The Guide to the NEBOSH National Diploma in Occupational Health and Safety, dated February 2010, is divided into four parts:
- Unit A: Managing health and safety
- Unit B: Hazardous agents in the workplace
- Unit C: Workplace and work equipment safety
- Unit D: Application of health and safety theory and practice.

This book provides material relevant to 'Managing health and safety'. Unit D is an assessment unit and this book provides some material that is relevant to this Unit. It does not contain material relevant to Units B and C.

Unit A: Managing health and safety

Overall learning outcome
On completion of this unit, candidates will be able to demonstrate their understanding of the domain of knowledge covered through:
1. the application of knowledge to familiar and unfamiliar situations; and
2. the critical analysis and evaluation of information presented in both quantitative and qualitative forms.

List of elements

Element	Title
A1	Principles of health and safety management
A2	Loss causation and incident investigation
A3	Measuring and reviewing health and safety performance
A4	Identifying hazards, assessing and evaluating risk
A5	Risk control
A6	Organisational factors
A7	Human factors
A8	Principles of health and safety law
A9	Criminal law
A10	Civil law

Note on legal matters (Elements A8, A9 and A10)
There are no chapters that deal specifically with legal matters. However, where there is UK legislation relevant to a health and safety management topic, this legislation is described in the appropriate chapter. The content of units A8 to A10 are not included in this Appendix.

Element title and learning outcomes	Relevant chapters of this book
A1: Principles of health and safety management On completion of this element, candidates should be able to demonstrate understanding of the content through the application of knowledge to familiar and unfamiliar situations and the critical analysis and evaluation of information presented in both quantitative and qualitative forms. In particular they should be able to: A1.1 Explain the moral, legal and economic reasons for the effective management of health and safety A1.2 Outline the societal factors which influence health and safety standards and priorities A1.3 Explain the principles and content of effective health and safety, quality, environmental, and integrated management systems with reference to recognised models and standards A1.4 Outline the role and responsibilities of the health and safety practitioner.	**A1: Principles of health and safety management** 3 Risk management – setting the scene 4 Key elements of risk management 8 Monitoring and measuring losses 9 Identifying causes and patterns 10 Monitoring and measuring conformity 11 Other elements of OH&SMSs 16 The human factors environment 18 Management systems 19 Measuring performance 23 Advanced audit and review
A2: Loss causation and incident investigation On completion of this element, candidates should be able to demonstrate understanding of the content through the application of knowledge to familiar and unfamiliar situations and the critical analysis and evaluation of information presented in both quantitative and qualitative forms. In particular they should be able to: A2.1 Explain theories of loss causation A2.2 Explain the quantitative analysis of accident/incident and ill-health data, limitations of their application, and their presentation in numerical and graphical form A2.3 Explain the statutory and the internal reporting and recording systems for injuries, ill-health, dangerous occurrences and near misses A2.4 Explain loss investigations; the requirements, benefits, the procedures, the documentation, and the involvement of and communication with relevant staff and representatives.	**A2: Loss causation and incident investigation** 8 Monitoring and measuring losses 9 Identifying causes and patterns 10 Monitoring and measuring conformity 19 Measuring performance 20 Advanced accident investigation and risk assessment

Element title and learning outcomes	Relevant chapters of this book
A3: Measuring and reviewing health and safety performance On completion of this element, candidates should be able to demonstrate understanding of the content through the application of knowledge to familiar and unfamiliar situations and the critical analysis and evaluation of information presented in both quantitative and qualitative forms. In particular they should be able to: A3.1 Explain the purpose of performance measurement in relation to health and safety objectives and arrangements A3.2 Explain the need for, and the objectives and limitations of, health and safety monitoring systems A3.3 Describe the variety of monitoring and measurement techniques A3.4 Explain the requirements for reviewing health and safety performance.	**A3: Measuring and reviewing health and safety performance** 3 Risk management – setting the scene 4 Key elements of risk management 8 Monitoring and measuring losses 9 Identifying causes and patterns 10 Monitoring and measuring conformity 11 Other elements of OH&SMSs 18 Management systems 19 Measuring performance 23 Advanced review and audit 24 Financial issues
A4: Identifying hazards, assessing and evaluating risk On completion of this element, candidates should be able to demonstrate understanding of the content through the application of knowledge to familiar and unfamiliar situations and the critical analysis and evaluation of information presented in both quantitative and qualitative forms. In particular they should be able to: A4.1 Describe how to use internal and external sources of information in the identification of hazards and the assessment of risk A4.2 Outline a range of hazard identification techniques A4.3 Explain how to assess and evaluate risk and to implement a risk assessment programme A4.4 Explain the principles and techniques of failure tracing methodologies with the use of calculations.	**A4: Identifying hazards, assessing and evaluating risk** 3 Risk management – setting the scene 4 Key elements of risk management 5 Risk assessment 7 Safe systems of work 20 Advanced accident investigation and risk assessment

Element title and learning outcomes	Relevant chapters of this book
A5: Risk control On completion of this element, candidates should be able to demonstrate understanding of the content through the application of knowledge to familiar and unfamiliar situations and the critical analysis and evaluation of information presented in both quantitative and qualitative forms. In particular they should be able to: A5.1 Outline common risk management strategies A5.2 Outline factors to be taken into account when selecting risk controls A5.3 Explain the development, main features and operation of safe systems of work and permit-to-work systems.	**A5: Risk control** 6 Risk control 7 Safe systems of work 15 The individual – psychology 20 Advanced accident investigation and risk assessment 21 Advanced risk control techniques 22 Emergency planning 24 Financial issues 27 Human error 28 Perception and decision making 29 External influences on human error 30 Improving human reliability
A6: Organisational factors On completion of this element, candidates should be able to demonstrate understanding of the content through the application of knowledge to familiar and unfamiliar situations and the critical analysis and evaluation of information presented in both quantitative and qualitative forms. In particular they should be able to: A6.1 Explain the internal and external influences on health and safety in an organisation A6.2 Outline the organisation as a system, the different types of organisation, their characteristics and relationship to individuals within them A6.3 Identify the various categories of third parties in a workplace – the relevant legislative requirements, responsibilities and controls A6.4 Explain the role, influences on and procedures for formal and informal consultation with employees in the workplace A6.5 Outline the development of a health and safety management information system, the relevant legal requirements, and the data it should contain A6.6 Explain health and safety culture and climate A6.7 Outline the factors which can both positively and negatively affect health and safety culture.	**A6: Organisational factors** 3 Risk management – setting the scene 4 Key elements of risk management 16 The human factors environment 18 Management systems 19 Measuring performance 21 Advanced risk control techniques 29 External influences on human error

Element title and learning outcomes	Relevant chapters of this book
A7: Human factors On completion of this element, candidates should be able to demonstrate understanding of the content through the application of knowledge to familiar and unfamiliar situations and the critical analysis and evaluation of information presented in both quantitative and qualitative forms. In particular they should be able to: A7.1 Outline psychological and sociological factors which may give rise to specific patterns of safe and unsafe behaviour in the working environment A7.2 Explain the nature of the perception of risk and its relationship to performance in the workplace A7.3 Explain the classification of human failure A7.4 Explain appropriate methods of improving individual human reliability in the workplace A7.5 Explain how organisational factors could contribute to improving human reliability A7.6 Explain how organisational factors could contribute to improving human reliability A7.7 Outline the principles, conditions and typical content of behavioural change programmes designed to improve safe behaviour in the workplace.	**A7: Human factors** 14 The individual – sensory and perceptual processes 15 The individual – psychology 16 The human factors environment 26 Individual differences 27 Human error 28 Perception and decision making 29 External influences on human error 30 Improving human reliability

2: Vocational standards knowledge requirements

Introduction

Structure of the units

The National Vocational Qualifications in occupational health and safety practice and their Scottish equivalents, collectively abbreviated to N/SVQs, are presented in a series of 13 units of competence.* Each unit starts with a general description of the role associated with the activities of the particular unit. The unit is then divided into two parts: performance criteria and knowledge requirements.

Performance criteria

These start with the phrase 'You must be able to': this is then followed by a list of activities that amplify the general description of the role.

Knowledge requirements

These are headed by the phrase 'You need to know and understand:'. In general, the requirements are then presented under three headings:

- 'The nature and role of health and safety' – this is followed by a phrase relevant to the particular unit;
- 'Principles and concepts'; and
- 'External factors' – this is then followed by a phrase relevant to the particular unit.

Unit catalogue

Unit	Title
PROHSP1	Develop, implement and review the organisation's health and safety strategy
PROHSP2	Promote a positive health and safety culture
PROHSP3	Develop and implement the health and safety policy
PROHSP4	Develop and implement effective communication systems for health and safety information
PROHSP5	Develop and maintain individual and organisational competence in health and safety matters
PROHSP6	Control health and safety risks
PROHSP7	Develop, implement and review proactive monitoring systems for health and safety
PROHSP8	Develop, implement and review reactive monitoring systems for health and safety
PROHSP9	Develop and implement a health and safety audit
PROHSP10	Develop and implement health and safety emergency response systems and procedures
PROHSP11	Develop and implement health and safety review systems
PROHSP12	Contribute to health and safety legal actions
PROHSP13	Influence and keep pace with improvements in health and safety practice

On the following pages, we have listed the various learning outcomes for each unit and where relevant material for each unit can be found in this book.

---

* At the time of the latest revision (April 2012), the National Occupational Standards were in the process of being updated by Proskills. When these have been finalised, we will outline how they relate to the N/SVQs at levels 3, 4 and 5.

Element title and learning outcomes	Relevant chapters of this book
PROHSP1: Develop, implement and review the organisation's health and safety strategy	**PROHSP1: Develop, implement and review the organisation's health and safety strategy**
The nature and role of the identification of health and safety hazards within the organisation	14 The individual – sensory and perceptual processes
K1 what is strategy and the purpose it?	15 The individual – psychology
K2 the internal factors, including organisational structures, strategies, and human and physical resources available	16 The human factors environment
K3 the key change factors impacting on the organisation	18 Management systems
K4 the key drivers of the current internal structure, internal opportunities and rigidities	26 Individual differences
K5 how to influence the organisation's strategies, policies and practices	27 Human error
K6 how to evaluate the organisation's business plan in relation to health and safety	28 Perception and decision making
	29 External influences on human error
Principles and concepts	
K7 information networks and sources	
K8 theories of motivation	
K9 where to obtain comprehensive, valid and reliable information on the external environment	
K10 sources of organisational performance data	
K11 how to assess organisational resources against any required changes	
K12 effective presentation of cases for change in structure and systems	
K13 how to encourage participation in, and feedback on, change strategies	
K14 change methodologies	
K15 objective setting	
K16 the change cycle and how to use it	
K17 the impact on the organisation when implementing the required changes	
K18 the appropriate communication channels for the change strategy	
K19 how to acknowledge, manage and resolve conflict	
K20 acceptable forms of compromise which maintain the integrity of the change process	

Element title and learning outcomes	Relevant chapters of this book
External factors impacting health and safety K21 key requirements of health and safety legislation and any other legal requirements in the workplace K22 how to respond to new legislation K23 benchmarking against current best practice K24 how to respond to technical developments	
PROHSP2: Promote a positive health and safety culture The nature and role of a positive health and safety culture within the organisation K1 the health and safety culture within the organisation K2 the organisation's communication system which can be used to promote the benefits of a positive health and safety culture K3 how people communicate K4 the people and groups who may be affected K5 how to engage people and groups who may be affected K6 the information needs of those people in the workplace affected K7 the available information sources for health and safety within the workplace K8 the importance of keeping people regularly informed and discussing their involvement K9 what problems may arise K10 which performance measures to utilise Principles and concepts K11 providing effective information, advice and guidance to others K12 external factors influencing a positive health and safety culture K13 other sources of expertise and advice on health and safety matters	**PROHSP2: Promote a positive health and safety culture** 14 The individual – sensory and perceptual processes 15 The individual – psychology 16 The human factors environment 21 Advanced risk control techniques 24 Financial issues 26 Individual differences 27 Human error 28 Perception and decision making 29 External influences on human error 30 Improving human reliability
PROHSP3: Develop and implement the health and safety policy The nature and role of a positive health and safety culture within the organisation K1 the health and safety culture within the organisation K2 the organisation's communication system which can be used to promote the benefits of a positive health and safety culture	**PROHSP3: Develop and implement the health and safety policy** 4 Key elements of risk management 16 The human factors environment 18 Management systems

Element title and learning outcomes	Relevant chapters of this book
K3 how people communicate K4 the people and groups who may be affected K5 how to engage people and groups who may be affected K6 the information needs of those people in the workplace affected K7 the available information sources for health and safety within the workplace K8 the importance of keeping people regularly informed and discussing their involvement K9 what problems may arise K10 which performance measures to utilise Principles and concepts K11 providing effective information, advice and guidance to others K12 external factors influencing a positive health and safety culture K13 other sources of expertise and advice on health and safety matters	
PROHSP4: Develop and implement effective communication systems for health and safety information Principles and concepts K1 the principles for effective communication K2 the barriers to effective communication K3 the formal and informal communications systems within an organisation K4 the different ways people can communicate K5 effective written and verbal communication External factors influencing effective communication systems for health and safety information K6 health and safety risk assessment, control procedures and practices, technical developments and best practice K7 proposed and new health and safety legislation, codes of practice and standards K8 health and safety promotional activities relevant to the needs of an organisation K9 health and safety statutory reporting requirements for an organisation K10 health and safety statutory information requirements for the products, services and waste of an organisation	**PROHSP4: Develop and implement effective communication systems for health and safety information** 4 Key elements of risk management 5 Risk assessment 6 Risk control 7 Safe systems of work 12 Communication and training 16 The human factors environment 18 Management systems 20 Advanced accident investigation and risk assessment 21 Advanced risk control techniques

Element title and learning outcomes	Relevant chapters of this book
K11 the health and safety standards and procedures of an organisation that are relevant to the contractors used by the organisation	

PROHSP5: Develop and maintain individual and organisational competence in health and safety matters

The nature and role of individual and organisational competence in health and safety matters within the organisation

K1　the structure of the organisation with respect to functions, activities, tasks and jobs

K2　the principles of competence, activity and task analysis

K3　health and safety competency training needs analysis

K4　the relationships between competencies, skills and qualifications

Principles and concepts

K5　principles of systematic training

K6　different learning styles

K7　training design and delivery

K8　the advantages and disadvantages of different methods of presentation

K9　training evaluation and validation

K10 preparing, delivering and marking tests and assignments

K11 effective written and verbal communication

External factors influencing individual and organisational competence in health and safety

K12 health and safety statutory requirements and industry best practice

K13 the quality management requirements for documentation

PROHSP5: Develop and maintain individual and organisational competence in health and safety matters

4　Key elements of risk management

7　Safe systems of work

12 Communication and training

14 The individual – sensory and perceptual processes

15 The individual – psychology

16 The human factors environment

18 Management systems

26 Individual differences

29 External influences on human error

30 Improving human reliability

PROHSP6: Control health and safety risks

The nature and role of the identification of health and safety hazards within the organisation

K1　health and safety hazards

K2　risk assessment techniques

K3　instruments and survey techniques which may be used to determine the exposure of people who may be affected

PROHSP6: Control health and safety risks

3　Risk management – setting the scene

4　Key elements of risk management

5　Risk assessment

6　Risk control

7　Safe systems of work

14 The individual – sensory and perceptual processes

Element title and learning outcomes	Relevant chapters of this book
Principles and concepts K4 the analysis techniques suitable for determining risks K5 methods for reviewing effectiveness **External factors influencing the identification of health and safety hazards** K6 health and safety statutory requirements K7 tolerability/acceptability of risk K8 quality management requirements for documentation **The nature and role of health and safety risk control measures within the organisation** K9 risk control measures, including safe systems of work K10 external factors influencing health and safety risk control methods K11 risk control hierarchies K12 the risk control measures required by relevant health and safety legislation and industry best practice	15 The individual – psychology 16 The human factors environment 20 Advanced accident investigation and risk assessment 21 Advanced risk control techniques 24 Financial issues 26 Individual differences 27 Human error 28 Perception and decision making 29 External influences on human error 30 Improving human reliability
PROHSP7: Develop, implement and review proactive monitoring systems for health and safety **The nature and role of active health and safety monitoring systems within the organisation** K1 proactive monitoring techniques for health and safety risk control measures K2 monitoring equipment K3 sampling routines K4 workplace inspections and activity observations **Principles and concepts** K5 effective written and verbal communication K6 how to respond to the needs of others **External factors influencing active health and safety monitoring systems** K7 health and safety statutory requirements and industry best practice for proactive monitoring systems and documentation K8 quality management requirements for documentation	**PROHSP7: Develop, implement and review proactive monitoring systems for health and safety** 3 Risk management – setting the scene 4 Key elements of risk management 8 Monitoring and measuring losses 9 Identifying causes and patterns 10 Monitoring and measuring conformity 18 Management systems 19 Measuring performance 23 Advanced review and audit 24 Financial issues

Element title and learning outcomes	Relevant chapters of this book
The nature and role of health and safety risk control measures within the organisation K9 risk control measures, including safe systems of work K10 external factors influencing health and safety risk control methods K11 risk control hierarchies K12 the risk control measures required by relevant health and safety legislation and industry best practice	
PROHSP8: Develop, implement and review reactive monitoring systems for health and safety **The nature and role of reactive health and safety monitoring systems within the organisation** K1 health and safety loss events K2 reporting forms and recording procedures for health and safety loss events K3 the health and safety loss events that require formal investigation K4 health and safety loss event investigation systems and procedures **Principles and concepts** K5 failure tracing methods and techniques K6 effective written and verbal communication K7 how to respond to the needs of others K8 statistical and epidemiological analyses of data K9 histograms, pie charts, and line graphs **External factors influencing reactive health and safety monitoring systems** K10 health and safety statutory requirements regarding loss events and investigations	**PROHSP8: Develop, implement and review reactive monitoring systems for health and safety** 3 Risk management – setting the scene 4 Key elements of risk management 8 Monitoring and measuring losses 9 Identifying causes and patterns 10 Monitoring and measuring conformity 19 Measuring performance 20 Advanced accident investigation and risk assessment 24 Financial issues
PROHSP9: Develop and implement a health and safety audit **The nature and role of the health and safety audit systems within the organisation** K1 health and safety management systems K2 health and safety management operational and technical standards and procedures	**PROHSP9: Develop and implement a health and safety audit** 4 Key elements of risk management 11 Other elements of OH&SMSs 18 Management systems 23 Advanced audit and review

Element title and learning outcomes	Relevant chapters of this book
K3 health and safety audit questionnaires K4 health and safety documentation **Principles and concepts** K5 the nature and scope of audits K6 how an audit system can be reviewed K7 the advantages and disadvantages of in-house and bought-in audits K8 the competency of auditors and companies offering an audit K9 effective written communication **External factors influencing health and safety audit systems** K10 the health and safety statutory requirements K11 industry best practice K12 quality management requirements for documentation	

PROHSP10: Develop and implement health and safety emergency response systems and procedures
 	
The purpose of emergency planning
K1   the potential causes of emergencies including natural and manmade both accidental and deliberate
 	
The nature and role of health and safety emergency response systems and procedures within the organisation
K2   emergency procedures for your organisation taking into account:
  • K2.1 relevant health and safety statutory requirements
  • K2.2 dealing with the ongoing effects of fatalities
  • K2.3 injury accidents
  • K2.4 dangerous occurrences
  • K2.5 fire and explosion
  • K2.6 toxic release
  • K2.7 major disaster
  • K2.8 environmental impact
  • K2.9 rescue and security alert
K3   how to manage simulated emergency procedures
 	
Principles and concepts
K4   press releases and media management
	PROHSP10: Develop and implement health and safety emergency response systems and procedures
	22 Emergency planning

Element title and learning outcomes	Relevant chapters of this book
External factors influencing health and safety emergency response systems and procedures K5 health and safety statutory requirements: • K5.1 first aid and medical service provision • K5.2 fire precautions • K5.3 emergency procedures • K5.4 major disasters • K5.5 ionising radiation incident • K5.6 environmental impact events • K5.7 the control of an emergency	
PROHSP11: Develop and implement health and safety review systems The nature and role of health and safety review systems within the organisation K1 how to make sure targets are specific, measurable, achievable, relevant and timely K2 health and safety management systems K3 the factors and features that are essential for the efficient and cost-effective working of a health and safety management system K4 efficiency and cost-effectiveness of a health and safety management system Principles and concepts K5 how to respond to the needs of others K6 electronic and paper record systems External factors influencing health and safety review systems K7 health and safety statutory requirements and industry best practice K8 quality management requirements for documentation	**PROHSP11: Develop and implement health and safety review systems** 4 Key elements of risk management 11 Other elements of OH&SMSs 18 Management systems 23 Advanced review and audit
PROHSP12: Contribute to health and safety legal actions The nature and role of health and safety legal actions within the organisation K1 health and safety criminal legislation K2 mitigating circumstances and/or defences in a health and safety criminal or civil case	**PROHSP12: Contribute to health and safety legal actions** There are no directly relevant chapters.

Element title and learning outcomes	Relevant chapters of this book
Principles and concepts K3 being a witness in court K4 using expert witnesses **External factors influencing health and safety legal actions** K5 criminal court procedure K6 the torts of negligence and breach of statutory duty as they apply to health and safety civil liability K7 defences in a health and safety civil claim K8 civil claims procedure	
PROHSP13: Influence and keep pace with improvements in health and safety practice **The nature and role of improvements in health and safety practice within the organisation** K1 sources of relevant information and advice on health and safety matters **Principles and concepts** K2 effective communication methods **External factors influencing improvements in health and safety practice** K3 appropriate professional, non-professional and technical groups and organisations K4 the level of relevant statutory duties and interpretation of 'so far as is reasonably practicable' when considering improvements to health and safety	**PROHSP13: Influence and keep pace with improvements in health and safety practice** 12 Communication and training 16 The human factors environment 29 External influences on human error

3: IOSH learning objectives

The material in this Appendix includes all the learning objectives set out in the IOSH booklet *Higher level qualification accreditation* (published in January 2011). However, not all of the learning objectives are relevant to this textbook. The introduction to the learning objectives begins with the following statement: 'The learning objectives specified here are designed to cover the core knowledge requirements for occupational safety and health practice.'

Section A – Basic knowledge requirements

This list covers the outline knowledge requirements for a structured progressive undergraduate programme. Most of the topics should be covered in year 1 and part of year 2 of the programme. Years 2 and 3 should build on the occupational safety and health subjects in section B. Students on master's programmes are expected to have covered the basic concepts in this list before entering a postgraduate programme. Any accredited prior learning or experiential learning which is used for entry to postgraduate programmes should cover the objectives in this section. Submissions for accreditation of postgraduate programmes will need to demonstrate how this requirement is met during their admission processes.

Element title and learning outcomes	Relevant chapters of this book
A1: Science and technology • Principles of chemistry and physics necessary to understand occupational safety and health hazards and effective controls • Principles of human physiology and microbiology necessary to understand health and safety hazards and consequences	**A1: Science and technology** 15 The individual – psychology
A2: Legal and regulatory systems • The origins and principles of legal frameworks and processes • The regulatory regimes in place and principles employed by enforcing agencies	**A2: Legal and regulatory systems** There are no directly relevant chapters
A3: Workplace health and safety • Principles of materials, structural and mechanical science with particular reference to the overall safety of work equipment and structures • Principles of engineering design, manufacturing, reliability and testing • Principles of ergonomics (human factors) -• Methods to identify the elements of a physical or organisational system that may give rise to unsafe or unhealthy conditions • Fire theory and principles of prevention • Building and working environment factors that influence safety and health	**A3: Workplace health and safety** 7 Safe systems of work 30 Improving human reliability

Element title and learning outcomes	Relevant chapters of this book
A4: Definitions of risk and risk management principles • Concepts of risk and risk management • How risk assessment, particularly for occupational safety and health, fits into the broader management concepts used in organisations • Social, political and cultural influences, intolerable and acceptable risks • Hierarchy of controls	**A4: Definitions of risk and risk management principles** All of Part 1.1 16 The human factors environment All of Part 2.1, except Chapter 22
A5: The principles and theories of health and safety management • Organisational behaviour and management theory • Role and practices of systematic health and safety management • Commonly applied development and implementation systems, procedures and performance standards • Loss control and the costs of accidents and ill health to an organisation and society • Causation theories relevant to health and safety • Fault and no-fault insurance processes	**A5: The principles and theories of health and safety management** 3 Risk management – setting the scene 4 Key elements of risk management 8 Monitoring and measuring losses 9 Identifying causes and patterns 10 Monitoring and measuring conformity 16 The human factors environment 18 Management systems 19 Measuring performance 24 Financial issues
A6: Occupational health and hygiene • The working environment, including organisational culture • The benefits of 'good work' • Work-related exposures and ill health • Toxicology and epidemiology • Principles of health hazard evaluation, monitoring and control • Relationship between occupational, environmental and public health • Workplace health promotion • Rehabilitation and workplace adjustments	**A6: Occupational health and hygiene** 8 Monitoring and measuring losses 19 Measuring performance
A7: Occupational psychology • Factors which influence perceptions, attitudes and behaviours • Formal and informal leadership • Risk perception and society	**A7: Occupational psychology** All of Parts 1.2 and 2.2
A8: Information technology (IT), literacy and numeracy • Computer-based software applications that are in frequent use within workplaces • Security and legal issues associated with IT systems • Producing technical and legal letters, reports and other documents • The numeracy necessary to undertake calculations relating to health and safety monitoring	**A8: Information technology (IT), literacy and numeracy** 19 Measuring performance 23 Advanced audit and review 24 Financial issues

Section B – Core professional learning objectives

All qualification programmes meeting the academic standard for IOSH Graduate membership must cover the following outline learning objectives at the depth appropriate for either undergraduate or postgraduate programmes. The core objectives should be modified in line with the academic policies and practices operating in the university or college and the overall aims of the programme. The academic descriptors and assessment of the core objectives must reflect an appropriate depth of study.

The learning objectives are grouped together in the stages of a safety management system, but this format does not need to be followed exactly. Institutions can design their own programme and modules can be developed that are not in this format. However, at least 80 per cent of the outline objectives must demonstrably be covered. Students successfully completing a course will have a level of knowledge and understanding at the appropriate academic depth of the following topics:

Element title and learning outcomes	Relevant chapters of this book
B1: Health and safety strategy, policy and culture • Health and safety culture in an organisation and its relationship to and integration with other management functions	**B1: Health and safety strategy, policy and culture** 4 Key elements of risk management 6 Risk control 16 The human factors environment 18 Management systems 21 Advanced risk control techniques
• The importance of engaging individuals and groups in an organisation's health and safety processes	4 Key elements of risk management 16 The human factors environment 18 Management systems
• The impact of practical and statutory health and safety requirements on the inputs, conversion processes and outputs of an organisation	There are no directly relevant chapters
• Management techniques to initiate, develop, promote, monitor and improve health and safety strategies, policies and organisational arrangements	4 Key elements of risk management 8 Monitoring and measuring losses 9 Identifying causes and patterns 10 Monitoring and measuring conformity 11 Other elements of OH&SMSs 16 The human factors environment 18 Management systems 19 Measuring performance 23 Advanced audit and review
• Potential barriers to developing effective policies and associated organisational arrangements	4 Key elements of risk management 16 The human factors environment 18 Management systems

Element title and learning outcomes	Relevant chapters of this book
• Internal, external and change factors affecting organisations' health and safety arrangements	16 The human factors environment
• The incorporation of relevant health and safety law into organisation's health and safety policies and management systems	4 Key elements of risk management 16 The human factors environment 18 Management systems
• Potential barriers to effective policies and positive health and safety cultures	4 Key elements of risk management 16 The human factors environment 18 Management systems
• Goals and performance targets in health and safety policy design and development	4 Key elements of risk management 19 Measuring performance
• Organisational arrangements for contractors and shared workplaces	There are no directly relevant chapters
B2: Communication and organisational competence • The theory and practice of organisational communication and its applicability to health and safety management systems	**B2: Communication and organisational competence** 4 Key elements of risk management 8 Monitoring and measuring losses 9 Identifying causes and patterns 10 Monitoring and measuring conformity 11 Other elements of OH&SMSs 12 Communication and training 16 The human factors environment 18 Management systems 19 Measuring performance 23 Advanced audit and review
• Organisational communication and its relationship to organisational culture and health and safety	10 Monitoring and measuring conformity 16 The human factors environment 18 Management systems 21 Advanced risk control techniques
• Identifying, locating and evaluating current sources of health and safety information	There are no directly relevant chapters
• Documentation and control systems to manage information and archives	11 Other elements of OH&SMSs
• Communication tools used to acquire and disseminate information	12 Communication and training

Element title and learning outcomes	Relevant chapters of this book
• Competence for organisations, teams and individuals, including documentation and assessments; competence development, prevention of decay	4 Key elements of risk management 12 Communication and training 18 Management
• Training analysis for health and safety needs	12 Communication and training
• Preparing, delivering and assessing the impacts of health and safety training programmes	12 Communication and training

B3: Identification, assessment and control of health and safety risks

B3: Identification, assessment and control of health and safety risks	**B3: Identification, assessment and control of health and safety risks**
• Legislation, control standards and guidance that affect work environments	4 Key elements of risk management 5 Risk assessment 6 Risk control 20 Advanced accident investigation and risk assessment 21 Advanced risk control techniques
• The factors that influence risk tolerability or acceptability	5 Risk assessment 14 The individual – sensory and perceptual processes 15 The individual – psychology 16 The human factors environment 20 Advanced accident investigation and risk assessment 28 Perception and decision making 29 External influences on human error
• Hazard identification and evaluation across a range of environments	5 Risk assessment 20 Advanced accident investigation and risk assessment
• Risk assessment techniques and the practical application of risk assessment across a range of environments	5 Risk assessment 20 Advanced accident investigation and risk assessment
• Risk control strategies across a range of environments	4 Key elements of risk management 5 Risk assessment 6 Risk control 7 Safe systems of work 18 Management systems 20 Advanced accident investigation and risk assessment 21 Advanced risk control techniques

Element title and learning outcomes	Relevant chapters of this book
• The principles and applicability of the tools and techniques available to measure risk	5 Risk assessment 20 Advanced accident investigation and risk assessment
• Safe systems of work and associated documents	7 Safe systems of work
• The use of suitable and appropriate analysis, assessment and recording techniques for risk control	5 Risk assessment 20 Advanced accident investigation and risk assessment
B4: Monitoring systems for health and safety • Active monitoring tools and their applicability in helping organisations meet their statutory and organisational needs, including employee feedback (climate)	**B4: Monitoring systems for health and safety** 8 Monitoring and measuring losses 9 Identifying causes and patterns 10 Monitoring and measuring conformity 19 Measuring performance 21 Advanced risk control techniques
• Reactive monitoring tools	8 Monitoring and measuring losses 9 Identifying causes and patterns
• Loss events, including their investigation from a legal perspective	8 Monitoring and measuring losses 9 Identifying causes and patterns 10 Monitoring and measuring conformity 19 Measuring performance 20 Advanced accident investigation and risk assessment
• Internal investigation techniques for actual and potential loss situations	8 Monitoring and measuring losses 9 Identifying causes and patterns 10 Monitoring and measuring conformity 19 Measuring performance 20 Advanced accident investigation and risk assessment
• The effectiveness of monitoring systems	8 Monitoring and measuring losses 9 Identifying causes and patterns 10 Monitoring and measuring conformity 19 Measuring performance
• Analysis techniques for monitoring data	8 Monitoring and measuring losses 9 Identifying causes and patterns 10 Monitoring and measuring conformity 19 Measuring performance

Element title and learning outcomes	Relevant chapters of this book
• Interpretation of data, effective use of the findings and production of clear findings and recommendations.	8 Monitoring and measuring losses 9 Identifying causes and patterns 10 Monitoring and measuring conformity 19 Measuring performance
B5: Audit and review • Health and safety review systems	**B5: Audit and review** 4 Key elements of risk management 11 Other elements of OH&SMSs 19 Measuring performance 23 Advanced audit and review
• Performance targets	4 Key elements of risk management 11 Other elements of OH&SMSs 19 Measuring performance 23 Advanced audit and review
• The development of action plans	11 Other elements of OH&SMSs 23 Advanced audit and review
• The concept of continual improvement in health and safety performance and the techniques for benchmarking	19 Measuring performance 23 Advanced audit and review
• Safety audits, their purposes, design, techniques and review systems, and associated international and national standards	11 Other elements of OH&SMSs 23 Advanced audit and review

References and
further reading

References and essential reading

The majority of the references listed below are available online – either free or for purchase. UK legislation is available at www.legislation.gov.uk, where it is provided in its original form and with no information on changes which may have been made to it. Note that essential reading references are printed in bold.

1 Health and Safety Executive. *Successful health and safety management*, HSG65 (2nd edition). HSE Books, 1997. Available online at www.hse.gov.uk/pubns/priced/hsg65.pdf.

2 British Standards Institution. *Guide to achieving effective OH&S performance*, BS 18004:2008. BSI, 2008. Note: BS 18004 has replaced BS 8800:2004.

3 British Standards Institution. *Occupational health and safety management systems – specification*, BS OHSAS 18001:2007. BSI, 2007. Notes: (a) This Standard replaced OHSAS 18001:1999 (as amended) and in this book 'BS OHSAS 18001' refers to the 2007 version unless otherwise stated; (b) There is a version of this Standard intended for international use, which is referred to as OHSAS 18001:2007; it differs from BS OHSAS 18001 only in having no National foreword.

4 Health and Safety Executive. *Costs to Britain of workplace injuries and work-related ill health: 2009/10 update*. HSE, 2011. Available online at www.hse.gov.uk/statistics/pdf/cost-to-britain.pdf.

5 British Standards Institution. *Quality management systems – requirements*, BS EN ISO 9001. BSI, 2008.

6 British Standards Institution. *Environmental management systems – requirements with guidance for use*, BS EN ISO 14001. BSI, 2004.

7 International Safety Management Organisation. *Safety management system Part 1. Requirements for safety and health management systems in manufacturing and services*. ISMO Ltd, 1994.

8 **Health and Safety Executive. *Successful health and safety management*, HS(G)65 (1st edition). HSE, 1991.**

9 International Labour Organization. *Guidelines on occupational safety and health management systems*, ILO-OSH 2001. ILO, 2001. Available online at www.ilo.org/wcmsp5/groups/public/---asia/---ro-bangkok/documents/publication/wcms_099129.pdf.

10 Health and Safety at Work etc Act 1974, Chapter 37. HMSO, 1974.

11 Employers' Health and Safety Policy Statements (Exception) Regulations 1975, SI 1975/1584 (SR 1980/269). HMSO, 1975.

12 National Examining Board in Occupational Safety and Health. *Guide to the NEBOSH National Diploma in Occupational Health and Safety*. NEBOSH, 2010.

13 Management of Health and Safety at Work Regulations 1999, SI 1999/3242 (SR 2000/388). HMSO, 1999.

14 Health and Safety Executive. *The Management of Health and Safety at Work Regulations 1999: Approved Code of Practice and Guidance*, L21. HSE Books, 1999. Available online at www.hse.gov.uk/pubns/priced/l21.pdf.

15 Manual Handling Operations Regulations 1992, SI 1992/2793 (SR 1992/535). HMSO, 1992.

16 Personal Protective Equipment at Work Regulations 1992, SI 1992/2966 (SR 1993/20). HMSO, 1992.

17 Health and Safety (Display Screen Equipment) Regulations 1992, SI 1992/2792 (SR 1992/513). HMSO, 1992.

18 Control of Noise at Work Regulations 2005, SI 2005/1643 (SR 2006/1). HMSO, 2005.

19 Control of Substances Hazardous to Health Regulations 2002, SI 2002/2677 (SR 2003/34). HMSO, 2002.

20 Control of Substances Hazardous to Health (Amendment) Regulations 2004, SI 2004/3386 (SR 2005/165). HMSO, 2004.

21 Control of Asbestos Regulations 2012:632, SI 2006/2739 (SR 2007/31). HMSO, 2006.

22 Control of Lead at Work Regulations 2002, SI 2002/2676 (SR 2003/35). HMSO, 2002.

23 Cullen, The Hon. Lord. *The Public Inquiry into the Piper Alpha disaster.* HMSO, 1990.

24 British Standards Institution. *Quality management systems – guidelines for performance improvement*, BS EN ISO 9004. BSI, 2009.

25 Health and Safety Executive. *Health and safety benchmarking*, INDG301. HSE Books, 1999. Available online at www.hse.gov.uk/pubns/indg301.pdf.

26 Work at Height Regulations 2005, SI 2005/735 (SR 2005/279). HMSO, 2005.

27 British Standards Institution. *Specification of common management system requirements as a framework for integration*, BS PAS 99:2006. BSI, 2006.

28 Hale A and Hale M. Accidents in perspective. *Occupational Psychology* 1970; 44: 115–122.

29 Heinrich H W. *Industrial accident prevention* (4th edition). McGraw-Hill, 1959.

30 Bird F E and Loftus R G. *Loss control management*. Institute Press, 1976.

31 Reporting of Injuries, Diseases and Dangerous Occurrences Regulations 1995, SI 1995/3163 (SR 1997/455). HMSO, 1995.

32 Social Security (Claims and Payments) Regulations 1979, SI 1979/628. HMSO, 1979.

33 British Standards Institution. *Guidelines for quality management system documentation*, ISO/TR 10013. BSI, 2001.

34 British Standards Institution. *Guidelines for auditing management systems*, BS EN ISO 19011:2011. BSI, 20011.

35 Evans D W. *Intermediate GNVQ core skills: communication*. Longman, 1996.

36 Evans D W. *Advanced GNVQ core skills: communication*. Longman, 1996.

37 Safety Representatives and Safety Committees Regulations 1977, SI 1977/500 (SR 1979/437). HMSO, 1977.

38 Health and Safety (Consultation with Employees) Regulations 1996, SI 1996/1513 (SR 1996/251). HMSO, 1996.

39 Health and Safety Information for Employees Regulations 1989, SI 1989/682 (SR 1991/105). HMSO, 1989.

40 Health and Safety Laboratory. Safety Climate Tool. HSL, 2010. More information at www.hsl.gov.uk/health-and-safety-products/safety-climate-tool.aspx.

41 Health and Safety Executive. *Managing competence for safety related systems. Part 1: key guidance.* HSE, 2007. Available online at www.hse.gov.uk/humanfactors/topics/mancomppt1.pdf.

42 Kroemer K H E and Grandjean E. *Fitting the task to the human* (5th edition). Taylor and Francis, 1997.

43 Health and Safety Executive. *Reducing error and influencing behaviour*, HSG48 (2nd edition). HSE Books, 1999. Available online at www.hse.gov.uk/pubns/priced/hsg48.pdf.

44 Health and Safety Executive. *How to tackle work-related stress*, INDG430. HSE Books, 2009. Available online at www.hse.gov.uk/pubns/indg430.pdf.

45 Jurgens H W, Aune I A and Piper G. *International data on anthropometry* (Occupational Safety and Health Series No. 65). ILO, 1990.

46 Duncan W J. *Organizational behavior* (2nd edition). Houghton Mifflin, 1981.

47 Taylor F W. *The principles of scientific management*. Originally published in 1911; reprinted by Norton Library, 1967. Original text available as a free ebook online at www.gutenberg.org/etext/6435.

48 Payne R L and Pugh D S. Organisations as psychological environments. In: Warr P B (ed.). *Psychology at work*. Penguin, 1971.

49 Arnold J, Silvester J, Patterson F, Robertson I T, Cooper C L and Burnes B. *Work psychology: understanding human behaviour in the workplace* (4th edition). FT Prentice Hall, 2004.

50 Handy C. *Understanding organisations*. Penguin, 1993.

51 British Standards Institution. *Quality management systems – fundamentals and vocabulary*, ISO 9000:2000. BSI, 2000.

52 British Standards Institution. *Environmental management standards*, ISO 14000:2004 series. BSI, 2004.

53 British Standards Institution. *Environmental management systems – general guidelines on principles, systems, and support techniques*, ISO 14004:2004. BSI, 2004.

54 Soanes C and Stevenson A (eds). *Concise Oxford Dictionary* (11th edition). Oxford University Press, 2004.

55 Martin C. *The health and safety coach*. Gower, 2001.

56 Det Norske Veritas. *International Safety Rating System* (7th edition). DNV, 2007.

57 HASTAM. *CHASE audit and evaluation system*. See www.hastam.co.uk/software/chase.html.

58 European Foundation for Quality Management. *Self-assessment – guidelines for companies*. EFQM, 1996.

59 Smith D. *IMS: implementing and operating*. BSI, 2002.

60 Department of Trade and Industry. *Total quality management and effective leadership – a strategic overview*. DTI, 1991.

61 Health and Safety Executive. *Total quality management and the management of health and safety*, Contract Research Report CRR153. HSE Books, 1997. Available online at www.hse.gov.uk/research/crr_pdf/1997/crr97153.pdf.

62 Institution of Occupational Safety and Health. *Joined-up working: an introduction to integrated management systems*. IOSH, 2012. Available online at www.iosh.co.uk/techguide.

63 Smith D. *Integrated management systems – the framework*, HB 10190:2001. BSI, 2001.

64 Hale A R. Conditions of occurrence of major and minor accidents. *Journal of the Institution of Occupational Safety and Health* 2002; 5 (1): 7–21.

65 Shipp P J. *Presentation and use of injury data*. British Iron and Steel Research Association, nd.

66 Boyle A J. Minor injury accident records – a case study, part 1. *Safety Surveyor* 1982; 9 (5): 23–28.

67 Boyle A J. Minor injury accident records – a case study, part 2. *Safety Surveyor* 1982; 9 (6): 8–11.

68 Haddon W. Energy damage and the ten countermeasure strategies. *Human Factors Journal*, August 1973.

69 Bird F E. *Management guide to loss control*. Institute Press, 1974.

70 Workplace (Health, Safety and Welfare) Regulations 1992, SI 1992/3004 (SR 1993/37). HMSO, 1992.

71 British Standards Institution. *Occupational health and safety management systems – specification*, OHSAS 18001. BSI, 1999.

72 Health and Safety Executive. *Five steps to risk assessment*, INDG163 (rev2). HSE Books, 2006. Available online at www.hse.gov.uk/pubns/indg163.pdf.

73 Department for Communities and Local Government. Fire safety risk assessment series. The Stationery Office, 2006. The series includes volumes covering offices and shops; factories and warehouses; sleeping accommodation; residential care premises; educational premises; small and medium places of assembly; large places of assembly; theatres, cinemas and similar premises; open air events and venues; healthcare premises; and transport premises and facilities.

74 Health and Safety Executive. *Essentials of health and safety at work*. HSE Books, 2006. Available online at www.hse.gov.uk/pubns/priced/essentials.pdf.

75 Health and Safety Executive. *Managing health and safety: five steps to success*, INDG275. HSE Books, 2003. Available online at www.hse.gov.uk/pubns/indg275.pdf.

76 Boyle A J. A fresh look at risk assessment. *The Safety and Health Practitioner*, February 1997; page 15.

77 Swain A D and Guttmann H E. *Handbook of human reliability analysis with emphasis on nuclear power plant applications*, Report NUREG/CR-1278. Sandia Laboratories, 1980.

78 Waring A. *Practical systems thinking*. Thomson Learning, 1996.

79 Health and Safety Executive. *Health and safety climate tool – process guidelines*. HSE Books, 1997 (out of print).

80 Institution of Occupational Safety and Health. *Promoting a positive culture – a guide to health and safety culture*. IOSH, 2012. Available online at www.iosh.co.uk/techguide.

81 Health and Safety Commission. *ACSNI Human Factors Study Group third report: organising for safety*. HMSO, 1993.

82 Fennel D. *Investigation into the King's Cross Underground fire*. HMSO, 1988.

83 Regulatory Reform (Fire Safety) Order 2005, SI 2005/1541 (SSI/2006/456; SR 2010/325). HMSO, 2005.

84 Confined Spaces Regulations 1997, SI 1997/1713 (SR 1999/13). HMSO, 1997.

85 Control of Major Accident Hazards Regulations 1999, SI 1999/743 (SR 2000/93). HMSO, 1999.

86 **British Standards Institution. *Occupational health and safety management systems – guidelines for the implementation of OHSAS 18001*, OHSAS 18002:2008. BSI, 2008.**

87 Maslow A H. A theory of motivation. *Psychological Review* 1943; 50: 370–396.

88 Vroom V H. *Work and motivation*. John Wiley, 1964.

89 Boring E G. Intelligence as the tests test it. *New Republic* 1923; 35: 35–37.

90 Binet A and Simon T; translated by Kite E S. *The development of intelligence in children (the Binet–Simon scale)*. Arno Press, 1916.

91 **Glendon A I, Clarke S G and McKenna E F. *Human safety and risk management* (2nd edition). CRC Press, 2006.**

92 Rasmussen J. Human errors: a taxonomy for describing human malfunctions in industrial installations. *Journal of Occupational Accidents* 1982; 4: 311–335.

93 Reason J. *Human error*. Oxford University Press, 1990.

94 Hale A R and Glendon A I. *Individual behaviour in the control of danger*. Elsevier, 1987. Note: This book is now out of print but it can be downloaded free of charge at www.hastam.co.uk/personnel/publications/hale_and_glendon.html.

95 Stranks J. *Human factors and safety*. Elsevier, 2007.

96 UK Commission for Employment and Skills. *National occupational standards: health and safety*. 2012. Available online at www.ukstandards.co.uk.

97 Skinner B F. *About behaviourism*. Knopf, 1974.

98 Roethlisberger F J and Dickson W J. *Management and the worker*. John Wiley, 1939.

99 Herzberg F. *Work and the nature of man*. World Publishing, 1966.

100 Sanders M S and McCormick E J. *Human factors in engineering design* (7th edition). McGraw-Hill, 1993.

101 Singleton W T. Current trends towards system design. *Applied Ergonomics* 1971; 2: 150–158.

102 Hartley C. *Health and safety: hazardous agents*. IOSH, 2004.

103 Turnbull N (chairman). *Internal control: guidance for directors on the Combined Code*. Institute of Chartered Accountants in England and Wales, 1999. An updated version was published in 2005 and is available online at www.frc.org.uk/images/uploaded/documents/Revised%20Turnbull%20Guidance%20October%202005.pdf.

Further reading

The texts below are suggested further reading for certain chapters.

Chapter 9
Ferry T S. *Modern accident investigation and analysis: an executive guide*. J Wiley and Sons, 1981 (out of print).

Chapter 12
Evans D W. *Intermediate GNVQ core skills: communication*. Longman, 1996.
Evans D W. *Advanced GNVQ core skills: communication*. Longman, 1996.
Suitable texts on listening skills, presentation skills and the skills required to run effective meetings.

Chapter 14
Beckett B S. *Illustrated human and social biology*. Oxford University Press, 1980.

Chapter 15
Most general textbooks on psychology will give further information on the topics covered in this chapter. However, textbooks focusing on psychology at work will be of most relevance. Searching for 'work psychology' or 'psychology and work' on any good bookshop's website will provide a selection of titles from which you can choose. Alternatively, visit a good bookshop, browse in the relevant section, and choose a book which meets your current level of knowledge and learning needs.

Health and Safety Executive. *Managing the causes of work related stress: a step-by-step approach using the Management Standards work*, HSG218. HSE Books, 2007.

Chapter 18
British Standards Institution. ISO 9000 series.
British Standards Institution. ISO 14000:2004 series.
Institution of Occupational Safety and Health. *Systems in focus – guidance on occupational safety and health management systems*. IOSH, 2011. Available online at www.iosh.co.uk/techguide.
Kirwan B and Ainsworth L K (eds). *A guide to task analysis*. Taylor and Francis, 1992.
Management Oversight and Risk Tree (MORT) Users' Manual. Note: The most up-to-date versions of MORT documentation are available on the internet, with a search using the full title (not 'MORT') being most effective.
Martin C. *The health and safety coach*. Gower, 2001.
Pybus R. *Safety management: strategy and practice*. Butterworth-Heinemann, 1996.
Selected items from the BSI Integrated Management System Series
> *IMS: a framework for integrated management systems. Background to PAS 99 and its application*, BIP 2119:2007
> *IMS: the excellence model*, BIP 2010:2004
> *IMS: continual improvement through auditing*, BIP 2011:2003
> *IMS. Corporate governance*, BIP 2012:2003
> *IMS: creating a manual*, BIP 2002:2003

Chapter 19

There are numerous textbooks on statistics ranging from the humorous to the incomprehensible. A selection is given below:

The humorous
Huff D. *How to lie with statistics*. Penguin, 1954.
Huff D. *How to take a chance*. Penguin, 1959.

General parametric statistics
Hannagan, T. *Mastering statistics*, Macmillan, 1997.

General non-parametric statistics
Siegel S and Castellan N J. *Non-parametric statistics for the behavioural sciences*. McGraw-Hill, 1988.

Specifically in the context of risk
Dickson G C A. *Risk analysis*. Witherby, 1991.
van der Schaaf T W and Hale A R (eds). *Near miss reporting as a safety tool*. Butterworth-Heinemann, 1991. A collection of papers discussing, *inter alia*, the difficulties associated with using near misses as a measure of occupational health and safety performance.

Chapter 20

Dickson G C A. *Risk analysis*. Witherby, 1991.
Fortune J and Peters G. *Learning from failure – the systems approach*. Wiley and Sons, 1997.
Kletz T A. *HAZOP and HAZAN*. Institution of Chemical Engineers, 1983.
Institution of Occupational Safety and Health. *Learning the lessons: how to respond to deaths at work and other serious incidents*. IOSH, 2012. Available online at www.iosh.co.uk/techguide.
British Standards Institution. *Hazard and operability studies (HAZOP studies) – application guide*, BS IEC 61882:2001. BSI, 2001.

Detailed, up-to-date information on MORT, ECFA, HAZOP, FMEA, ETA and FTA is available on the internet. Search using the full title, rather than the acronym, for the best results.

Chapter 21

Bignell V and Fortune J. *Understanding systems failures*. Manchester University Press, 1984.
Checkland P and Scholes J. *Soft systems methodology in action*. Wiley, 1999.
Fortune J and Peters G. *Learning from failure – the systems approach*. Wiley and Sons, 1997.
Institution of Occupational Safety and Health. *Promoting a positive culture – a guide to health and safety culture*. IOSH, 2012. Available online at www.iosh.co.uk/techguide.
Kirwan B and Ainsworth L K (eds). *A guide to task analysis*. Taylor and Francis, 1992.
Turner B A and Pidgeon N F. *Man-made disasters*. Butterworth-Heinemann, 1997.
Waring A. *Practical systems thinking*. International Thomson Business Press, 1996.
Wilkinson S. Physical control of risk. Witherby, 1992.

Chapter 23

Asbury S and Ashwell P. *Health and safety, environment and quality audits*. Elsevier, 2007.

Chapter 24

Much of the literature on Total Loss Control is relevant in this area. For example, see:
Bird F E and Germain G L. *Practical loss control leadership*. International Loss Control Institute, 1985.

More general texts include:
Bernstein P L. *Against the gods*. Wiley and Sons, 1998.
Deacon S. *Measuring business value in health and safety*. Technical Communications (Publishing) Ltd, 1995.
See also the 'Business benefits' topic on the HSE website, www.hse.gov.uk.

Chapter 26

There are many general texts on the psychology of individual differences, for example:
Arnold J, Silvester J, Patterson F, Robertson I T, Cooper C L and Burnes B. *Work psychology: understanding human behaviour in the workplace*. Pearson Higher Education, 2004.
Eysenck M W. *Cognitive psychology: a student's handbook*. Taylor and Francis, 2005.
McKenna E F. *Business psychology and organisational behaviour*. Taylor and Francis, 2006.

Chapter 27

Toft B and Reynolds S. *Learning from disasters: a management approach*. Perpetuity Press, 1997.

Chapter 28

There are many general texts on human perception. For example, see:
Sekuler R and Blake R. *Perception* (5th edition). McGraw-Hill, 2005.

For further information on the perception of risk, see, for example:
Glendon A I, Clarke S G and McKenna E F. *Human safety and risk management* (2nd edition). CRC Press, 2006.

Chapters 29 and 30

Kroemer K H E and Grandjean E. *Fitting the task to the human*. Taylor and Francis, 1997.
McKenna E F. *Psychology and organisational behaviour*. Taylor and Francis, 2006.

Index